Handbook of Research on Wireless Security

Yan Zhang
Simula Research Laboratory, Norway

Jun Zheng
City University of New York, USA

Miao Ma
Hong Kong University of Science and Technology, Hong Kong

Volume II

Information Science REFERENCE

INFORMATION SCIENCE REFERENCE

Hershey · New York

Acquisitions Editor:	Kristin Klinger
Development Editor:	Kristin Roth
Senior Managing Editor:	Jennifer Neidig
Managing Editor:	Sara Reed
Copy Editor:	Ashlee Kunkel, Holly J. Powell
Typesetter:	Jamie Snavely, Carole Coulson
Cover Design:	Lisa Tosheff
Printed at:	Yurchak Printing Inc.

Published in the United States of America by
Information Science Reference (an imprint of IGI Global)
701 E. Chocolate Avenue, Suite 200
Hershey PA 17033
Tel: 717-533-8845
Fax: 717-533-8661
E-mail: cust@igi-global.com
Web site: http://www.igi-global.com

and in the United Kingdom by
Information Science Reference (an imprint of IGI Global)
3 Henrietta Street
Covent Garden
London WC2E 8LU
Tel: 44 20 7240 0856
Fax: 44 20 7379 0609
Web site: http://www.eurospanonline.com

Library of Congress Cataloging-in-Publication Data

Handbook of research on wireless security / Yan Zhang, Jun Zheng, and Miao Ma, editors.

p. cm.

Summary: "This book combines research from esteemed experts on security issues in various wireless communications, recent advances in wireless security, the wireless security model, and future directions in wireless security. As an innovative reference source for students, educators, faculty members, researchers, engineers in the field of wireless security, it will make an invaluable addition to any library collection"--Provided by publisher.

Includes bibliographical references and index.

ISBN 978-1-59904-899-4 (hardcover) -- ISBN 978-1-59904-900-7 (ebook)

1. Wireless communication systems--Security measures. I. Zhang, Yan, 1962- II. Zheng, Jun, Ph.D. III. Ma, Miao. IV. Title.

TK5102.85.H35 2008

005.8--dc22

2007036301

British Cataloguing in Publication Data
A Cataloguing in Publication record for this book is available from the British Library.

Editorial Advisory Board

Table of Contents

Detailed Table of Contents

to enabling secure and private service discovery in these environments, including the dynamic population of participants, the lack of a universal trust mechanism, and the limited capabilities of the devices. To ensure secure service discovery while addressing privacy issues, trust-based models are inevitable. We survey secure service discovery in the broadband wireless environment. We include case studies of two protocols which include a trust mechanism, and we summarize future research directions.

The recent developments in the mobile technology (mobile phones, middleware, wireless networks, etc.) created a need for new methods of protecting the code transmitted through the network. The oldest and the simplest mechanisms concentrate more on the integrity of the code itself and on the detection of unauthorized manipulation. The newer solutions not only secure the compiled program, but also the data that can be gathered during its "journey," and even the execution state. Some other approaches are based on prevention rather than detection. In the chapter we present a new idea of securing mobile agents. The proposed method protects all components of an agent: the code, the data, and the execution state. The proposal is based on a zero-knowledge proof system and a secure secret sharing scheme, two powerful cryptographic primitives. Next, the chapter includes security analysis of the new method and its comparison to other currently most widespread solutions. Finally, we propose a new direction of securing mobile agents by straightening the methods of protecting integrity of the mobile code with risk analysis and a reputation system that helps avoiding a high-risk behavior.

The broad aim of identity management (IdM) is to manage the resources of an organization (such as files, records, data and communication infrastructure, and services) and to control and manage access to those resources in an efficient and accurate way. Consequently, identity management is both a technical and process orientated concept. The concept of IdM has begun to be applied in identities related applications in enterprises, governments, and Web services since 2002. As the integration of heterogeneous wireless networks becomes a key issue in towards the next generation (NG) networks, IdM will be crucial to the success of NG wireless networks. A number of issues, such as mobility management, multioperator, and securities require the corresponding solutions in terms of user authentication, access control, and so forth. IdM in NG wireless networks is about managing the digital identity of a user and ensuring that users have fast, reliable, and secure access to distributed resources and services of an NGN and the associated service providers, across multiple systems and business contexts.

Wardriving is the practice of searching wireless networks while moving. Originally, it was explicitly referred to people searching for wireless signals by driving on vans, but nowadays it generally identifies people searching for wireless accesses while moving. Despite the legal aspects, this "quest for connectivity" spawned a quite productive underground community, which developed powerful tools, relying on cheap and standard hardware. The knowledge of these tools and techniques has many useful aspects. First, when designing the security framework of a wireless LAN (WLAN), the knowledge of the vulnerabilities exploited at the basis of wardriving is a mandatory step, both to avoid penetration issues and to detect whether attacks are ongoing. Second, hardware and software developers can design better devices by avoiding common mistakes and using an effective suite for conducting security tests. Lastly, people who are interested in gaining a deeper understanding of wireless standards can conduct experiments by simply downloading software running on cost effective hardware. With such preamble, in this chapter we will analyze the theory, the techniques, and the tools commonly used for wardriving IEEE 802.11-based wireless networks.

The broadcast nature of wireless networks and the mobility features created new kinds of intrusions and anomalies taking profit of wireless vulnerabilities. Because of the radio links and the mobile equipment features of wireless networks, wireless intrusions are more complex because they add to the intrusions developed for wired networks, a large spectrum of complex attacks targeting wireless environment. These intrusions include rogue or unauthorized access point (AP), AP MAC spoofing, and wireless denial-of-service and require adding new techniques and mechanisms to those approaches detecting intrusions targeting wired networks. To face this challenge, some researchers focused on extending the deployed approaches for wired networks while others worked to develop techniques suitable for detecting wireless intrusions. The efforts have mainly addressed (a) the development of theories to allow reasoning about detection, wireless cooperation, and response to incidents, and (b) the development of wireless intrusion and anomaly detection systems that incorporate wireless detection, preventive mechanisms, and tolerance functions. This chapter aims at discussing the major theories, models, and mechanisms developed for the protection of wireless networks/systems against threats, intrusions, and anomalous behaviors. The objectives of this chapter are to (a) discuss security problems in wireless environment, (b) to present the current research activities, (c) study the important results already developed by researchers, and (d) to discuss

A lot of networks today are behind firewalls. In peer-to-peer networking, firewall-protected peers may have to communicate with peers outside the firewall. This chapter shows how to design peer-to-peer systems to work with different kinds of firewalls within the object-oriented action systems framework by combining formal and informal methods. We present our approach via a case study of extending a Gnutella-like peer-to-peer system (Yan et al, 2003) to provide connectivity through firewalls.

An ubiquitous access and pervasive computing concept is almost intrinsically tied to wireless communications. Emerging next-generation wireless networks enable innovative service access in every situation. Apart from many remote services, proximity services will also be widely available. People currently rely on numerous forms of identities to access these services. The inconvenience of possessing and using these identities creates significant security vulnerability, especially from network and device point of view in wireless service access. After explaining the current identity solutions scenarios, the chapter illustrates the on-going efforts by various organizations and the requirements and frameworks to develop an innovative, easy-to-use identity management mechanism to access the future diverse service worlds. The chapter also conveys various possibilities, challenges, and research questions evolving in these areas.

This chapter provides a survey of privacy enhancing techniques and discusses their effect using a scenario in which a charged location-based service is used. We introduce four protection levels and discuss an assessment of privacy enhancing techniques according to these protection levels.

This chapter shows that the security challenges posed by the 802.11 wireless networks are manifold and it is therefore important to explore the various vulnerabilities that are present with such networks.

Converging networks and mobility raise new challenges towards the existing authentication, authorisation, and accounting (AAA) systems. Focus of the research is towards integrated solutions for seamless service access of mobile users. Interworking issues between mobile and wireless networks are the basis for detailed research on handover delay, multidevice roaming, mobile networks, security, ease-of-use, and anonymity of the user. This chapter provides an overview over state-of-the-art in authentication for mobile systems, and suggests extending AAA-mechanisms to home and community networks, taking into account security and privacy of the users.

Wireless networks are gaining popularity that comes with the occurrence of several networking technologies raising from personal to wide area, from centralized to distributed, and from infrastructure-based to infrastructure-less. Wireless data link characteristics such as openness of transmission media make these networks vulnerable to a novel set of security attacks, despite those that they inherit from wired networks. In order to ensure the protection of mobile nodes that are interconnected using wireless protocols and standards, it is essential to provide an in-depth study of a set of mechanisms and security models. In this chapter, we present the research studies and proposed solutions related to the authentication, privacy, trust establishment, and management in wireless networks. Moreover, we introduce and discuss the major security models used in a wireless environment.

In the design and analysis of authentication and key agreement protocols, provable secure formal methods play a very important role, among which the Canetti-Krawczyk(CK) model and the universal composable(UC) security model are very popular at present. This chapter focuses on these two models and consists mainly of three parts. (1) There is an introduction to the CK model and the UC model. (2) There is also a study of these two models, which includes an analysis of the CK model and an extension of the UC security model. The analysis of the CK model presents its security analysis, advantages, and disadvantages, and a bridge between this formal method and the informal method (heuristic method) is established; an extension of the UC security model gives a universally composable anonymous hash certification model. (3) The applications of these two models are also presented. With these two models, the four-way handshake protocols in 802.11i and Chinese WLAN security standard WAPI are analyzed.

In a wireless environment, multimedia transmission is often affected by the error rate, delaying, terminal's power or bandwidth, and so forth, which brings difficulties to multimedia content protection. In the past decade, wireless multimedia protection technologies have been attracting more and more researchers. Among them, wireless multimedia encryption and watermarking are two typical topics. Wireless multimedia encryption protects multimedia content's confidentiality in wireless networks, which emphasizes improving the encryption efficiency and channel friendliness. Some means have been proposed, such as the format-independent encryption algorithms that are time efficient compared with traditional ciphers, the partial encryption algorithms that reduce the encrypted data volumes by leaving some information unchanged, the hardware-implemented algorithms that are more efficient than software based ones, the scalable encryption algorithms that are compliant with bandwidth changes, and the robust encryption algorithms that are compliant with error channels. Compared with wireless multimedia encryption, wireless multimedia watermarking is widely used in ownership protection, traitor tracing, content authentication, and so forth. To keep costs low, a mobile agent is used to partition some of the watermarking tasks. To counter transmission errors, some channel encoding methods are proposed to encode the watermark. To keep robust, some means are proposed to embed a watermark into media data of low bit rate. Based on both watermarking and encryption algorithms, some applications arise, such as secure multimedia sharing or secure multimedia distribution. In this chapter, the existing wireless multimedia encryption and watermarking algorithms are summarized according to the functionality and multimedia type, their performances are analyzed and compared, the related applications are presented, and some open issues are proposed.

In this chapter, a system-on-chip design of the newest powerful standard in the hash families, named Whirlpool, is presented. With more details, an architecture and two VLSI implementations are presented. The first implementation is suitable for high-speed applications while the second one is suitable for applications with constrained silicon area resources. The architecture permits a wide variety of implementation tradeoffs. Different implementations have been introduced and each specific application can choose the appropriate speed-area trade-off implementation. The implementations are examined and compared in the security level and in the performance by using hardware terms. Whirlpool with RIPEMD, SHA-1, and SHA-2 hash functions are adopted by the International Organization for Standardization (ISO/IEC) 10118-3 standard. The Whirlpool implementations allow fast execution and effective substitution of any previous hash families' implementations in any cryptography application.

The fourth generation of mobile networks (4G) will be a technology-opportunistic and user-centric system combining the economic and technological advantages of different transmission technologies to provide a context-aware and adaptive service access anywhere and at any time. Security turns out to be one of the major problems that arise at different interfaces when trying to realize such a heterogeneous system by integrating the existing wireless and mobile systems. Indeed, current wireless systems use very different and difficult to combine proprietary security mechanisms, typically relying on the associated user and infrastructure management means. It is generally impossible to apply a security policy to a system consisting of different heterogeneous subsystems. In this chapter, we first briefly present the security of candidate 4G access systems, such as 2/3G, WLAN, WiMax and so forth. In the next step, we discuss the arising security issues of the system interconnection. We namely define a logical access problem in heterogeneous systems and show that both the technology-bound low-layer and the overlaid high-layer access architectures exhibit clear shortcomings. We present and discuss several proposed approaches aimed at achieving an adaptive, scalable, rapid, easy-to-manage, and secure 4G service access independently of the used operator and infrastructure. We then define general requirements on candidate systems to support such 4G security.

The integration of heterogeneous mobile/wireless networks using an IP-based core network materializes the beyond 3G (B3G) mobile networks. Along with a variety of new perspectives, the new network model raises new security concerns, mainly because of the complexity of the deployed architecture and the heterogeneity of the employed technologies. In this chapter, we examine and analyze the security architectures and the related security protocols, which are employed in B3G networks focusing on their functionality and the supported security services. The objectives of these protocols are to protect the involved parties and the data exchanged among them. To achieve these, they employ mechanisms that provide mutual authentication as well as ensure the confidentiality and integrity of the data transferred over the wireless interface and specific parts of the core network. Finally, based on the analysis of the security mechanisms, we present a comparison of them that aims at highlighting the deployment advantages of each one and classifies the latter in terms of (a) security, (b) mobility, and (c) reliability.

This chapter analyzes the security architecture designed for the protection of the universal mobile telecommunication system (UMTS). This architecture is built on the security principles of 2G systems with improvements and enhancements in certain points in order to provide advanced security services. The main objective of the 3G security architecture is to ensure that all information generated by or relating to a user, as well as the resources and services provided by the serving network and the home environment, are adequately protected against misuse or misappropriation. Based on the carried analysis, the critical points of the 3G security architecture, which might cause network and service vulnerability, are identified. In addition, the current research on the UMTS security and the proposed enhancements that aim at improving the UMTS security architecture are briefly presented and analyzed.

Motivated by the requirements for higher data rate, richer multimedia services, and broader radio range, wireless mobile networks are currently in the stage evolving from the second-generation (2G), for example, global system for mobile communications (GSM), into the era of third-generation (3G) or beyond 3G or fourth-generation (4G). Universal mobile telecommunications system (UMTS) is the natural successor of the current popular GSM. Code division multiple access 2000 (CDMA2000) is the next generation version for the CDMA-95, which is predominantly deployed in the North America and North Korea. Time division-sychrononous CDMA (TD-SCDMA) is in the framework of 3GPP2 and is expected to be one of the principle wireless technologies employed in China in the future. It is envisioned that each of three standards in the framework of international mobile telecommunications-2000 (IMT-2000) will play a significant role in the future due to the backward compatibility, investment, maintenance cost, and even politics. In all of the potential standards, access security is one of the primary demands as well as challenges to resolve the deficiency existing in the second generation wireless mobile networks such as GSM, in which only one-way authentication is performed for the core network part to verify the user equipment (UE). Such access security may lead to the "man-in-middle" problem, which is a type of attack that can take place when two clients that are communicating remotely exchange public keys in order to initialize secure communications. If both of the public keys are intercepted in the route by someone, that someone can act as a conduit and send in the messages with a fake public key. As a result, the secure communication is eavesdropped on by a third party.

The global system for mobile communications (GSM) is the most popular standard that implements second generation (2G) cellular systems. 2G systems combined with general packet radio services (GPRS) are often described as 2.5G, that is, a technology between the 2G and third (3G) generation of mobile systems. GPRS is a service that provides packet radio access for GSM users. This chapter presents the security architecture employed in 2.5G mobile systems, focusing on GPRS. More specifically, the security measures applied to protect the mobile users, the radio access network, the fixed part of the network, and the related data of GPRS, are presented and analyzed in details. This analysis reveals the security weaknesses of the applied measures that may lead to the realization of security attacks by adversaries. These attacks threaten network operation and data transfer through it, compromising end-users and network security. To defeat the identified risks, current research activities on the GPRS security propose a set of security improvements to the existing GPRS security architecture.

Security measures of mobile infrastructures have always been important from the early days of the creation of cellular networks. Nowadays, however, the traditional security schemes require a more fundamental approach to cover the entire path from the mobile user to the server. This fundamental approach is so-called end-to-end (E2E) security coverage. The main focus of this chapter is to discuss such architectures for IEEE 802.16e (Mobile-WiMAX) and major 3G cellular networks. The end-to-end implementations usually contain a complete set of algorithms and protocol enhancements (e.g., mutual identification, authentications, and authorization), including the VLSI implementations. This chapter discusses various proposals at the protocol level.

This chapter outlines how cellular authentication can be utilized for generic application security. It describes the basic concept of the generic bootstrapping architecture that was defined by the 3rd generation partnership project (3GPP) for current networks and outlines the latest developments for future networks.The chapter will provide an overview of the latest technology trends in the area of generic application security.

Network mobility (NEMO) enables seamless and ubiquitous Internet access while on board vehicles. Even though the Internet Engineering Task Force (IETF) has standardized the NEMO basic support protocol as a network layer mobility solution, few studies have been conducted in the area of the authentication, authorization, and accounting (AAA) framework that is a key technology for successful deployment. In this chapter, we first review the existing AAA protocols and analyze their suitability in NEMO environments. After that, we propose a localized AAA framework to retain the mobility transparency as the NEMO basic support protocol and to reduce the signaling cost incurred in the AAA procedures. The proposed AAA framework supports mutual authentication and prevents various threats such as replay attack, man-in-the-middle attack, and key exposure. Performance analysis on the AAA signaling cost is carried out. Numerical results demonstrate that the proposed AAA framework is efficient under different NEMO environments.

Section III
Security in Ad Hoc and Sensor Networks

Mobile ad hoc network (MANET) is a self-configuring and self-maintaining network characterized by dynamic topology, absence of infrastructure, and limited resources. These characteristics introduce security vulnerabilities, as well as difficulty in providing security services to MANETs. To date, tremendous research has been done to develop security approaches for MANETs. This work will discuss the existing approaches that have intended to defend against various attacks at different layers. Open challenges are also discussed in the chapter.

Providing privacy is often considered a keystone factor for the ultimate take up and success of mobile ad hoc networking. Privacy can best be protected by enabling anonymous communication and, therefore, this chapter surveys existing anonymous communication mechanisms for mobile ad hoc networks. On the basis of the survey, we conclude that many open research challenges remain regarding anonymity provisioning in mobile ad hoc networks. Finally, we also discuss the notorious Sybil attack in the context of anonymous communication and mobile ad hoc networks.

The pervasiveness of wireless communication recently gave mobile ad hoc networks (MANET) significant researchers' attention, due to its innate capabilities of instant communication in many time and mission critical applications. However, its natural advantages of networking in civilian and military environments make it vulnerable to security threats. Support for anonymity in MANET is orthogonal to a critical security challenge we faced in this chapter. We propose a new anonymous authentication protocol for mobile ad hoc networks enhanced with a distributed reputation system. The main objective is to provide mechanisms concealing a real identity of communicating nodes with an ability of resistance to known attacks. The distributed reputation system is incorporated for a trust management and malicious behavior detection in the network.

The increasing diffusion of wireless portable devices and the emergence of mobile ad hoc networks promote anytime and anywhere opportunistic resource sharing. However, the fear of exposure to risky interactions is currently limiting the widespread uptake of ad hoc collaborations. This chapter introduces to the challenge of identifying and validating novel security models/systems for securing ad hoc collaborations by taking into account the high unpredictability, heterogeneity, and dynamicity of envisioned wireless environments. We claim that the concept of trust management should become a primary engineering design principle, to associate with the subsequent trust refinement into effective authorization policies, thus calling for original and innovative access control models. The chapter overviews the state-of-the-art solutions for trust management and access control in wireless environments, by pointing out both the need for their tight integration and the related emerging design guidelines (e.g., exploitation of context awareness and adoption of semantic technologies).

Security has become a primary concern in mobile ad hoc networks (MANETs). The characteristics of MANETs pose both challenges and opportunities in achieving security goals, such as confidentiality, authentication, integrity, availability, access control, and nonrepudiation. Cryptographic techniques are widely used for secure communications in wired and wireless networks. Most cryptographic mechanisms, such as symmetric and asymmetric cryptography, often involve the use of cryptographic keys. However, all cryptographic techniques will be ineffective if the key management is weak. Key management is also a central component in MANET security. The purpose of key management is to provide secure procedures for handling cryptographic keying materials. The tasks of key management include key generation, key distribution, and key maintenance. Key maintenance includes the procedures for key storage, key update, key revocation, key archiving, and so forth. In MANETs, the computational load and complexity for key management are strongly subject to restriction by the node's available resources and the dynamic nature of network topology. A number of key management schemes have been proposed for MANETs. In this chapter, we present a survey of the research work on key management in MANETs according to recent literature.

Mobile-IP ad hoc networks (MANETs) have gained popularity in the past few years with the creation of a variety of ad hoc protocols that specifically offer quality of service (QoS) for various multimedia traffic between mobile stations (MSs) and base stations (BSs). The lack of proper end-to-end security coverage, on the other hand, is a challenging issue as the nature of such networks with no specific infrastructure is prone to relatively more attacks, in a variety of forms. The focus of this chapter is to discuss a number of attack scenarios and their remedies in MANETs including the introduction of two entities, ad hoc key distribution center (AKDC) and decentralize key generation and distribution (DKGD), which serve as key management schemes.

The integration of wireless communication and embedded video systems is a demanding and interesting topic which has attracted significant research efforts from the community of telecommunication. This chapter discusses the challenging issues in wireless video surveillance and presents the detailed design for a novel highly-secure video surveillance system over ad hoc wireless networks. To this end, we explore the state-of-the-art in the cross domains of wireless communication, video processing, embedded systems, and security. Moreover, a new media-dependent video encryption scheme, including a reliable data embedding technique and real-time video encryption algorithm, is proposed and implemented to enable the system to work properly and efficiently in an open and insecure wireless environment. Extensive experiments are conducted to demonstrate the advantages of the new systems, including high security guarantee and robustness. The chapter would serve as a good reference for solving the challenging issues in wireless multimedia and bring new insights on the interaction of different technologies within the cross application domain.

Intrusion detection in ad hoc networks is a challenge because of the inherent characteristics of these networks, such as, the absence of centralized nodes, the lack of infrastructure, and so forth. Furthermore, in addition to application-based attacks, ad hoc networks are prone to attacks targeting routing protocols, which is a novel problem. Issues in intrusion detection in ad hoc networks are addressed by numerous research proposals in literature. In this chapter, we first enumerate the properties of ad hoc networks which hinder intrusion detection systems. Second, significant intrusion detection system (IDS) architectures and methodologies proposed in the literature are elucidated. Strengths and weaknesses of these works are then studied and explained. Finally, the future directions, which will lead to the successful deployment of intrusion detection in ad hoc networks, are discussed.

In this chapter we present the growing challenges related to security in wireless sensor networks. We show possible attack scenarios and evidence the ease of perpetrating several types of attacks due to the extreme resource limitations that wireless sensor networks are subjected to. Nevertheless, we show that security is a feasible goal in this resource-limited environment. To prove that security is possible we survey several proposed sensor network security protocols targeted to different layers in the protocol stack. The work surveyed in this chapter enable several protection mechanisms vs. well documented network attacks. Finally, we summarize the work that has been done in the area and present a series of ongoing challenges for future work.

The applications of wireless sensor networks (WSNs) are continuously expanding. Recently, consistent research and development activities have been associated to this field. Security ranks at the top of the issues that should be discussed when deploying a WSN. This is basically due to the fact that WSNs are, by nature, mission-critical. Their applications mainly include battlefield control, emergency response (when a natural disaster occurs), and healthcare. This chapter reviews recent research results in the field of WSN security.

Since routing is a fundamental operation in all types of networks, ensuring routing security is a necessary requirement to guarantee the success of routing operations. A securing routing task gets more challenging as the target network lacks an infrastructure-based routing operation. This infrastructure-less nature that invites a multihop routing operation is one of the main features of wireless sensor networks that raises the importance of secure routing problem in these networks. Moreover, the risky environment, application

The electromagnetic energy source used by wireless communication devices in a vehicle can cause electromagnetic compatibility problems with the electrical and electronic equipment on board. This work is focused on the radiated susceptibility – EMS – issue and proposes a method for quantifying the electromagnetic influence of wireless RF transmitters on board vehicles. The key to the analysis is the evaluation of the relation between the electrical field emitted by a typical Bluetooth device operating close to the automobile's electrical and electronic systems and the field level specified by the EMC directive 2004/104/EC for radiated susceptibility tests. The chapter includes the model of a closed circuit structure emulating an automobile's electric wire system and the simulation of its behavior under electromagnetic fields' action. According to this a physical structure is designed and implemented, which is used for laboratory tests. Finally, simulated and experimental results are compared and the conclusions obtained are discussed.

The great promise of wireless LAN will never be realized unless there is an an appropriate security level. From this point of view, various security protocols have been proposed to handle WLAN security problems that are mostly due to the lack of physical protection in WLAN or because of the transmission on the radio link. The purpose of this chapter is (1) to provide the reader with a sample background in WLAN technologies and standards, (2) to give the reader a solid grounding in common security concepts and technologies, and (3) to identify the threats and vulnerabilities of WLAN communications.

Wireless local area networks (WLAN) are rapidly becoming a core part of network access. Supporting user mobility, more specifically, session continuation in changing network access points, is becoming an integral part of wireless network services. This is because of the popularity of emerging real-time streaming applications that can be commonly used when the user is mobile, such as voice-over-IP and Internet radio. However, mobility introduces a new set of problems in wireless environments because of handoffs between network access points (APs). The IEEE 802.11i security standard imposes an authentication delay long enough to hamper real-time applications. This chapter will provide a comprehensive

Preface

Wireless networks have been seen unprecedented growth in the past few years. Wireless technologies provide users with a variety of benefits like portability, flexibility, increased productivity, and lower installation costs. Various wireless technologies, from wireless local area network (WLAN) and Bluetooth to WiMAX and third generation (3G) have been developed. Each of these technologies has its own unique applications and characteristics. For example, a WLAN can provide the wireless users with high bandwidth data communication in a restricted and dense area (hotpot). Ad hoc networks, like those enabled by Bluetooth, allow data synchronization with network systems and application sharing between devices. WiMAX can provide high-speed, high bandwidth efficiency, and high-capacity multimedia services for residential as well as enterprise applications.

However, any wireless technology is inherently risky. It has the same risks as the wired networks as well as new risks brought by the wireless connectivity. There have been many reports of security weaknesses and problems related to different wireless technologies, which make wireless security quite a hot research topic recently, both in the academia and industry.

Wireless security is a very broad area as there are so many different wireless technologies existing. Each wireless technology has its own architecture, algorithms, and protocols. Different wireless technologies have their own application areas and different security concerns, requirements, and solutions. To this end, we want to bring up the *Handbook of Research on Wireless Security* to serve as a single comprehensive reference in the field of wireless security.

In this book, the basic concepts, terms, protocols, systems, architectures, and case studies in the wireless security are provided. It identifies the fundamental problems, key challenges, and future directions in designing secure wireless systems. It covers a wide spectrum of topics in a variety of wireless networks, including attacks, secure routing, encryption, decryption, confidentiality, integrity, key management, identity management, and also security protocols in standards.

The chapters of this book are authoritatively contributed by a group of internationally renowned experts on wireless security. They are organized in four sections:

- Section I: Security Fundamentals
- Section II: Security in 3G/B3G/4G
- Section III: Security in Ad Hoc and Sensor Networks
- Section IV: Security in Wireless PAN/LAN/MAN

Section I introduces the basic concepts and fundamental mechanisms of wireless security. This section is able to provide the necessary background for readers and introduce all the fundamental issues on wireless security without previous knowledge on this area. Section II discusses all the security aspects in 3G/B3G/4G. It is well known that 3G mobile systems offer mobile users content rich services, wireless broadband access to Internet, and worldwide roaming. Future 4G mobile communication networks are expected to provide all IP-based services for heterogeneous wireless access technologies, assisted by mobile IP to provide seamless Internet access for mobile users. However the broadcast nature of the wireless communication and increased popularity of wireless devices introduce serious security vulnerabilities. A variety of security issues regarding 3G/B3g/4G will be introduced and addressed with effective solutions (e.g., identity management, confidentiality and integrity mechanisms, evaluation of the current 3G/B3G/4G security protocols, analysis of the impact of security deployment upon the network performance, etc.). Section III explores the security in ad hoc and sensor networks. In recent years, tremendous technological advances have been made in the areas of wireless ad hoc and sensor networks. Such networks have a significant impact on a variety of applications including scientific, military, medical, industrial, office, home, and personal domains. However, these networks introduce new security challenges due to their dynamic topology, severe resource constraints, and absence of a trusted infrastructure. Many aspects of security issues regarding the ad hoc and sensor networks will be covered, including key management, cryptographic protocols, authentication and access control, intrusion detection and tolerance, secure location services, privacy and anonymity, secure routing, resilience against different types of attacks, and so forth. Section IV exploits the security problems in wireless PAN/LAN/MAN. Nowadays we have continuously growing markets for the wireless PANs, wireless LANs, and wireless MANs, but there is a big black hole in the security of this kind of network. Diverse aspects of the security issues on these types of networks will be introduced. For instance, the threats and vulnerabilities in wireless LANs, access control in wireless LANs, evaluating security mechanisms in wireless PANs, the protocols and mechanisms to enhance the security of wireless LANs/MANs, security issues in WiMAX, and so forth are discussed. Practical examples will also be introduced to enhance the understanding.

This book can serve as an essential and useful reference for undergraduate and graduate students, educators, scientists, researchers, engineers, and research strategists in the field of wireless security.

We hope that by reading this book the reader can not only learn the basic concepts of wireless security but also get a good insight into some of the key research works in securing the wireless networks. Our goal is to provide an informed and detailed snapshot of this fast moving field. If you have any feedback or suggestion, please contact the editors.

Yan Zhang, Jun Zheng, and Miao Ma

Acknowledgment

The editors would like to acknowledge the help of all involved in the collation and review process of the handbook, without whose support the project could not have been successfully completed.

Deep appreciation and gratitude is first due to Editorial Advisory Board, whose suggestions and comments have greatly enhanced the quality of the book. Most of the authors of the chapters included in this handbook also served as referees for chapters written by other authors. We would like to thank them for their time, valuable comments, and hard work in reviewing the peers' work. Thanks also go to all the external reviewers who provided constructive and comprehensive reviews. Their critical suggestions and comments ensure the quality of the book.

Special thanks also go to the publishing team at IGI Global Inc., whose contributions throughout the whole process from inception of the initial idea to final publication have been invaluable. In particular to Kristin Roth, who continuously prodded via e-mail for keeping the project on schedule, to Jessica Thompson, whose support, patience, and professionalism during this project, and to Nicole Dean, for enhancing the book marketability. We are grateful for the staffs for the great efforts during the typesetting period. Last but not least, a special thank to the families and friends for their constant encouragement, patience, and understanding throughout this project.

In closing, we wish to thank all of the authors for their insights, excellent contributions, and professional cooperation to this handbook.

Co-Editors for *Handbook of Research on Wireless Security*

Yan Zhang, Ph.D.
Simula Research Laboratory, Norway

Jun Zheng, Ph.D.
CUNY, USA

Miao Ma, Ph.D.
HKUST

May 2007

Section III
Security in Ad Hoc and Sensor Networks

Chapter XXVI
Security in Mobile Ad Hoc Networks

Bin Lu
West Chester University, USA

ABSTRACT

Mobile ad hoc network (MANET) is a self-configuring and self-maintaining network characterized as dynamic topology, absence of infrastructure, and limited resources. These characteristics introduce security vulnerabilities, as well as difficulty in providing security services to MANETs. Up to date, tremendous research has been done to develop security approaches to MANETs. This work will discuss the existing approaches that have intended to defend against various attacks at different layers. Open challenges are also discussed in the chapter.

INTRODUCTION

A mobile ad hoc network (MANET) is a self-configuring and self-maintaining network composed of mobile nodes that communicate over wireless channels (Perkins, 2001). MANETs are characterized as infrastructure-less with rapid topology change, high node mobility, and stringent resource constraints. A MANET is usually used in situations such as military battles, disaster recovery, and emergent medical situations. While applications in these areas still dominate the research needs for MANETs, commercial applications (such as home networking and personal area networks) have also been brought to attention with the rapid research progress in mobile telephony and personal digital assistants.

Early research in MANETs assumed a cooperative and trusted environment, which unfortunately is not always true. In an unfriendly environment, a variety of attacks can be launched, ranging from passive eavesdropping to active interference. The attacks could target a number of devices or services in MANETs, such as wireless channels, routing protocols, high-level applications, or even security mechanisms themselves. A misbehaving node can be *selfish* or *malicious*, based on their intensions. A selfish node can simply deviate from network protocols in order to maximize its own profit, while a malicious node may intend to corrupt some services or bring down some other nodes. Both selfish and malicious misbehaviors are dangerous in that they could cause degradation in the network performance, or even paralyzation of the entire network. Therefore security has become a primary concern, especially for security-sensitive applications in a noncooperative or hostile environment.

However, introducing security features to MANETs is not a trivial task. The lack of a fixed infrastructure determines that MANETs do not have a clear physical line of defense, unlike their wired counterparts, who can deploy security defense mechanisms (e.g., firewalls) at network devices such as gateways or routers. The decentralized manner of operations also implies that a central administration point is not realistic for MANETs. Moreover, all security services come with a price. The security mechanisms will share with other services the precious communication and computation resources, which may consequently affect the performance of the node, or even the entire network. Performance is also a basic concern for ad hoc networks, which means a tradeoff has to be made between security and other services such as computation and communication. Therefore, minimum consumption of resources is one of the most important requirements for security solutions in MANETs.

This chapter will discuss security issues in MANETs, including security attacks, security requirements, security solutions, and their advantages and weakness.

The remainder of this chapter is organized as follows: the following section will discuss the security vulnerabilities, security services, and security challenges for MANETs; the third section will focus on the security solutions that have been proposed for MANETs. The security mechanisms to protect MAC (medium access control) layer communications and routing protocols will be described. Intrusion detections, authentication, and key management will be also discussed in this section. In the last section we will discuss the open research issues for MANET security and then we will conclude the chapter.

VULNERABILITIES, SECURITY SERVICES, AND CHALLENGES

MANETs Vulnerabilities

MANETs suffer from all the vulnerabilities that their wired counterparts encountered. An adversary may launch various attacks ranging from *passive* eavesdropping to *active interference* such as traffic jamming, packet modification and fabrication, message replay, denial-of-service (DoS), and so forth. Some of these vulnerabilities are aggravated in a wireless context due to the characteristics of MANETs, such as the lack of a clear line of defense and the in-the-air communications.

Besides, ad hoc networks are susceptible to vulnerabilities that are inherent to wireless networks, which reside in their routing and autoconfiguration mechanisms. The MAC (medium access control) protocols (such as IEEE [1999] 802.11 series) and most of the routing protocols for MANETs are designed with the assumption that all the nodes will cooperate and would not intentionally deviate from the protocols. However, this is not always true, especially in an autonomous network where nodes belong to different self-profited organizations.

Eavesdropping is generally easier in MANETs than in the Internet due to the open nature of the communication medium in MANETs. Passive attacks are by nature difficult to detect, not mentioning in MANETs where many mobile devices support promiscuous mode. Like in the wired networks, cryptographic operations are used to prevent ad hoc networks from eavesdropping.

The MAC protocols in MANETs are vulnerable to traffic jamming, which is caused by nodes who fail to follow the protocols in order to maximize their own profit or simply to disrupt network operations. A node can obtain an unfair share of the bandwidth by transmitting without waiting its turn, or interrupt signal transmissions by injecting bogus signals into the network. Communication channels in MANETs are open and shared, therefore it is difficult to prevent and detect this kind of attacks. Moreover, ad hoc nodes are usually battery-powered, which makes energy a precious resource in MANETs. An adversary could launch a new type of DoS attack, namely "sleep deprivation torture" attack (Stajano & Anderson, 1999), by forcing a node to relay packets.

Ad hoc routing requires the participation of all the nodes in the network. MANETs are *peer-to-peer*, namely all the nodes play the same roles as end hosts and routers as well. However, some selfish nodes may refuse to forward data packets or routing requests for other nodes to save energy or communication resources. Some more dramatic attacks by malicious nodes include dissemination of false routing information, sending frequent routing updates to achieve denial-of-service, and deviating traffic from legal route.

Like in the traditional wired networks, attacks can target the security mechanisms as well. For examples, cryptographic operations can be at risk if a secret key is intercepted and compromised, or a trusted authority is brought down. These attacks are not intrinsic to wireless networks, but they are difficult to prevent and detect in the context of MANETs.

Security Services

The services that should be provided in MANETs are the same as those in the wired networks, which include *availability*, *authentication*, *integrity*, *confidentiality*, and *nonrepudiation*.

- **Availability** ensures that network services are provided as supposed to be. In an ad hoc network without protection of proper security mechanisms, its service performance and availability can be easily compromised. For example, signal jamming at the physical and media access control layers can seriously interfere with communications or even bring down the physical channels. A malicious or selfish node can also disrupt routing services, which may result in network partition. To solve the problem, some economic models have been proposed to stimulate cooperation among nodes. Monitoring techniques are also used to ensure proper provision of network services. For instance, a node in promiscuous mode can monitor the communications in the vicinity.
- **Authentication** ensures that the identity of a node in communication is indeed the entity it declares to be. Authentication can prevent identity masquerade and unauthorized access to resource or information. Authentication is usually provided by digital signature or possession of a secret (such as a key). Due to stringent resource constraint of MANETs, the authentication protocols for the traditional Internet are not applicable because these protocols consume too much computational resources. Some authentication approaches that use one-way hash function, which proves to be faster than other cryptographic operations, have drawn much attention because of their efficiency.
- **Integrity** ensures that a message in transmission has not been maliciously altered or corrupted. A message can be corrupted due to presence of malicious attacks, or communication failures, which may be common on the lossy channels of ad hoc networks. In addition to the traditional approaches for the Internet, some researchers proposed that a node could perform integrity check by overhearing the next hop when this next hop forwards the packet on along the path. This overhearing technique can be easily used in ad hoc networks because of the open nature of the communication channels.
- **Confidentiality** guarantees that sensitive information is not disclosed to unauthorized entities. Encryption used in wired networks is also used for MANETs.
- **Nonrepudiation** ensures that the origin of a message cannot deny having sent the message. Nonrepudiation allows a malicious node who has sent false information to be accused by legitimate users, and therefore is important in intrusion detection. Asymmetric key cryptography has been used to provide nonrepudiation for both the Internet and MANETs.

Other security services for MANETs include *authorization* and *accounting*. But to our best of knowledge, not much research work has been published on authorization or accounting especially for MANETs.

Security Challenges

Security is an important issue for mobile ad hoc networks, especially for those in security-sensitive environments. However, the unique characteristics of MANETs, such as absence of infrastructure, rapid and unpredictable change of topology, open and shared wireless medium, and stringent resource constraints, have posed nontrivial challenges to security designs.

First, use of open shared medium makes an ad hoc network susceptible to attacks such as eavesdropping, signal jamming, impersonation, message distortion, and message injection. A malicious node is able to impersonate other nodes even without gaining physical access to the victims.

Second, absence of infrastructure and frequent change of topology and membership has tremendously raised the probability of a network being compromised. Unlike the traditional wired networks, MANETs do not have dedicated routers to form a clear line of defense where traffic monitoring or access control mechanisms can be deployed. In addition, each mobile node functioning as a router and participating in routing and packet forwarding may lead to significant vulnerability since a malicious node en route can tamper the routing and data packets. In an ad hoc network, it is also difficult to introduce a central administrative entity to security solutions in that such an entity can easily become a target of attacks, which may then cause a failure of the entire network.

Third, due to constraints of resources (such as power, bandwidth, CPU capacity and memory), security mechanisms for MANETs must be lightweight in terms of communication overhead, computation complexity, and storage overhead. Asymmetric cryptography is usually considered too expensive for MANETs. Therefore symmetric cryptographic algorithms and one-way functions are commonly used to protect data integrity and confidentiality.

Last, an ad hoc network may consist of a great number of nodes, which renders *scalability* another main concern for security solutions.

The security mechanisms or approaches should be adapted to the characteristics of MANETs. Yang, Luo, Ye, Lu, and Zhang (2004) propose that the security solutions for MANETs should accommodate the following needs. First, the prevention and detection mechanisms should be fully distributed through the network. They should collect security information from individual nodes to secure the entire network. The security devices on each node are able to work alone in local prevention and detection with limited computational resources and battery power. Second, the security mechanism on different layers of the protocol stack should all cooperate and contribute to a line of defense. Also, all the three components of intrusion prevention, detection, and response, should be used to provide security. Finally, the security solutions should be able to adapt to the highly dynamic topology. It is difficult to accommodate these needs.

SECURITY SOLUTIONS FOR MANETS

Several security techniques or mechanisms have been commonly used to provide security features to MAC layer and ad hoc routing.

Digital signature can protect integrity of data packets or nonmutable fields in routing packets. However, public key cryptography is much slower than symmetric key cryptography, especially on devices with limited resources (such as CPU power and memory space), and has been considered unacceptable for MANETs by many researchers. Moreover, it is vulnerable to DoS attacks of flooding network with bogus packets for which signature verification is required. But public key cryptography is still adopted in many security mechanisms because of its superiority in key distribution and its effectiveness in providing integrity and nonrepudiation services.

Hash function is much faster than public key cryptography and therefore well suits the requirement of low overhead for MANETs. Hash chain is usually used to protect authentication for mutable fields in neighboring communications. Hash chain is built by applying a one-way hash function repeatedly. To create a one-way hash chain, a node should choose a random value and then generate a list of hash values, $h_0, h_1, h_2, ..., h_n$, from the random value, where $h_{i+1} = H(h_i)$ for $0 \leq i < n$, where H is the hash function. To use a one-way hash chain for authentication, h_n should be distributed first. Consecutive element, h_i, can be authenticate by applying H to previously distributed element, h_j $(j > i)$, for $(j - i)$ times.

Monitoring technique has been proved an effective way to provide availability to routing advertisement or data packet forwarding, and to promote fair share of bandwidth at MAC layer. To monitor, nodes turn on *promiscuous* mode to listen to communication of neighboring nodes in order to ensure proper transmission of frames or packets.

Reputation mechanisms have been used together with *cooperatio*n mechanisms to enhance security in routing and MAC layer protocols. It will be discussed in "Cooperation" topic in a later section.

Wireless MAC Security

MAC protocols for wireless networks such as IEEE 802.11 (1999) use a contention resolution mechanism for sharing the open communication channel. This resolution mechanism is fully distributed and requires cooperation among all the participating nodes. The participating nodes are expected to perform a random *backoff* before transmission to reduce contention and to ensure a reasonably fair share of the channel.

However, in an untrusted network environment where selfish or malicious nodes may be included, cooperation cannot always be guaranteed. A *selfish* node may intentionally deviate from MAC protocols to maximize its throughput by obtaining an unfair share of the bandwidth. A *malicious* node may intend denial-of-service (DoS) attacks by injecting frames on the wireless medium continuously, or intermittently with the intention of conserving its own energy. The injection may cause radio collisions and transmission jamming, and thus repeated backoffs among legitimate nodes. A malicious node can also transmit strong noise signals to prevent messages in the victim vicinity from being received.

No matter a node is selfish or malicious, the consequences of their misbehaviors can be severe and disastrous, and therefore should be addressed as security problems with essential concerns. The security solution is to detect misbehaviors and to locate the misbehaving nodes in a timely and reliable manner. This is not a trivial task due to the random nature of the MAC protocols and the shared and volatile medium. It is especially difficult to differentiate between misbehavior and an occasional deviation caused by impairment of wireless link.

Several approaches have been proposed to handle selfish and malicious misbehaviors at the MAC layer[1].

One approach is to address selfish misbehaviors by using game theoretic techniques to find a state where the misbehaving nodes cannot gain any advantage over the well-behaved nodes (Cagalj, Ganeriwal, Aad, & Hubaux, 2004; Konorski, 2001, 2002; Mackenzie & Wicker, 2000, 2003; Michiardi & Molva, 2002b). This approach has also been used at network layer to secure routings.

Konorski (2001, 2002) proposes a game theoretic model that targets *selfish* nodes who fail to adhere to MAC protocols by waiting for smaller backoff intervals than supposed to be. By applying the noncooperative game model (Jones, 2000), the approach modifies the backoff algorithm using blackbursts and leads the game to a Nash equilibrium point (Nash, 1950). The approach requires accurate measurement of the duration, which is difficult to grant in MANETs. Cagalj et al. (2004) developed a strategy that employs two Markov chains (Jha, Tan, & Maxion, 2001) to derive from contention windows the access possibilities of the misbehaving nodes and the well-behaved nodes, respectively. The approach can reach the Nash equilibrium with multiple selfish nodes.

Another approach, which has been mostly used, is to monitor the neighboring node by overhearing and then penalize the identified misbehaving nodes (Gupta, Krishnamurthy, & Faloutsos, 2002; Kyasanur & Vaidya, 2005; Radosavac, Baras, &

Koutsopoulos, 2005; Radosavac, Cardenas, Baras, & Moustakides, 2006; Raya, Hubaux, & Aad, 2004; Xu, Trappe, Zhang, & Wood, 2005).

Raya et al. (2004) deals with MAC misbehaviors in wireless hot-spot communities, such as intentionally scramble frames or illegal manipulation of backoff intervals also. A sequence of observations is required to detect misbehaviors based on the extent to which MAC protocol parameters are manipulated.

Kyasanur and Vaidya (2005) propose modifications to IEEE 802.11, such as letting the receiver of the particular transmission decide whether the sender has deviated from the protocol. It is proposed to use additional nodes in the vicinity to detect collusions between the receiver and the sender. The authors also present a diagnosis scheme, which uses a moving window and thresh to capture the misbehaving nodes. A scheme for punishing a selfish node is also presented. Simulation results show that the detection and penalty schemes are effective in handling selfish MAC misbehaviors.

Radosavac et al. (2005) propose to let a node compute the backoff values of its neighboring node based on the RTS (request-to-send), CTS (clear-to-send), or ACK (acknowledgement) messages. The problem is cast into a "minimax robust detection framework," in which the worst-case instance of attack will be identified and a detection rule of optimum performance is generated with uncertain information. The approach requires clock synchronization, which is considered not realistic by some researchers. A recently published work by Radosavac et al. (2006) is an advanced version of the published work of Radosavac et al. in 2005. The work studies the single-node attacks as well as colluding attacks.

Gupta et al. (2002) and Xu et al. (2005) studied the DoS attacks at MAC layer and analyzed different attack models with their traffic patterns. Gupta et al. (2002) demonstrate simulation of IEEE 802.11 protocol as well as emulation of a perfectly fair MAC (FAIRMAC) protocol in order to show how the employment of MAC layer fairness can prevent or alleviate the effect of the DoS attacks. The authors also show that many other factors such as location of the malicious node, availability of other compromised nodes, availability of routing information, together with the fairness, determine the efficacy of the DoS attacks. Xu et al. (2005) also provide interesting insights into jamming attacks at MAC layer. They proposed four jamming attack models that can be used by an adversary who intend DoS attacks: *constant*, *deceptive*, *random*, and *reactive jamming*. The effectiveness of the four jammer strategies is evaluated by implementation of a prototype using Berkeley Motes platform. Different measurements for detecting jamming attacks are proposed. The authors found that not a single measurement is sufficient to conclusively differentiate malicious attacks from link impairment.

To reliably detect misbehaviors at MAC layer, accurately and reliably monitoring the transmission pattern from a node is a critical factor and still worth further investigation.

Secure Routing Protocols

Routing protocols for MANETs are very different from those existing Internet protocols, because MANETs are self-organized and the protocols need to cope with frequent topology change, open shared medium, and resource restrictions. In addition, all the nodes also serve as routers, participating in route discovery, route maintenance, and packet delivery. These characteristics have introduced significant difficulty to routing security in MANETs.

In 1996, The Internet Engineering Task Force (IETF) established a MANET workgroup (Macker & Chakeres, 2006), which goal is "to standardize of the IP routing protocol functionality suitable for wireless routing applications." Since then, some routing protocols have been proposed particularly for MANETs.

AODV (ad hoc on-demand vector) (Perkins, Belding-Royer, & Das, 2003) is a reactive routing protocol. In AODV, the node who needs to establish a route to another node will broadcast a route request (RREQ) message to its neighbors. Each node that receives the message establishes a reverse link toward the originator of the RREQ, unless such a link has already existed. *Dynamic source routing* (DSR) (Johnson, Maltz, & Hu, 2004) is a protocol that uses *source routing* technique, in which the sender constructs a "source route" in the packet's header that gives the hosts on the path. *Destination-sequenced distance-vector* (DSDV) (Perkins & Bhagwat, 1994) is a proactive routing protocol which maintains a routing table that lists all possible destinations in the network as well as metric and next hop to the destination.

These protocols are designed without security concern in mind, and therefore are susceptible to various attacks.

Attacks on MANET Routing

A selfish or malicious node can disrupt routing services *passively* or *actively*. Their purposes include selfish conservation of own resource, disruption of routing, excessive resource consumption, and so forth.

A selfish node may refuse to participate in routing by simply discarding routing packets. This attack is usually not defended against secure routing protocols in that the node can still fail to forward data packets even if a path including the selfish node has been established. To prevent this attack, some cooperation mechanisms have been proposed, which will be discussed later.

A malicious node can maliciously advertise falsified routing information by tampering fields such as source, destination, metric, and so forth. For example, an attacker can claim falsified short distance information by advertising zero or a very small metric in order to attract and later drop the traffic originally destined to other nodes (*blackhole* attack), or in order to include itself on the path so that it can analyze the communications. Another example is that an attacker can use forged routing packets to create a routing loop, causing packets to circulate in the network without reaching their destinations. This malicious attack should be distinguished from nodes unknowingly providing incorrect or obsolete routing information, which may result from topology change. This is not a trivial task due to the nature of ad hoc networks.

Another type of attack, *wormhole* attack (Hu, Perrig, & Johnson, 2003a), happens when two malicious nodes establish a link via private network connection and forward all the received traffic to each other. In this type of attack the normal flow of routing packets will be short-circuited, and a virtual vertex cut of nodes can be created in the network that the attackers control.

An adversary can also mount a *replay* attack by sending an old advertisement in an attempt to get other nodes to update its routing table with stale routes. Sequence number is usually used to prevent packets from being repeatedly passed on.

Denial-of-service (DoS) attack can be attempted by injecting packets into the network which may cause excessive consumption of resources. One special type of DoS attacks, *jellyfish attacks* (Aad Hubaux, & Knightly, 2004), is to hold packets unnecessarily for some amount of time before forwarding them. The jellyfish attack can cause high end-to-end delay and delay jitter. *Rushing attacks* (Hu, Perrig, & Johnson, 2003b) takes advantage of the suppression mechanisms that are used by on-demand routing protocols to prevent duplicate routing requests from being spread. The suppression mechanism processes only the first request while skipping the duplicate ones. All these attacks are difficult to detect in MANETs due to the inherent volatility of the communication channels.

Besides failing to follow routing protocols, which is sometimes referred as *routing attacks*, an attacker may also target the data messages traversing an established path. A misbehaving node may maliciously alter or drop data packets in transit, which is called *packet forwarding attacks*. These two types of attacks are different due to the differences of routing and data packets. Usually, routing packets are altered as they circulate around the network (such as in *metric* field that states the shortest distance to destination). Thus routing packets are *mutable*, and called *hop-by-hop* transmission. The data packets are *nonmutable*, because the data are not changed during transmission (except for some particular fields in the header) and therefore is *end-to-end* transmission. The integrity of the data packets can be protected by traditional cryptographic operations, while routing packets are hard to protect.

Secure Routing Protocols

Sanzgiri, Dahill, Levine, Shields, and Royer (2002) propose the authenticated routing for ad hoc networks (ARAN), which is a secure protocol that provides authentication and nonrepudiation to route discovery and maintenance. ARAN requires that each node have a certificate signed by a trusted certificate server. It introduces much overhead by requiring every node that forwards route request to sign the certificate, and therefore is vulnerable to DoS attacks.

Papadimitratos and Haas (2002) propose the secure routing protocol (SRP), which can be applied to DSR. SRP requires a security association between the source and destination nodes and uses the association to authenticate route request and route reply messages. Malicious modifications of the routing messages will be detected at the destination. SRP does not attempt to secure route error messages, therefore the messages are subject to forgery.

Ariadne was developed by Hu, Johnson, and Perrig (2002) based on DSR. Ariadne can authenticate routing messages using one of the three schemes: shared secret keys between all pairs of nodes, shared secret keys between communicating nodes combined with broadcast authentication, or digital signatures. Ariadne uses symmetric cryptography primitives, with TESLA(timed efficient stream loss-tolerant authentication) (Perrig, Canetti, Song, & Tygar, 2001; Perrig, Canetti, Tygar, & Song, 2002), a broadcast authentication scheme that requires time synchronization. Some researchers argue time synchronization is an unrealistic requirement for ad hoc networks.

Hu, Perrig, and Johnson (2002) designed a secure efficient ad hoc distance vector routing protocol (SEAD) for DSDV to prevent from attacks of DoS, replay attacks, and wormhole attacks. SEAD also uses hash chains to authenticate metric and sequence numbers. SEAD does not use asymmetric cryptography operations thus the authentication overhead is maintained at a reasonable level.

Zapata (2006) proposed secure AODV (SAODV). SAODV is a secure extension of the AODV routing protocol that can be used to protect the route discovery. SAODV uses a digital signature to authenticate in an end-to-end manner and to protect the integrity of the nonmutable fields in routing messages (such as source, destination, sequence number, etc.). Hash chain is used to authenticate in a hop-by-hop manner the hop-count information, which is the only mutable field in the messages. A signature extension is added to the original AODV RREQ and RREP messages for authentication with signature and hash chain.

Hu et al. (2003) also designed a mechanism, called *packet leash*, to defend against wormhole attack. However, the mechanism requires clock synchronization. Song, Qian, and Li (2005) therefore proposed a statistical approach that eliminates the need of clock synchronization. An approach to defend against rushing attacks has also been proposed (Hu et al., 2003).

Cooperation in MANETs

The presence of selfish nodes that do not respect the routing protocols or MAC protocols can cause performance degradation or even network partition. This subsection will discuss the approaches that have been proposed to solve this problem. Most of these approaches are used at the network layer, but some approaches can also be applied to MAC layer with proper modifications.

1. **One of the approaches is to detect misbehaving nodes and then avoid such nodes in routing.**

Marti Giuli, Lai, and Baker (2000) propose two techniques, *watchdog* and *pathrater*, to detect the presence of nodes that have agreed to forward packets but fail to do so. The watchdog, run by each node on a path, can identify the misbehaving node by monitoring the next hop to ensure that the packets are timely passed on. Although it can detect misbehaviors at the forwarding level, the watchdog might not be able to detect in the presence of collisions, colluding attacks, and partial dropping. The pathrater can help to avoid the misbehaving nodes. Each node maintains a rating for every other node in the network. The rating will be incremented if the node is on an actively used path, and decremented on a broken path. A node calculates a path metric by averaging the node ratings in the path.

2. **Another approach is to design protocols that stimulate cooperation by penalizing misbehavior or rewarding behavior of forwarding for other nodes' benefit.**

Buttyan and Hubaux (2000, 2003) propose a protocol that can stimulate packet forwarding. It requires a node to pass all packets to its security module, which maintains a counter called *nuglet counter*. The counter is decreased whenever the node sends a packet as the originator, and increased when the node forwards a packet for another node. Since the value of the counter must remain positive, a node needs to maintain a balance on the counter by forwarding packets for the benefits of others to have its own packets to be sent. To prevent a node from illegitimately increasing its own counter, the counter is required to be maintained by a trusted and tamper resistant hardware module (such as a Smart card).

CONFIDANT (cooperation of nodes fairness in dynamic ad-hoc networks) (Buchegger & Boudec, 2001, 2002a, 2002b) was proposed to detect, discourage and stop selfish misbehaviors. CONFIDANT consists of four components: a *monitor* to observe the neighborhood; a *trust manager* to deal with incoming and outgoing warning messages; a *reputation system* to maintain reputation records based on own experiences, vicinity observations, and reported records; and a *path manager* for nodes to adapt their behavior according to the reputation of a node or a path. CONFIDANT takes into consideration the problem of nodes providing false information to gain good reputation. With a proper weight system and a modified Bayesian estimation procedure, the second-hand information can still speed up the detection while suppressing false positives and negatives. The simulation results show that the network performance can still be good even when half of the network population misbehaves.

CORE is a collaborative reputation mechanism (Michiardi & Molva, 2002a). Similarly to CONFIDANT, CORE also differentiates between observations and reports by other nodes. It applies different weights to subjective reputation (observations), indirect reputation (positive reputation reported by others), and functional reputation (the subjective and indirect reputation calculated with respect to different functions). At each node, reputation values are stored in a reputation table, and a watchdog mechanism is used to detect misbehaving nodes.

Sprite is a cheat-proof and credit-based system (Zhong, Chen, & Yang, 2003), which also requires that nodes receive enough credits by forwarding for other nodes to send their own packets. To prove a node has received or forwarded a message, the node keeps a receipt of the message and uploads the receipt to a credit clearance service (CCS). To motivate nodes to report receipts, CCS gives more credits to a node that forwards a message than to a node that does not. Proper actions are taken to prevent the cheating action. If a message is not received by the destination, the credits to the intermediate nodes will be greatly reduced, and therefore the benefit of falsely reporting a receipt by an intermediate node will be reduced too. The approach needs a centralized trusted entity, which is hard for MANETs.

Some other interesting approaches that use punishment or rewarding systems can be found by Mohan and Joiner (2004) and Salem, Buttyan, Hubaux, and Jakobsson (2003).

3. **Game-theoretic techniques (Jones, 2000) have also been used to develop protocols for stimulating cooperation (Anderegg & Eidenbenz, 2003; Srinivasan, Nuggehalli, Chiasserini, & Rao, 2003).**

These techniques assume that all nodes are selfish and *rational*, that is, they only do things that are beneficial to themselves and their purpose is to maximize their own utility. Usually *noncooperative* game model is used in these approaches. By means of imposing suitable costs on network operation, the game reaches a stable state called "Nash equilibrium" (Nash, 1950), where a selfish node cannot gain an advantage over well-behaved nodes.

Anderegg and Eidenbenz (2003) provide a game theoretic approach, which goal is to achieve truthfulness and cost-efficiency for routing protocols in MANETs. The approach pays the forwarding nodes a premium over their actual costs for forwarding data packets. The authors show that the total overpayment is relatively small.

Although protocols developed with game-theoretic techniques may be resilient to misbehavior, they may not achieve the same performance of protocols developed under the assumption that all nodes are well-behaved.

Authentication and Key Management in MANETs

Authentication and key management are essential problems for MANET security.

Authentication in MANETs

Up to date, a number of authentication protocols have been proposed for MANETs (Balfanz, Smetters, Stewart, & Wong, 2002; Lu & Pooch, 2005; Perrig, Canetti, Tygar, & Song, 2000; Venkatraman & Agrawal, 2000; Weimerskirch & Thonet, 2001).

Stajano and Anderson (1999) propose an approach for ad hoc network of wireless devices: *secure transient association*. The purpose of the approach is to provide transient association between the controller and the peripheral, which is essential for ad hoc authentication. The idea came from the biology fact that a duckling emerging from its egg will recognize the first moving object it sees as its mother. Similarly, the approach defines that a device will recognize the first entity that sends it a secret key as its owner. As soon as this ownership has been established, the relationship will last for the rest of the nodes' life.

Perrig et al. (2001, 2002) propose a broadcast authentication scheme, TESLA, which uses a one-way key chain with delayed key disclosure. TESLA first bootstraps an authentic key from a one-way key chain between the sender and its receivers, and then broadcasts authentications with delayed key disclosure. The delayed key disclosure can prevent a malicious node from tampering a node that has delays in receiving the newest key, by means of using the newest key to forge packets with valid authentication information. Authentication techniques that use one-way hash chain keys can tolerate packet loss and have the advantage of low overhead. TESLA has been adopted by many approaches to authenticate *neighboring communications* in MANETs.

Zhu, Xu, Setia, and Jajodia (2003) propose a light-weight hop-by-hop authentication protocol (LHAP), in which every node authenticates all the packets received from neighbors before forwarding it. LHAP also uses one-way hash chain, like TESLA, but it does not use delayed key disclosure. LHAP uses TRAFFIC chain (a one-way hash chain) to authenticate packets, and uses TESLA chain to authenticate TRAFFIC keys. Security properties and performance is analyzed. The analysis shows that LHAP is lightweight and practical.

Key Management in MANETs

Key management is an essential cryptographic primitive that is the basis of the other security primitives. In the traditional wired networks, centralized key management approaches are usually used. However, an ad hoc network is peer-to-peer and does not have a central administration point. In addition, a central authority may become a single point of failures in case of heavy workload, as well as an easy target of malicious attacks. Therefore, recent research has been focused on looking for key management approaches that are not only efficient, but also well functional on a dynamic network topology, and tolerant to link failures.

A partially or fully distributed certificate authority is commonly used for key management in MANETs.

Zhou and Haas (1999) propose a fully distributed public-key management service for ad hoc networks. It is assumed that the communication channels are reliable, and all nodes in the system know the public key and trust any certificates signed using the corresponding private key. A $(n, t+1)$ *threshold cryptography* scheme is used for distribution of the private key, where the key is divided into n shares. Therefore, n parties are allowed to share the ability to perform a cryptographic operation (e.g., creating a digital signature), and any $t + 1$ parties can perform the operation jointly. To sign a certificate, each server produces a partial signature for the certificate using its share and submits the partial signature to a combiner that can generate the entire signature. In this way, the system can tolerate a certain number ($t < n$) of compromised servers.

A similar approach proposed by Kong, Zerfos, Luo, Lu, and Zhang (2001) provide a more fair distribution by allowing *each* node to carry a secret share. Any $t + 1$ nodes in the vicinity of the requesting node can jointly provide complete service, which increases availability and scalability of the service. However, this scheme is not secure if an attacker can compromise arbitary $t + 1$ nodes and thus can collect enough shares and reconstruct the system's private key.

According to Zhou and Haas (1999) and Kong et al. (2001), a trusted authority is needed for initialization of the first $t + 1$, which is difficult in MANETs. In addition, it is still not clear how to determine the number t initially and adapt t based on n.

Capkun, Buttyan, and Hubaux (2003) propose a fully self-organized public-key management system that does not require use of any trusted authority even in the system initialization phase. Like PGP (pretty good privacy)(Zimmermann, 1995), the scheme allows a node to create public and private keys by itself. But the keys and certificates are not stored in centralized certificate repositories. Instead, they can be stored at the nodes in a fully distributed manner. When a node wants to obtain the public key of another node, it acquires a chain of valid public-key certificates. The first certificate of the chain can be directly verified by using a trusted public key. Then each sequential certificate can be verified using the public key contained in the previous certificate of the chain. The last certificate contains the public key of the target user. The system allows the nodes in the network to perform key authentication based only on their local information.

However, Chan (2004) argues that although some protocols are fully distributed and self-organized without needing any trusted third party (TTP), they are not robust to dynamic topology or sporadic links because they need the routing structure that has been established initially.

Chan (2004) proposes a distributed symmetric key management scheme for MANETs, which uses a fully distributed and self-organized key pre-distribution scheme (DKPS) without relying on TTPs or infrastructure support. The DKPS scheme has three phases, namely *distributed key selection* (DKS), *secure shared-key discovery* (SSD), and *key exclusion property testing* (KEPT). In the DKS phase, each node randomly picks keys from the publicly known universal set to form its key ring, in which *exclusion property* will be ensured to avoid collision. As soon as each node shares a common key with any other node, it enters the SSD phase and broadcasts its key identifiers to others. To guarantee that the nodes can let each other know which keys they are having in common without revealing the keys to others, the author proposes MRS (modified Rivest's scheme) and built SSD upon MRS. MRS is based on the work of Rivest, Adleman, and Dertouzos (1978), and is a special class of encryption functions that allow operations on the encrypted data without needing knowledge of the decryption functions. In KEPT phase, a node tests whether its set of keys satisfy the exclusion property.

Crepeau and Davis (2003) provide a certificate revocation scheme that can defend against attacks of maliciously accusing other nodes and using revoked certificate to access network services.

Many researchers are still making efforts to find a secure yet cost-efficient key distribution approach.

Intrusion Detection Systems (IDS) for MANETs

In the traditional Internet, network devices such as routers, switches, and gateways can be used to monitor the traffic. Due to the lack of these network devices and a fixed infrastructure, intrusion detection in MANETs is more challenging than that in the Internet. Moreover, the restriction of resources again brings more difficulty to data analysis, which usually plays an important role in intrusion detection. A comprehensive survey on IDS for MANETs can be found by Avantvalee and Wu (2006).

An IDS for MANETs not only has the same requirements as in the wired networks (such as reliability, minimal false positive and false negative rates, transparency to system and users, etc.), but also requires low usage of system and network resources. Therefore, the design and development of IDS for MANETs is not a trivial task.

A simple solution for IDS in MANETs is that each host relies on itself for detection, where the audit data are gathered and processed locally. Some IDS proposed for MANETs use this solution of letting individual nodes to determine intrusions independently in case the local evidence is strong. But many systems also allow a node to request complementary information from others so that cooperation can be reinforced in case of weak or inconclusive local evidence.

Albers, Camp, Percher, Jouga, Me, and Puttini (2002) propose a local IDS (LIDS), which uses several mobile agents on each node. All the LIDS in a community can collaborate to alert each other of intrusions. These data are independent from operating system and need no additional resources for local information. A LIDS has several data collecting agents of different types: a *local agent* that locally detects intrusions and responds to intrusions; a collection of *mobile agents* that collect and process data from remote hosts; and a *local MIB agent* that collects MIB (management information base) variables for the mobile agents or the local LIDS agent. The implementation of prototypes was claimed by the authors, but the results are not demonstrated in the publication.

A distributed intrusion detection model was later proposed by Zhang, Lee, and Huang (2003). The model of the IDS agent is composed of six modules: a *local data collection module* that collects real-time audit data; a *local detection engine* that performs local anomaly detection; a *cooperative detection engine* that helps collaboration and collects broader data sets from other agents; a *local response module* that triggers local response actions; a *global response module* that coordinates responses among neighboring nodes; and a *secure communication module* that provides secure communication channels among IDS agents. On the anomaly detection model, two classification techniques, RIPPER(repeated incremental prunig to produce error reduction) and SVM (support vector machine) light, are applied to compute classifiers as anomaly detectors. The classifiers are used to detect anomaly updates to routing tables. The performances are evaluated and compared through simulations. The authors find that protocols with strong traffic correlation tend to have better detection performance.

Kachirski and Guha (2003) propose an agent-based IDS that uses multiple mobile sensors to determine intrusions. The system assigns functional tasks different agents: a *network monitoring* agent to monitor network packets (only on certain nodes to preserve resources); a *host monitoring* agent on every node to monitor system and applications level activities; a *decision-making* agent on every node to determine intrusions based on host-level information, and on certain nodes to determine network-level intrusions; and an *action* agent on every node to respond to intrusions. Similarly to the two IDS described above, this system makes intrusion decisions based on both independent and collaborative monitoring, and the level of the monitoring can be adapted according to the availability of the computational and network resources.

Another intrusion detection technique is the dynamic hierarchical intrusion detection architecture proposed by Sterne, Balasubramanyam, Carman, Wilson, Talpade, Ko et al. (2005). The system requires every node to monitor, log, analyze, and respond to detected intrusions. It also uses clustering to form a hierarchical structure. Different nodes (e.g., leaf nodes and clusterhead nodes in the structure) may perform different functions in intrusion detections. This hierarchical structure is advantageous in monitoring end-to-end traffic and thus can help detect end-to-end attacks. The system does not use promiscuous listening, which is arguably unrealistic for MANETs. However, some researchers have also argued that a hierarchical architecture may not be suitable to MANETs either, due to the rapid topology change of MANETs and the high overhead introduced by organizing the hierarchy.

Sun, Wu, and Pooch (2003) propose a zone-based IDS (ZBIDS). ZBIDS divides the network into nonoverlapping zones. The nodes are categorized into two types based on their locations to a zone: intrazone nodes (within a zone and not connected to nodes in another zone) and interzone nodes (within a zone and connected to nodes in another zone). Intrazone nodes are responsible for local detection and broadcast in case of alerts. Interzone nodes perform aggregation and correlation of these local detection results. The system can limit the detection cooperation in a zone, which may reduce the overhead by the broadcast and aggregation. However, the system requires that each node know its physical location, which needs prior design setup. The management of zones is not a trivial task either.

Intrusion detection has been a challenging task for MANETs, mainly due to the distribution nature and resource constraints of ad hoc networks. To determine intrusions with local or incomplete information and with low overhead has been a major concern for researchers.

OPEN CHALLENGES AND CONCLUSION

Challenges

The research in MANET security is still in its early stage. Some areas that are interesting but little explored include accounting, trust management, authentication, and key management.

Accounting provides the method for collecting the information used for billing, auditing, and reporting. Accounting mechanisms can track the services that users are accessing as well as the amount of network resources they are consuming. Accounting is a challenging problem due to the distributed and ephemeral nature of MANETs.

The characteristics of MANETs also bring difficulty to *trust management*. In MANETs, the trustworthiness is evaluated based on the information or evidence provided by peers, not by trusted authorities or a central administration point (as in the Internet or wireless networks with base-stations). Additionally, the gathering of the trust evidence may be difficult due to the small bandwidth, and therefore local information has to be relied on. Evaluation with uncertain and incomplete trust evidence certainly poses challenges to trust management.

Research progress has been made on authentication and key management. But finding cryptographic mechanisms that consume less computational resources and impose lower time complexity is still a major research concern in MANET security.

Another problem for MANET security is to find an effective and efficient approach for *intrusion response*. Many publications simply mentioned that proper actions should be taken to react to intrusions, which may include alarming the other nodes in the network, isolating the compromised nodes, or re-establishing the trust relationship for the entire network. But the problem of *how to* locate and then isolate the compromised nodes is not discussed in details. The location and isolation could be even more difficult when distributed attacks are launched from multiple sources. Eliminating the compromised nodes by rekeying or rebuilding the trust could be an effective solution. However, it is certainly not efficient taking into account the computation and communication overhead it may cause.

Some other unexplored research problems include the tradeoff between privacy (such as identity anonymity and location privacy) and other security services (such as accounting and intrusion detection), and the tradeoff between security strengths and network performance.

Yang et al. (2004) argue that MANET security needs a "multifence security solution," namely *resiliency-oriented* security design. They argue that the existing proposals are attack-oriented because the protocols target some specific attack that has been identified first. These protocols therefore may not work well in the presence of unanticipated attacks. They propose that a security solution is needed that can be embedded into every component or every layer in the network. The solution can offer multiple lines of defense against many both known and unknown security threats.

Besides problems described above, how to adapt the security mechanisms in a large-scale wireless network is also an interesting problem. The scalability of security mechanisms and the compromise between security and network scalability are certainly topics worth further research study.

Conclusion

With the rapid proliferation of wireless networks and mobile computing applications, MANETs have received increased attention. Security is an important feature for ad hoc networks, especially in untrustworthy environments such as battlefields. Development of security solutions for ad hoc networks has therefore become a major research concern.

However, the characteristics of ad hoc networks have not only introduced vulnerabilities to malicious attacks varying from passive eavesdropping to active interfering, but also imposed difficulty and challenges in introducing security features to MANETs.

This book chapter has discussed the security vulnerabilities, challenges, and security solutions for MANETs. A variety of attacks and their countermeasures have been identified for different network operations, mechanisms, and network layers. Existing research efforts as well as the open challenges were discussed in the chapter.

REFERENCES

Aad, I., Hubaux, J.-P., & Knightly, E.W. (2004). Denial of service resilience in ad hoc networks. In *Proceedings of the ACM International Conference on Mobile Computing and Networking (MobiCom 2004),* Philadelphia, (pp. 202-215).

Albers, P., Camp, O., Percher, J., Jouga, B., Me, L., & Puttini, R. (2002). Security in ad hoc networks: A general intrusion detection architecture enhancing trust based approaches. In *Proceedings of the 1st International Workshop on Wireless Information Systems (WIS-2002)* (pp. 1-12).

Anderegg, L., & Eidenbenz, S. (2003). Routing and forwarding: Ad hoc-VCG: A truthful and cost-efficient routing protocol for mobile ad hoc networks with selfish agents. In *Proceedings of the 9th Annual International Conference on Mobile Computing and Networking (MobiCom 05),* San Diego, (pp. 245-259). ACM Press.

Avantvalee, T., & Wu, J. (2006). A survey on intrusion detection in mobile ad hoc networks. In Y. Xiao, X. Shen, & D. -Z. Du (Eds.), *Wireless/mobile network security* (pp. 170-196).

Balfanz, D., Smetters, D.K., Stewart, P., & Wong, H.C. (2002). *Talking to strangers: Authentication in ad-hoc wireless networks.* Paper presented at the Symposium on Network and Distributed Systems Security (NDSS '02), San Diego.

Buchegger, S., & Boudec, J.L. (2001). *The selfish node: Increasing routing security in mobile ad hoc networks* (IBM Research Report: RR 3354).

Buchegger, S., & Boudec, J.L. (2002a) Nodes bearing grudges: Towards routing security, fairness, and robustness in mobile ad hoc networks. In *Proceedings of the Tenth Euromicro Workshop on Parallel, Distributed and Network-based Processing,* Canary Islands, Spain, (pp. 403-410). IEEE Computer Society.

Buchegger, S., & Boudec, J.L. (2002b). Performance analysis of the CONFIDANT protocol: Cooperation of nodes - fairness in dynamic ad-hoc networks. In *Proceedings of IEEE/ACM Symposium on Mobile Ad Hoc Networking and Computing (MobiHoc),* Lausanne, CH, (pp. 226-236). ACM Press.

Buttyán, L., & Hubaux, J.-P. (2000). Enforcing service availability in mobile ad-hoc WANs. In *Proceedings of Workshop on Mobile Ad-hoc networking and Computing (MobiHOC),* Boston, *(pp. 87-96).*

Buttyán, L., & Hubaux, J.-P. (2003). Stimulating cooperation in self-organizing mobile ad hoc networks. *Mobile Networks and Applications,* *8*(5), 579-592.

Cagalj, M., Ganeriwal, S., Aad, I., & Hubaux, J.-P. (2004). *On cheating in CSMA/CA ad hoc networks* (Tech. Rep. IC/2004/27, EPFL-DI-ICA). Lausanne, Switzerland: Swiss Federal Institute of Technology Lausanne.

Capkun, S., Buttyan, L., & Hubaux, J.-P. (2003). Self-organized public-key management for mobile ad hoc networks. *IEEE Transactions on Mobile Computing, 2*(1), 52-64.

Chan, A.C.-F. (2004). Distributed symmetric key management for mobile ad hoc networks. In *Proceedings of the 23rd Annual Joint Conference of the IEEE Computer and Communications Societies (INFOCOM),* Hong Kong, China, (pp. 2414-2424). IEEE.

Crepeau, C., & Davis, C. R. (2003). A certificate revocation scheme for wireless ad hoc networks. In *Proceedings of the 1st ACM Workshop Security of Ad Hoc and Sensor Networks*, Fairfax, Virginia, (pp. 54-61). ACM Press.

Gupta, V., Krishnamurthy, S., & Faloutsos, M. (2002). Denial of service attacks at the MAC layer in wireless ad hoc networks. In *Proceedings of MILCOM.*

Hu, Y.C., Johnson, D., & Perrig, A. (2002). SEAD: Secure efficient distance vector routing for mobile wireless ad hoc networks. In *Proceedings of the 4th IEEE Workshop on Mobile Computing Systems and Applications (WMCSA '02),* Callicoon, New York, (pp. 3-13).

Hu, Y.C., Perrig, A., & Johnson, D. (2002). Ariadne: A secure on-demand routing protocol for ad hoc networks. In *Proceedings of the 8th ACM International Conference on Mobile Computing and Networking (MobiCom)*, Atlanta, Georgia, (pp. 12-23). ACM Press.

Hu, Y.C., Perrig, A., & Johnson, D. (2003a). Packet leashes: A defense against wormhole attacks in wireless ad hoc networks. In *Proceedings of the Twenty-Second Annual Joint Conference of the IEEE Computer and Communications Societies (INFOCOM 2003)* (pp. 1976-1986). IEEE.

Hu, Y.C., Perrig, A., & Johnson, D. (2003b). Rushing attacks and defense in wireless ad hoc network routing protocols. In *Proceedings of ACM WiSe 2003*, San Diego, (pp. 30-40). ACM Press.

IEEE. (1999). *Standard for wireless LAN-medium access control and physical layer specification, P802.11.*

Jha, S., Tan, K., & Maxion, R. (2001). Markov chains, classifiers, and intrusion detection. In *Proceedings of the 14th IEEE Computer Security Foundations Workshop,* Cape Breton, Nova Scotia, Canada, (pp. 206-219).

Johnson, D.B., Maltz, D.A., & Hu, Y. (2004). The dynamic source routing protocol for mobile ad hoc networks (DSR). *INTERNET DRAFT, MANET working group.* Retrieved November 17th, 2006, from http://www.ietf.org/internet-drafts/draft-ietf-manet-dsr-10.txt

Jones, A. (2000). *Game theory: Mathematical models of conflict* (pp. 210-236). Horwood Publishing.

Kachirski, O., & Guha, R. (2003). Effective intrusion detection using multiple sensors in wireless ad hoc networks. In *Proceedings of the 36th Annual Hawaii International Conference on System Sciences (HICSS'03)* (pp. 57.1-57.8). IEEE.

Kong, J., Zerfos, P., Luo, H., Lu, S., & Zhang, L. (2001). Providing robust and ubiquitous security support for mobile ad hoc networks. In *Proceedings of the 9th International Conference on Network Protocols (ICNP)* (pp. 251 - 260). ACM Press.

Konorski, J. (2001). *Protection of fairness for multimedia traffic streams in a non-cooperative wireless LAN setting.* Paper presented at PROMS (LNCS 2213, pp. 116-129). Springer.

Konorski, J. (2002). Multiple access in ad-hoc wireless LANs with noncooperative stations. *Networking* (LNCS 2345, pp. 1141-1146). Springer.

Kyasanur, P., & Vaidya, N.H. (2005). Selfish MAC layer misbehavior in wireless networks. *IEEE Transactions on Mobile Computing, 4*(5), 502-516.

Lu, B., & Pooch, U.W. (2005). A lightweight authentication protocol for mobile ad hoc networks. In *Proceedings of the International Conference on Information Technology: Coding and Computing (ITCC'05),* Las Vegas, (pp. 546-551). ACM Press.

Mackenzie, A.B., & Wicker, S.B. (2000). Game theory and the design of self-configuring, adaptive wireless networks. *IEEE Communications Magazine, 39*(11), 126-131.

Mackenzie, A.B., & Wicker, S.B. (2003). Stability of multipacket slotted aloha with selfish users and perfect information. In *Proceedings of Infocom 2003*, San Francisco, (pp. 1583 -1590). IEEE.

Macker, J., & Chakeres, I. (2006). *Mobile ad-hoc networks (MANET).* Retrieved November 17th, 2006, from http://www.ietf.org/html.charters/manet-charter.html

Marti, S., Giuli, T., Lai, K., & Baker, M. (2000). Mitigating routing misbehavior in mobile ad hoc networks. In *Proceedings of the 6th ACM International Conference on Mobile Computing and Networking (MobiHoc'05),* Urbana Champaign, IL, (pp. 255- 265). ACM Press.

Michiardi, P., & Molva, R. (2002a). *CORE: A collaborative reputation mechanism to enforce node cooperation in mobile ad hoc networks.* Paper presented at the Sixth IFIP Conference on Security Communications, and Multimedia (CMS 2002), Portoroz, Slovenia.

Michiardi, P., & Molva, R. (2002b). *Game theoretic analysis of security in mobile ad hoc networks* (Tech. Rep. RR-02-070). Institut Eurecom.

Mohan, M., & Joiner, L.L. (2004). Solving billing issues in ad hoc networks. In *Proceedings of ACMSE '04,* Huntsville, AL, (pp. 31-36). ACM Press.

Nash, J. (1950). The bargaining problem. *Econometrica, 18*, 155-162. The Econometric Society.

Papadimitratos, P., & Haas, Z.J. (2002). *Secure routing for mobile ad hoc networks.* Paper presented at the SCS Communication Networks and Distributed Systems Modeling and Simulation Conference (CNDS 2002), San Antonio, TX.

Perkins, C.E. (Ed.). (2001). *Ad hoc networks.* Upper Saddle River, NJ: Addison-Wesley.

Perkins, C.E., Belding-Royer, E.M., & Das, S.R. (2003). Ad hoc on-demand distance vector (AODV) routing. *Internet request for comments RFC 3561.* Retrieved November 17th, 2006, from http://www.ietf.org/rfc/rfc3561.txt.

Perkins, C.E., & Bhagwat, P. (1994). *Highly dynamic destination-sequenced distance-vector routing (DSDV) for mobile computers.* Paper presented at the ACM Conference on Communications Architectures, Protocols and Applications (SIGCOMM '94) London, (pp. 234-244). ACM Press.

Perrig, A., Canetti, R., Song, D., & Tygar, D. (2001). Efficient and secure source authentication for multicast. In *Proceedings of Network and Distributed System Security Symposium (NDSS'01),* San Diego, CA, (pp. 35-46).

Perrig, A., Canetti, R., Tygar, D., & Song, D. (2000) Efficient authentication and signing of multicast streams over lossy channels. In *Proceedings of IEEE Symposium on Security and Privacy,* Berkeley, CA, (pp. 56-73). IEEE

Perrig, A., Canetti, R., Tygar, D., & Song, D. (2002, Summer). The TESLA broadcast authentication protocol. *RSA CryptoBytes, 5*, 2-13.

Radosavac, S., Baras, J.S., & Koutsopoulos, I. (2005). *A framework for MAC protocol misbehavior detection in wireless networks.* Paper presented at the Wireless Security Workshop (WiSe '05), Cologne, Germany, (pp. 33-42).

Radosavac, S., Cardenas, A., Baras, J.S., & Moustakides, G. (2006). Detecting IEEE 802.11 MAC layer misbehavior in ad hoc networks: Robust strategies against individual and colluding attacker. *Journal of Computer Security: Special Issue on Security of Ad Hoc and Sensor Networks 15*(2007), 103-128.

Raya, M., Hubaux, J.-P., & Aad, I. (2004). DOMINO: A system to detect greedy behavior in IEEE

802.11hotspots. In *Proceedings of the Second International Conference on Mobile Systems, Applications, and Services (MobiSys '04),* Boston, MA, (pp. 84-97).

Rivest, R.L., Adleman, L., & Dertouzos, M.L. (1978). On data banks and privacy homomorphisms (pp. 169-179). *Foundations of secure computation.* Academic Press.

Salem, N.B., Buttyan, L., Hubaux, J.-P., & Jakobsson, M. (2003). A charging and rewarding scheme for packet forwarding in multi-hop cellular networks. In *Proceedings of MobiHoc'03,* Annapolis, MD, (pp. 13-24). ACM Press.

Sanzgiri, K., Dahill, B., Levine, B.N., Shields, C., & Royer, E.M. (2002). A secure routing protocol for ad hoc networks. In *Proceedings of the 10th IEEE International Conference on Network Protocols (ICNP'02),* Paris, (pp. 78-87). IEEE.

Song, N., Qian, L., & Li, X. (2005). Wormhole attacks detection in wireless ad hoc networks: A statistical analysis approach. In *Proceedings of 19th IEEE International Parallel and Distributed Processing Symposium (IPDPS '05),* Denver, CO, (pp. 289-296).

Srinivasan, V., Nuggehalli, P., Chiasserini, C.F., & Rao, R.R. (2003). Cooperation in wireless ad hoc networks. In *Proceedings of IEEE INFOCOM,* San Francisco, (pp. 808-817).

Stajano, F., & Anderson, R.J. (1999). The resurrecting duckling: Security issues for ad-hoc wireless networks. In B. Christiano, B. Crispo, & M. Roe (Eds.), *Security Protocols, 7th International Workshop Proceedings* (LNCS, vol. 1796, pp. 172-194).

Sterne, D., Balasubramanyam, P., Carman, D., Wilson, B., Talpade, R., Ko, C., et al. (2005). A general cooperative intrusion detection architecture for MANETs. In *Proceedings of the 3rd IEEE International Workshop on Information Assurance (IWIA '05),* Oahu, HI, (pp. 57-70).

Sun, B., Wu, K., & Pooch, U.W. (2003). Alert aggregation in mobile ad hoc networks. In *Proceedings of the 2003 ACM Workshop on Wireless Security (WiSe '03) in conjunction with the 9th Annual International Conference on Mobile Computing and Networking (MobiCom '03),* San Diego, (pp. 69-78). ACM Press.

Venkatraman, L., & Agrawal, D. (2000). *A novel authentication scheme for ad hoc networks.* Paper presented at the IEEE Wireless Communications and Networking Conference (WCNC 2000), Chicago, IL, (Vol. 3, pp. 1268-1273). IEEE.

Weimerskirch, A., & Thonet, G. (2001). A distributed light-weight authentication model for ad-hoc networks. In *Proceedings of 4th International Conference on Information Security and Cryptology (ICISC 2001),* Seoul, Korea, (pp. 341-354). ACM Press.

Xu, W., Trappe, W., Zhang, Y., & Wood, T. (2005). The feasibility of launching and detecting jamming attacks in wireless networks. In *Proceedings of the Sixth ACM International Symposium on Mobile Ad Hoc Networking and Computing (MobiHoc '05),* Urbana Champaign, IL, (pp. 48-57). ACM Press.

Yang, H., Luo, H., Ye, F., Lu, S., & Zhang, L. (2004). Security in mobile ad hoc networks: Challenges and solutions. *IEEE Wireless Communications, 11*(1), 38-47.

Zapata, M.G. (2006). Secure ad hoc on-demand distance vector (SAODV) routing. *INTERNET DRAFT, MANET working group.* Retrieved December 12th, 2006, from http://www.ietf.org/internet-drafts/draft-guerrero-manet-saodv-06.txt.

Zhang, Y., Lee, W., & Huang, Y. (2003). Intrusion detection techniques for mobile wireless networks. *Wireless Networks Journal (ACM WINET), 9*(5), 545-556. ACM/Kluwer Press.

Zhong, S., Chen, J., & Yang, Y.R. (2003). Sprite: A simple, cheat-proof, credit-based system for mobile ad-hoc networks. In *Proceedings of IEEE Infocom,* San Francisco, (pp. 1987-1997). IEEE.

Zhou, L., & Haas, Z. (1999). Securing ad hoc networks. *IEEE Network, 6*(13), 24-30.

Zhu, S., Xu, S., Setia, S., & Jajodia, S. (2003). LHAP: A lightweight hop-by-hop authentication protocol for ad-hoc networks. In *Proceedings of 23rd International Conference on Distributed Computing Systems Workshops (ICDCSW '03)*, Providence, RI, (pp. 749-755). IEEE.

Zimmermann, P. (1995). *The official PGP user's guide.* MIT Press.

KEY TERMS

Authentication: The processes of verifying the identity of an entity if it is indeed the entity it declares to be.

Intrusion Detection: The techniques or processes of detecting inappropriate, incorrect, or anomalous activities.

Key Management: The techniques or processes of creating, distributing, and maintaining a secret key, which will be used to protect the secrecy of communications or to ensure the original data are not maliciously altered.

MANET (mobile ad hoc network): An infrastructure-less, self-organizing network of mobile hosts connected with wireless communication channels. A MANET does not have a fixed topology because all the hosts can move freely, which results in rapid and unpredictable topology change.

Medium Access Control (MAC): A sublayer of the data link layer specified in the seven-layer OSI (open systems interconnection) model. It addresses problems of moving data frames across a shared channel.

Routing: The process of selecting paths in a network along which to send data packets.

Security: The concepts, measures, or processes of protecting data from unauthorized access or disruption.

END NOTE

[1] Signal jamming can also be launched at physical layer, but it is not within the scope of this chapter because it is more related to electrical engineering than computer security.

Chapter XXVII
Privacy and Anonymity in Mobile Ad Hoc Networks

Christer Andersson
Combitech, Sweden

Leonardo A. Martucci
Karlstad University, Sweden

Simone Fischer-Hübner
Karlstad University, Sweden

ABSTRACT

Providing privacy is often considered a keystone factor for the ultimate take up and success of mobile ad hoc networking. Privacy can best be protected by enabling anonymous communication and, therefore, this chapter surveys existing anonymous communication mechanisms for mobile ad hoc networks. On the basis of the survey, we conclude that many open research challenges remain regarding anonymity provisioning in mobile ad hoc networks. Finally, we also discuss the notorious Sybil attack in the context of anonymous communication and mobile ad hoc networks.

INTRODUCTION

The quest for privacy in today's increasingly pervasive information society remains a fundamental research challenge. In the traditional (wired) Internet, one essential means for protecting privacy is *anonymous communication*. Being anonymous usually implies that a user remains unlinkable to a set of items of interest (e.g., communication partners, messages) from an attacker's perspective (Pfitzmann & Hansen, 2006). The capabilities of the attacker are usually modeled by an *attacker model*, which can, for instance, include a rogue communication partner or an observer tapping the communication lines. Further, more advanced applications can be deployed on top of anonymous communication mechanisms, to, for instance, enable pseudonymous applications.

This chapter investigates how anonymous communication can be enabled in *mobile ad hoc networks* (Corson & Macker, 1999); networks constituted by mobile platforms that establish on-the-fly wireless connections among themselves and ephemera networks without central entities to control it. They are of great importance as they constitute a basic core functionality needed for deploying *ubiquitous computing*. In short, ubiquitous computing would allow for computational environments providing information instantaneously through "invisible interfaces," thus allowing unlimited spreading and sharing of information. If realized, ubiquitous computing could offer an invaluable support for many aspects of our society and its institutions. However, if privacy aspects are neglected, there is a great likelihood that the end product will resemble an Orwellian nightmare.

In this chapter, we study how privacy and anonymity issues are tackled today in mobile ad hoc networks by surveying existing anonymous communication mechanisms adapted for mobile ad hoc networks[1]. Only recently, a number of such proposals have been suggested. In the survey, we evaluate some of these approaches against a set of general requirements (Andersson, Martucci, & Fischer-Hübner, 2005), which assess to which degree these approaches are suitable for mobile ad hoc networks. We also discuss Sybil attacks (Douceur, 2002) in the context of anonymous communication and mobile ad hoc networks.

This chapter is structured as follows. First, an introduction to privacy, anonymity, and anonymity metrics is provided in "Background." Then, existing approaches for enabling anonymity in ad hoc networks are described in "Anonymous Communication in Mobile Ad Hoc Networks." In "Survey of Anonymous Communication Mechanisms for Ad Hoc Networks" these approaches are evaluated against the aforementioned requirements. Then, Sybil attacks in the context of anonymous communication and mobile ad hoc networks are discussed in "Future Trends." Finally, conclusions are drawn in "Conclusions."

BACKGROUND

In this section, the concepts of privacy and anonymity and their relation are introduced. Methods for quantifying anonymity are also discussed.

Definitions of Anonymity and Related Concepts

Pfitzmann and Hansen (2006) define *anonymity* as "the state of being not identifiable within a set of subjects, the *anonymity set*" (p. 6). The anonymity set includes all possible subjects in a given scenario, such as possible senders of a message.

Related to anonymity is *unlinkability*, where unlinkability of two or more items of interest (IOIs, e.g., subjects, messages, events, actions, etc.) means that within the system (comprising these and pos-

Figure 1. Unlinkability between a user in the anonymity set and an item of interest

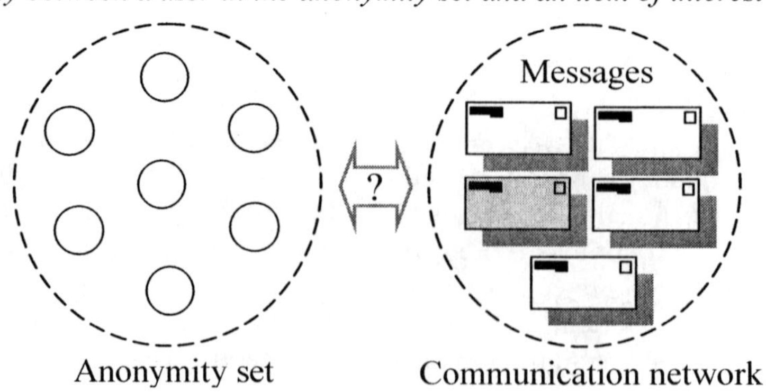

Figure 2. Setting a path between A and D (through B and C) using layered encryption; PK_B and PK_C are the public keys of B and C. K_{AB} and K_{AC} are shared symmetric keys. D is an external receiver

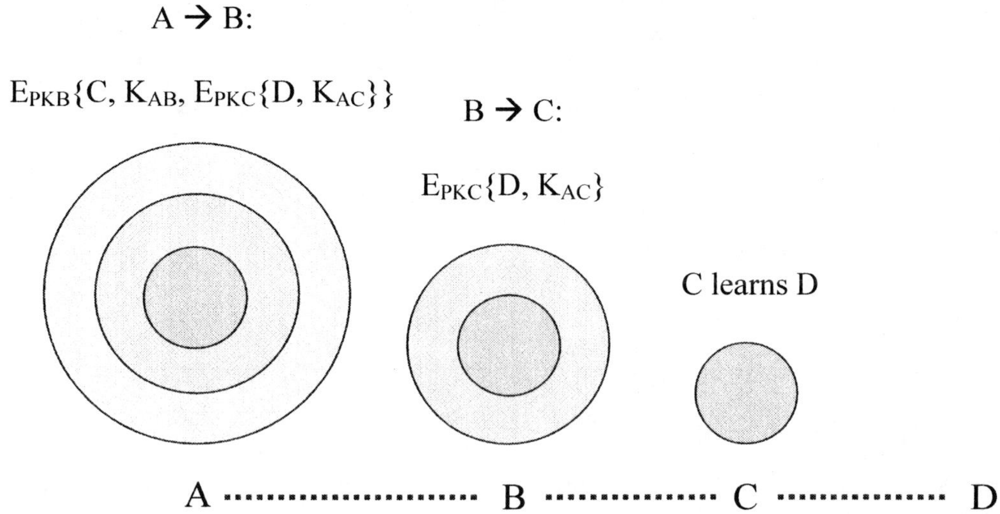

sibly other items), from the attacker's perspective, these items of interest are no more and no less related after his observation than they are related concerning his a-priori knowledge. (Pfitzmann & Hansen, 2006, p. 8)

Anonymity can be defined in terms of unlinkability: *sender anonymity* entails that a message cannot be linked to the sender, while *receiver anonymity* implies that a message cannot be linked to the receiver (see Figure 1).

In traditional networks, such as the Internet, anonymous communication is often realized by *anonymous overlay networks*, which establish *virtual paths* consisting of one or more intermediary nodes, along which packets are transmitted. Using methods described below, the anonymous overlay network constructs the paths in such a manner that the correlation between the sender and receiver, and possibly also the identity of the sender and/or the receiver, is hidden.

A classic method enabling anonymity, where the sender determines the full path, is *layered encryption*[2]: a message is wrapped into several encryption layers. As the message propagates the network, these layers are sequentially decrypted by each successive node in the path, until the receiver decrypts the final layer. Each layer usually includes the identity of the next node in the path and a symmetric key shared with the initiating node (see Figure 2). In this way, expensive public key encryption is only used for constructing the path; for data delivery symmetric encryption is used. Messages encrypted in layers are often denoted *message onions*. Layered encryption enables anonymity as intermediary nodes do not know whether their predecessor and successor nodes are the sender or receiver, respectively.

An alternative approach, first applied in Crowds (Reiter & Rubin, 1997), is to let the sender select its successor randomly, which in turn flips a biased coin to decide whether it should end the path and connect to the receiver, or extend the path to a random node. The flipping of the biased coin is repeated until a node decides to connect to the receiver (see Figure 3). In this approach, link-to-link encryption between intermediary hops in the path is usually combined with end-to-end encryption. This approach enables sender anonymity towards network nodes and the receiver, as neither of these nodes can deduce if the previous node in the path is the sender.

Another method specifically tailored for providing receiver anonymity is *invisible implicit addressing* (Pfitzmann & Waidner, 1987). Invisible implicit addressing hides the identity of the receiver by first encrypting a message (or a part of it) with

Figure 3. "Crowds-like" path setting between the sender and receiver

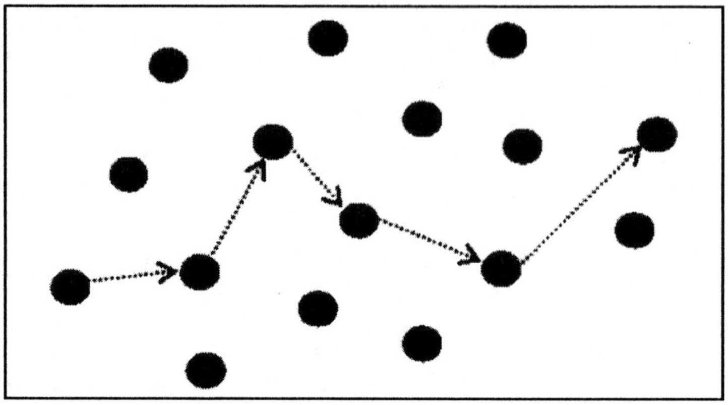

the receiver's public key (or a shared symmetric key). Instead of sending the message directly to the receiver, the message is then *broadcasted* to all nodes in the network, which all must try to decrypt the message. However, only the intended receiver will be able to successfully decrypt the message.

On the Relation between Privacy and Anonymity

Privacy is recognized either explicitly or implicitly as a fundamental human right by most constitutions of democratic societies. Privacy can be defined as the right to *informational self-determination*, that is, individuals must be able to determine for themselves when, how, to what extent, and for what purpose personal information about them is communicated to others.

In Europe, the right for privacy of individuals is protected by the by a legal framework mainly consisting of the EU Data Protection Directive 95/46/EC, which defines general privacy requirements, and the E-Communications Privacy Directive 2002/58/EC, which specifically applies for personal data processing within the electronic communication sector.

An important privacy principle is *data minimization*, stating that the collection and processing of personal data should be minimized. Clearly, the less personal data are collected or processed, the less the right to informational self-determination is affected. Art. 6 (1) of the EU Data Protection Directive 95/46/EC embodies the principle of data minimization by stating that personal data should be limited to data that are adequate, relevant, and not excessive, and by requiring that data should only be kept in a form that permits identification of data subjects for no longer than it is necessary for the purpose for which the data were collected or for which they are further processed. Consequently, technical tools such as privacy-enhancing technologies should be available to contribute to the effective implementation of these requirements by providing anonymity and/or pseudonymity for the users and other concerned individuals.

More specific legal requirements for anonymization can also be found in the E-Communications Privacy Directive 2002/58/EC: Pursuant to Art.9 of the Directive: location data may only be processed when they are made anonymous, or with the consent of the user or subscriber to the extent and for the duration necessary for the provision of a value-added service.

On Measuring Anonymity

This section discusses *anonymity metrics*, which quantify the degree of anonymity in a given scenario in the following manner. First, the given attacker model, together with the properties of the anonymous communication mechanism, are passed

Table 1. A summary of anonymity metrics

Anonymity set size
A classic indicator of anonymity is the size of the anonymity set. This metric is appropriate for mechanisms in which all users are equally likely to be the sender of a particular message, as in the DC-networks (Chaum, 1988) or Crowds, regarding the Web server (Reiter & Rubin, 1997).
K-anonymity
If a mechanism provides *k*-anonymity (Sweeney, 2002), *k* constitutes a lower bound of the anonymity set size *n*. For example, $k = 3$ implies that an attacker cannot exclude more than $(n - 3)$ users from the anonymity set.
Crowds-based metric
In the Crowds-based metric[3] (Reiter & Rubin, 1997), anonymity is measured on a continuum, including the points *possible innocence* (the probability that a user is not the sender is not negligible), *probable innocence* (the probability that a user is a sender ≥ 1/2), and *beyond suspicion* (the user is not more likely than any other user to be the sender). The analysis is based on the communication patterns in Crowds, and the result is a probability depending on the anonymity set size and the number of corrupted users.
Entropy-based metrics
In entropy-based metrics (Diaz, Seys, Claessens, & Preneel, 2002; Serjantov & Danezis, 2002), each user is first assigned with a probability of being the sender of a message. The entropy regarding which user sent the message is then calculated using Shannon's theories (Shannon, 1948). The resulting degree is system-wide and may change depending on, for example, changes in the attacker's knowledge. Diaz et al. solely bases their analysis on the probability distributions (equally distributed probabilities → max degree of anonymity), while in Serjantov and Danezis metric, a large anonymity set contribute positively to the degree of anonymity.

as input to the anonymity metric. Then, the metric determines the degree of anonymity based using for example, analysis or by simulation, depending on the metric at hand. In Table 1, we summarize the most common anonymity metrics.

Although the metrics listed above differs in many respects, the main parameters contributing to the degree of anonymity in all metrics are *size of anonymity set* (anonymity set size and *k*-anonymity), *probability distributions* (entropy-based metric by Diaz et al.), and both (entropy-based metric by Serjantov and Danezis and the Crowds-based metric).

Anonymous Communication in Mobile Ad Hoc Networks

In *proactive* routing protocols (Perkins, 2001), each node always maintains routes to all other nodes, including nodes to which no packets are being sent. Standard proactive protocols do not enable anonymity as all nodes know significant amounts of information about other nodes.

In *reactive* routing protocols (Perkins, 2001), routes between nodes are established on demand, meaning that less packets are circulated in the network, for example, for status sensing. Also standard reactive routing protocols fail to enable anonymity. As a proof of concept, consider the reactive protocols dynamic source routing (DSR) (Johnson & Maltz, 1996) and ad hoc on-demand distance vector routing (AODV) (Perkins & Royer, 1999).

- In DSR, during route discovery[4] the route request (RREQ) includes the IP addresses of the sender and receiver in plain. The IPs are also disclosed by the route reply (RREP) message. During data transfer, the path between the sender and receiver is included in plain in the packet headers.
- Also in AODV, the RREQ and RREP messages disclose the sender and receiver IP addresses. Also, routing data at each node in an active path discloses the receiver IP.

This situation applies for virtually any standard routing protocol. So far, two methods for enabling anonymous communication in mobile ad hoc networks have been proposed: *anonymous routing protocols* and *anonymous overlay networks*. They are explained in the next sections.

Anonymous Routing Protocols

An anonymous routing protocol replaces the standard routing protocol with a protocol preserving anonymity (see Figure 4). Anonymous routing protocols normally include building blocks for *anonymous neighborhood authentication*, *anonymous route discovery*, and *anonymous data transfer*. The first phase is not always included; instead many approaches assume that other mechanisms offer this service.

During anonymous neighborhood authentication, nodes establish trust relationships with their *neighbors* (i.e., nodes within one-hop distance). "Trust" implies that the nodes prove mutual possession of some valid identifiers, such as certificates, pseudonyms, public/private key-pairs, or combinations thereof.

The task of anonymous route discovery is to establish an anonymous path between the sender and receiver. Sender anonymity is often achieved through layered encryption. Sometimes, receiver anonymity is enabled by invisible implicit addressing, meaning in this context that a challenge is included in the RREQ that only the receiver can decrypt[5].

The main disadvantage with invisible implicit addressing is that all nodes receiving the RREQ must try to decrypt the challenge, resulting in considerable overhead (especially as the RREQ reaches all nodes). When the RREP is propagated back to the sender on the path created by the corresponding RREQ message, *visible implicit addressing* (Pfitzmann & Waidner, 1987) is often used to hinder nodes other than the sender from matching RREP messages with corresponding RREQ messages. This is often enabled by including sequence numbers in the RREP and RREQ so that only the sender can conclude that the sequence number of a given RREP corresponds to an earlier sent out RREQ.

During anonymous data transfer, data messages are sent along the paths created during route discovery. Only protocols that use *source routing* can apply layered encryption, as the sender in this case needs to decide the full path. Else, link-to-link encryption, possibly combined with end-to-end encryption, is normally used.

Figure 4. Anonymous routing protocol

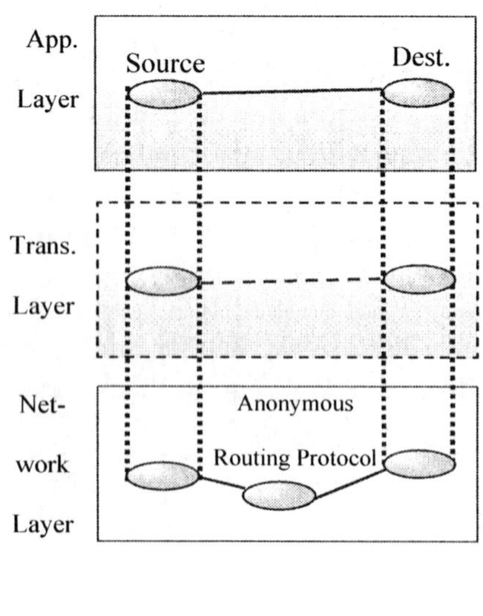

Figure 5. Anonymous overlay network

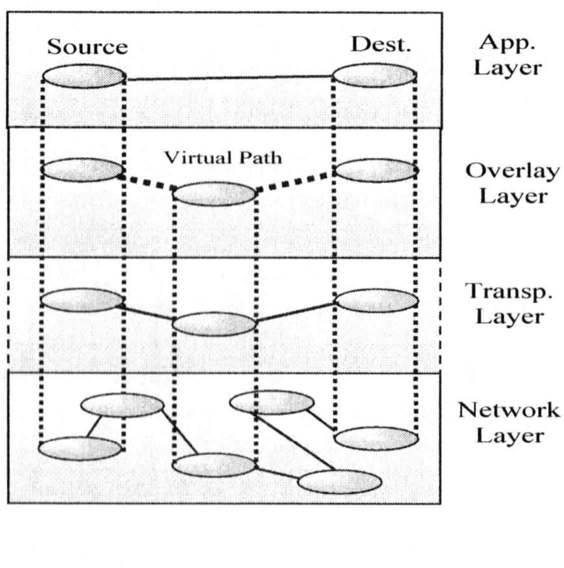

Anonymous Overlay Networks

In mobile ad hoc networks, anonymous overlay networks are normally deployed above the routing or transport layer (see Figure 5), where they can use services from the standard routing protocol (e.g., finding a route to the next node in the path) or the transport layer (e.g., reliable data delivery).

Anonymous overlay networks can be divided into the following phases: *group buildup*, *path construction*, and *data transfer*.

During group buildup, the user base of the overlay network is populated. One strategy for group buildup is to assign this task to one or more *directory servers*, where a set of nodes (or at least one node) must act as a directory server (Martucci, Andersson, & Fischer-Hübner, 2006). Similarly as in anonymous routing protocols, virtual path setting and data transfer are either based on layered encryption, or link-to-link encryption combined with end-to-end encryption.

Comparison between Anonymous Routing Protocols and Anonymous Overlay Networks

In Table 2 we summarize the respective pros and cons with anonymous routing protocols and anonymous overlay networks.

Survey of Anonymous Communication Mechanisms for Ad Hoc Networks

The survey is divided into two parts: one part for anonymous routing protocols and one for anonymous overlay networks[6]. Before the survey, however, we list the evaluation criteria against which the mechanisms included in the survey are evaluated.

Evaluation Criteria

Six requirements were defined by Andersson et al. (2005) that an anonymous overlay network should meet to be suitable for mobile ad hoc networks. These requirements are general enough to be suitable for providing the criteria against which the mechanisms surveyed in this chapter are evaluated. They are listed below:

R1. The anonymous communication mechanism must scale well. It should perform well also with a large number of participants.

R2. The anonymous communication mechanism must provide strong anonymity properties. We examine how the studied approaches resist an attacker model including a global observer[7], path insiders, other network nodes, and the receiver.

Table 2. Pros and cons with anonymous routing protocols and anonymous overlay networks

Advantages with Anonymous Routing Protocols
They make it possible to control already on the routing level what information is being disclosed during routing. Yet, this does not exclude the possibility that additional efforts may be needed in upper layers. Also, most approaches use the shortest path between the sender and receiver.
Disadvantages with Anonymous Routing Protocols
The replacement of the standard routing protocol; this will likely decrease the user base, which degrades anonymity according to many metrics. Besides, nodes may be exposed if a connection-oriented transport layer is used above the anonymous routing protocol, as they establish direct connections between nodes.
Advantages with Anonymous Overlay Networks
Flexibility; an anonymous overlay network is independent of the routing protocol and, further, compatible with applications expecting services from for example, a reliable transport layer.
Disadvantages with Anonymous Overlay Networks
The performance can be expected to be slightly worse as messages are detoured through a set of overlay nodes, instead of being transmitted on the shortest route between the sender and recipient.

R3. The anonymous communication mechanism must be fair regarding the distribution of workload among the nodes. The workload should be equally distributed (and nodes should not be forced to spend a lot of resources on behalf of others). Else, incentives should be given for accepting a higher workload.

R4. The anonymous communication mechanism must provide acceptable performance. It should be lightweight (e.g., generate few messages and avoid public key operations). We evaluate whether the studied approaches presents arguments indicating a good performance. We also evaluate whether there are strong assumptions that could hamper performance.

R5. The anonymous communication mechanism must employ a peer-to-peer paradigm (P2P) model. There should be no dependence on central hardware/services, or at least, it should be minimized. We also study whether there are some implicit requirements for centralized services that are hidden by strong assumptions.

R6. The anonymous communication mechanism must handle a dynamic topology. It must tolerate that nodes are frequently entering or leaving the network.

In the survey, we grade the approaches according to which degree they satisfy these requirements: ●●● = the requirement is satisfied to a high degree; ●● = ... is satisfied to a medium degree; ● = ... is satisfied to a low degree; and ○ = ... is violated. Regarding the grading of R2, the approaches are graded according to which degree they provide anonymity against each item in the assumed attacker model (see R2).

Survey of Anonymous Routing Protocols

In this section, we survey a variety of prominent anonymous routing protocols proposed in recent years. The ratings of the mechanisms are listed in table-form in the next section.

Anonymous Dynamic Source Routing Protocol (AnonDSR)

AnonDSR (Song, Korba, & Yee, 2005) is a source routing protocol using invisible implicit addressing for route discovery. The RREP is created as a message onion. Both the sender and recipient know the intermediary nodes in the path. Data messages are sent as message onions on bidirectional paths. AnonDSR includes a security parameter establishment (SPE) protocol for exchanging security parameters prior to route discovery, which contains a major flaw (see R2).

R1. As the SPE protocol is used to establish shared secrets between sender and receivers, the issues regarding it (see R2) may hamper scalability.

R2. The SPE protocol broadcasts the IDs of the senders and receivers in plain. If used, AnonDSR provides merely confidentiality. If not used, AnonDSR provides sender and receiver anonymity against observers, path insiders, and network nodes.
AnonDSR changes the message appearance at intermediary hops. Yet, a global observer may correlate the RREQ sizes or trace data flows in the network.

R3. During route discovery, nodes spend energy to assess whether they are the intended receiver. Intermediary nodes must perform public key encryptions.

R4. The range of the nodes and the network size is not specified in the performance simulation of AnonDSR, and only route discovery is evaluated while data transfer and node mobility are not considered. Also, as implicit addressing with public key cryptography is used, AnonDSR cannot be expected to provide high performance.

R5. No special nodes needed, and thus AnonDSR adheres well to the P2P paradigm.

R6. AnonDSR does not support rebuilding of broken paths. Also, the insecurities in the SPE protocol may cause problems for new nodes joining the network that wish to establish security parameters with existing nodes.

Secure Distributed Anonymous Routing Protocol (SDAR)

SDAR (Boukerche, El-Khatib, Xu, & Korba, 2004) is a source routing protocol enabling a system for managing trust: nodes associate their neighbors with a trust level based on past behavior. Invisible implicit addressing is used to hide the receiver identity in the RREQ. The RREP and data messages are sent as message onions.

R1. SDAR can be expected to scale badly as every node in the network must perform three public key operations per received RREQ message.

R2. SDAR offers sender and receiver anonymity against observers and other network nodes. SDAR alters messages appearance and applied padding to thwart global observers. Still, only nodes assumed to forward RREQ/RREP packets do so, others drop them.

R3. It is not specified whether the certificate authority (CA) is a central service or distributed among the nodes. When processing RREQ packets, all nodes must perform one public key encryption, one public key decryption, and one signature generation.

R4. There are serious performance issues in SDAR. For instance, every node must perform must perform three public key operations for each RREQ it forwards.

R5. The existence of a CA (or similar) is assumed for distributing public keys. It is not specified how it would be implemented.

R6. We predict that the trust management system in SDAR would suffer in a dynamic topology; it would be difficult for nodes to be highly trusted as they would be • punished for leaving the network in the midst of a communication. Also, path rebuilding in case of broken paths is not considered.

MASK

MASK (Zhang, Liu, & Lou, 2005) does not use source routing. Prior to route discovery, MASK performs anonymous neighborhood authentication, and nodes know each other by temporal pseudonyms. For performance reasons, MASK avoids invisible implicit addressing during route discovery; instead, the receiver identity is disclosed in the RREQ. After route discovery, a sender may have multiple active paths to the receiver. End-to-end and/or link-to-link encryption is employed during data transfer, depending on the application at hand.

R1. MASK can be expected to scale well as it avoids the usage of implicit addressing. Yet, an increased node density (i.e., more neighbor nodes) may degrade performance during anonymous neighborhood authentication.

R2. MASK offer sender anonymity against path insiders, network nodes, and observers, but no receiver anonymity. MASK uses altered message appearance, random choice of paths, and per-hop message delay to harden traffic analysis during low traffic. No node forwards RREQ/RREP messages more than once.

R3 The avoidance of implicit addressing bears a positive impact on fairness.

R4. Simulation results indicate that MASK provides good performance. However, the mutual authentication between neighboring nodes was shown to be the most costly operation and in scenarios where the transmission range is small compared to the network size,
this may affect performance negatively.

R5. A trusted authority (TA) is used during the bootstrapping phase of the network.

R6. Broken paths are handled by broadcasting error packets in case of a broken path. Still, the tight synchronization scheme between neighboring nodes may lead to problems in some situations where neighboring nodes leave and join often.

Anonymous On-Demand Routing (ANODR)

ANODR (Kong, Hong, Sanadidi, & Gerla, 2005) is a source routing protocol aiming to protect privacy by avoiding persistent identifiers. Invisible implicit addressing based on symmetric encryption is used to hide the receiver identity during route discovery.

The RREP is created as a message onion. During data transfer, it is not specified whether or not the data payload is encrypted.

R1. It is unclear how senders and receivers share symmetric keys. Given that they share a key, to solve the challenge in the RREQ, the receiver
may have to try all keys shared with other nodes (see R4). Further, other network nodes must try all their shared keys to conclude that they are not the intended receiver.

R2. ANODR offers sender and receiver anonymity against observers, path insiders, and networks nodes. Senders and receivers are not mutually anonymous. ANODR uses traffic mixing to thwart observers, where messages are independently and randomly delayed. Yet, traffic patters are leaked as only nodes assumed to forward the RREP does so. Further, as the payload of data messages is not altered at intermediary hops, it is trivial for a global observer to trace data traffic.

R3. Each node must spend considerable resources when forwarding RREQ packets.

R4. There are serious performance issues in ANODR (see R1).
Although ANODR has performed reasonably well in a simulation scenario, problems can be expected in a real world scenario.

R5. No special nodes are needed, and thus ANODR adheres well to the P2P paradigm.

R6. ANODR supports path rebuilding in case of broken paths. However, it is unclear how new nodes should share symmetric keys with old nodes

Discount Anonymous On-Demand Routing (Discount ANODR)

Discount ANODR (Yang, Jakobsson, & Wetzel, 2006) is a low-latency source routing protocol that avoids invisible implicit addressing. A random time to live counter is used for RREQ/RREP messages to confuse observers (implemented by flipping a biased coin). Data are sent as message onions along unidirectional paths (i.e., a new path must be build for the reply).

R1. Discount ANODR can be expected to scale well. However, the bias of the coin flipping may have to be adapted if the geographical size of the network increases.

R2. Discount ANODR provides sender anonymity against local observers, as the coin flipping and random padding during route discovery confuse observers to a certain degree. No receiver anonymity.
Data messages are padded with random bits.

R3. There are no special nodes and no public encryption on behalf of other nodes.

R4. Discount ANODR avoids public key encryption and invisible implicating addressing.
The coin flipping may degrade performance as nodes on the shortest path may drop the RREQ, resulting in nonoptimal paths. Also, RREP packets can be lost for the same reason. Unidirectional paths also hamper performance.

R5. The nodes have to collectively administrate two values determining the bias of the coins deciding whether a node should forward a RREQ and a RREP, respectively.

R6. Discount ANODR rebuilds broken paths, but does not discuss how to collectively adapt the bias of the coin flipping when the network characteristics change.

Anonymous Routing Protocol for Mobile Ad Hoc Networks (ARM)

ARM (Seys & Preneel, 2006) aims to foil global observers by using random time-to-live values and padding for all messages. Senders and receivers share one-time pseudonyms. Invisible implicit addressing hides the receiver by including the secret pseudonym in the RREQ. The RREP is created as a message onion. Link-to-link encryption is used for data transfer.

R1. As a tight synchronization scheme is used between sender and recipients, it is assumed that senders shares keys and pseudonyms with a limited set of receivers.

R2. ARM offers sender and receiver anonymity against networks nodes, path insiders, and observers. Senders and receivers have an a-priori relationship. In ARM, data messages have a uniform size, RREQ/RREP messages are randomly padded, and RREQ/RREP/data messages are propagated using random time-to-live values. The effectiveness of this limited dummy traffic is not formally proven.

R3. While no nodes perform public key operations, the amount of nodes forwarding RREQ/RREP and data messages increases due to the random time-to-life values.

R4. If assuming a static environment, there are no conclusive arguments orthogonal to performance. However, all nodes in ARM generate overhead traffic.
ARM has not yet been simulated to assess the performance.

R5. There are no special nodes in ARM. In a real world scenario, central infrastructure may be required to realize the assumption that each node should possess a unique identifier; it is unclear how this would clash with the P2P paradigm.

R6. The assumption that each node establishes a broadcast key with its neighbors is problematic when considering dynamic topologies. Further, ARM does not consider path rebuilding in case of broken paths.

Distributed Anonymous Secure Routing Protocol (ASRP)

ASRP (Cheng & Agrawal, 2006) is a routing protocol not based on source routing where nodes are known by dynamic random pseudonyms. Invisible implicit addressing (based on public encryption) is used for both RREQ and RREP packets. Data messages are link-to-link and end-to-end encrypted. It is not specified whether the paths are bidirectional or unidirectional.

R1. All nodes in the network must perform two public key operations per RREQ (one private key decryption and one public key generation). This hampers scalability as the more nodes in the network, the more generated RREQ packets.

R2. Senders and receivers are not mutually anonymous as they have an a-priori relationship. Anonymity is offered against path insiders and network nodes, and ASRP alters message appearance and maintains a uniform message size to confuse attackers.

R3. All nodes spend significant resources when forwarding RREQ and RREP packets. For the RREQ, see R1. For propagation of RREP packets, all nodes on the path must perform three public key operations (one private key decryption and two public key encryptions).

R4. The performance of ASRP has not been simulated. Route discovery can be expected to offer a low performance, as public key encryption is extensively used.

R5. No special nodes are needed, and thus ASRP adheres to the P2P paradigm.

R6. Path rebuilding in case of broken paths is not considered. This means that the expensive route discovery process has to be initiated for each case of path failure.

Privacy Preserving Routing (PPR)

PPR (Capkun, Hubaux, & Jakobsson, 2004) is a proactive protocol for communication between ad hoc networks interconnected by fixed access points (AP). Nodes know each other by temporal pseudonyms. In the sender network, nodes maintain the shortest path to the AP. In the receiver's network, the AP maintain the shortest paths to the nodes. Routing consists of three parts: *uplink* (distance vector protocol), *inter-station*, and *downlink* (source routing). In uplink, a sender sends a message that reaches the AP as a message onion. In downlink, the receiver's AP send an onion to the receiver.

R1. The AP and the CA are the major points of workload aggregation in PPR, but as these are centrally offered services, PPR can be expected to scale well.

R2. PPR offers sender and receiver anonymity

against observers, network nodes, and path insiders. There are no countermeasures against global observers in the senders or receivers networks, except message alteration at intermediary hops. Anonymity is quantified using the entropy-based anonymity metric (see section "On Measuring Anonymity"). There is no anonymity against the AP.

R3. Nodes do not perform special roles or execute public key operations on behalf of others.

R4. Public key encryption is only used for establishing trust relationships among neighboring nodes. The performance of PPR has not yet been simulated.

R5. PPR violates the P2P model as the existence of a CA and several AP is assumed.

R6. The existence of the AP facilitate the handling of trust and security issues in a dynamic topology. The uplink protocol is the most vulnerable part regarding routing, but it can be expected to handle dynamic topologies well.

Summary of Survey Results for Anonymous Routing Protocols

The survey results for all requirements (except R2) are summarized in Table 3. The survey results for R2 are summarized in Table 4.

SURVEY OF ANONYMOUS OVERLAY NETWORKS

In this section, we study two anonymous overlay networks for ad hoc networks: Chameleon (Martucci et al., 2006) and MRA (Jiang, Vaidya, & Zhao, 2004).

Chameleon

Chameleon can be described as a variant of Crowds adapted for mobile ad hoc networks. In Chameleon, the nodes share the responsibility of being directory servers during group buildup. Node authentication is based on certificates (the existence of a TCP (transmission control protocol)/SSL (secure

Table 3. Summary of survey results (except R2)

Requirement	*ARM*	*AnonDSR*	*ANODR*	*SDAR*	*Discount ANODR*	*ASRP*	*MASK*	*PPR*
R1: Scalability	•	••	•	•	•••	••	••	•••
R3: Fairness	••	••	•	••	•••	•	•••	••
R4: Performance	••	•	•	•	••	•	••	••
R5: P2P	••	•••	•••	••	••	•••	••	○
R6: Dyn. Top.	•	•	••	•	••	•	••	•••

Table 4. Summary of anonymity requirement R2

Attacker model	*ARM*	*AnonDSR*	*ANODR*	*SDAR*	*Discount ANODR*	*ASRP*	*MASK*	*PPR*[8]
Sender – observer	••	•	•	•	•	•	••	•
Send. – path insider	•••	•••	•••	•••	•••	•••	•••	•••
Sender – net. node	•••	•••	•••	•••	•••	•••	•••	•••
Sender – receiver	○	○	○	○	○	○	○	○
Rec. – observer	••	•	•	•	○	•	○	•
Rec. - path insider	•••	•••	•••	•••	○	•••	○	•••
Rec. – net. node	•••	•••	•••	•••	○	•••	○	•••

socket layer) layer is assumed). Data messages are end-to-end and link-to-link encrypted or only link-to-link encrypted.

R1. The load on each node is approximately constant as the size of the network grows. However, if too few directory servers are used, this may put a limit on scalability.

R2. Chameleon offers sender anonymity against receivers and sender and receiver anonymity against local observers and malicious nodes.
The degree of anonymity is quantified by the Crowds-based metric (see "On Measuring Anonymity").

R3. A small subset of the nodes must act as directory servers. It is suggested that nodes take turns in acting as the directory servers.

R4. Chameleon is based on light-weight encryption.
However, the performance of Chameleon has not yet been assessed through simulation.

R5. Chameleon generally follows the P2P paradigm. However, nodes are assumed to possess certificates obtained in advance
and the global probability deciding the expected path length has to be administrated collectively by the nodes

R6. Chameleon repairs broken paths at the point of breach, rather that rebuilding the whole path. Without redundancy, vanishing directory servers may be a problem.

Mix Route Algorithm (MRA)

MRA applies traffic mixing[9] (Chaum, 1981) in a mobile ad hoc scenario. A subset of the nodes acts as mixes, which constitute the virtual paths. Each node assigns a mix as its *dominator mix*. A RREQ is sent to the receiver via the sender's dominator mix, triggering the receiver to register at its dominator mix with a DREG (dominator registration) message. Each mix periodically broadcasts RUPD (route update) messages containing its registered receivers and a path field, which is updated as the RUPD propagates through the network. When it reaches the sender, it contains the path to the receiver.

R1. Scalability may be hampered if the mix set is static in a growing network.

R2. As the min path length is one, a mix may learn the identity of both the sender and receiver. The first mix always learns the sender ID. Receiver anonymity is in doubt as all mixes broadcast information in the network about which receivers it is currently providing services for (i.e., the RUPD messages).

R3. Incentives for the costly operating of mixes are left as a future research problem.

R4. MRA is based on pubic-key cryptography. Basing MRA on symmetric cryptography is left as future research. Results from a performance simulation are presented, but only different mix settings are compared.

R5. No central services are needed. Still, establishing trust between mixes and other nodes are left as future research. This may require aid from external trusted nodes.

R6. If the sender or dominator mix move, the sender may have to switch dominator mix. If the mix set is small, problems may arise regarding the mix advertisement as nodes only retransmit advertisement messages from their dominator mixes.

Summary of Survey Results for Anonymous Overlay Networks

The results from the survey are summarized in Table 5.

DISCUSSION

From the survey, we can make the following observations:

1. **It is difficult to protect against a global eavesdropper.** None of the studied approaches implement powerful and proven countermeasures against global observers. We believe that it is an open research problem regarding how to enable such countermeasures while at the same time offering an

Table 5. Summary of survey results (left) and summary of anonymity requirement R2 (right)

Requirement	*Chameleon*	*MRA*
R1: Scalability	●●	●●
R3: Fairness	●●	●
R4: Performance	●●	●
R5: P2P	●●	●●
R6: Dyn. Top.	●●	●●

Attacker model	*Chameleon*	*MRA*
Sender – observer	●	●●
Send. – path insider	●●●	●●●/○[10]
Sender – net. node	●●●	●●●
Sender – receiver	●●●	○
Rec. – observer	●	●
Rec. - path insider	○	●/○[11]
Rec. – net. node	●●●	●

acceptable level of performance in mobile ad hoc networks[12].

2. **It is difficult to implement invisible implicit addressing efficiently.** There is a clear trade-off between on the one hand enabling receiver anonymity by using invisible implicit addressing and on the other hand satisfying the fairness, dynamic, and scalability requirements. The proposals using invisible implicit addressing either use costly public key cryptography (e.g., AnonDSR, SDAR, ASRP) or avoid public key operations at the cost of including strong assumptions regarding in beforehand mutual distribution of secrets (e.g., ARM, ANODR).
3. **It is straightforward to hide the identity of the sender from other network nodes.** This is probably because most of the approaches use classical techniques for hiding the identity of the sender, such as layered encryption, that have been used before in other contexts.
4. **No anonymous routing protocol implements sender anonymity towards the receiver.** Hiding the sender identity during route discovery would require a mechanism for hiding the propagation of the RREP messages similar (and equally costly as) to the invisible implicit addressing schemes used for hiding the propagation of the RREQ messages.

FUTURE TRENDS

A *Sybil attack* (Douceur, 2002) implies one attacker forging multiple identifiers in the network to control an unbalanced portion of the network. Sybil attacks can undermine security in, for instance, mobile ad hoc networks based on reputation schemes or threshold cryptography (Piro, Shields, & Levine, 2006). Douceur has showed that *preventing* Sybil attacks is practically impossible as it requires a TTP (trusted third party) to manually assert that each identity corresponds to only one logical entity in the network. Yet, during the years, and recently also for mobile ad hoc networks, many approaches for *detecting* Sybil attacks have been proposed. In this section, we discuss why Sybil attacks threaten anonymity in ad hoc networks, and discuss some proposed countermeasures.

The Sybil Attack in Mobile Ad Hoc Networks

Mobile ad hoc networks are highly susceptible to Sybil attacks because of, for instance, the lack of reliable network or data link identifiers, and the absence of a trusted entity capable of vouching for the one-to-one binding between physical devices and logical network identifiers. This may give the impression that ad hoc nodes are naturally anonymous as nodes could confuse observers by regularly changing their {IP, MAC} pairs. Although this may prevent long-term tracking, other problems may arise. For instance, when there is a need to identify a node offering a specific service, a rouge node could easily impersonate this service. The absence of reliable network identifiers may also disrupt routing, as a rouge user could announce false information using multiple {IP, MAC} pairs. Also, as senders and receivers establish direct connections, they are still vulnerable to traffic analysis and physical layer oriented attacks (Capkun et al., 2004).

However, the Sybil attack also poses a threat against anonymous routing protocols and anonymous overlay networks. For both approaches, the anonymity set denotes the user base. There are some differences though. In an anonymous overlay network, the anonymity set is used as a pool of nodes serving as an input parameter to the path creation algorithm. Polluting the anonymity set with many Sybil identities might yield a path only containing Sybil identities. If this happens, the attacker can easily break anonymity by linking the sender to the receiver. In an anonymous routing protocol, however, each node only stepwise extends the path to another node within a single-hop distance, until the receiver is reached. Thus, the locations of the nodes play a more important role here, and as all Sybil identities share the same location, it is difficult for the attacker to force the creation of paths in which it controls all nodes.

Thus, the Sybil attack poses a greater threat to anonymity in anonymous overlay networks compared to anonymous routing, although it still poses a great threat to other security properties for anonymous routing.

Mechanisms for Detecting the Sybil Attack in mobile Ad Hoc Networks

In this section, we describe two recent proposals for thwarting Sybil attacks in mobile ad hoc networks.

- The fact that Sybil nodes in mobile ad hoc networks naturally travel together in clusters can be used for detecting Sybil attacks (Piro et al., 2006). Piro et al. propose a detection mechanism in which each node records all encountered {IP, MAC} pairs. If a user repeatedly observes a set of {IP, MAC} pairs sharing the same location, there is an increased likelihood that these {IP, MAC} pairs represent Sybil nodes. One drawback with this strategy is that it is unclear how to prevent a detected attacker from generating new {IP, MAC} pairs and relaunch a new attack later, as there is no underlying long-term identity that can be blocked from the system.
- Another strategy is to cryptographically guarantee a one-to-one mapping between all temporal network identifiers seen in a particular network and corresponding certified long-term identifiers (Martucci et al., 2008). To tailor this approach for ad hoc networks, the nodes must be able to assert the validity of the temporal identifiers without having to interact with the TTP. Further, to protect privacy, only the TTP should be able to link a temporal identifier to the corresponding long-term identifier and there should be unlinkability between temporal identifiers used in different contexts. The fact that you need reliable identifiers to protect against the Sybil attack and to provide reliable anonymous communication has been labeled as the *identity-anonymity paradox* (Martucci et al., 2006).

CONCLUSION

In mobile ad hoc networks, anonymous communication can either be enabled by anonymous routing protocols or anonymous overlay networks. Currently, anonymous routing is the most popular approach, although future requirements, such as flexibility regarding the applications, may raise the need for anonymous overlay networks.

We evaluated commonly proposed anonymous routing protocols and anonymous overlay networks for mobile ad hoc networks against a set of evaluation criteria and showed that a number of research challenges remain. For instance, it is difficult to offer receiver anonymity without using a complex and performance-hampering invisible implicit addressing scheme, and it is further difficult to protect against global observers.

Finally, we introduced Sybil attacks, a notorious

threat to all computer networks, including mobile ad hoc networks. We expect that the area of enabling reliable identifiers in a privacy-friendly manner is an interesting future research area.

REFERENCES

Andersson, C., Martucci, L. A., & Fischer-Hübner, S. (2005). Requirements for privacy: Enhancements in mobile ad hoc networks. In *Proceedings of the 3rd German Workshop on Ad Hoc Networks (WMAN 2005)* (pp. 344-348). Gesellschaft für Informatik (GI).

Boukerche, A., El-Khatib, K., Xu, L., & Korba, L. (2004). A novel solution for achieving anonymity in wireless ad hoc networks. In *Proceedings of the 7th ACM International Symposium on Modeling, Analysis and Simulation of Wireless and Mobile Systems* (pp. 30-38).

Capkun, S., Hubaux, J. P., & Jakobsson, M. (2004). *Secure and privacy-preserving communication in hybrid ad hoc networks* (EPFL-IC Tech. Rep. No. IC/2004/10). Lausanne, Switzerland: Laboratory for Computer Communications and Applications (LCA)/Swiss Federal Institute of Technology Lausanne (EPFL).

Chaum, D. (1981). David Chaum: Untraceable electronic mail, return addresses, and digital pseudonyms. *Communications of the ACM, 24*(2), 84-88.

Cheng, Y., & Agrawal, D. P. (2006). *Distributed anonymous security routing protocol in wireless mobile ad hoc networks.* Paper presented at the OPNETWORK 2005.

Corson, M. S., & Macker, J. (1999). *Mobile ad hoc networking (MANET): Routing protocol performance issues and evaluation considerations* (RFC-2501), Internet RFC/STD/FYI/BCP Archives.

Dıaz, C., Seys, S., Claessens, J., & Preneel, B. (2002). Towards measuring anonymity. In *Proceedings of the Workshop on Privacy Enhancing Technologies (PET 2002)* (LNCS 2482). Springer-Verlag.

Douceur, J. R. (2002). The Sybil attack. In P. Druschel, F. Kaashoek, & A. Rowstron (Eds.), *Peer-to-peer Systems: Proceedings of the 1st International Peer-to-Peer Systems Workshop (IPTPS)* (pp. 251-260). Springer-Verlag.

Goldschlag, D. M., Reed, M. G., & Syverson, P. F. (1996). Hiding routing information. *Information hiding* (LLNCS 1174, pp. 137-150). Springer-Verlag.

Jiang, S., Vaidya, N. H., & Zhao, W. (2004). A mix route algorithm for mix-net in wireless mobile ad hoc networks. In *Proceedings of the 1st IEEE International Conference on Mobile Ad Hoc and Sensor Systems (MASS 2004).*

Johnson, D. B., & Maltz, D. A. (1996). Dynamic source routing in ad hoc wireless networks. In *Computer Communications Review: Proceedings of the ACM SIGCOMM'96 Conference on Communications Architectures, Protocols and Applications.*

Kong, J., Hong, X., Sanadidi, M. Y., & Gerla, M. (2005). Mobility changes anonymity: Mobile ad hoc networks need efficient anonymous routing. In *Proceedings of the 10th IEEE Symposium on Computers and Communications (ISCC 2005).*

Levine, B. N, Shields, C., & Margolin, N. B. (2006). *A survey of solutions to the Sybil attack* (Tech. Rep. 2006-052). Amherst, MA: University of Massachusetts Amherst.

Martucci, L. A., Andersson, C., & Fischer-Hübner, S. (2006). Chameleon and the identity-anonymity paradox: Anonymity in mobile ad hoc networks. *In Short-Paper Proceedings of the 1st International Workshop on Security (IWSEC 2006)* (pp. 123-134).

Martucci, L., Kohlweiss, M., Andersson, C.,& Panchenko, A. (2008). Self-certified Sybil-free pseudonyms. In *1st ACM Conference on Wireless Network Security* (WiSec 2008).

Perkins, C. E. (2001). *Ad hoc networking.* Addison-Wesley Professional.

Perkins, C. E., & Royer, E. M. (1999). Ad-hoc on demand distance vector routing. In *Proceedings of the 2nd IEEE Workshop on Mobile Computing Systems and Applications (WMCSA '99).*

Pfitzmann, A., & Hansen, M. (2006) *Anonymity, unlinkability, unobservability, pseudonymity, and identity management: A consolidated proposal for terminology v0.27.* Retrieved April 25, 2007, from http:// dud.inf.tu-dresden.de/literatur/Anon_Terminology_v0.28.doc

Pfitzmann, A., & Waidner, M. (1987). Networks without user observability. *Computers and Security, 6*(2), 158-166.

Piro, C., Shields, C., & Levine, N. L. (2006). Detecting the Sybil attack in mobile ad hoc networks. In *Proceedings of the IEEE/ACM International Conference on Security and Privacy in Communication Networks (SecureComm).*

Reiter, M., & Rubin, A. (1997). *Crowds: Anonymity for Web transactions.* Technical report No. 97-15, DIMACS (pp. 97-115).

Serjantov, A., & Danezis, G. (2002). Towards and information theoretic metric for anonymity. In *Proceedings of the Workshop on Privacy Enhancing Technologies (PET 2002)* (LNCS 2482) .Springer-Verlag.

Seys, S., & Preneel, B. (2006). ARM: Anonymous routing protocol for mobile ad hoc networks. In *Proceedings of International Workshop on Pervasive Computing and Ad Hoc Communications (PCAC '06).*

Shannon, C. E. (1948). A mathematical theory of communication. *The Bell System Technical Journal, 27,* 379-423.

Song, R., Korba, L., & Yee, G. (2005). AnonDSR: Efficient anonymous dynamic source routing for mobile ad-hoc networks. In *Proceedings of the 2005 ACM Workshop on Security of Ad Hoc and sensor Networks (SASN 2005)* (pp. 32-42). Alexandria.

Sweeney, L. (2002). k-Anonymity: A model for protecting privacy. *International Journal on Uncertainty, Fuzziness and Knowledge-based Systems, 10*(5), 557-570.

Yang, L., Jakobsson M., & Wetzel, S. (2006). Discount anonymous on demand routing for mobile ad hoc networks. In *Proceedings of SecureComm 2006*, Baltimore, MD.

Zhang, Y., Liu, W., & Lou, W. (2005). Anonymous communication in mobile ad hoc networks. In *Proceedings of the 24th Annual Joint Conference of the IEEE Communication Society (INFOCOM 2005)*, Miami.

KEY TERMS

Anonymity: The state of being not identifiable within a set of subjects.

Anonymity Metrics: Metrics for quantifying the degree of anonymity in a scenario.

Mobile Ad Hoc Network: Networks constituted of mobile devise which may function without the help of central infrastructure or services.

Privacy: The right to informational self-determination, that is, individuals must be able to determine for themselves when, how, to what extent, and for what purpose personal information about them is communicated to others.

Receiver Anonymity: Implies that a message cannot be linked to the receiver.

Sender Anonymity: Means that a message cannot be linked to the sender.

Unlinkability: If two items are unlinkable, they are no more or less related after an attacker's observation than they are related concerning the attacker's a-priori knowledge.

END NOTES

1 As devices in ad hoc networks are responsible for their own services, including security and routing, protocols for anonymous communication for wired networks are not suitable for ad hoc networks, not even those based on the

peer-to-peer paradigm (P2P) (Andersson et al., 2005).

[2] This method is sometimes also called telescope encryption. A public key based version of the method was initially introduced by Chaum (1981). Onion Routing, which only uses public key encryption for setting the path, and then relies on symmetric encryption, was later proposed by Goldschlag, Reed, and Syverson (1996).

[3] The Crowds-based metric was developed for Crowds, but has since been used in other contexts.

[4] This denotes the process of setting a path between the sender and a receiver. First, the sender floods a route request (RREQ) into the network, which triggers the sending of a route reply (RREP) from the receiver to the sender. During the propagation of the RREQ and RREP, respectively, the path is interactively formed.

[5] In the context of mobile ad hoc networks, this method is often referred to as a global trapdoor.

[6] In the survey, we omit approaches relying on the existence of either a positioning device (e.g., GPS) in the mobile devices or a location server in the mobile ad hoc network.

[7] A global observer is an observer that is capable of observing all networks traffic in the whole network.

[8] Note that no anonymity is provided against the access points (not included in attacker model).

[9] Batching and reordering traffic to hide the correlation between incoming and outgoing traffic.

[10] No sender anonymity if path length is one.

[11] No receiver anonymity against last mix on the path.

[12] It is commonly believed that omnipresent protection against a global observer can only be achieved if all nodes transmit a constant flow of traffic, requiring massive usage of dummy traffic.

Chapter XXVIII
Secure Routing with Reputation in MANET

Tomasz Ciszkowski
Warsaw University, Poland

Zbigniew Kotulski
Warsaw University, Poland

ABSTRACT

The pervasiveness of wireless communication recently gave mobile ad hoc networks (MANET) significant researchers' attention, due to its innate capabilities of instant communication in many time and mission critical applications. However, its natural advantages of networking in civilian and military environments make it vulnerable to security threats. Support for anonymity in MANET is orthogonal to a critical security challenge we faced in this chapter. We propose a new anonymous authentication protocol for mobile ad hoc networks enhanced with a distributed reputation system. The main objective is to provide mechanisms concealing a real identity of communicating nodes with an ability of resistance to known attacks. The distributed reputation system is incorporated for a trust management and malicious behaviour detection in the network.

INTRODUCTION

The contemporary information society extensively takes advantage of wireless communication using several specific network technologies. This continuously evolving area provides a flexible and convenient way for improving work standards in business, home, education, or rescue applications. Thanks to the pervasiveness of private unlicensed spectrum technologies such as Bluetooth and IEEE 802.11 family protocols, the communication between personal and handheld electronic devices is easier, comfortable, and mobile. Through the years, a lot of researches' efforts were devoted to the functional and network performance improvements, covering existing standards designed for fully cooperative environments. However, many first pioneering deployments of wireless networks quickly turned out its several vulnerabilities they suffer from. Since that time substantially more attention has been paid to the security as a supplementary service protecting and supporting performance in wireless communication. The specific and unique characteristics of mobile ad hoc networks (MANET) such as a multihop routing and highly dynamic topology impose a new type of security concerns that we present in this chapter.

In response to the vulnerabilities being identified in several MANET protocols a set of security considerations have taken place in a number of extensions to existing nonsecure approaches. Even though, the strong security requirements are met in many MANET protocol designs, only few of them address anonymity and privacy guaranties (Boukerche, 2004; Ciszkowski & Kotulski, 2006; Kong & Hong, 2003; Zhang, Liu, & Lou, 2005), which are treated as an orthogonal to security critical challenge we discuss in this chapter. On the example of a novel anonymous authentication protocol (ANAP) for mobile ad hoc networks (Ciszkowski & Kotulski, 2006) we present an enhanced distributed reputation system designed for efficient and secure routing in MANET. The main objective of this work is to provide protocol with mechanisms concealing the real identity of the communicating nodes maintaining the resistance to known attacks (Chaum, 1981; Pfitzmann & Hansen, 2005). The distributed reputation system is incorporated in order to build and manage mutual trust of the communicating nodes. The trust knowledge reflects a trustworthy and malicious activity in the network, effectively improving secure routing in MANET by means of anonymous authentication and path discovery phases. ANAP delivers links for secure exchange of data, taking advantage of an on-demand routing approach (Hu et al., 2002; Perkins & Royer, 1999; Royer & Toh, 1999).

The following sections present related work and protocol designs focusing on the distributed reputation system improving secure and anonymous routing in MANET. Two last sections cover some concluding remarks and further research directions.

BACKGROUND

MANET is a set of mobile nodes which operates wirelessly in an environment with a devoid of fixed network structure enforced by self-configuring and self-organizing mechanisms. All its nodes are free to move, join, or leave the network in ad hoc manner, while the end-to-end communication between nodes being beyond its radio range is performed in a multihop fashion. This specific feature demands for additional requirements to every node that, apart from sending and receiving data, must act as an interconnecting router. Since every node may be obliged to perform data forwarding, appropriate routing algorithms were developed to meet such a requirement. The main objective of routing protocols for ad hoc network is creating an up-to-date multihop communication path in a dynamically changing network topology. The appropriate and specific path discovery and path maintenance algorithms have already been developed which characterizes particular routing protocols. One can distinguish two groups of protocols designed for MANET: reactive (on-demand) and proactive (table-driven). The first type tries to resolve a path to a destination node on the source node demand, whereas the second approach is more preventive and continuously keeps routing tables up to date by monitoring the nearest neighbourhood. The detailed description and comparison of both classes of routing protocols for ad hoc networks may be found in works by Hu et al. (2002), Johnson (1994), and Royer et al. (1999).

At the moment several applications apart from strict MANET paradigm take advantage of the dynamic ad hoc routing phenomenon and make use of it in an akin to MANET wireless environments such as wireless mesh networks or vehicular ad hoc networks (VANET). This increasing application potential gives the MANET's security a primary concern for researching communities.

For MANETs there are several solutions considering multilayer defence against known attacks, mainly focusing on provided services such as authentication, anonymity, confidentiality, and integrity based on the network layer security. Most of them extend existing protocols for which the scope of security considerations was significantly limited, such as AODV, DSR, DSDV, and LSP (Royer et al., 1999). A complete protection from various attacks in MANET takes into account actions such as prevention, detection, and reaction. The prevention is usually achieved by means of secure routing protocols (path discovery and maintenance), while the abnormal behaviour detection is performed by monitoring end-to-end communication or by overhearing the local neighbourhood. In both cases the security extensively employs different cryptographic primitives authenticating routing messages (Hu & Perrig, 2004; Yang, Luo, Ye, Lu, & Zhang, 2004). The main difference between authentication schemes was characterized in the following list:

- First method takes advantage of a single shared session key widely distributed in the group of MANET nodes, for example, used in secure routing protocol (SRP) (Papadimitratos & Haas, 2002). This approach does not provide anonymity and is vulnerable to single node compromise.
- Second method assumes sharing pair-wise keys between all nodes in the networks and is used for HMAC in Ariadne (Hu et al., 2002). It suffers from lack of scalability and in this case of N nodes; N(N-1)/2 keys are required priory to allow communication.
- Third approach makes use of scalable public key cryptography where digital certificates and signatures allow the mobile nodes to be mutually authenticated (Sanzgiri, Dahill, Levine, Shields, & Belding-Royer, 2002; Zapata & Asokan, 2002). It is a principle method for secure routing in ARAN, SEAD, and SAODV. Although this method is linearly scalable when a node's number increases, its flexibility and efficiency may suffer from vulnerability to denial-of-service (DoS) attacks and computation overhead.

All of aforementioned methods assume existence of trusted authority TA (certifying authority CA) dealing with a key setup phase. The online and distributed TA supports key management and should assure a self-configurable principle of MANET, which is an important and challenging task for the nowadays research (Hu & Perrig, 2004; Mangipudi, Katti, & Fu, 2006; Yang et al., 2004).

Even though the many solutions for mobile ad hoc networks ensure secure communications (Hu & Perrig, 2004; Yang et al., 2004), very few of them address privacy guaranties as complementary to a strict security approach (Kong, Hong, & Gerla, 2005). The main objective of the anonymous communication in mobile ad hoc network is to provide privacy for all of its users represented by the nodes. Demanding for anonymity imposes a series of requirements for protocol construction and creates the new types of attacks. A set of formal notions of the anonymity and its related properties can be found by Chaum (1981) and Pfitzmann and Hansen (2005), where the authors characterize a general anonymous system with identity management designed for fixed network topology but easily applicable for MANET. ANODR protocol (Kong & Hong, 2003) delivers a full irrevocable anonymity based on symmetric cryptography primitives. Note that the pure anonymity of the users limits the accountability for malicious activity in the network. In order to avoid such a restriction a revocable anonymity was introduced which is based on pseudonyms managed by trusted third party, as it was proposed in ANAP (Ciszkowski & Kotulski, 2006), MASK (Zhang et al., 2005), and SDAR (Boukerche, 2004). Authors of SDAR identified that anonymous MANET is an environment suffering from the lack of personal and direct incentives to be well cooperative between nodes. In order to overcome this important vulnerability, SDAR was proposed: a node's trust management system providing a scoring of node's local activity whereby during the communication the most trustful nodes were promoted.

In the anonymous communication in the mobile ad hoc networks the trust management is used for misbehaviour detection and nodes evaluation. This evaluation creates a long-term node assessment called a reputation. A reputation is known from e-market mostly thanks to online auctioning systems, but in terms of MANET, it was defined by

Buchegger (2005) as a means for providing incentives for good behaviour of nodes and metric used for identifying the most truthful node. The main purpose of reputation management is a detection of any untrustworthy behaviour in the network interfering to ordinary and regular functions such as routing (next hop finding), forwarding (packets relaying), and, finally, isolating originators of such an activity. This goal of reputation management motivates nodes to be cooperative and improves the network security and performance.

Blaze, Feigenbaum, and Lacy (1996) argue for enhancing the existing trust management systems that incorporate PGP (Zimmermann, 1994) and PKI infrastructure of X.509. They propose a flexible and independent security system, PolicyMaker, which deals with policy management. They point out that trust transitiveness usually depends on the context of service, for example, e-main, link capacity, and quality of service in data forwarding, while PGP and X.509 support a trust transiting considering only guaranties of an association of public key with its owner's identity. One of the first solutions addressing the trust transitiveness in MANETs was protocol's Confidant (Buchegger & Le Boudec, 2002) and SDAR. Confidant incorporates the reputation system maintaining a node trust, path rates, and shared reputation information; however its construction does not provide the anonymous communication. The protocol SDAR supports a secure and anonymous communication but the proposed reputation system is limited in its efficiency because it supports only three levels of permissible trust that a node may obtain.

In the next section we present an introduction to the reputation-based secure routing, defining concepts of trust and reputation in relation to the mechanism of its modelling and managing. In the main part of the following section, we introduce an example of a distributed reputation model implemented in an anonymously authentication protocol for mobile ad hoc network (Ciszkowski & Kotulski, 2006).

REPUTATION-BASED SECURE ROUTING IN MANET

In many secure routing protocols for mobile ad hoc networks (Hu & Perrig, 2004; Yang et al., 2004) it can be found only few proposals considering anonymity as critical service assuring privacy for MANET's customers ANODR, MASK, SDAR and ANAP. The demand for anonymity in an open environment, such as mobile ad hoc networks, decreases node accountability and may be a source of unexpected behaviour coming from unknown network identities. It is postulated that by incorporating a distributed reputation system into secure routing protocol we can detect and avoid cooperation of hostile and anonymous nodes. The reputation system is an essential part of routing in MANET, which facilitates a prediction of node's behaviour and improves a performance of an anonymous communication. The reputation system provides a set of mechanisms and polices that when applied locally, similar to PolicyMaker (Blaze et al., 1996), allow assigning of a value of trust to a particular action performed in the network. Such an approach maintains trust knowledge of local neighbourhood activity without revealing its real identity. This information acts as a probability of future intentions and behaviours in the network and may be used to enforce the path discovery process by choosing the communication path containing only trusted nodes.

Trust and Reputation Modelling

In the literature it can be found that many interchangeable cases of the use of reputation and trust, even though, in popular understanding, they are not synonyms. In order to avoid any mistakes in this chapter we correspond with Hussain, Chang, and Dillon's (2004) definitions, and by trust we mean a subjective probability of a one peer (trustee), so that the particular actions of another peer (trusted) they are willing and capable to perform will be done according to the trustee's expectations in the given context and time. The defined trust is asymmetrical and usually represented by knowledge gathered during direct interactions and observations. The reputation is a perceived grade of trustworthiness to a particular peer created by their historical behaviour during observations and interactions with third party peers in the given context and time. This definition describes a concept of the peer's reputation expressed by a level of aggregated trust exchanged and shared between other peers. The main difference between the reputation and trust is that the subjective trust is usually created by the direct and own experience while the reputation is established combining other's trust knowledge. Considering the stated explanation, the reputation tends to represent a generalized opinion in a local group of peers. As the reputation may comprise an aggregated trust of several network nodes, it becomes a very valuable metric that supports the routing process. The measure expressing the level of trustworthiness for a particular node is also an important incentive for cooperation and good behaviour in the anonymous ad hoc network.

In terms of trust aggregation, sharing, and assessment, we can distinguish several types of modelling and management of the reputation in collaborative environments such as P2P networks, auctioning systems, and in particular, MANET. In one of the leading concepts of modelling, the distributed reputation assumes probabilistic representation of trust and introduces importance or uncertainty of shared knowledge. Lee, Hwang, Lee, and Kim (2006) apply the fuzzy set theory for the trust modelling. This approach defines multiple evaluating criteria with different importance factors. It allows building of trust by classifying the different types of observations and aggregating them with different importance weights. Since every criterion is strictly related to a class of observations, the interpretation of shared reputation depends on own preferences. Jøsang (2002) introduces the subjective logic where the trust is represented by a probability vector named an opinion, which is composed of belief, disbelief, uncertainty, and atomicity function. The atomicity function additionally determines the rate of the uncertainty in expected value of trust. Based on this approach, Huang, Hu, and Wang, (2006) propose a system similar to Lee et al.'s (2006) weighted evaluation trust, but the weights are dynamically modified by a feedback control unit according to the trust-policing module. The related probabilistic trust management methods are very attractive, but they usually suffer from complexity of probability logic incorporated into calculations.

Very interesting and especially suitable for MANET trust model is an enhanced reputation model for mobile ad hoc networks proposed by Liu and Issarny (2004). It takes advantage of the monitoring system as a principle module of misbehaviour detection (Buchegger & Le Boudec, 2002), which was effectively improved by exchanging the second-hand information (Buchegger, 2005). This work considers self-experience of nodes, time, and context dependency and introduces the definitions of services and recommendation reputation. In SDAR for anonymous MANET a three level of community management trust was introduced, in which every node may act as a central node of the community consisting of the rest nearest one-hop neighbours. The reputation is created locally, based on detected packet drops and modifications. In the network one distinguishes three classes of trust, which are assigned to every node. Trustworthy behaviour usually promotes nodes to the higher trust level but finally it may depend on local polices. A route discovery process is conducted considering a node's class membership. The main drawback of this approach is a low granularity of trust classes and considering only the local own experience.

Distributed Reputation for Secure MANET

In this section we present a distributed reputation system, which extends Liu's and Issarny (2004) reputation model incorporated in ANAP (Ciszkowski & Kotulski, 2006). We propose a new method of evaluating recommendation reputation considering past experience and recommendation reputation of voters (recommends). We define two types of second-hand information, related to the immediate nodes and cumulative reputation describing aggregated reputation of immediate nodes' neighbourhood. Second-hand information is exchanged on demand of interested nodes. In order to detect malicious activity and any anomalies in information exchange we incorporated a second-hand recommendation validation by a statistical correlation approach.

The reputation depends on time, own past experience, second-hand information, and is expressed by a level of trust. These input features are organized with a reputation dynamic evaluation scheme providing a node assessment. A proposed model consists of the following definitions and assumptions:

- $\mathbf{SR_n^B(A)}$ – service reputation held by **B** expressing a level of trust to node **A** in time **n**, and is taken into account whenever **B** is going to interact with **A**
- $\mathbf{IR_n^B(C)}$ – information reputation held by **B** expressing the level of trust for second-hand information $\mathbf{V_n^{BC}}$ received from node **C** at time **n**, and is used for evaluating received the second-hand information from **C**
- $\mathbf{V_n^{BC}(A)}$ – second-hand information (vote) coming from **C** to **B**; contains a recommendation of trust to node **A**, for an honest node is equal $\mathbf{SR_n^C(A)}$
- $\mathbf{CR_n^B(A)}$ – cumulative reputation expressing an aggregated grade for **A**'s neighbourhood which is unreachable by **B** at time **n**, it acts as a reputation of path going from **B** through node **A**
- $\mathbf{PR_n^B(A)}$ – path reputation is a product of service and cumulative reputation and expresses the trust of a path going from node **B** through the node **A** at time **n**,
- $\mathbf{STE^B(A)}$ – satisfaction degree of node **B** during elementary interaction with the node **A**
- $\mathbf{ST_n^B(A)}$ – average satisfaction degree of node **B** during several elementary interactions with the node **A** in time **n**
- $\mathbf{OE_n^B(A)^t}$ – own experience of node **B** based on history of interactions with the node **A**
- Nodes exchange **V** information with truthful neighbours
- Aforementioned parameters vary in range <-1,1>, where the most positive value reflects to the trustworthiest parameter

The building process of key reputation metrics is illustrated in Figure 1.

We introduced a virtual discrete time **n** in order to make the event-based nature of building reputation independent on real time. The virtual time depends on a number of events and its quantum consists of constant **Q** elementary interactions $\mathbf{STE_i}$. Every $\mathbf{STE_i}$ is expressed by Equation (1) and

Figure 1. Model of distributed reputation system providing the following vector metrics: own experience OEB(A), votes VB(A), service reputation SRB(A), cumulative reputation CRB(A) and path reputation PRB(A)

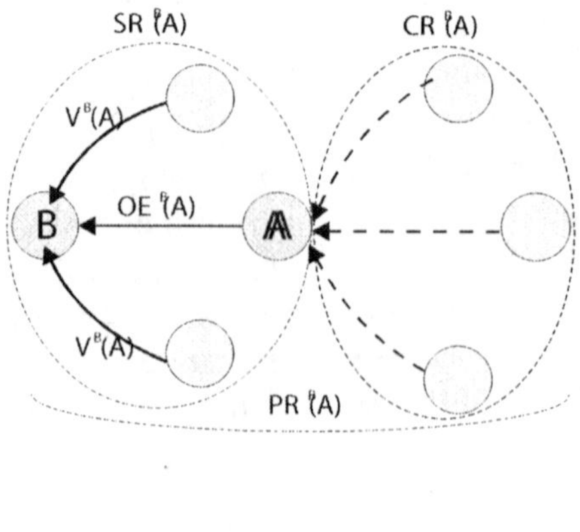

depends on a set of weighted metrics **m** monitored by a node during network packets exchanging. A metrics vector corresponds to all kinds of detectable observations such as every overheard packet modifications, attacks (DoS, reply attack, etc.), and network quality of service (QoS) parameters, for example, transmission delay and packet drops. This set of direct measurement is evaluated by expectation function **E**, which allows the assigning of different importance factor to a particular type of its arguments. The $\mathbf{STE_i}$ building process should take into account all observable misbehaviour defined at the end of this section.

$$STE_i^B(A) = \sum_{j=0}^{L-1} w_j E_i(m_j) \tag{1}$$

After every **Q** interactions between **B** and **A**, the time value **n** is incremented and aggregated $\mathbf{STE_i}$ updates satisfaction degree **ST**:

$$ST_n^B(A) = \frac{1}{Q} \sum_{i=0}^{L-1} STE_i \tag{2}$$

The proposed model takes advantage of past experience with a finite **L**-length memory, where every reputation measure with time **n - L** becomes the oldest value and is forgotten.

An own experience of node **B** at time n is based on the history of interactions with the node **A** and is evaluated as weighted average of **ST**:

$$OE_n^B(A) = \frac{\sum_{j=0}^{L-1} \gamma_j ST_{n-j}^B(A)}{\sum_{j=0}^{L-1} \gamma_j},$$

where

$$\gamma_n = \begin{cases} \rho, & n = 0 \\ (1-\rho)^{n+1}, & n > 0 \end{cases} \tag{3}$$

γ_n is an exponential fading function and depends on time, where $0<\rho<1$.

Whenever the own experience (**OE**) is updated or the second-hand information **V** is obtained the service reputation (**SR**) is modified. **SR** consists of own experience and weighted average of votes taking into account the information reputation (**IR**) of recommending nodes. Considering a set **GV** of voting nodes on **A**, the node **B** takes into account only nodes with positive **IR**. Own information is usually more valuable (Kong et al., 2005; Buchegger, 2005), hence scaling factor $\alpha \in <0,1>$ is introduced to the formula:

$$SR_n^B(A) = \alpha OE_n^B(A) + (1-\alpha) \frac{\sum_{p \in GV \setminus B} IR_n^B(p) V_n^{Bp}(A)}{\sum_{p \in GV \setminus B} IR_n^B(p)}, \qquad IR_n(p) > 0 \tag{4}$$

Note that nodes cooperating rarely have small service reputation **SR** and are less trustworthy.

In order to evaluate a credibility of recommendation **V** obtained from neighbouring nodes, it is required to update the information reputation (**IR**). In our model we propose a formula, which considers close relations between node's experiences (**OE**) with particular node, say **A**, as well as other voter's **IR**:

$$IR_n^B(C) = \beta OE_n^B(C) - \frac{\sum_{p \in GV \setminus B} IR_n^b(p) \left| V_n^{Bp}(C) - OE_n^B(C) \right|}{2 \sum_{p \in GV \setminus B} IR_n^B(p)}, \quad IR_n(p) > 0 \tag{5}$$

where $\boldsymbol{\beta} \in <0,1>$ is a scaling factor for own experience. It is recommended to let the service reputation (**SR**) evolve more dynamically then an information reputation (**IR**), which is equivalent to $\boldsymbol{\beta < \alpha}$. This allows nodes to rehabilitate their service reputation faster than their recommendation credibility. It means the nodes providing only a good service are able to rebuild already lost the information reputation.

Revealing all node identities on the communication path one could provide an easy way for building a global node reputation and evaluate a reputation along path, from the source to the destination. However, in ANAP for a communication path longer than three hops a physical identity of nodes lasts pure anonymous. Therefore we evaluate a cumulative reputation (**CR**) for the intermediate node, as an aggregated trust for its neighbour nodes, which are unreachable by the source node. This metric reflects a reputation along a path and does not break the anonymity:

$$CR_n^B(A) = \frac{\sum_{p \in GAV \setminus GBV} V_n^{Bp}(A)}{|GAV \setminus GAB|}, \qquad IR_n(p) > 0 \tag{6}$$

where **GBV** and **GAV** are sets of nodes being in neighbourhood of respectively node **B** and **A.**

Every time a source node wants to send data to an immediate node it calculates a path reputation (**PR**) as a product of service reputation, combining own and immediate nodes experience and cumulative reputation:

$$PR_n^B(A) = SR_n^B(A)CR_n^B(A) \tag{7}$$

A set of path with the highest **PR** is selected for communication. For path, which **PR** falls below zero, a communication channel should be closed.

A trust history evolution of own and immediate nodes is stored at every node in appropriate reputation parameters, represented by **L**-length vectors. This set of information is used for validating incoming second-hand votes by means of correlation analysis. Note, the own experience vector **OE** is a weighed moving average process (**MA**) of the order **L**. The γ function coefficients determine the dynamic of changes in **MA** process and make the all observation jointly dependent. As it was defined in Equations 4 and 5 the correlation is propagated to the service reputation (**SR**) and information reputation (**IR**) (indirectly by **IR** influence to **CR** and **PR**).

The correlation validation is based on autocorrelation functions of own experience $\hat{R}_n^V$ second-hand information $\hat{R}_n^O$ and distance functions, respectively **O** and **V**:

$$D_O = \sum_{i=0}^{L-1} (\hat{R}_i^O - \hat{R}_{i+1}^V)^2 \tag{8}$$

$$D_V = \sum_{i=0}^{L-1} (\hat{R}_i^V - \hat{R}_{i+1}^V)^2 \tag{9}$$

where $\hat{R}_n^O$ and $\hat{R}_n^V$ are estimators of autocorrelation function know as a convolution time series evaluated for a linear and stationary system, such as a reputation system:

$$\hat{R}_i^O = \sum_{n=0}^{L-1} OE_n OE_{n-i} \tag{10}$$

$$\hat{R}_i^V = \sum_{n=0}^{L-1} V_n V_{n-i} \tag{11}$$

Defined distance functions for every new vote **V** provide a measure of correlation change between own experience and already known history of votes from a particular node. Any attempt of voting unrelated to the historical observation will be observed by substantial increase of distance function $\mathbf{D_V}$. Additionally we can observe how much our own experience differ from the received by neighbour nodes analyzing value of $\mathbf{D_O}$. These two metrics should be taken into account if their values exceed some threshold separately defined for each of them $\mathbf{Th_O}$, $\mathbf{Th_V}$. In the case where the abnormal behaviour is detected following actions may be undertaken:

- $\mathbf{D_O > Th_O}$ and $\mathbf{D_V > Th_V}$ - the votes from misbehaved nodes are rejected and their information reputation is arbitrary decreased.
- $\mathbf{D_O < Th_O}$ and $\mathbf{D_V > Th_V}$ - the votes are accepted but service reputation (SR) is updated with higher scaling factor a. This less restrictive approach gives an ability to react to dynamically changing reputation but prevents too fast malicious attacks from targeting too discrediting nodes.
- $\mathbf{D_O > Th_O}$ and $\mathbf{D_V < Th_V}$ - in this case a node reaction should be similar to the previous anomaly, however, it may be symptomatic of long term attack against reputation service by nodes being in collusion.

Every node during messages exchanging collects its own experience of elementary interactions **STE**. The following list describes types of behaviour that can be taken into consideration during reputation building:

- **Forwarding:** During network operations nodes are able to verify integrity of messages anonymously forwarded in behalf of them by overhearing the first intermediate node. Every message tampering, delays, double relays, and dropping are detected as a malicious behaviour.

- **Receiving:** Every obtained message that could not be successfully verified, repeated messages and break down paths without error message notification coming form involved immediate node should be treated as untrustworthy.
- **Anonymous path establishing:** In case of ANAP, an anonymous path establishing a three-pass process and in every phase multilayered operations are performed. By default every request packet **REQ** should be forwarder only once by every node. In the case of detection of behaviour inconsistent with this rules or obtaining multiple copies of reply **REP** or error **ERR** messages, the reputation system should be informed.
- **Recommendation exchanging:** Sharing a reputation between nodes allows to compare an own experience with a given by recommending nodes. In the case when the one of the votes differs much from the rest voters there exists presumption of node discrediting. Additional statistical cross-validation (Hildebrand, Laing, & Rosenthal, 1977) methods may be used for this case evaluation.

The interaction of the presented reputation system with the anonymous authentication protocol is performed ensuring the purely anonymous communication. The reputation information is exchanged between nodes in on-demand manner of interested node, encrypted by public key of message originator. This ensures that recommendation sharing is hidden and may be read only by legitimated recipients.

FUTURE TRENDS

In the contemporary information society the mobile ad hoc networks is a promising and very attractive alternative for wireless access networks. Proposed in the last section, a solution for managing routing in secure MANETs is based on the distributed reputation system. We expect in the near future a class of new attacks will appear focusing on the reputation system. Keeping in mind that the security of the every system depends on its weakest point, the potential vulnerabilities of the reputation system may be treated as an important challenge for the future research. Two interesting forms of attacks for the reputation system may be Sybil and Collusion attack. In the case of first, the attacker takes advantage of using multiple identities by adversary's node, while in the second several malicious nodes are in collusion. In both cases it is highly possible that own experience and shared reputation may be affected by these attacks. Proposed by us, autocorrelation analysis for anomaly detection in reputation recommendations may not be sufficiently sensitive to cope with mentioned attacks. Now, a statistical method validation of recommendation, such as the cross-validation (Hildebrand et al., 1977), has been proposed and is being developed. It is a very promising direction of research, since the cross-validation is very flexible and easily applicable for complex data. On the other hand, the method is mathematically rigorous, so the obtained results are verifiable and easy to implement.

Another interesting area is the secure routing in MANET enforced by an ontology-based reputation system (Caballero, Botia, & Gomez-Skarmeta, 2006). A conceptual-based reputation may be identified as a reputation created for different types of services provided in MANET with an ability of creating a similarity measures between them. This approach in a natural way improves the model of incentives for the ad hoc communication giving ability to treat MANET networks as a service oriented.

At the moment several applications apart from strict MANET paradigm take advantage of the dynamic ad hoc routing phenomenon and make use of it in an akin to MANET wireless environments such as wireless mesh networks or vehicular ad hoc networks (VANET). This example shows that researching in the MANET's area may bear unlimited applications.

CONCLUDING REMARKS

In this chapter we presented a new approach of distributed reputation-based secure routing mechanism in MANET. In the background section the main concepts of secure and anonymous mobile ad hoc networks were presented. The overview of applied authentication schemes in secure MANET was analyzed giving an introduction to trust management and reputation basis as a mean for detecting misbehaviour and improving the routing performance.

In the main part of this chapter we focused on a new proposal of a distributed reputation system, which was an extension of the Liu and Issarny (2004) model and which was introduced in the anonymous authentication protocol for mobile ad hoc networks (Ciszkowski & Kotulski, 2006). We emphasized in the proposal the method of evaluating recommendation reputation considering the past experience and recommendation reputation of voters. We defined two types of the second-hand information, related to the immediate nodes and cumulative reputation, describing aggregated reputation of immediate nodes' neighbourhood. Second-hand information is exchanged on demand of interested nodes. In order to detect the malicious activity and any anomalies in the information exchange we incorporated the second-hand recommendation validation by the statistical correlation approach.

We pointed out the the security in MANET is a primary concern for researchers, in particular this becomes a very important issue since several applications apart from strict MANET communication model take advantage of the dynamic ad hoc routing phenomenon.

REFERENCES

Blaze, M., Feigenbaum, J., & Lacy, J. (1996). Decentralized trust management. In *Proceedings of the IEEE Symposium on Security and Privacy* (p. 164). IEEE Xplore.

Boukerche, A., El-Khatiba, K., Xua, L., & Korba, L. (2005). An efficient secure distributed anonymous routing protocol for mobile and wireless ad hoc network. *Computer Communications, 28*(10), 1193-1203.

Buchegger, S. (2005). Self-policing mobile ad hoc networks by reputation systems. *IEEE Communications Magazine, 43*(7), 101-107.

Buchegger, S., & Le Boudec, J.-Y. (2002, June). Performance analysis of the CONFIDANT protocol: Cooperation of nodes fairness in dynamic ad-hoc networks. In *Proceedings of IEEE/ACM Symposium on Mobile Ad Hoc Networking and Computing (MobiHOC),* Lausanne, Switzerland, (pp. 226-236).

Caballero, A., Botia, J. A., & Gomez-Skarmeta, A. F. (2006). A new model for trust and reputation management with an ontology based approach for similarity between tasks. In K. Fischer, I. J. Timm, E. André, & N. Zhong (Eds.), *Multi-agent System Technologies, 4th German Conference, MATES 2006*, Erfurt, Germany, (LNCS 4196, pp. 172-183). Berlin: Springer.

Chaum, D. (1981). Untraceable electronic mail, return addresses, and digital pseudonyms. *Communications of the ACM, 24*(2), 84-88.

Ciszkowski, T., & Kotulski, Z. (2006). *ANAP: Anonymous authentication protocol in mobile ad hoc networks.* Paper presented at the 10th Domestic Conference on Applied Cryptography ENIGMA, Warsaw, Poland, (pp. 191-203).

Hildebrand, D. K., Laing, J. D., & Rosenthal, H. (1977). *Prediction analysis of cross classification.* New York: John Wiley & Sons.

Hu, Y., & Perrig, A. (2004). A survey of secure wireless ad hoc routing. *IEEE Security & Privacy Magazine, 2*(3), 28-39.

Hu, Y.-C., Perrig, A., & Johnson, D. B. (2005). Ariadne: A secure on-demand routing protocol for ad hoc networks. *Wireless Networks, 11*(1-2), 21-38.

Huang, C., Hu, H., & Wang, Z. (2006, September 3-6). A dynamic trust model based on feedback control mechanism for P2P applications. In L. T. Yang, H. Jin, J. Ma, & T. Ungerer (Eds.), *Proceedings of Third International Conference on Autonomic and Trusted Computing (ATC 2006),* Wuhan, China, (LNCS 4158, pp. 312-321). Berlin: Springer.

Hussain, F., Chang, E., & Dillon, T. S. (2004, March). Classification of trust in logistic peer-to-peer communication. In *Proceedings of the IEEE International Conference on Sciences of Electronic, Technologies of Information and Telecommunications (SETIT 2004),* Tousse, Tunisia.

Johnson, D. B. (1994). Routing in ad hoc networks of mobile hosts. In *Proceedings of IEEE Workshop on Mobile Computing Systems and Applications* (pp. 158-163). IEEE Press.

Jøsang, A. (2002, July). Subjective evidential reasoning. In *Proceedings of the 9th International Conference on Information Processing and Management of Uncertainty in Knowledge-Based Systems (IPMU 2002),* Annecy, France.

Kong, J., & Hong, X. (2003). ANODR: Anonymous on demand routing with untraceable routes for mobile ad-hoc networks. In *Proceedings of the 4th ACM International Symposium on Mobile Ad Hoc Networking & Computing (MobiHoc03),* Annapolis, MD, (pp. 291-302).

Kong, J., Hong, X., & Gerla, M. (2005). Mobility changes anonymity: Mobile ad hoc networks need efficient anonymous routing. In *Proceedings of 10th IEEE Symposium on Computers and Communications (ISCC 2005)* (pp. 57-62). IEEE Computer Society.

Lee, K.-M., Hwang, K.-S., Lee, J.-H., & Kim, H. J. (2006, September). A fuzzy trust model using multiple evaluation criteria. In L. Wang, L. Jiao, G. Shi, X. Li, & J. Liu (Eds.), *Proceedings of Third International Conference on Fuzzy Systems and Knowledge Discovery (FSKD 2006)* (LNCS 4223, pp. 961-969). Berlin: Springer.

Liu, J., & Issarny, V. (2004, March 29-April 1). Enhanced reputation mechanism for mobile ad hoc networks. In C. Jensen, S. Poslad, & T. Dimitrakos (Eds.), *Proceedings of Second International Conference on Trust Management (iTrust 2004),* Oxford, UK, (LNCS 2995, pp. 48-62). Berlin: Springer.

Mangipudi, K., Katti, R., & Fu, H. (2006). Authentication and key agreement protocols preserving anonymity. *International Journal of Network Security, 3*(3), 259-270.

Nilsson, N. J. (1986). Probabilistic logic. *Artificial Intelligence, 28*(1), 71-87.

Papadimitratos, P., & Haas, Z. (2002, January 27-31). Secure routing for mobile ad hoc networks. In *Proceedings of the SCS Communication Networks and Distributed Systems Modelling and Simulation Conference (CNDS 2002),* San Antonio, (pp.192-204).

Perkins, C., & Royer, E. (1999, February). Ad hoc on-demand distance vector routing. In *Proceedings of the 2nd IEEE Workshop on Mobile Computing Systems and Applications,* New Orleans, (pp. 90-100).

Pfitzmann, A., & Hansen, M. (2005). *Anonymity, unobservability, pseudonymity, and identity management: A proposal for terminology.* Retrieved October 4, 2007, from http://dud.inf.tu-dresden.de/Literatur_V1.shtml

Royer, E., & Toh, C. (1999, April). A review of current routing protocols for ad hoc mobile wireless networks. *IEEE Personal Communications, 6*(2), 46-55.

Sanzgiri, K., Dahill, B., Levine, B. N., Shields, C., & Belding-Royer, E. M. (2002, November). A secure routing protocol for ad hoc networks. In *Proceedings of 10th IEEE International Conference on Network Protocols* (pp. 78-87). IEEE Press.

Yang, H., Luo, H., Ye, F., Lu, S., & Zhang, L. (2004). Security in mobile ad hoc networks: Challenges and solution. *IEEE Wireless Communications, 11*(1), 38-47.

Zapata, Z. G., & Asokan, N. (2002). Securing ad hoc routing protocols. In *Proceedings of ACM Workshop on Wireless Security (WiSe 2002)* (pp. 1-10). ACM Press.

Zhang, Y., Liu, W., & Lou, W. W. (2005). Anonymous communications in mobile ad hoc networks (INFOCOM 2005). In *Proceedings of 24th Annual Joint Conference of the IEEE Computer and Communications Societies* (Vol. 3, pp. 1940-1951). Proceedings IEEE.

Zimmermann, J. (1994). *PGP user's guide*. Cambridge: MIT Press.

KEY TERMS

Anonymity: Aims at hiding an entity's identity completely.

Anonymous Authentication: A method of proving that someone has rights to certain actions or resources without disclosing the user's real identity.

Attacks: Attacks on MANET can destroy availability of nodes (attacks on routing) and contest reputation of nodes.

Authentication: A method of proving someone's identity, especially if that someone is an authorized user of processes or resources.

Collusion Attack: If a number of adversary nodes make a coalition against reputation of other nodes.

Cross-validation: A statistical method derived from cross-classification which main objective is to detect the outlying point in a population set. It is a candidate method for anomalies detection in the reputation sharing (recommendations) and regular communication in MANET. **Denial-of-Service (DoS) attack:** An attempt of keeping an access to computer resources (nodes) unavailable, especially by generating dummy traffic from one source **(DoS)** or a large number of sources **(distributed DoS [DDoS])**.

MANET: Mobile ad hoc network is a self-configuring network of freely moving nodes connected by wireless links that can constitute a path joining two arbitrary nodes of the network.

Privacy: The ability of keeping secret someone's identity, resources, or actions. It is realized by anonymity and pseudonymity.

Pseudonymity: Hides the user's real identity behind some virtual identity called a pseudonym.

Reputation: Perceived grade of trustworthiness to a particular peer created by their historical behaviour during observations and interactions with third party peers in the given context and time

Routing: A method of selecting a path (a chain of links between neighbouring nodes) from a source node to a destination node. One can distinguish two groups of protocols designed for MANET: reactive (on-demand) and proactive (table-driven). The first type tries to resolve a path to a destination node on the source node demand, whereas the second approach is more preventive and continuously keeps routing tables up to date by monitoring the nearest neighbourhood.

Security: Security of a system means that the system does exactly what it is designed to do and nothing else, even in a case of attack. Secure MANET enables reliable routing: privacy of communication with immediate degree of authentication of the parties of the information exchange process.

Sybil Attack: When one adversary node uses several identities to multiply its ability of rating other nodes in MANET.

Trust: A subjective probability of a one peer (trustee) so that particular actions of another peer (trusted) they are willing and capable to perform will be done according to trustee's expectations in the given context and time

VANET: A form of mobile ad hoc network, to provide communications among nearby vehicles and between vehicles and nearby fixed equipment, usually described as roadside equipment.

Chapter XXIX
Trust Management and Context-Driven Access Control

Paolo Bellavista
University of Bologna, Italy

Rebecca Montanari
University of Bologna, Italy

Daniela Tibaldi
University of Bologna, Italy

Alessandra Toninelli
University of Bologna, Italy

ABSTRACT

The increasing diffusion of wireless portable devices and the emergence of mobile ad hoc networks promote anytime and anywhere opportunistic resource sharing. However, the fear of exposure to risky interactions is currently limiting the widespread uptake of ad hoc collaborations. This chapter introduces the challenge of identifying and validating novel security models/systems for securing ad hoc collaborations, by taking into account the high unpredictability, heterogeneity, and dynamicity of envisioned wireless environments. We claim that the concept of trust management should become a primary engineering design principle, to associate with the subsequent trust refinement into effective authorization policies, thus calling for original and innovative access control models. The chapter overviews the state-of-the-art solutions for trust management and access control in wireless environments by pointing out both the need for their tight integration and the related emerging design guidelines, that is, exploitation of context awareness and adoption of semantic technologies.

INTRODUCTION

Wireless telecommunication systems and the Internet are converging towards an integrated distributed environment that permits users to access/share services and to collaborate anytime and anywhere even when they are on the move. The increasing diffusion of portable devices with wireless connectivity and the emergence of mobile ad hoc networks (MANET) further promote opportunistic and temporary resource sharing by enabling mobile users in physical proximity of each other to spontaneously form ad hoc communities without the need to rely on the availability of a fixed network infrastructure. Mobile file sharing, mobile e-campus, emergency response, and vehicle coordination are just few collaborative application examples that illustrate the novel opportunities leveraged by envisioned and converged wired-wireless networks of the future. Hereinafter we will indicate this integrated network computing scenario formed by fixed Internet hosts, wireless terminals and wireless access points in between, as well as by collections of wireless mobile hosts forming MANET without the aid of any established fixed infrastructure, with the comprehensive term of *wireless Internet*.

However, the fear of exposure to risky interactions (possibly compromising confidentiality, availability, and integrity of both data and services) is currently limiting the widespread uptake of anywhere and anytime collaboration. To some extent, the above risk is present in any traditional distributed collaborative setting, but the wireless Internet exacerbates the perception of that risk because of the complex security challenges arising from the increased degree of openness and dynamicity of the scenario. Collaborating participants often cannot be statically preidentified; they usually change frequently due to high mobility and/or occasional failures, forming continuously varying ad hoc coalitions with entities entering and leaving groups dynamically. At the same time, roaming participants are often interested in establishing opportunistic collaborations with dynamically discovered partners, without having previous knowledge or long-term pre-established relationships with them. One of the most difficult security challenge in these environments is how to decide who to trust in the plethora of opportunistically discovered entities. In addition, MANET introduce a further level of complexity to secure collaborative applications: differently from traditional fixed networks where dedicated nodes support basic networking functions, for example, routing, in MANET these functions are carried out by available peers in the network, and there is no reason to assume that these peers will all cooperate uniformly. For instance, because network operations consume energy, some nodes may exhibit a selfish behavior and deny their cooperation, thus leading to severe degradation of network performance and functioning.

To protect and/or provide incentives for anywhere and anytime collaborations, there is the need for appropriate security models/systems that should follow novel design guidelines to take into account the high unpredictability, heterogeneity, and dynamicity of wireless Internet environments. In those scenarios where identities/roles of collaborating entities are difficult to be a-priori established, we claim that the concept of trust should become a primary design principle for the engineering of secure collaborative applications (Cahill, Gray, Seigneur, Jensen, Yong, Shand, *et al.*, 2003; Capra, 2004; Kagal, Finin, Joshi, 2001; Ruohomaa & Kutvonen, 2005). Trust provides a means to reduce the exposure to risky transactions in unfamiliar environments with no possibility to offer absolute protection against potential dangers. Trust solutions allow entities to decide whether to accept or refuse the dangers presumably associated with interactions with other entities. How to access resources and to whom to grant permissions should depend on the trust degree that collaborating entities mutually have.

Using trust as the basis to support secure ad hoc collaborations requires the design of novel trust management frameworks that enable entities to form, maintain, and evolve trust opinions in highly dynamic wireless environments. In fact, the wireless Internet deployment scenario poses complex issues to trust management and requires rethinking traditional solutions based on assumptions that are unacceptable in these environments (Cahill et al., 2003; Capra, 2004). In fact, in traditional distributed systems trust decisions can be delegated to centralized and trusted third parties with full visibility and control over the whole trust management domain (most entities are fixed and statically known). On the contrary, in the wireless Internet the lack of both a globally available trust management infrastructure and clearly defined administrative boundaries calls for fully decentralized and self-organized trust solutions. Moreover, trust management solutions are effective as far as it is possible to bind trust opinions to security decisions. We claim that trust management should be considered as the key starting point for subsequent refinement of trust into security policies related to authorization and security management. In particular, authorization can be seen as the outcome of the refinement of trust relationships among strangers (Grandison & Sloman, 2000).

Therefore, the issue of access control is also crucial for the provisioning of anytime and anywhere collaborative applications, and raises challenges similar to trust management, thus calling for novel access control models. Only few proposals are starting to emerge in that research area, by addressing two main needs. A primary requirement is to design/develop access control solutions that take into account heterogeneity and dynamicity of available services, computing devices, and user characteristics. Along this direction, the emerging design guideline for novel access control solutions advocates a paradigm shift from subject-centric access control models to context-centric ones (Covington, Long, Srinivasan, Dey, Ahamad, & Abowd, 2001; Corradi, Montanari, & Tibaldi, 2004; Ko, Won, Shin, Choo, & Kim, 2006; Toninelli, Montanari, Kagal, & Lassila, 2006). Hereinafter, at a high abstraction level, the term "context" is defined as any information that is useful for characterizing the state or the activity of an entity or the world where this entity operates (Dey, Abowd, & Salber, 2001). Differently from subject-centric solutions where context is an optional element of policy definition, simply used to restrict the applicability scope of the permissions assigned to the subject, in context-centric solutions context is the first-class principle that explicitly guides both policy specification and enforcement; it is not possible to define a policy without the explicit specification of the context making the policy valid. The second main requirement is the full integration of novel trust models/solutions with trust-dependent (possibly context-aware) access control policies. That integration represents the most significant goal in the state-of-the-art research in security for ad hoc wireless collaborations, with currently only a very few proposals at an early stage.

The achievement of secure, open, and dynamic wireless collaborations requires not only proper trust and access control models, but also shared and interoperable vocabularies for trust and access control specifications to avoid inconsistent interpretations. Some initial research efforts tend to propose the adoption of ontological technologies as a significant guideline toward common policy understanding (Kagal, Finin, & Joshi, 2003; Tonti et al., 2003; Uszok, Bradshaw, & Jeffers, 2004). Semantically rich representations of trust and access control policies permit resource/context descriptions at different levels of abstraction and enable reasoning about both structure and properties of entities, context, and operations, thus enabling flexible opportunities for policy analysis, conflict detection, and harmonization. It is worth noticing that current security solutions for wireless Internet collaborations represent interesting steps forward, but are still more proof-of-concept prototypes of single aspects rather than comprehensive methodological and technical reference guides.

The goal of the chapter is to survey the most relevant support solutions in the literature by considering the two primary research directions emerging in the area, that is, trust management and semantic context-driven access control. In particular, examples of solutions in each category will be presented in the Trust Management section and the Semantic Context-driven Access Control section, respectively. The COMITY Framework section will focus on the main design choices of our trust-dependent context-aware middleware proposal, with the aim of exemplifying the main concerns and solution guidelines about the integration of trust and access control management. Primary open issues and expected directions of evolution end the chapter.

TRUST MANAGEMENT

The adoption of the concept of trust as the basis for engineering secure collaborative applications is currently attracting relevant research interests. Trust has always been an important element in the establishment of relationships in many fields. Humans use trust daily to promote interaction and to accept risk in situations where they have only partial information (Cahill et al., 2003). In computing, the need for trust models and support systems has recently grown with the widespread Internet usage where transactions involve entities spanning a range of domains and organizations, not all of which may be trusted to the same extent. Recently, trust issues have taken on more urgency due to wireless environments of emerging relevance populated by a plethora of unknown and anonymous users/devices. Entities can interact as far as they are able to autonomously assess trust and to use this as the basis for automated decision making, for example, whether to use a service or whether to permit access to resources.

Incorporating trust in wireless Internet systems is important because trust can be an enabling technology for application provisioning in open and dynamic environments in situations where we are given up complete control because traditional security solutions are inadequate or even inapplicable. For instance, certificate-based authentication and authorization mechanisms exhibit several limitations when deployed over ad hoc wireless scenarios. First, they impose too much computational overhead (especially due to certificate validation), often intolerable for mobile devices with limited computational resources. Second, the transient nature of ad hoc collaborations does not justify the efforts of going through the laborious and expensive certificate issuance process. Finally, the lack of central authority and network infrastructure in MANET, coupled with the dynamic nature of the network topology, complicates the adoption of certificate-based authentication and authorization mechanisms.

Trust-related research has been carried out along several different directions and has proposed many approaches for trust definition, formation, evolution, and management, but has not yet achieved universally accepted techniques/tools, as detailed in the following.

Trust Definition and Properties

Trust is a complex and multifaceted notion relating to belief in the honesty, truthfulness, competence, and reliability of a trusted person or service (Grandison & Sloman, 2000). Currently there is no consensus in the literature on the meaning of trust though several research activities recognize its importance. Due to the fact that trust is an integral part of human nature, it is normally treated as an intuitive and universally understood concept. However, by realizing that it is unwise to assume it is an intuitive, universal, and well-understood concept, many researchers have proposed different definitions of trust and the importance of trust standardization is widely recognized (Frank & Peters, 1998; Gambetta, 2001; Marsh, 1994; Staab, 2004). However, trust definitions vary depending on researcher background and on addressed application domain.

Despite these differences, most proposals result in having common basic properties. Trust is usually specified in terms of a relationship between two entities that specifies the expectation of one trust-assigning entity, called the trustor, about the actions of another entity (object of a trust estimation), that is, the trustee, within a specified context (Grandison & Sloman, 2000). Entities bound by a trust relationship may be completely or partially unknown to each other.

Trust relationships may differentiate depending on the number of entities involved. They include one-to-one relationships between two entities, one-to-many in the case of one entity that needs to trust a group, many-to-many in the case, for example, of a committee, or many-to-one in the case of departments trusting a head branch. In any case, trust relationship is asymmetric: trustor and trustee do not need to have similar trust in each other even if they exploit the same information as their basis to establish their trust relationship. This derives from the observation, common to all trust definition proposals, that trust is a subjective notion (Cahill et al., 2003).

A crucial characteristic of trust, especially in wireless collaborative environments, is that trust is context-specific, that is, trust attributes depend on the context where trust is evaluated. For instance, honesty might be more significant for financial applications, whereas competence could be relevant for medical applications. In addition, the trust level determined in one context does not directly transfer to another application domain.

Trust is also inherently linked to risk, typically with an approximate inverse relationship, where risk is the probability of loss with respect to an interaction (English, Terzis, & Wagealla, 2004; Josang & Presti, 2004; Marsh, 1994; Sloman, 2004). The riskier an activity is, the higher is the trust level required to engage in the activity. The analysis of the exact relationship between risk and trust is a key issue for enabling cooperation, but there is still little work on risk analysis within trust management models.

Trust Management Systems

Trust management is the activity of collecting, codifying, analyzing, evaluating, and reevaluating evidence that relates to trust attributes with the purpose of making assessments and decisions about trust relationships. Several solutions have been proposed, each tailored to specific computing environments and focusing only on a subset of trust management problems (Ruohomaa & Kutvonen, 2005; Srinivasan, Teitelbaum, Liang, Wu, & Cardei, in press). The aim of this section is not to provide an exhaustive survey of all trust management systems, but to overview some exemplar state-of-the-art solutions along their historical evolution to point out how and to what extent they can address the heterogeneity and dynamicity of targeted wireless environments. The section also examines the recent trust-related MANET research work and outlines the novel research directions in the field of trust management that provide useful guidelines for the design of appropriate models to secure wireless ad hoc collaborations.

The issue of trust has been initially studied in the area of distributed systems where it has been faced in close association with authentication and authorization. Trust between entities is typically established by means of credentials, such as digital certificates, that act as proofs of either the identity of credential owners or the membership of credential owners to a trusted group. For instance, a digital certificate issued by a certification authority proves that a public key is owned by a particular entity. The certification authority vouches for the authenticity of the key owner's identity. Credential-based trust management solutions are designed to verify the authenticity of credentials and to determine whether certain credentials are sufficient for performing a certain action, that is, to decide how much to trust a given credential or its issuer/owner (Blaze, Feigenbaum, & Keromytis, 1998; Blaze, Feigenbaum, & Lacy, 1996; Chu, Feigenbaum, LaMacchia, Resnick, & Strauss, 1997). These approaches, however, have some limitations. They do not precisely define how trust is built and do not provide any model/tool to support trust formation and evolution. For instance, PolicyMaker and KeyNote focus on access control issues based on credential attributes rather than on trust evolution and reasoning issues. In addition, these trust management solutions usually assume a static form of trust. Moreover, traditional approaches can be only deployed in centrally administered settings with mostly fixed and known entities where a centralized trusted authority stores information about involved entities (Blaze et al., 1996, 1998).

To overcome the aforementioned limitations, trust solutions have recently evolved to better take into account the characteristics of open and dynamic environments. The Sultan trust management framework provides a wider notion of trust and allows the specification, analysis, and management of complex trust relationships. In particular, it includes a language for describing trust and recommendation relationships (Grandison & Sloman, 2003). In the Sultan model, a trustor can specify whether a trusted entity can perform (or not) actions when associated with a specific trust level within a specific context. The trust level is a measure of belief in the honesty, competence, security, and dependability of the trustee; the context defines the conditions to satisfy to establish the trust relationship. The proposed approach requires keeping track of historical information about entity behaviors and is based on a central server where trust information is stored and used for decision making and analysis.

A step further to better address the open and dynamic nature of wireless collaborative scenarios is represented by trust management solutions specifically designed to deal with incomplete knowledge and uncertainty. They all exploit the concepts of recommendation and reputation to enable trust decision making. The approach described by Abdul-Rahman and Hailes (2000) improves traditional solutions by proposing a distributed trust model that allows entities to autonomously reason about trust, without relying on a central authority. Based on direct experiences and recommendations, an entity derives its own trust evaluations. Mui et al. (2001) introduce a trust computational model, based on a Bayesian formalization of distributed rating processes, that takes into account the concept of reputation. The SECURE (secure environments for collaboration among ubiquitous roaming entities) project proposes another trust management solution, based on recommendations, that addresses how entities in unfamiliar environments can overcome initial suspicion to provide secure collaboration (Cahill et al., 2003). The model dynamically builds trust from local trust policies, based on past observations and recommendations. A different approach for mobile environments not relying on any third-party trusted entity is presented by Capra (2004): the proposed model enables nodes to form their trust opinions on the basis of aggregated trust information, mainly based on direct experiences regarding past interactions with neighbor nodes.

MANET push to the extreme the dynamicity, heterogeneity, and uncertainty about the execution environment, thus requiring further improvements to the design of trust management solutions to secure collaborative applications. The previously sketched solutions rely on historical reputation and recommendation to derive appropriate trust levels. In MANET, entities are likely to have only one-shot encounters, thus making difficult to collect sufficient historical reputation data about their past behavior for an appropriate trust level evaluation. In addition, due to the high variability of MANET scenarios, even if two entities re-encounter and collaborate, the collected historical data about one entity's reputation may refer to a completely different deployment context, thus making historical data less relevant to be the only aspect to consider for calculating the appropriate trust level for the new interaction context.

Other solutions in the literature have recently proposed trust models specifically targeted to MANET, in particular to deal with MANET node selfishness. For example, nodes can make trust-guided decisions to choose the appropriate relay node for forwarding packets. Several relevant activities have focused on MANET reputation, defined as the perception of one node regarding the performance of another node during the execution of a protocol (Srinivasan et al., 2007). Several reputation schemes have been proposed to mitigate selfishness. CONFIDANT (cooperation of nodes - fairness in distributed ad-hoc. networks) proposes a reputation system for misbehavior detection in MANET based on a modified Bayesian estimation approach (Buchegger & Le Boudec, 2002): everyone maintains a reputation rating and a trust rating about everyone else of interest. The reputation value of a node is calculated based on direct observation and trusted second-hand reputation messages. If a node observes another node not participating correctly, it reports this observation to other nodes that then perform actions to avoid being affected and to punish the "bad" node, for example, by refusing to forward its traffic. The approach is fully distributed and no static agreement is necessary. CORE (collaborative repuation) proposes another reputation mechanism to enforce node cooperation in MANET (Michiardi & Molva, 2002). CORE uses a collaborative monitoring technique and measures reputation as a node contribution to network operations. The OCEAN (observation-based cooperation enforcement in ad hoc networks) proposal uses only direct first-hand observations of other nodes behavior in contrast to CONFIDANT and CORE (Bansal & Baker, 2003). OCEAN discards second-hand reputation information because such information is considered subject to false accusations and requires maintaining trust relationships with other nodes. Liu, Joy, and Thompson (2004) allows a MANET node to form and maintain a trust level value with each other node, based on node behavior over time. The model relies on a collaborative scheme among nodes and exploits the trust value to determine a secure routing path for message forwarding and delivery. Parker et al. (2006) propose a solution that applies only to a laboratory setting, and relies on a reputation management system to give MANET nodes the ability to independently evaluate trust of nodes which they interact with. The reputation model neither relies on any wired infrastructure nor assumes continuous connectivity among all devices; the model only assumes that every device can assign an accuracy degree to any information provided to other peers and that every node maintains trust degrees for a subset of their peers.

Partially related to trust management MANET solutions, there is also the relevant issue of intrusion detection, which is however out of the scope of this chapter. For a comprehensive survey on the various interesting proposals available in the literature about MANET intrusion detection, please refer to Anantvalee and Wu's (in press) work.

To end the section with extremely recent and just opened research directions in trust management, some activities are focusing on risk analysis aspects relating to trust. Integrated management of risk and trust is still at an early stage with only few proposals. The SECURE project incorporates an explicit risk model for assessing collaborations (Cahill et al., 2003; Carbone, Nielsen, & Sassone, 2003; Dimmock, 2003; Josang & Presti, 2004), analyzes the relationship between risk and trust, and derives a computational model integrating the two concepts; risk is used to deduce trust. An opposite approach is taken by English et al. (2004) and Quercia, Hailes, and Capra (2006) where it is trust that drives the determination of risk: based on trust assessment input, the proposed decision model estimates the probability of potential risks associated with an action, and consequently decides whether to carry out requested actions. Dimmock, Bacon, Ingram, and Moody (2005) propose a risk model based on utility theory and evaluates it in a peer-to-peer collaborative scenario to drive trust-based access control.

SEMANTIC CONTEXT-DRIVEN ACCESS CONTROL

The design and deployment of wireless Internet collaboration services impose new challenges to distributed resource retrieval and operation, undermining several assumptions of traditional collaborative solutions. Whereas traditional collaboration relies on a static characterization of context where changes in the set of both service clients (users/devices) and accessible resources are relatively rare and predictable, user/device mobility causes frequent changes in physical user location, accessible resources, and the visibility/availability of collaborating partners. Therefore, traditional subject-centric access control solutions exhibit some limitations when applied to wireless ad hoc collaborations. The tight coupling of principal identities/roles with operating conditions, needed in subject-centric access control, would require security administrators to foresee all contexts where each collaborating principal/role is likely to operate. In wireless environments where entities are typically unknown and where operating conditions frequently change even unpredictably, that traditional approach may lead to a combinatorial explosion of the number of policies to write, force a long development time, and induce potential bugs. The traditional subject-centric approach also lacks flexibility: new access control policies need to be designed and implemented from scratch for any new principal and for any principal when new context situations occur.

The emerging guideline proposed in several recent security solutions is to tightly bind context awareness and access control models. By drawing inspiration from role-based access control (RBAC), which exploits the role concept as a mechanism for grouping subjects based on their properties, context can provide a level of indirection between entities requesting resource access and their permitted set of actions. Instead of assigning permissions directly to subjects and defining in which contexts these permissions are valid, for each resource system administrators should define the context where to enable operations on it. When an entity operates in a specific context, it automatically acquires the ability to perform the set of actions active for the current context; when context changes, entities are instantaneously granted/denied access to resources accordingly. For instance, let us suppose that a printer can only be accessed by people located in the same room the printer is. Then, whenever a person leaves that room, the person looses the permission.

The integration of access control with context has two main characteristics. First, it is an example of an active access control model (Georgiadis, Mavridis, Pangalos, & Thomas, 2001). Active security models are aware of the context associated with an ongoing activity; that distinguishes the passive concept of permission assignment from the active concept of context-based permission activation. Second, the exploitation of context as a mechanism for grouping policies and for evaluating applicable ones simplifies access control management by both increasing policy reuse and simplifying policy update and revocation.

The explicit consideration of context for access control decisions is a very recent research direction that is attracting increasing interest, but only a few context-dependent policy model proposals have currently emerged. Covington et al. (2001) strongly point out the relevance of context. The proposal allows policy designers to represent contexts through a new type of role, called *environment role*. Environment roles capture relevant environmental conditions, used to restrict and regulate user privileges. Permissions are assigned to roles (both traditional and environment ones) and role activation/deactivation regulates resource access. However, the proposed model considers a very limited set of context information represented by environment properties and states. Environment roles do not provide any means for representing other kinds of contexts, such as current collaborative activities, or user movements.

An attempt of extending these kinds of contexts is proposed by Neumannb and Strembeck (2003). They present a context-based security approach using special purpose RBAC constraints. A context constraint is defined as a dynamic RBAC constraint that checks the actual values of one or more context attributes. Accordingly, context constraints are used to define conditional permissions, that is, RBAC permissions constrained by context situations. In this security model, access rights are granted to users if corresponding context constraints are satisfied. This approach also presents an engineering process for context constraints, based on goal-oriented requirement engineering: the first step is to fetch the model of the current deployment scenario, by identifying goals and obstacles; each goal is then examined to derive the context attributes needed to describe that goal; then, goals are used to specify context constraints. Although context is considered crucial, the proposal continues to follow a subject-centric approach where context conditions act simply as constraints on privilege applicability.

Another effort of exploiting context for access control decisions is presented by McDaniel (2003). Conditions in access control policy statements are not defined as expressions over a fixed set of attributes, but viewed as general purpose programs used to measure context (every condition is a parametric function). This extended condition view embraces the environment dynamicity and allows expression of complicated relationships between context conditions. However, in this approach the policy infrastructure only evaluates policies in response to the trigger of user-requested operations; the access control process is still subject-based.

Corradi et al. (2004) propose a context-based access control management system for ubiquitous environments that dynamically determines the contexts of mobile users and rules the access to resources by taking into account user contexts. In particular, access control policies are defined as association rules between a set of permissions and a set of contexts.

By specifically focusing on access control in spontaneous wireless coalitions, Liscano and Wang (2005) propose a delegation-based approach, where users participating in a communication session can delegate a set of their permissions to a temporary *session role*, in order to enable transitory access to the resources of all participants in the session. In particular, one user assigns the session role to the entities the user is willing to communicate with. Context is used to define the conditions that must hold for the assignment to take place. Only a limited set of contextual information can be specified and no semantic technologies are exploited to represent either session role or context constraints. In addition, security problems may arise whenever an entity delegated to play the session role leaves the collaborative session. In fact, unless the user explicitly states she is leaving the session, there is no way for the framework to be aware that the session role must be revoked.

Semantic Access Control Policies

Access control policies can be specified in many different ways and multiple approaches have been proposed in different application domains (Tonti et al., 2003). Anyway, there are some general requirements that any policy representation should satisfy regardless of its field of applicability: expressiveness to handle the wide range of possible management requirements, simplicity to ease policy definition tasks for administrators with different degrees of expertise, enforceability to ensure a mapping of policy specifications into implementable policies for various platforms, scalability to ensure adequate performance, and analyzability to enable reasoning about policies. The challenge is to achieve a suitable balance among the objectives of expressiveness, computational tractability, and ease of use.

An important principle that is currently emerging for novel access control models for wireless Internet environments is the adoption of semantically-rich policy representations to improve expressiveness, analyzability, and interoperability. A semantic-based approach allows description of contexts and associated policies at a high level of abstraction, in a form that enables their classification and comparison. This feature is essential, for instance, in order to detect conflicts between policies before their enforcement. In addition, semantic techniques can provide the reasoning features needed to deduce new information from existing knowledge. This ability may be exploited by policy frameworks when faced with unexpected situations to react in a contextually appropriate way. For example, let us consider an access control policy granting access to the projector in the meeting room of a company. Suppose that meetings are normally scheduled until 5 p.m. and that access to the projector is consequently not allowed after that time. Let us now suppose that a meeting goes beyond its scheduled end time. In this example, we need to "instruct" the system such that, if certain context conditions hold, the meeting is still taking place and access to the projector should therefore continue to be allowed.

The importance of adopting a semantic approach to the specification of all security policy building elements (subjects, actions, context, ect.) has recently emerged in well-known policy frameworks, such as KAoS (knowledgeable agent-oriented system) (Uszok et al., 2004) and Rei (Kagal et al., 2003), which have adopted semantic technologies due to their rich expressiveness and tractability.

However, the integration of semantic technologies within context-driven access control is still at its infancy with only a few proposals. In fact, a major drawback of the approaches mentioned in the previous section is that they do not exploit semantic information describing contexts, which could provide various advantages to context-aware access control systems. On the contrary, Toninelli et al. (2006) present a novel semantic-based access control model that exploits context awareness to control resource access and to enable dynamic policy adaptation in response to context changes, while adopting semantic technologies for context/policy specification to enable reasoning about context and policies. Ko et al. (2006) propose an approach that allows to overcome the semantic gap between contexts specified in policies at design time and context data dynamically collected in the execution environment. For instance, consider an authorization policy stating that a doctor working in a pediatrics ward can access information about the infants' parents. If Doctor Green is currently located in Room 209, and Room 209 belongs to a pediatrics ward, then the authorization policy should apply to Doctor Green. Semantic-based policy rules are therefore used to explicitly state the semantic relationship between the condition expressed in the policy and the condition revealed by sensors, for example, the doctor's location.

THE COMITY FRAMEWORK

COMITY (context-based middleware for trustworthy services) is our original proposal of trust-dependent access control framework that exploits semantic-based context descriptions for collaborative applications over MANET. In particular, COMITY allows different actors, sharing little or no prior knowledge about each other, to establish trustworthy relationships and to define proper access control policies governing operations on shared resources depending on mutual trust relationships. Let us note that COMITY largely exploits context awareness to model trust and access control policies. In addition, it adopts semantic-based context/policy descriptions to provide expressive representation and reasoning over trust relationships and access control policies.

We claim that the COMITY framework is a good exemplification of an emerging trend for novel security solutions based on semantic technologies, on context awareness, and on the trust concept to properly handle highly dynamic environments. To point out those emerging solution guidelines, the following section provides a rapid overview of the main characterizing COMITY features.

COMITY Context Model

COMITY defines context as any characterizing information about the controlled entities and their surrounding environment that is relevant for establishing trust relationships and trust-dependent access control policies. The COMITY context model distinguishes between the concepts of entity and context. An entity "bears" one or more contexts and a context "inheres in" one or more entities. COMITY contexts may be either primary or secondary (see Figure 1). The primary context identifies the set of information about a collaborating entity/device (and its resources/services) to determine an initial level of trust; this set of information should be always acquirable, without the need to rely on a fixed network infrastructure for its validation. The secondary context includes context data that may not be always available, but that can be used, if available, to refine the initial trust level, such as credentials and historical data about one entity's behavior. For instance, credentials, such as digital certificates, are often available in traditional execution environments, but may not be verifiable in MANET, in that case becoming useless to determine the first level of trust. The idea underpinning our choice of distinguishing primary and secondary context is to base initial trust decisions mainly on context information that can be always acquired, for example, user characteristics, device properties, and offered services. In addition, in wireless ad hoc collaborations, users tend to discover partners of interest and to create group aggregations on the basis of user preferences/interests, offered resources/services, and access terminal capabilities, that is, on the basis of primary context.

More in details, as depicted in Figure 1, one user's primary context includes data about the user, the user's offered services, and the characteristics of the currently used access device. In particular, the User Context represents user characteristics and includes various types of context element parts. The Identifier part represents the unique system-level user identifier, for example, a personal identifier that does not convey information qualifying user identity. The Social Attitude part represents the user willingness to cooperate, that is, the user's attitude to share information with other entities. A user may declare oneself *selfish* when the user denies cooperation, for instance by not providing a service/information to save device battery degradation; *ignorant* when the user disposition is not declared; and *cooperative* when the user is willing to cooperate. The Team part specifies the team, that is, a group of users with common characteristics, if any, that the user belongs to. For instance, in the case of temporary unavailability of one collaborating service provider, this context information facilitates entities to discover other entities possibly replacing it within the team. The Distance part represents the physical distance between two entities. In particular, distance is the number of network hops required for an entity to route a message to the collaborating entity which the distance context part is associated with. The need for this context information relies on the fact

Figure 1. The COMITY model for primary and secondary contexts

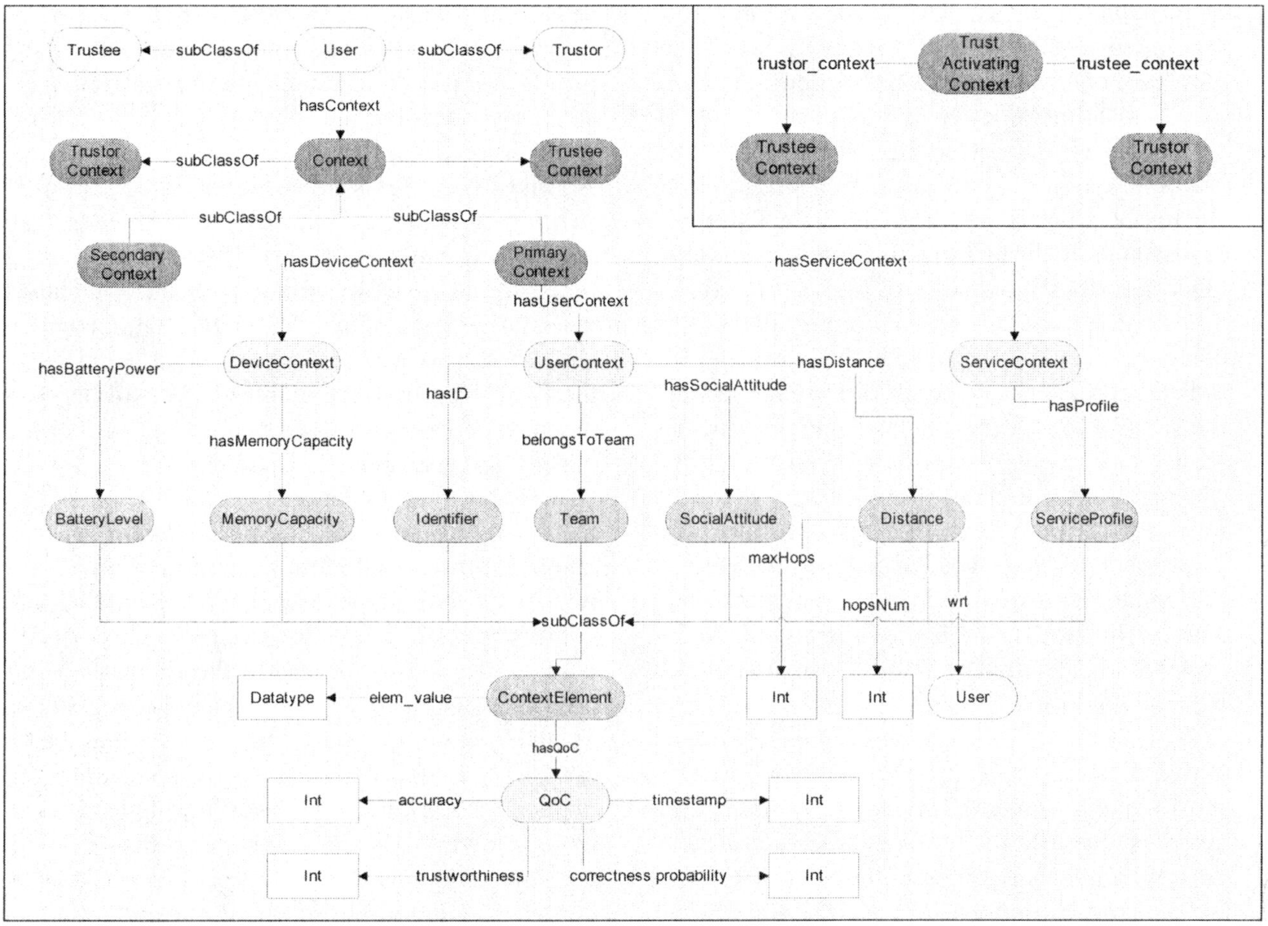

that MANET users tend to prefer collaborating with close peers due to typical link instability and high error rates in message delivery in long routing paths.

The secondary context includes all other types of information characterizing users/devices and provided services that are considered relevant, but not necessary, for determining the initial trust level. For instance, user identities/roles, digital certificates, reputations, and recommendations are all information that COMITY takes into account as secondary context. Differently from primary context, COMITY does not provide any static prebuilt classification for secondary context elements; the classification of secondary context data is usually tailored to application-specific requirements and network deployment conditions at runtime.

It is worth noticing that in highly dynamic MANET scenarios there are several problems arising in context determination and processing. A primary complexity is due to frequent context changes. Due to dynamic context variations, not only the value of a context attribute is of interest, but also the moment when the value was determined (van Sinderen et al., 2006). In addition, context sources are generally restricted in their ability to estimate the precise value of a context attribute. Another factor of complexity stems from the possible unreliability of the context information provided by collaborating entities. An entity may purposely provide incorrect information (malicious entity), may be unable to guarantee a reliable level of provided information (ignorant entity), may provide correct information (cooperative entity), or may have correct information but denies to make it available (uncooperative).

To deal with such aspects, COMITY adopts the notion of quality of context (QoC), as defined by Buchholz, Kupper, and Schiffers (2003) and van Sinderen et al. (2006), which is metainformation "that describes the quality of information that is used as context information." In particular, each context element is associated with a parameter that indicates the accuracy (i.e., the difference between the value of the context attribute and the aspect of reality it represents), the probability of correctness (the probability, estimated by the context source, to be the correct value), the trustworthiness (the probability, estimated by the context receiver, that the context source provided the correct value), and the freshness (the age of the context value, typically represented by a timestamp). Let us note that the determination of the correct QoC levels to apply in highly dynamic wireless environments is a challenging issue, in particular to achieve the most suitable tradeoff between minimum intrusiveness and accuracy. These mechanisms are currently under investigation in COMITY where we are experimenting and validating novel highly decentralized and localized techniques based on collaborative game theory.

COMITY Trust Model

There are different forms of trust in the literature relating to whether access is being provided to trustor's resources, the trustee is providing a service, trust concerns authentication, or it is being delegated (Grandison & Sloman, 2000). The COMITY trust model focuses on two specific aspects: an entity A (trustor) trusts an entity B (trustee) i) to use resources that A owns or controls, and ii) to provide a service not requiring access to A's resources.

COMITY defines trust relationships as one-to-one associations between contexts and trusted actions, with each relationship characterized by a specific trust degree. Contexts act as intermediaries between A and B and the set of operations for which A trusts B. COMITY users exploit the primary and secondary context described in the previous section to define their own trust relationships. In particular, the context involved in a trust relationship definition is the intersection between A's and B's contexts (hereinafter called *trust activating context*, see Figure 1). This means that an action is considered trusted if executed within the scope of specific conditions of the applicable trust activating context. Which trustor and which trustee are involved in the trust relationship is directly derived from the identifier parts of the two primary contexts involved in the trust activating context intersection.

COMITY trust relationships have all an associated trust degree, which reflects the trustor opinion about the trustworthiness of executing an action within the scope of its associated trust activating context. We are currently integrating COMITY with existing trust models in the literature suitable to derive trust degree values in decentralized MANET environments (Capra, 2004; Quercia et al., 2006). In general, trust degree determination depends on various factors. For instance, the need perceived by A of letting B to perform the specified action on either A's or B's resources influences trust degree calculation; that need may be quantified as the expected loss deriving from not executing the specified action. In addition to need, it is necessary to take into account QoC levels and the risk of the loss deriving from performing critical actions in an incorrect way.

COMITY Trust-Dependent Access Control Model

COMITY uses trust-dependent access control policies to govern operations on resources when A trusts B to use resources that A owns or controls. In this case, A has to define suitable authorization policies. On the contrary, in the case A trusts B to provide a service that does not involve access to A's resources, B's usage of the service is outside A's control, but depends on the access control policies defined by B. Both A and B follow the COMITY trust-dependent access control model to define proper access control policies.

The COMITY trust-dependent access control model extends our approach proposed by Toninelli et al. (2006). In particular, the access control model is context-centric: contexts are associated with allowed actions that represent all and only the conditions that enable access to the resources. By drawing inspiration from Java protection domains (Gong, 1999), we call these contexts as *protection contexts*: they are determined by policies and provide users with a controlled visibility of the performable resource access actions.

The COMITY extension we are currently working on consists in exploiting the trust degree as a specific protection context with possibly associated actions. In other words, it is possible to allow/deny an action as far as the resource access requestor has a proper trust degree. More in details, at any resource access request, COMITY verifies whether a trust activating context has been defined for the requested operation. In that case, COMITY checks whether the trust activating context is currently in effect (*active context*), that is, the context defining conditions match the operating conditions of the requesting entity, requested resource, and environment. In particular, there is the need to verify whether the requestor is the trustee indicated in the trust activating context. Then, the corresponding trust degree associated with the activating trust context and with the requested operation is used to select the proper access control policy.

Context and Trust Relationship Implementation

The COMITY context model is based on an underlying system model that describes interactions by using the concepts of entities and actions. An entity represents any actor/resource in the system and is logically characterized by a number of properties expressed as attribute-value pairs. An action describes an activity that an actor is able to perform on another entity. An interaction is the association of an entity and an action.

COMITY adopts description logics (DL) and associated inference mechanisms to model entities, actions, trust activating contexts, trust relationships, and access control policies. In particular, we use Web ontology language (OWL)-based ontologies, as shown in Figure 2a (for the sake of simplicity, we use hereinafter a compact DL notation). A trust activating context is defined as a subclass of a generic context. Each generic context consists of several context elements, with each element characterized by at least an identity property and a location property defining the physical or logical position of an entity, and eventually by other additional properties. A trust activating context is defined by restriction on the trustor and trustee primary contexts (possibly by considering also secondary contexts when available). Each primary context consists of several context elements, grouped within a specific context category, with each element characterized by a value and at least a QoC attribute.

Figure 2a shows an example of a trust activating context, where the trustee is within two-hop distance from the trustor and has a device with a full battery; both trustor and trustee are cooperative. Each information associates with values of trustworthiness and/or freshness. Figure 2b shows an example of trust-dependent access control policy. The trust relationship defines an association between the trust activating context shown in Figure 2a (Trust_Activating_Context_1) and the action of the trustee accessing the files included in folder(A). COMITY assigns a level of trust (70%) to this association. The access control policy states that access to folder(A) files is granted to a trustee if the trust level associated to that action reaches the minimum threshold of 60%. Therefore, Trust_Activating_Context_1 is effective and the trustee is granted with the requested permission.

The DL adoption in context modeling and reasoning has well-known benefits. For instance, when considering activating contexts as classes and a set of sensor inputs as individuals, DL-based reasoning allows one to determine which activating contexts are in effect by verifying which trust activating context classes the current state is an instance of, and to figure out how activating contexts relate to each other, for example, via nesting [6]. For instance, let us consider the case of a user

Figure 2. COMITY specifications for trust relationship and trust-based access control policy

Trustor Primary Context Specification

```
Trustor_Primary_Context_1 ≡ Primary_Context ⊓
                    ∃has_User_Context.Trustw_Coop_UserCtx
Trustw_Coop_UserCtx ≡ UserContext ⊓ ∃hasSocialAttitude.Cooperative_T10
Cooperative_T10 ≡ Social_Attitude ⊓ ∃elem_value.{"Cooperative"} ⊓ ∃trustworthiness.{10}
```

Trustee Primary Context Specification

```
Trustee_Primary_Context_1 ≡ Primary_Context ⊓ ∃hasUserContext.Trustw_Coop_2Hops_UserCtx
                  ∃hasDeviceContext.FullBattery_DeviceCtx
Trustw_Coop_2Hops_UserCtx ≡ UserContext ⊓ ∃hasSocialAttitude.Cooperative_T8 ⊓
∃hasDistance.Max2Hops
Max2Hops ≡ Distance ⊓ ∃maxHops.{2} ⊓ ∃wrt.Trustor
Cooperative_T8 ≡ Social_Attitude ⊓ ∃elem_value.{"Cooperative"} ⊓ ∃trustworthiness.{8}
FullBattery_DeviceCtx ≡ DeviceContext ⊓ ∃hasBatteryPower.FullBattery_T8_TS67
FullBattery__T8_TS67 ≡ BatteryLevel ⊓ ∃elem_value.{"Full"} ⊓ ∃trustworthiness.{8} ⊓
```

Trust Activating Context Specification

```
Trust_Activating_Context_1 ≡ Trust_Activating_Context ⊓ ∃trustor_context.Trustor_Primary_Context_1
                    ⊓ ∃trustee_context.Trustee_Primary_Context_1
```

A)

Context-based Trust Relationship Specification

```
ileA_Access_Trust_Rel ≡ Trust_Association ⊓ ∃activating_context.Trust_Activating_Context_1 ⊓
                 ∃trusted_action.FileA_Sharing_Action ⊓ ∃trust_degree.TrustDegree_70%
ileA_Access_Action ≡ AccessAction ⊓ ∃target.FolderA_Files ⊓ ∃performedBy.Trustee
rustDegree_70% ≡ TrustDegree ⊓ ∃hasTrustValue.{70%}
```

Trust-dependent Access Control Policy Specification

```
ileA_Access_Policy ≡ Access_Control_Policy ⊓ ∃controls. FileA_Access_Action ⊓
                 ∃protection_context.MinTrustDegree_60%
inTrustDegree_60% ≡ Trust_Degree ⊓ ∃hasMinValue.{60%}
```

B)

who has a PDA with full battery, is self-defined as cooperative, and is one-hop far from a trustor. By means of DL-based reasoning, COMITY can automatically recognize that this enables the trustee part of the trust activating context defined in Figure 2a.

CONCLUSIONS AND OPEN RESEARCH ISSUES

Securing ad hoc wireless collaborations is a complex management issue that calls for novel security paradigms suitable for high heterogeneity and dynamicity. In this perspective, trust is starting to be recognized as an enabling principle for promoting cooperation among (partially) unknown entities. However, trust is insufficient for securing wireless Internet collaborations and there is the need to associate it with proper resource access control policies.

In addition, in the specific field of trust management, novel solutions have started to emerge specifically designed to deal with uncertainty and incomplete knowledge that are usual in dynamic wireless environments. However, despite the widespread interest, the field is still immature. Sound terminological and methodological frameworks are still missing, there is not even a commonly agreed definition of the concept of trust, and several aspects still require investigation, such as the proper evaluation and combination of trust and risk.

Also the relationship between context and trust has not received the needed attention whereas it is crucial to form trust opinions and to reason about recommendation and reputation. The relationship is twofold. First, trust, recommendation, and reputation opinions can be context-dependent. Second, context elements at the time when trust, recommendation, and reputation opinions are created should be compared with the trustor's context where interactions currently take place. Moreover, there is the need to integrate trust management solutions with run-time monitoring both to support self-adaptation of trust opinions to dynamic variations and to detect (and possibly react to) violations to the initially estimated trust level.

Another open research issue is about mechanisms and strategies to promote cooperation among possibly unknown entities in dynamic environments: credit systems on top of trust management solutions could reward the good behavior of cooperating peers and punish selfish/cheating nodes. A very relevant research challenge at the moment is the full integration of novel trust-based management models/solutions with trust-dependent (possibly context-aware) access control policies. Finally, there is the need for standardization efforts, based on semantic technologies, to open new possibilities for entities to represent contexts, trust relationships, and access control policies in a fully interoperable way, which is crucial in open and dynamic deployment scenarios.

ACKNOWLEDGMENT

Work supported by the MIUR PRIN MOMA and the CNR IS-MANET Projects.

REFERENCES

Abdul-Rahman, A., & Hailes, S. (2000). Supporting trust in virtual communities. In *Proceedings of the 33rd Hawaii International Conference on System Sciences* (Vol. 6, p. 6007). Washington D.C.: IEEE Computer Society Press.

Anantvalee, T., & Wu, J. (in press). A survey on intrusion detection in mobile ad hoc networks. In Y. Xiao, X. Shen, & D.Z. Du (Eds.), *Wireless/mobile network security.* New York: Springer-Verlag.

Bansal, S., & Baker, M. (2003). *Observation-based cooperation enforcement in ad hoc networks.* Stanford, CA: Stanford University. Retrieved May 15, 2007, from http://stanford.edu/~sbansal/pubs/ocean03.pdf

Blaze, M., Feigenbaum, J., & Keromytis, A.D. (1998). KeyNote: Trust management for public-key infrastructures. In B. Christianson et al. (Eds.), *Proceedings of the 6th International Workshop on Security Protocols* (LNCS 1550, pp. 59-63). Cambridge, UK: Springer-Verlag.

Blaze, M., Feigenbaum, J., & Lacy, J. (1996). Decentralized trust management. In *Proceedings of IEEE Symposium on Security and Privacy* (pp. 164-173). Oakland, CA: IEEE Computer Society Press.

Buchegger, S., & Le Boudec, J.Y. (2002). Performance analysis of the CONFIDANT protocol: Cooperation of nodes - fairness in dynamic ad-hoc networks. In *Proceedings of the 3rd ACM International Symposium on Mobile Ad Hoc Networking & Computing* (pp. 226-236). Lausanne, Switzerland: ACM Press.

Buchholz, T., Kupper, A., & Schiffers, M. (2003). *Quality of context: What it is and why we need it.* Paper presented at the 10th HP–OVUA Workshop, Geneva, Switzerland.

Cahill, V., Gray, E., Seigneur, J.M., Jensen, C.D., Yong C., Shand, B., et al. (2003). Using trust for secure collaboration in uncertain environments. *IEEE Pervasive Computing, 2*(3), 52-61.

Capra, L. (2004). Engineering human trust in mobile system collaborations. In *Proceedings of the 12th ACM International Symposium on Foundations of Software Engineering* (pp. 107-116). Newport Beach, CA: ACM Press.

Carbone, M., Nielsen, M., & Sassone, V. (2003). A formal model for trust in dynamic networks. In *Proceedings of the 1st International Conference on*

Software Engineering and Formal Methods (pp. 54-63). Brisbane, Australia: IEEE Press.

Chu, Y.H., Feigenbaum, J., LaMacchia, B., Resnick, P., & Strauss, M. (1997). REFEREE: trust management for Web applications. *Computer Networks and ISDN Systems, 29*(8-13), 953-964.

Corradi, A., Montanari, R., & Tibaldi, D. (2004). Context-based access control management in ubiquitous environments. In *Proceedings of the 3rd IEEE International Symposium on Network Computing and Applications* (pp. 253-260). Cambridge, MA: IEEE Computer Society Press.

Covington, M.J., Long, W., Srinivasan, S., Dey, A.K., Ahamad, M., & Abowd, G.D. (2001). Securing context-aware applications using environmental roles. In *Proceedings of the 6th ACM Symposium on Access Control Models and Technologies* (pp. 10-20). Chantilly, VA: ACM Press.

Dey, A., Abowd, G., & Salber, D. (2001). A conceptual framework and a toolkit for supporting the rapid prototyping of context-aware applications. *Human-Computer Interaction, 16*(2-4), 97-166.

Dimmock, N., Bacon, J., Ingram, D., & Moody, K. (2005). Risk models for trust-based access control. In P. Herrmann, S. Shiu, V. Issarny (Eds.), *Proceedings of the 3rd Annual Conference on Trust Management* (LNCS, Vol. 3477, pp. 364-371). Rocquencourt, France: Springer-Verlag.

Dimmock, N., Bacon, J., Ingram, D., & Moody, K. (2005). Risk models for trust-based access control. In P. Herrmann et al. (Eds.), *Proceedings of the 3rd Annual Conference on Trust Management* (LNCS 3477, pp. 364-371). Rocquencourt, France: Springer-Verlag.

English, C., Terzis, S., & Wagealla, W. (2004). Engineering trust-based collaborations in a global computing environment. In C. Jensen, S. Poslad, T. Dimitrakos (Eds.), *Proceedings of the 2nd International Conference on Trust Management, Lecture Notes in Computer Science* (Vol. 2995, pp. 120-134). Oxford, UK: Springer-Verlag.

Frank, N.M., & Peters, L. (1998). Building trust: The importance of both task and social precursors. In *Proceedings of the International Conference on Engineering and Technology Management: Pioneering New Technologies - Management Issues and Challenges in the Third Millennium* (pp. 322-327). San Juan, PR: IEEE Computer Society Press.

Gambetta, D. (2001). Can we trust trust? In D. Gambetta (Ed.), *Trust making and breaking cooperative relations* (pp. 213-237), Oxford, UK: University of Oxford. Retrieved December 29, 2006, from http://www.sociology.ox.ac.uk/papers/gambetta213-237.pdf

Georgiadis, C.K., Mavridis, I., Pangalos, G., & Thomas, K.R. (2001). Flexible team-based access control using contexts. In *Proceedings of the 6th ACM Symposium on Access Control Models and Technologies* (pp. 21-27). Litton-TASC, Chantilly, VA: ACM Press.

Gong, L. (1999). *Inside Java 2 platform security.* Boston: Addison Wesley.

Grandison, T., & Sloman, M. (2000). A survey of trust in internet applications. *IEEE Communications and Surveys, 3*(4). IEEE Communications Society Press.

Grandison, T., & Sloman, M. (2003). Trust management tools for internet applications. In P. Nixon & S. Terzis (Eds.), *Proceedings of the 1st International Conference on Trust Management* (LNCS 2692, pp. 91-107). Heraklion, Crete, Greece: Springer-Verlag.

Josang, A., & Presti, S.L. (2004). Analysing the relationship between risk and trust. In C. Jensen, S. Poslad, T. Dimitrakos (Eds.), *Proceedings of the 2nd International Trust Management, Lecture Notes in Computer Science* (Vol. 2995, pp. 135-145). Oxford, UK: Springer-Verlag.

Kagal, L., Finin, T., & Joshi, A. (2001). Trust-based security in pervasive computing environments. *IEEE Computer, 34*(12), 154-157.

Kagal, L., Finin, T., & Joshi A. (2003). A policy language for pervasive computing environment. In *Proceedings of the 4th IEEE International Workshop on Policy for Distributed Networks and Systems* (pp. 63-74). Lake Como, Italy: IEEE Computer Society Press.

Ko, H.J., Won, D.H., Shin, D.R., Choo, H.S., & Kim, U.M., (2006). A semantic context-aware access control in pervasive environments. In M. Gavrilova et al. (Eds.), *Proceedings of the International Conference on Computational Science and its Applications* (LNCS, Vol. 3981, pp.165-174). Glasgow, Scotland: Springer-Verlag.

Liscano, R. & Wang, K. (2005). A SIP-based architecture model for contextual coalition access control for ubiquitous computing. In *Proceedings of the 2nd Annual Conference on Mobile and Ubiquitous Systems* (pp. 384-392). San Diego: IEEE Computer Society Press.

Liu, Z., Joy, A.W., & Thompson, R.A. (2004). A dynamic trust model for mobile ad hoc networks. In *Proceedings of the 10th IEEE International Workshop on Future Trends of Distributed Computing Systems* (pp. 80-85). Suzhou, China: IEEE Computer Society Press.

Marsh, S.P. (1994). Formalising trust as a computational concept. *Computing science and mathematics* (p. 170). Stirling, Scotland: University of Stirling.

McDaniel, P. (2003). On context in authorization policy. In *Proceedings of the 8th ACM Symposium on Access Control Models and Technologies* (pp. 80-89). Lake Como, Italy: ACM Press.

Michiardi, P., & Molva, R. (2002). Core: A collaborative reputation mechanism to enforce node cooperation in mobile ad hoc networks. In B. Jerman-Blazic & T. Klobucar (Eds.), *Proceedings of the 6th IFIP TC6/TC11 Joint Working Conference on Communications and Multimedia Security: Advanced Communications and Multimedia Security* (Vol. 228, pp. 107-121). Deventer, The Netherlands: Kluwer.

Mui, L., Mohtashemi, M., Cheewee, A., Szolovits, P., & Halberstadt, A.: (2001). *Ratings in distributed systems: A bayesian approach.* Paper presented at the 11th Workshop on Information Technologies and Systems, New Orleans, LA.

Neumann, G., & Strembeck, M. (2003). An approach to engineer and enforce context constraints in an RBAC environment. In *Proceedings of the 8th ACM Symposium on Access Control Models and Technologies* (pp. 65-79). Como, Italy: ACM Press.

Parker , J., Patwardhan, A., Perich, F., Joshi, A., & Finin, T.: (2006). Trust in pervasive computing. *Mobile Middleware, 21*, 473-496. CRC Press.

Quercia, D., Hailes, S., & Capra, L. (2006). B-trust: Bayesian trust framework for pervasive computing. In K. Stølen W. H.Winsborough, F. Martinelli, & F. Massacci (Eds.), *Proceedings of the 4th International Conference on Trust Management* (LNCS 3986, pp. 298-312). Pisa, Italy: Springer-Verlag.

Ruohomaa, S., & Kutvonen L. (2005). Trust management survey. In P. Herrmann, S. Shiu, V. Issarny (Eds.), *Proceedings of the 3rd Annual Conference on Trust Management* (LNCS 3477, pp. 77-92). Rocquencourt, France: Springer-Verlag.

Sloman, M. (2004). Trust management in Internet and pervasive systems. *IEEE Intelligent Systems, 19*(5), 77-79.

Srinivasan, A., Teitelbaum, J., Liang, H., Wu, J., & Cardei, M. (in press). Reputation and trust-based systems for ad hoc and sensor networks. In A. Boukerche (Ed.), *Algorithms and protocols for wireless ad hoc and sensor networks.* Wiley & Sons.

Staab, S. (2004). The pudding of trust. *IEEE Intelligent Systems, 19*(5), 74.

Toninelli, A., Montanari, R., Kagal, L., & Lassila, O. (2006). A semantic context-aware access control framework for secure collaborations in pervasive computing environments. In I. Cruz et al. (Eds.), *Proceedings of the 5th International Semantic Web Conference* (LNCS 4273, pp. 473-486). Athens, GA: Springer-Verlag.

Tonti, G., et al. (2003). Semantic Web languages for policy representation and reasoning: A comparison of KAoS, Rei, and Ponder. In D. Fensel et al. (Eds.), *Proceedings of the 2nd International*

Semantic Web Conference (LNCS 2870, pp. 419-433). Sanibel Island, FL: Springer-Verlag.

Uszok, A., Bradshaw, J.M., & Jeffers, R. (2004). KAoS: A policy and domain services framework for grid computing and semantic Web services. In C. Jensen, S. Poslad, T. Dimitrakos (Eds.), *Proceedings of the 2nd International Conference on Trust Management*, (LNCS 2995, pp. 16-26). Oxford, UK: Springer-Verlag.

van Sinderen, M.J., van Halteren, A.T., Wegdam, M., Meeuwissen, H.B., & Eertink, E.H.: (2006). Supporting context-aware mobile applications: An infrastructure approach. *IEEE Communications Magazine, 44*(9), 96-104.

KEY TERMS

Access Control: The ability to limit and control the actions/operations that a legitimate user of a computer system can perform. Access control constrains what a user can do directly, as well what programs executing on behalf of a user are allowed to do.

Context: Many definitions of context are available in the literature (Dey et al., 2001). The most accepted one defines "context" as any information useful for characterizing the state or the activity of an entity or the world in which this entity operates. An entity is a person, place, or object that is considered relevant to the interaction between a user and an application, including the user and applications themselves.

Middleware for Distributed Systems: A distributed software support layer which abstracts over the complexity and heterogeneity of the underlying distributed environment with its multitude of network technologies, operating systems, and implementation languages. The primary role of middleware is to ease the task of developing, deploying, and managing distributed applications by providing a simple, consistent, and integrated distributed programming environment.

Recommendation: A communicated opinion about the trustworthiness of a third party entity.

Reputation: The opinion that one entity builds about another entity. In particular, reputation refers to the general expectation about the future actions of an entity based upon past actions. Reputation can be exploited in trust management by providing one relevant element to consider for the trustor to assess the prospective trustee's trustworthiness.

Semantic Technologies: Technologies that permit to add semantic metadata to information resources. Semantic metadata allow to effectively process data, for instance via automated inferences, that is, understanding what a data resource is and how it relates to other data independently of its name and syntax.

Trustee: The entity (individual, access terminal, resource/service component) that is the object of a trust evaluation by a trustor (see the following definition).

Trust Management: The collection, maintenance, and processing of the information required to make a trust relationship decision, to evaluate the criteria related to trust relationships, and to monitor and re-evaluate existing trust relationships. The term was first defined by Blaze et al. (1996) as a unified approach to specifying and interpreting security policies, credentials, and relationships in order to allow direct authorization of security-critical actions.

Trustor: An individual who sets up a trust or, in other words, the subject that has the possibility/responsibility to trust a target entity.

Chapter XXX
A Survey of Key Management in Mobile Ad Hoc Networks

Bing Wu
Fayetteville State University, USA

Jie Wu
Florida Atlantic University, USA

Mihaela Cardei
Florida Atlantic University, USA

ABSTRACT

Security has become a primary concern in mobile ad hoc networks (MANETs). The characteristics of MANETs pose both challenges and opportunities in achieving security goals, such as confidentiality, authentication, integrity, availability, access control, and nonrepudiation. Cryptographic techniques are widely used for secure communications in wired and wireless networks. Most cryptographic mechanisms, such as symmetric and asymmetric cryptography, often involve the use of cryptographic keys. However, all cryptographic techniques will be ineffective if the key management is weak. Key management is also a central component in MANET security. The purpose of key management is to provide secure procedures for handling cryptographic keying materials. The tasks of key management include key generation, key distribution, and key maintenance. Key maintenance includes the procedures for key storage, key update, key revocation, key archiving, and so forth. In MANETs, the computational load and complexity for key management are strongly subject to restriction by the node's available resources and the dynamic nature of network topology. A number of key management schemes have been proposed for MANETs. In this chapter, we present a survey of the research work on key management in MANETs according to recent literature.

INTRODUCTION

Mobile Ad Hoc Networks (MANETs)

In areas where there is little communication infrastructure or the existing infrastructure is inconvenient to use, wireless mobile users may still be able to communicate through the formation of *mobile ad hoc networks* (Perkins, 2001). A mobile ad hoc network, or simply MANET, is a collection of wireless mobile hosts that form a temporary network without the aid of any centralized administration or support. In such a network, each mobile node operates not only as a host but also as a router, forwarding packets for other mobile nodes in the network that may be multiple hops away from each other.

Possible applications of MANETs include: soldiers relaying information for situational awareness on the battlefield; business associates sharing information during a meeting; attendees using laptop computers to participate in an interactive conference; and emergency disaster relief personnel that are coordinating efforts at sites of fires, hurricanes, or earthquakes.

A routing protocol is necessary in such an environment, since two hosts that wish to communicate may not be able to exchange packets directly. Figure 1 shows a simple example of a MANET. Host *w* is not within the range of host *u*'s wireless transmitter and vice versa. If *u* and *w* wish to exchange packets, they may depend on the services of host *v* to forward packets for them because *v* is within the overlap between *u* and *w*'s transmission range. Although the number of hops for a host to reach another is likely to be small, the routing problem in a real MANET will still be complicated due to the inherent nonuniform propagation characteristics of wireless transmissions, and the highly dynamic topology of the networks.

Characteristics of MANETs

A MANET is an autonomous system of mobile nodes. The system may operate in isolation, or may have gateways to an interface with a fixed network. Its nodes are equipped with wireless transmitters/receivers using antennas that may be omnidirectional (broadcast), highly directional (point-to-point), or some combination thereof. At a given time, the system can be viewed as a random graph due to the movement of the nodes and their transmitter/receiver coverage patterns, the transmission power levels, and the cochannel interference levels (Karygiannis & Owens, 2002; Ravi, Raghunathan, & Potlapally, 2002; Stallings, 2002). The network topology may change with time as the nodes move or adjust their transmission and reception parameters. Thus, ad hoc networks have several salient characteristics:

Figure 1. An example of a mobile ad hoc network

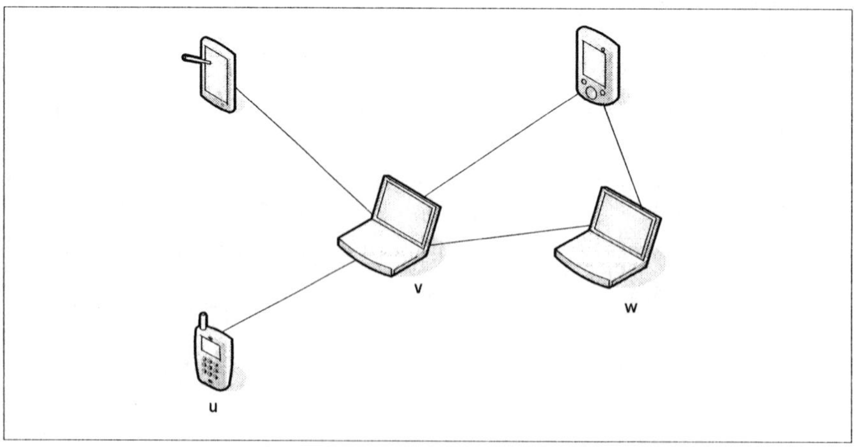

- **Dynamic topologies:** The network topology may change randomly and rapidly at unpredictable times, and may consist of both directional and unidirectional links. Nodes freely roam in the network, join or leave the network at their own will, and fail occasionally.
- **Resource constraints:** The wireless links have significantly lower capacity than wired links. The computation and energy resources of a mobile device are limited.
- **Infrastructure-less:** There is no well-defined infrastructure, or access point, or some other central control point available. Moreover, the wireless medium is accessible by both legitimate nodes and attackers. There is no clear boundary to separate the inside network from the outside world.
- **Limited physical security:** Portable devices are generally small with weak protection. The physical devices could be stolen or compromised.

Security Challenges Overview

Security Attacks

While MANETs can be quickly and inexpensively setup as needed, security is a more critical issue compared to wired networks or other wireless counterparts. Many *passive* and *active* security attacks could be launched from the outside by malicious hosts or from the inside by compromised hosts (Lou & Fang, 2003; Murthy & Manoj, 2005; Wu, Chen, Wu, & Cardei, 2006).

- **Passive attacks:** In passive attacks, an intruder captures the data without altering them. The attacker does not modify the data and does not inject additional traffic. The goal of the attacker is to obtain information that is being transmitted, thus violating the message confidentiality. Since the activity of the network is not disrupted, these attacks are difficult to detect. A powerful encryption mechanism can alleviate these attacks, making it difficult to read the transmitted data.
- **Active attacks:** In active attacks, an attacker actively participates in disrupting the normal operation of the network services. An attacker can create an active attack by modifying packets or by introducing false information. Active attacks can be further divided into internal and external attacks; internal attacks are from compromised nodes that were once a legitimate part of the network. Since the adversaries are already part of the network as authorized nodes, they are much more severe (Wu et al., 2006) and difficult to detect compared to external attacks. External attacks are carried by nodes that are not a legitimate part of the network. Such attacks are often prevented through firewalls or some authentication and encryption mechanisms.

Security Goals

Security services include the functionality that is required to provide a secure networking environment. It comprises *authentication, access control, confidentiality, integrity, nonrepudiation,* and *availability* (Ilyas, 2003; Nichols & Lekkas, 2002; Wu et al., 2005; Yang, Luo, Ye, Lu, & Zhang, 2004). Authentication is the ability to verify that a peer entity in an association is the one it claims to be, or can be used for the determination of data origins. Availability ensures the survivability of the network service despite denial-of-service attacks. Confidentiality ensures that certain information is never disclosed to unauthorized entities. Integrity guarantees that a message being transferred is not corrupted. Nonrepudiation ensures that the origin of a message cannot deny having sent the message. Access control is the ability to limit and control access to devices and/or applications via communication links. The main security services can be summarized as follows:

- **Authentication**: The function of the authentication service is to verify a user's identity and to assure the recipient that the message is from the source that it claims to be from. First, at the time of communication initiation, the service assures that the two parties are authentic, that each is the entity it claims to be. Second, the service must assure that a third party does not interfere by impersonating one of the two legitimate parties for the purpose of authorized transmission and reception.

- **Confidentiality**: Confidentiality ensures that the data/information transmitted over the network is not disclosed to unauthorized users. Confidentiality can be achieved by using different encryption techniques such that only legitimate users can analyze and understand the transmission.
- **Integrity:** The function of integrity control is to assure that the data are received exactly as sent by an authorized party. That is, the data received contain no modification, insertion, deletion, or replay.
- **Access control**: This service limits and controls the access of a resource such as a host system or application. To achieve this, a user trying to gain access to the resource is first identified (authenticated) and then the corresponding access rights are granted.
- **Nonrepudiation:** This is related to the fact that if an entity sends a message, the entity cannot deny that it sent that message. If an entity gives a signature to the message, the entity cannot later deny that message. In public key cryptography, a node A signs the message using its private key. All other nodes can verify the signed message by using A's public key, and A cannot deny the message with its signature.
- **Availability:** This involves making network services or resources available to the legitimate users. It ensures the survivability of the network despite malicious incidences.

Security Mechanisms

Cryptography is an important and powerful tool for secure communications. It transforms readable data (plaintext) into meaningless data (ciphertext). Cryptography has two dominant categories, namely symmetric-key (secret-key) and asymmetric-key (public-key) approaches (Saloma, 1996). In symmetric-key cryptography, the same key is used to encrypt and decrypt the messages, while in the asymmetric-key approach, different keys are used to convert and recover the information. Although the asymmetric cryptography approaches are versatile (can be used for authentication, integrity, and privacy) and are simpler for key distribution than the symmetric approaches, symmetric-key algorithms are generally more computation-efficient than the asymmetric cryptographic algorithms (Tanenbaum, 2003). There are varieties of symmetric and asymmetric algorithms available, including data encyrption standard (DES), advanced encryption standard (AES), international data encryption algorithm (IDEA), RSA, and ElGamal (Menezes, Oorschot, & Vanstone, 1996). Threshold cryptography is another cryptographic technique that is quite different from the above two approaches. In Shamir's (k, n) secret sharing scheme, secret information is split into n pieces according to a random polynomial. Meanwhile, the secret could be recovered by combining any threshold k pieces based on Lagrange interpolation. These cryptographic algorithms are the security primitives that are widely used in wired and wireless networks. They can also be used in MANETs and help to achieve the security in its unique network settings.

Key Management

As in the above description, cryptography is a powerful tool in achieving security. However, most cryptosystems rely on the underlying secure, robust, and efficient key management subsystem. In fact, all cryptographic techniques will be ineffective if the key management is weak. Key management is a central part of the security of MANETs. In MANETs, the computational load and complexity for key management are strongly subject to restriction by the node's available resources and the dynamic nature of network topology. Some asymmetric and symmetric key management schemes (including group key) have been proposed to adapt to the environment of MANETs. Key management deals with key generation, key storage, distribution, updating, revocation, deleting, archiving, and using keying materials in accordance with security policies. In this chapter, we present a comprehensive survey of research work on key management in MANETs based on recent literature. This chapter is organized as follows: The introduction gives an introduction of MANETs. The key management section discusses key and trust models in wired networks and MANETs. The asymmetric key management section presents the asymmetric key management schemes in MANETs. The symmetric key management section in MANETs presents the symmetric key management schemes in MANETs. The group key management section presents the group key management in the infrastructure networks with proper extension being necessary for MANETs. We conclude the chapter and discuss possible future work in the open challenges and future directions section. We include the reference and key terms at the end of the chapter.

KEY MANAGEMENT IN MANETS

Key management is a basic part of any secure communication. Most cryptosystems rely on some underlying secure, robust, and efficient key management system. Secure network communications normally involve a key distribution procedure between communication parties, in which the key may be transmitted through insecure channels. A framework of trust relationships needs to be built for authentication of key ownership in the key distribution procedure. While some frameworks are based on a centralized trusted third party (TTP), others could be fully distributed. For example, a certification authority (CA) is the TTP in asymmetric cryptosystems, a key distribution center (KDC) is the TTP in the symmetric system, and in pretty good privacy (PGP) no TTP is assumed. According to recent literature, the centralized approach is regarded as inappropriate for MANETs because of the dynamic environment and the transient relationships among mobile nodes. Most researchers prefer the decentralized trust model for MANETs. Several decentralized solutions have been proposed in recent papers with different implementations, such as how the CA's responsibility is distributed to all nodes, or to a subset of nodes.

Fundamentals of Key Management

Cryptographic algorithms are security primitives that are widely used for the purposes of authentication, confidentiality, integrity, and nonrepudiation. Most cryptographic systems require an underlying secure, robust, and efficient key management system. Key management is a central part of any secure communication and is the weakest point of system security and the protocol design.

A key is a piece of input information for cryptographic algorithms. If the key was released, the encrypted information would be disclosed. The secrecy of the symmetric key and private key must always be assured locally. The key encryption key (KEK) approach (Burnett & Paine, 2001) could be used at local hosts to protect the secrecy of keys. To break the cycle (use key to encrypt the data, and use key to encrypt key) some noncryptographic approaches need to be used, for example, smart card, biometric identity (such as fingerprint), and so forth.

Key distribution and key agreement over an insecure channel are at high risk and suffer from potential attacks. In the traditional digital envelop approach, a session key is generated at one side and is encrypted by the public-key algorithm. Then it is delivered and recovered at the other end. In the Diffie-Hellman (DH) scheme (Burnett & Paine, 2001), the communication parties at both sides exchange some public information and generate a session key on both ends. Several enhanced DH schemes have been invented to counter man-in-the-middle attacks. In addition, a multiway challenge response protocol, such as Needham-Schroeder (Tanenbaum, 2002), can also be used. Kerberos, which is based on a variant of Needham-Schroeder, is an authentication protocol used in many real systems, including Microsoft Windows. However, in MANETs, the lack of a central control facility, the limited computing resources, dynamic network topology, and the difficulty of network synchronization all contribute to the complexity of key management protocols.

Key integrity and ownership should be protected from advanced key attacks. Digital signatures, hash functions, and the hash function-based message authentication code (HMAC) (Menezes et al., 1996) are techniques used for data authentication and/or integrity purposes. Similarly, the public key is protected by the public-key certificate, in which a trusted entity called the certification authority in public key infrastructure (PKI) vouches for the binding of the public key with the owner's identity. In systems lacking a TTP, the public-key certificate is vouched for by peer nodes in a distributed manner, such as PGP (Burnett & Paine, 2001). In some distributed approaches, the system secret is distributed to a subset or all of the network hosts based on threshold cryptography. Obviously, a certificate cannot prove whether an entity is "good" or "bad." However, it can prove ownership of a key. Certificates are mainly used for key authentication.

A cryptographic key could be compromised or disclosed after a certain period of usage. Since the key should no longer be usable after its disclosure, some mechanism is required to enforce this rule. In PKI, this can be done implicitly or explicitly. The certificate contains the lifetime of validity; it is not useful after expiration. However, in some cases, the private key could be disclosed during the valid period, in which case the CA needs to revoke a certificate explicitly and notify the network by posting it onto the certificate revocation list (CRL) to prevent its usage.

Key management for large dynamic groups is a difficult problem because of scalability and security. Each time a new member is added or an old member is evicted from the group, the group key must be changed to ensure backward and forward security. Backward security means that new members cannot determine any past group key and discover the previous group communication messages. Forward security means that evicted members cannot determine any future group key and discover the subsequent group communication information. The group key management should also be able to resist against colluded members.

Trust Models

Centralized Trust Model

For the centralized trust model, there is a well-trusted entity known as a TTP (Kaufmanet, Perlman, & Speciner, 2002; Menezes et al., 1996). A TTP is an entity trusted by all users in the system, and it is often used to provide key management services. Depending on the nature of their involvement, TTPs can be classified into three categories: inline, online, or off-line. See Figure 2 for an illustration. An inline TTP participates actively in between the communication path of two users. An online TTP participates actively but only for

Figure 2. Categories of trust third parties

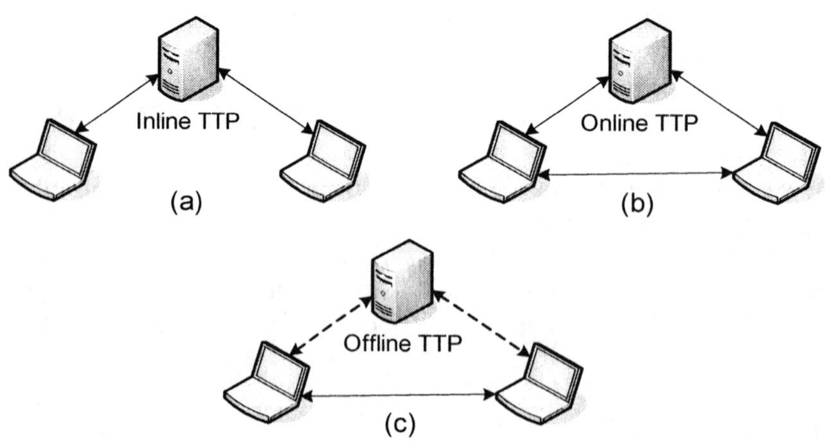

management purposes, as the two parties communicate with each other directly. An off-line TTP communicates with users prior to the setting up of communication links and remains off-line during network operation.

TTPs in Symmetric Key Management Systems

TTPs have been implemented in both symmetric and asymmetric key management systems. KDC and key translation centers (KTC) (Oppliger, 1998) are TTPs in symmetric cryptographic key management systems and the CA is the TTP in public-key management systems. KDC and KTC simplify the symmetric key management since each user does not have to share a secret key with every other user. Instead, it only needs to share one key with the TTP. This reduces the total number of keys that need to be managed from $\frac{n(n-1)}{2}$ to n, where n is the total number of users. Figure 3 illustrates the protocols by implementing KDC or KTC.

1. Alice requests to share a secret key with Bob. If the TTP is KDC, it generates a key to use. Otherwise, Alice provides it. The message is encrypted using the secret key shared between Alice and the TTP.
2. The TTP encrypts the session key with the key it shares with Bob and returns it to Alice.
3. Alice sends the encrypted session key to Bob, who can decrypt it and thereafter use it to communicate securely with Alice.

Public Key Infrastructure (PKI)

The use of public key cryptography requires the authenticity of public keys. Otherwise, it is easy to forge or spoof someone's public key. Some trusted framework must be present to verify the ownership of a public key. A straightforward solution is to have any two users that wish to communicate exchange their public keys in an authenticated manner. It would require the initial distribution of *n(n-1)* public keys. Obviously, this solution is not scalable for a large network and has the same problems we discussed in the symmetric key management system. However, by having a trusted third party issue certificates to each of the users, every user only needs to hold the public key of the TTP, which significantly simplifies the authentication process for users' public keys. Actually, there are two dominating trust models in PKI, namely, centralized and web-of-trust trust models (Wu et al., 2006; Yi & Kravets, 2004). For network scalability, the centralized trust model could be a hierarchical trust structure instead of a single CA entity. Multiple CA roots could be necessary for a large

Figure 3. Establishment of session key using KDC or KTC

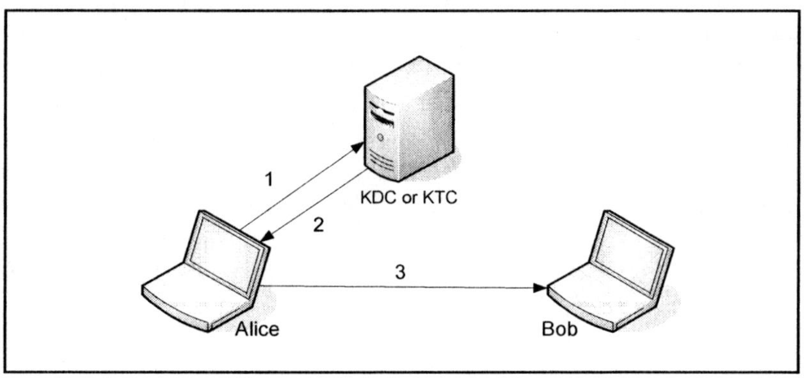

Figure 4. Components of a PKI

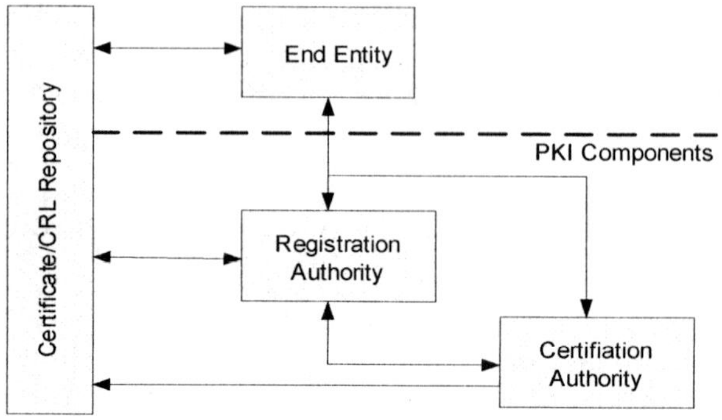

network, such as the Internet. We will discuss the fully distributed or web-of-trust model later.

A PKI provides the mechanisms needed to manage certificates, and normally consists of the components illustrated in Figure 4.

In this diagram, the CA is the component responsible for issuing and revoking certificates, while the registration authority (RA) is responsible for establishing the identity of the subject of a certificate and the mapping between the subject and its public key. The RA and CA can be implemented as one component; therefore, RA is an optional component. PKI components provide basic services, such as registration, initialization, certification, key update, revocation, key recovery, cross-certification, and so forth.

Web-of-Trust Model

The web-of-trust model is also called certificate chaining. PGP (Tanenbaum, 2002) is an example built on this trust model. In the web-of-trust model there is no TTP that is well-trusted by all network nodes. Instead, peer nodes can issue certificates to each other and populate the certificate graph. Certificates can be authenticated through certificate chaining. Compared with the centralized trust model, the web-of-trust model does not require a heavy infrastructure or complex bootstrapping procedures, and every node plays an identical role and shares the same responsibility. Although the web-of-trust model has the above advantages, it has two major limitations. First, a certificate graph may not populate enough to provide certificate chains for a given pair of nodes, so it is difficult to predict whether any given authentication request can be fulfilled. Second, without relying on a TTP, any trust relationship relies on the goodwill and the correct behaviors of all participants. Obviously, that cannot always be assumed. However, since there is no clear way to tell if a certificate chain includes any misbehaving nodes, the overall confidence for the certificate is relatively low.

Decentralized Trust Model

In MANETs, a framework for key management built on a fully centralized mode is not feasible, not only because of the difficulty of maintaining such a globally trusted entity but also because the central entity could become a hotspot of attacks. Thus, this network suffers from a security bottleneck. Meanwhile a completely distributed model may not be acceptable because there is no well-trusted security anchor available in the whole system. One feasible solution is to distribute the central trust to multiple entities (or the entire network) based on a secret sharing scheme. In the decentralized public key management scheme, the system public key is distributed to the entire network, while the system private key is split to multiple pieces and distributed to a subset (or all) of the nodes. The subset of group nodes creates a view of a CA and functions as a CA in combination.

Hybrid Trust Model

This scheme takes advantage of the positive aspects of two different trust systems. The basic idea is to incorporate a TTP into the certificate graph. Here, the TTP is a virtual CA node that represents all nodes that comprise the virtual CA. Some authentication metrics, such as confidence value, are introduced in order to "glue" two trust systems (Yi & Kravets, 2004). While this model is theoretically sound, it is difficult to "glue" two different trust systems since there is no clear way to assign a value of confidence level.

Overview of Key Management Schemes in MANETs

Asymmetric Key Management Schemes

Recently, research papers have proposed different key management schemes for MANETs. Most of them are based on public-key cryptography. The basic idea is to distribute the CA's functionality to multiple nodes. Zhou and Hass (1999) present a secure key management scheme by employing *(t, n)* threshold cryptography. The system can tolerate t-1 compromised servers. Luo et al., (2004, 2001) propose a localized key management scheme in which all nodes are servers and the certificate service can be performed locally by a threshold number of neighboring nodes. Yi, Naldurg, and Kravets (2002) put forward a similar scheme. The difference is that their certificate service is distributed to a subset of nodes, which are physically more secure and powerful than the others. Wu, Wu, and Dong, (in press) also introduce a scheme that is similar to Yi, in which server nodes form a mesh structure and a ticket scheme is used for efficiency. Capkun, Buttyan, and Hubaux (2003) consider a fully distributed scheme that is based on the same idea of PGP. Yi and Kravets (2004) provide a composite trust model. Their idea is to take advantage of the positive aspects of both the central and fully distributed trust models.

Symmetric Key Management Schemes

There are research papers that are based on the symmetric-key cryptography for securing MANETs. For instance, some symmetric key management schemes are proposed for sensor nodes that are assumed to be incapable of performing costly asymmetric cryptographic computations. Pair-wise keys can be preloaded into nodes, or based on the random key distribution in which a set of keys is preloaded. Chan (2004) introduces a distributed symmetric key distribution scheme for MANETs. The basic idea is that each node is preloaded with a set of keys from a large key pool (Chan, Perrig, & Song, in press; Du, Deng, Han, & Varshney, 2003). The key pattern should satisfy the property that any subset of nodes can find at least one common key, and the common key should not be covered by a collusion of a certain number of other nodes outside the subset. Chan and Perrig (2005) introduce a symmetric key agreement scheme for the sensor nodes. The basic idea of their approach is that each node shares a unique key with a set of nodes vertically and horizontally (in 2-dimensions). Therefore, any pair of nodes can rely on at least one intermediate node to establish the common key.

Group Key Management Schemes

Collaborative and group-oriented applications in MANETs are going to be active research areas. Group key management is one of the basic building blocks in securing group communications. However, key management for large dynamic groups is a difficult problem because of scalability and security (Rafaeli & Hutchison, 2003). For instance, each time a new member is added or an old member is evicted from a group, the group key must be changed to ensure backward and forward security.

ASYMMETRIC KEY MANAGEMENT SCHEMES IN MANETS

Secure Routing Protocol (SRP)

SRP is a decentralized public key management protocol proposed by Zhou and Hass (1999) by employing *(t, n)* threshold cryptography (Shamir, 1979; Wong, Wang, & Wing, 2002) in their research paper called "Securing Ad Hoc Networks." In the system, there are *n* servers, which are responsible for public-key certificate services. Therefore, the system can tolerate t-1 compromised servers. Servers can proactively refresh the secret shares using the proactive secret sharing (PSS) (Herzberg, Jarecki, Krawczyk, & Yung, 1995) techniques or by adjusting the configuration structure based on share redistribution techniques to handle compromised servers or system failure. Since the new shares are independent of the old ones, mobile adversaries would have to compromise a threshold number of servers in a very short amount of time, which obviously increases the difficulty of the success of adversaries. The system configuration of this scheme is illustrated in Figure 5. The system public key K is distributed to all nodes in the network, whereas the private key S is split to *n* shares s1, s2, s3, ..., sn, one share for each server according to a random polynomial function.

In this scheme, the system model is such that *n* servers are special nodes, each with its own public/private key pair and the public key of every node in the network. This is a critical issue in a large network. However, this scheme does not describe how a node can contact *t* servers securely and efficiently in case the servers are scattered in a large area. A share-refreshing scheme is proposed to counter mobile adversaries. The update of secret shares does not change the system public/private key pairs. Therefore, nodes in the network can still use the same system public key to verify a signed certificate so that the share-refreshing is transparent to all nodes. However, a method of distributing these updated subshares to all nodes securely and efficiently in the network is not addressed.

Ubiquitous and Robust Access Control (URSA)

URSA is a localized key management scheme proposed by Luo et al. (2004, 2001) in their paper "URSA: Ubiquitous and Robust Access Control for Mobile Ad Hoc Networks". The URSA protocol is also based on threshold cryptography as in SRP (Zhou & Haas 1999). The difference between URSA and SRP is that in URSA, all nodes are servers and are capable of producing a partial certificate, while in SRP only server nodes can produce certificates. Thus, certificate services are distributed to all nodes in the network. *URSA* also proposes a distributed self-initialization phase that allows a newly joined node to obtain secret shares by contacting a coalition of *k* neighboring nodes without requiring the existence of an online secret share dealer. The basic idea is to extend the PSS technique by shuffling the partial shares instead of shuffling the secret sharing polynomials. The purpose of this shuffling process is to prevent deducing the original secret share from a resulting share.

Figure 5. Illustration of SRP scheme

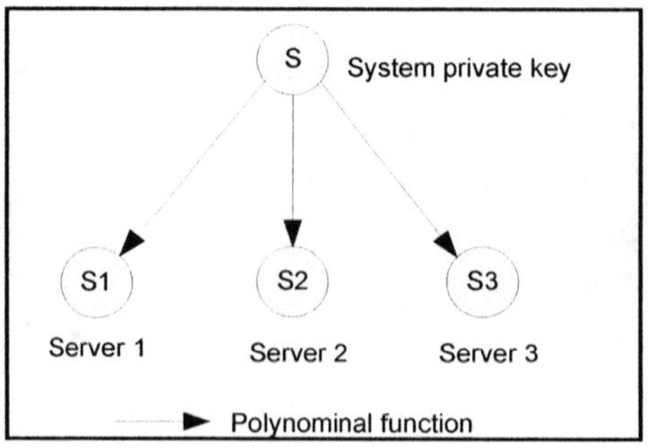

In URSA, every node should periodically update its certificate. To update its certificate, a node must contact its 1-hop neighbors, and request partial certificates from a collection of threshold k number of nodes. It can combine partial certificates into a legitimistic certificate. This will introduce either communication delays or cause search failures. It could potentially utilize services from 2-hop neighboring nodes.

The advantage of this scheme is efficiency and secrecy of local communications, as well as system availability since the CA's functionality is distributed to all network nodes. On the other hand, it reduces system security, especially when nodes are not well-protected because an attack can easily locate a secret holder without much searching and identifying effort. One problem is that in a sparse network where a node has a small number of neighbors, the threshold k is much larger than the network degree d and a node that wants to have its certificate updated needs to move around in order to find enough partial certificate "producers." The second critical issue is the convergence in the share-updating phase. Another critical issue is that too great an amount of off-line configuration is required prior to accessing the networks.

Mobile Certificate Authority (MOCA)

MOCA is a decentralized key management scheme proposed by Yi et al. (2002) in their paper "Key Management for Heterogeneous Ad Hoc Wireless Networks". In this approach, a certificate service is distributed to mobile certificate authority nodes. MOCA nodes are chosen based on heterogeneity if the nodes are physically more secure and computationally more powerful. In cases where nodes are equally equipped, they are selected randomly from the network. The trust model of this scheme is a decentralized model since the functionality of CA is distributed to a subset of nodes. A service-requesting node can locate $k + \alpha$ MOCA nodes either randomly, based on the shortest path, or according to the freshest path in its route cache. However, the critical question is how nodes can discover those paths securely since most secure routing protocols are based on the establishment of a key service in advance.

Self-Organized Key Management

Capkun et al. (2002) consider a fully distributed key management scheme in their paper "Self-organized Public Key Management for Mobile Ad Hoc Networks". This scheme is based on the web-of-trust model that is similar to PGP. The basic idea is that each user acts as its own authority and issues public key certificates to other users. A user needs to maintain two local certificate repositories. One is called the nonupdated certificate repository and the other one is called the updated certificate repository. The reason a node maintains a nonupdated certificate repository is to provide a better estimate of the certificate graph. Key authentication is performed via chains of public key certificates that are obtained from other nodes through certificate exchanging, and are stored in local repositories.

The fully distributed, self-organized certificate chaining has the advantage of configuration flexibility and it does not require any bootstrapping of the system. However, this certificate chaining requires a certain period to populate the certificate graph. This procedure completely depends on the individual node's behavior and mobility. One the other hand, this fully self-organized scheme lacks any trusted security anchor in the trust structure that may limit its usage for applications where high security assurance is demanded. In addition, many certificates need to be generated and every node should collect and maintain an up-to-date

Figure 6. An example of certificate chain

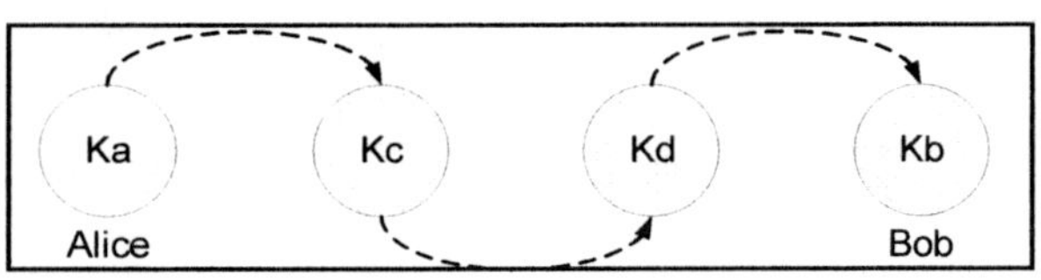

certificate repository. The certificate graph, which is used to model this web-of-trust relationship, may not be strongly connected, especially in the mobile ad hoc scenario. In that case, nodes within one component may not be able to communicate with nodes in different components. Certificate conflicting is another potential problem in this scheme.

Composite Key Management

Recently, Yi and Kravets (2004) provided a composite key management scheme in their paper "Composite Key Management for Ad Hoc Networks". In their scheme, they combine the centralized trust and the fully distributed certificate chaining trust models. This scheme takes advantage of the positive aspects of two different trust systems. The basic idea is to incorporate a TTP into the certificate graph. Here, the TTP is a virtual CA node that represents all nodes that comprise the virtual CA. Some authentication metrics, such as confidence value, are introduced in order to "glue" two trusted systems. A node certified by a CA is trusted with a higher confidence level. However, properly assigning confidence values is a challenging task. An example of a composite key management model is shown in Figure 7.

Secure and Efficient Key Management (SEKM)

SEKM is a decentralized key management scheme proposed by Wu and Wu (2007) in their paper "Secure and Efficient Key Management in Mobile Ad Hoc Networks". It is based on the decentralized virtual CA trust model. All decentralized key management schemes are quite similar in that the functionality of the CA is distributed to a set of nodes based on the techniques of threshold cryp-

Figure 7. An example of composite key management scheme

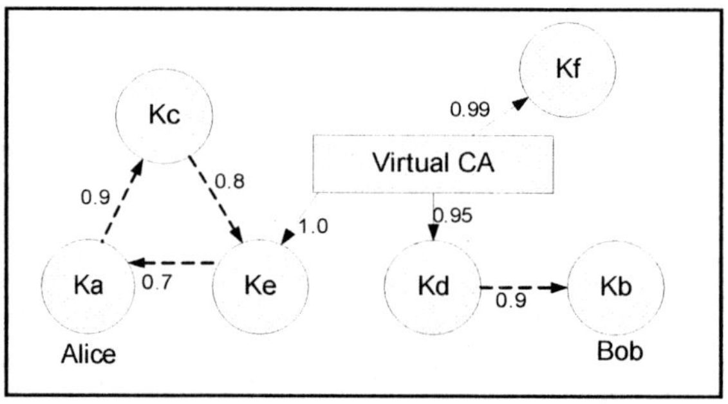

Figure 8. Server group structure in SEKM

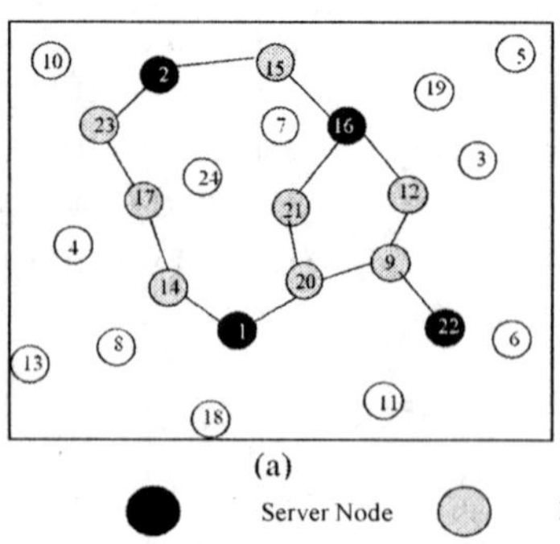

Server 16's Table

Server	Next Hop
1	12, 21
2	15
16	16
22	12, 21

(b)

tography. However, no schemes except for SEKM present detailed, efficient, and secure procedures for communications and cooperation between secret shareholders that have more responsibilities. In SEKM, all servers that have a partial system private key are to connect and form a server group. The structure of the server group is a mesh structure as shown in Figure 8. Periodic beacons are used to maintain the connection of the group so servers can efficiently coordinate with each other for share updates and certificate service. The problem with SEKM is that, for a large network with highly dynamic mobility, maintaining the structure server group can be costly.

SYMMETRIC KEY MANAGEMENT SCHEMES IN MANETS

Distributed Key Predistribution Scheme (DKPS)

DKPS is a distributed symmetric key management scheme proposed by Chan (2004) in the paper "Distributed Symmetric Key Management for Mobile Ad Hoc Networks". It is aimed at the network settings where mobile nodes are not assumed to be capable of performing computationally intensive public key algorithms and the TTP is not available. The basic idea of the DKPS scheme is that each node randomly selects a set of keys in a way that satisfies the probability property of cover-free family (CFF). Any pair of nodes can invoke the secure shared key discovery procedure (SSD). The theory behind the SSD is the additive and scalar multiplicative homomorphism of the encryption algorithm as well as the property of nontrivial zero encryption. To discover the common secret key, one side of the two parties can form a polynomial and send the encrypted polynomial to the other side. The coefficients of the polynomial are encrypted with the sender's secret key. The other side will send back the encrypted polynomial multiplied by a random value. Because of the homomorphism and nontrivial zero encryption properties, either side can only discover the common secret key, without disclosing the other noncommon keys.

Peer Intermediaries for Key Establishment (PIKE)

PIKE is another symmetric key management scheme proposed by Chan and Perrig (2005) in their paper "PIKE: Peer Intermediaries for Key Establishment in Sensor Networks". It is a random key predistribution scheme. The basic idea of PIKE is to use sensor nodes as trusted intermediaries to establish shared keys. Each node shares a unique secret key with a set of nodes. In the case of 2-dimension, a node shares a unique secret with each of the $O(\sqrt{n})$ nodes in the horizontal and vertical dimensions. Therefore, any pair of nodes can have a common secret with at least one intermediate node. This key predistribution scheme can be extended to three or more dimensions. Figure 9 shows the basic idea of the PIKE scheme. Dark lines connect

Figure 9. Illustration of PIKE scheme

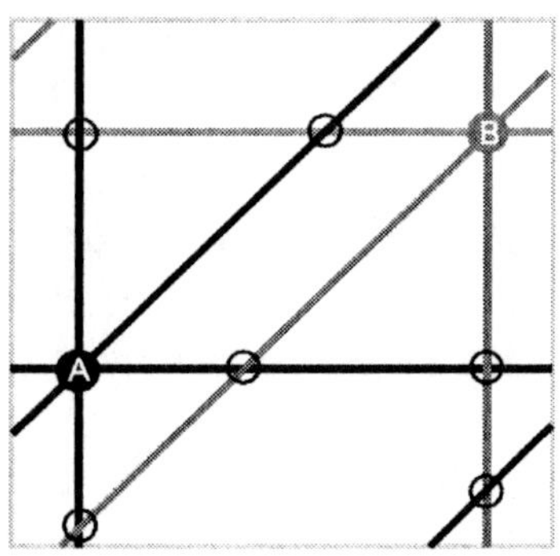

the nodes that share a unique key with node A, and light lines connect nodes that share a unique key with node B. There are six nodes that each share a unique key with node A and node B.

GROUP KEY MANAGEMENT APPROACHES

The messages are protected by encryption using the chosen key, which in the context of group communication is called the group key. Only those who know the current group key are able to recover the original message. Group key establishment means that multiple parties want to create a common secret to be used in the secure exchange of information. Two people who did not previously share a common secret can create one common secret with a DH key exchange protocol. The 2-party DH protocol can be extended to a generalized version of the n-party DH key-exchange model. Research efforts have been put into the design of group key agreement protocols to achieve better scalability, efficiency, and storage saving, such as the introduction of a tree structure and hash function. Furthermore, the group key management also needs to address the security issue related to membership changes. The modification of membership could require the group key to be refreshed (e.g., periodic rekey). The change of group keys when old members leave or new members join ensures backward and forward security. Therefore, a group key scheme must provide a scalable and efficient mechanism to rekey the group.

Group key management protocols can be roughly classified into three categories, namely, centralized, decentralized, and distributed (Rafaeli & Hutchison, 2003). In centralized group key protocols, a single entity is employed to control the whole group and is responsible for rekeying and distributing group keys to group members. In the decentralized approaches, a set of group managers is responsible for managing the group as opposed to a single entity being held responsible. In the distributed method, group members themselves contribute to the formation of group keys and are equally responsible for the rekeying and distribution of group keys. Recently, collaborative and group-oriented applications in MANETs have become an active research area. Obviously, group key management is a central building block in securing group communications in MANETs. However, group key management for large and dynamic groups in MANETs is a difficult problem because of the requirement of scalability and security under the restrictions of nodes' available resources and unpredictable mobility.

The literature presents several approaches to group key management. In this section, we give an overview of those protocols. Most of the following group key protocols are designed for the infrastructure networks. However, with the proper extension, some of them could be utilized and adapted to the MANET environment, or could serve as a hint for the design of MANET-specific group key management protocols. For instance, group Diffie-Hellman (GDH) and logical key hierarchy (LKH) have been extended into the MANETs. Wu, Wu, and Dong (in press) propose a simple and efficient group key management scheme, called SEGK, for MANETs. The basic idea of SEGK is that a physical multicast tree is formed in MANETs for efficiency. Group members take turns acting as group coordinator to compute and distribute intermediate key materials to group members. The keying materials are delivered through the tree links. The coordinator is also responsible for maintaining the connection of the multicast group. All group members can compute the group key locally in a distributed manner.

Logical Key Hierarchy (LKH)

LKH is a centralized group key management scheme proposed by Wallner, Harder, and Agee (1998) (Wong, Gouda, & Lam, 1998). It is based on the tree structure with each user (group participant) corresponding to a leaf and the group initiator as the root node. The tree structure will significantly reduce the number of broadcast messages and storage space for both the group controller and group members. The operation of this scheme is outlined below.

Figure 10. A sample tree structure of LKH

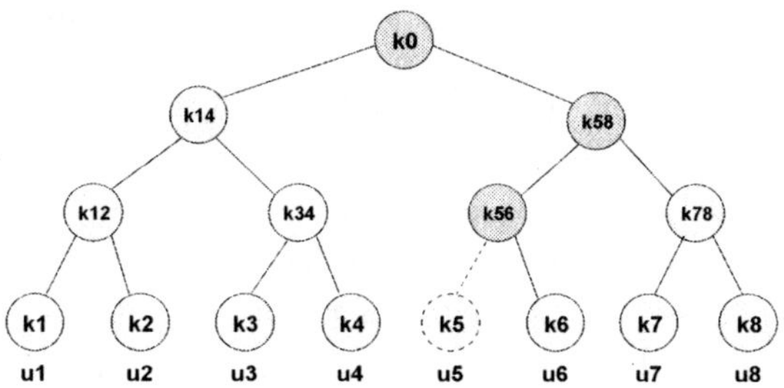

Each leaf node shares a pair-wise key with the root node as well as a set of intermediate keys from it to the root. So, for a balanced binary tree, each group member stores at most $d+1$ keys, where $d = \log_2 n$ is the height of the tree, and n is the total number of group members. See Figure 10: U_5 stores *k5, k56, k58,* and *k0.*

When a member joins the group, the rekey procedure will be started. A rekey message is generated containing the new set of keys encrypted with its respective node's children key. Figure 10

Figure 11. Illustration of joining member U5

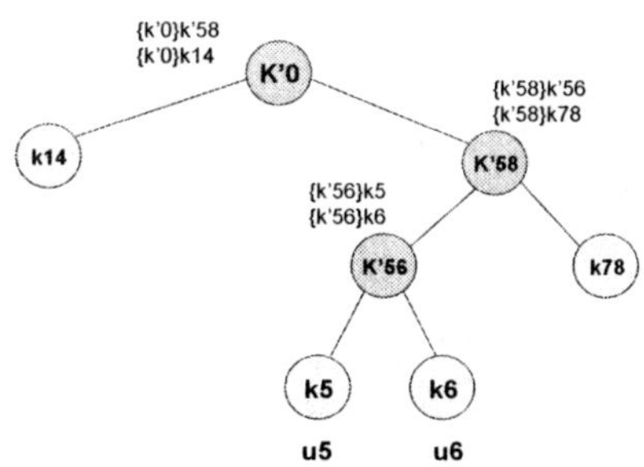

Figure 12. Illustration of leaving member U6

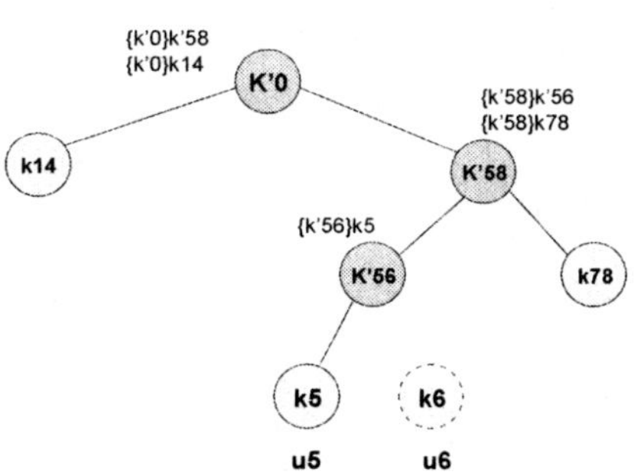

shows keys that are affected. The new member U_5 receives a secret key *k5* and attaches the intermediate node k56 logically. The keys *k56, k58,* and *k0,* which are in the path from *k5* to *k0,* need to be refreshed. New keys, *k'56, k'58*, and *k'0*, are generated as illustrated in Figure 11. These keys are encrypted with their respective node's children's key, for example, one instance of *k'56* is encrypted by *k5* and the other copy is encrypted by *k6* (see Figure 11). The removal of a member follows a similar procedure. For instance, when member U6 leaves the group, *k56, k58,* and *k0* should be changed and the new set of keys *k'56, k'58*, and *k'0* are encrypted with their respective children's key. See Figure 12 for an illustration of a member leave.

One-Way Function Trees (OFT)

OFT is another centralized group key management scheme proposed by Sherman and McGrew (2003). It is based on the tree structure that is similar to the above LKH scheme. However, all keys in the OFT scheme are functionally related according to a one-way hash function. The idea is that the keys held by a node's children are blinded using a one-way hash function and then combined together using a mixing function, such as a bitwise exclusive-or operation. Each group user receives blind keys from its sibling set as well as the blind key of its own sibling. Based on collected blinded keys, the group users can deduce each key of its ancestor set. See Figure 13 for an illustration. *k6* is the key of U5's sibling. *k56, k58,* and *k0* are the keys of U5's ancestor set. *k78* and *k14* are the keys of U5's sibling set.

Figure 13. A sample tree structure of OFT

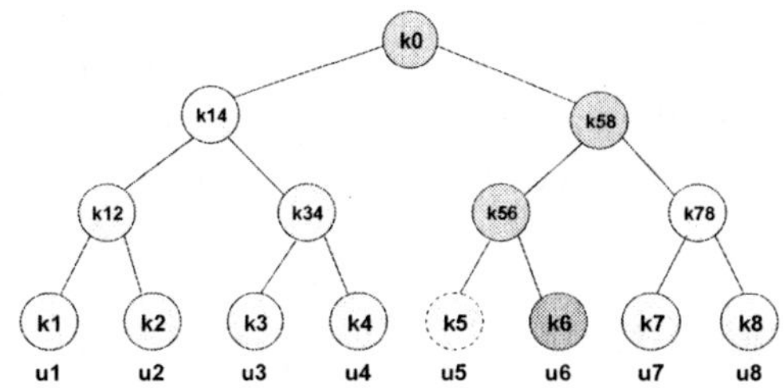

A group user still needs to store $d+1$ keys, where $d = \log_2 n$ is the height of the tree, and n is the total number of group members. The scheme has the same complexity as the LKH scheme for a balanced tree structure, but in the rekeying process, the size of keying materials reduces from $2 \cdot \log_2 n$ to $\log_2 n$.

Figure 14. Illustration of join member U5 in OFT

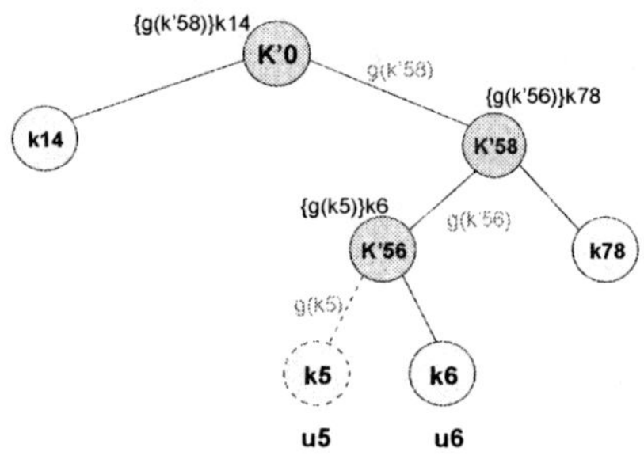

The message size reduction is achieved because in the OFT scheme. The blinded key changed in a node is encrypted only with the key of its sibling node while in LKH scheme the new key must be encrypted with its two children's keys. See Figure 14.

Tree-Based Group Diffie-Hellman (TGDH)

TGDH is a group key management scheme proposed by Kim, Perrig, and Tsudik (2000a, 2000b). It is a tree-based group DH scheme. The basic idea is to combine the efficiency of the tree structure with the contributory feature of DH.

The basic operation of this scheme is as follows. Each group member contributes its (equal) share to the group key, which is computed as a function of all the shares of current group members. As the group grows, new members' shares are factored into the group key but old members' shares remain unchanged. As the group shrinks, departing members' shares are removed from the new key and at least one remaining member changes its share. All protocol messages are signed by the sender using RSA.

In TGDH, a sponsor takes a special role that can involve computing keys and broadcasting the blinded keys to the group during events of member join, leave, partition, and merge. Any member in the group can take on this responsibility. Figure 15 illustrates the operation of member join. When M_4 joins the group, sponsor M_3 will rename node <1, 1> to <2, 2>, generate a new intermediate node <1, 1> and new member node <2, 3>, and promote <1, 1> as the parent node of <2, 2> and <2, 3>. Sponsor M_3 knows blinded key $BK_{<2,3>}$ (the blind key of newly joined member) and $BK_{<1,0>}$, so M_3 can compute the new group key $K_{<0,0>}$ as it can

Figure 15. Illustration of join member in TGDH

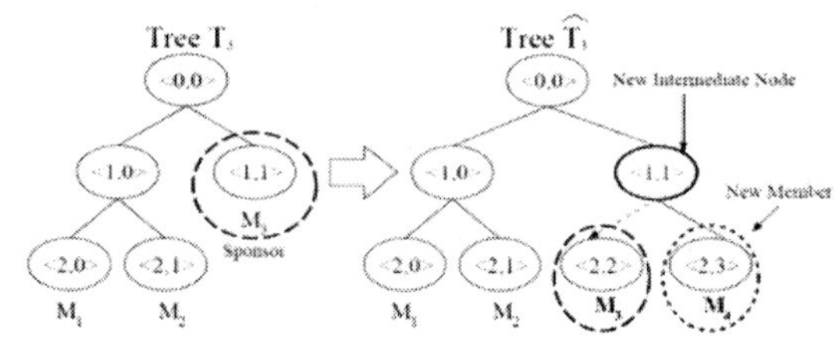

Figure 16. Illustration of leaving member in TGDH

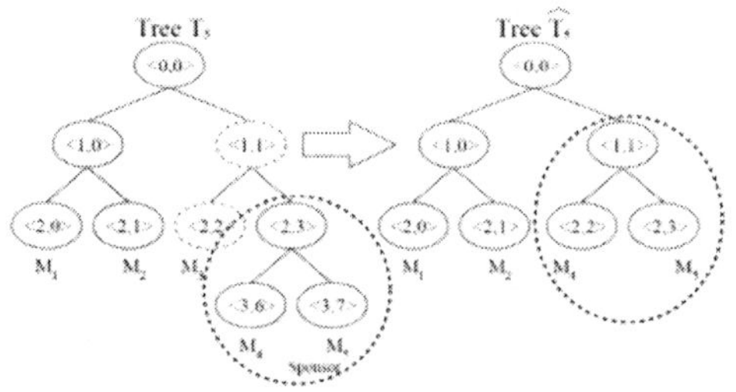

compute the intermediate key $K_{<1,0>}$. Any other member can also compute the new group key after sponsor M_3 publishes the blinded key of $K_{<1,0>}$. The leave operation is quite similar. See Figure 16 for an illustration.

Group Diffie-Hellman (GDH)

GDH is a group key distribution scheme proposed by Steiner, Tsudik, and Waidner (1998). GDH actually contains three key distribution schemes that are extended from the DH protocols. In this chapter, we only give the algorithm of GDH.3 and ignore GDH.1 and GDH.2 since these two protocols need a total of $O(n^2)$ exponentiations. The first stage involves collecting contributions from all group members (upflow). At the end of this stage, user U_{n-1} obtains $g^{\prod\{N_k | k\in[1,n-1]\}}$ and broadcasts this value to all other group members at the second stage. At the third stage, every user $U_i (i \neq n)$ factors out its own exponent and forwards the result to the last user U_n. At the final stage, U_n collects all inputs from the previous stage, raises every one of them to the power of N_n and broadcasts the resulting $n-1$ values to the rest of the group. In the end, every group member has a value of the form $g^{\prod\{N_k | k\in[1,n] \wedge k\neq i\}}$ and can easily compute the group key K_n. Member addition and deletion can be handled easily in this scheme.

A simple example is shown below to illustrate the operation of this scheme for a group of four members, A, B, C, and D:

Stage 1: A→ {B}: g^a ; B→{C}: g^{ab}
Stage 2: C →{A, B, D}: g^{abc}
Stage 3: A→{D}: g^{bc} ; B→{D}: g^{ac} ; C→{D}: g^{ab}
Stage 4: D→{A, B, C}: g^{bcd}, g^{acd}, g^{abd}, $\{g^{abc}\}$
Stage 5: $K = g^{abcd}$

The total number of exponentiations of GDH.3 is $5n-6$, the total number of rounds is $n+1$, and the number of messages is $2n-1$.

Burmester-Desmedt (BD)

BD is a distributed group key management scheme proposed by Burmester and Desmedt (1994). It is an extension of the Diffie-Hellman key distribution system. The core algorithm of this scheme is as follows:

Step 1: Each group member U_i selects a random exponent r_i, and then computes and broadcasts $z_i = g^{r_i} \bmod p$.

Step 2: Each group member U_i computes and broadcasts $X_i = (\frac{z_{i+1}}{z_{i-1}})^{r_i} \bmod p$

Step 3: Each group member U_i computes the common secret: $k_i = (z_{i-1})^{nr_i} \cdot X_i^{n-1} \cdot X_{i+1}^{n-2} \ldots X_{i-2} \bmod p$.

That is, each group user will come up with the same secret $k = g^{r_1 r_2 + r_2 r_3 + \ldots + r_n r_1} \bmod p$ which is the group key shared by all group members.

In BD scheme, each group member needs to perform $n+1$ exponentiations. It also requires a total number of $2n$ broadcast messages. Considering a simple example with a group of four users, A, B, C, and D, in the group, user B can compute $k = (g^a)^{4b} \cdot (g^{cb}/g^{ab})^3 \cdot (g^{dc}/g^{bc})^2 \cdot (g^{ad}/g^{cd})^1 = g^{ab+bc+cd+da}$. Obviously, it can be verified that other users A, C, and D can compute the same key as B.

Skinny Tree (STR)

STR is a simple group key management scheme proposed by Steer, Strawczynski, Diffie, and Wiener (1990). It is also extended from the DH. STR requires group users to be ordered in a chain. The outline of the algorithm is the following:

Step 1: Every user generates a random number r_i, and broadcasts $g^{r_i} \bmod p$.
Step 2: Users are ordered as a chain. The first and the second user can calculate the value

$$k = g^{(r_n g^{(r_{n-1} g^{(\ldots g^{(r_3 g^{(r_1 r_2)})})})})}$$

However, users 3 to n require further information to calculate k. The detailed algorithm as given by the author is skipped here, which can be referred by Steer et al. (1990). A simple example of four users A, B, C, and D is shown is Figure 17. This scheme takes two rounds and four modular exponentiations, which makes it suited for adding new group members. However, member exclusion is relatively difficult.

OPEN CHALLENGES AND FUTURE DIRECTIONS

Security is an important feature that determines the success and degree of deployment of MANETs. Cryptography is a powerful tool to defend against a variety of attacks and helps to achieve a variety of security goals. Most cryptographic

Figure 17. Illustration of STR

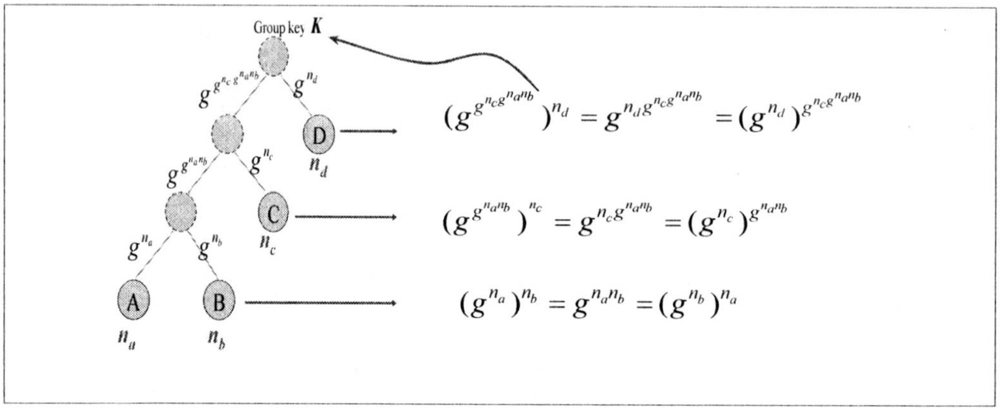

algorithms require the use of keying materials. If the cryptographic key is disclosed, then there is no security at all. Obviously, key management is in the central part of any secure communication and is the weakest point of the security. However, ensuring the security of MANETs is more challenging because of the host mobility, shared wireless medium, resource constraint of physical devices, and most seriously, lack of a fixed and trustable control point in MANETs. Designing and building an underlying secure, robust, and scalable key management system is a difficult problem that has received increased attention recently. The current research on key management in MANETs is still at its early stage.

Research on key management in MANETs goes in three directions according to the trust models, which are centralized, decentralized, and fully distributed. While centralized approaches are of least interest in MANETs, decentralized approaches have gained a lot of research attention. The fully distributed trust model is also favored for MANETs. Interestingly, a hybrid approach that combines the centralized model with the distributed scheme has been proposed recently.

Key management in MANETs can also be roughly classified into unicast and multicast key management according to the communication type. Previously, most research focused on the secure pair-wise communications, and key management focus was on how to distribute or establish a session key between a pair of communication parties. Currently, secure group communications, such as dynamic conferencing or multicasting in MANETs, is becoming an active research area. The security of group communication involves the management of group keys. For efficiency, tree-based structures are utilized when a central or virtual central control entity is available. Most contributory group key distributions are based on DH protocol with different implementations. Meanwhile, key management can also be classified into symmetric and asymmetric key management depending on the underlying cryptographic algorithms used. Currently, most key management schemes are based on asymmetric cryptosystems. However, for some specific types of MANETs, such as sensor networks, the symmetric key management scheme is dominant. An example of a symmetric approach is the random key predistribution in sensor networks.

In summary, based on different assumptions, many key management protocols have been proposed for MANETs. All key management approaches are subject to various restrictions such as the mobile device's available resources, the network bandwidth, and MANETs dynamic nature. An efficient key management protocol for MANETs is an ongoing hot research area.

REFERENCES

Burmester, M., & Desmedt, Y. (1994). A secure and efficient conference key distribution system.

In A. De Santis (Ed.), *Advances in Cryptology – EUROCRYPT '94* (No. 950).

Burnett, S., & Paine, S. (2001). *RSA security's official guide to cryptography.* RSA Press.

Capkun, S., Buttya, L., & Hubaux, P. (2003). Self-organized public key management for mobile ad hoc networks. *IEEE Transactions on Mobile Computing, 2*(1), 52-64.

Chan, A. (2004). Distributed symmetric key management for mobile ad hoc networks. In *Proceedings of IEEE INFOCOM.*

Chan, H., & Perrig, A. (2005). PIKE: Peer intermediaries for key establishment in sensor networks. In *Proceedings of IEEE INFOCOM.*

Chan, H., Perrig, A., & Song, D. (in press). Random key pre-distribution schemes for sensor networks. In *Proceedings of the IEEE Security and Privacy Symposium.*

Du, W., Deng, J., Han, Y., & Varshney, P. (2003). A pairwise key pre-distribution scheme for wireless sensor networks. In *Proceedings of the 10th ACM Conference on Computer and Communications Security (CCS)*, Washington, D.C.

Herzberg, A., Jarecki, S., Krawczyk, H., & Yung, H. (1995). Proactive secret sharing or: How to cope with perpetual leakage. *Proceedings of Crypto'95, 5*, 339-52.

Ilyas, M. (2003). *The handbook of ad hoc wireless networks.* CRC Press.

Karygiannis, T., & Owens, L. (2002). *Wireless network security-802.11, bluetooth and handheld devices* (pp. 800-848) (special publication). National Institute of Standards and Technology, Technology Administration, U.S Department of Commerce,.

Kaufman, C., Perlman, R., & Speciner, M. (2002). *Network security private communication in a public world.* Prentice Hall PTR.

Kim, Y., Perrig, A., & Tsudik, G. (2000a). *Simple and fault-tolerant key agreement for dynamic collaborative groups* (Tech. Rep. 2/USC Tech. Rep. 00-737).

Kim, Y., Perrig, A., & Tsudik, G. (2000b). *Simple and fault-tolerant key agreement for dynamic collaborative groups.* Paper presented at the 7th ACM Conference on Computer and Communications Security (pp. 235-244). ACM Press.

Lou, W., & Fang, Y. (2003). A survey of wireless security in mobile ad hoc networks: Challenges and available solutions. In X. Chen, X. Huang, & D. Du (Eds.), *Ad hoc wireless networks* (pp. 319-364). Kluwer Academic Publishers.

Luo, H., & Lu, S. (2004). URSA: Ubiquitous and robust access control for mobile ad hoc networks. *IEEE/ACM Transactions on Networking, 12*(6), 1049-1063.

Luo, H., Zerfos, P., Kong, J., Lu, S., & Zhang, L. (2001). Providing robust and ubiquitous security support for mobile ad-hoc networks. In *Proceeding of the 9th International Conference on Network Protocols.*

Menezes, A., Oorschot, P., & Vanstone, S. (1996). *Handbook of applied cryptography.* CRC Press.

Murthy, C., & Manoj, B. (2005). *Ad hoc wireless networks: Architectures and protocols.* Prentice Hall PTR.

Nichols, R., & Lekkas, P. (2002). *Wireless security-models, threats, and solutions.* McGraw Hill.

Oppliger, R. (1998). *Internet and Intranet security.* Artech House.

Perkins, C. (2001). *Ad hoc networks.* Addison-Wesley.

Rafaeli, S., & Hutchison, D. (2003). A survey of key management for secure group communication. *ACM Computing Surveys, 35*(3), 309-329.

Ravi, S., Raghunathan, A., & Potlapally, N. (2002). Secure wireless data: System architecture challenges. In *Proceedings of the International Conference on System Synthesis.*

Saloma, A. (1996). *Public-key cryptography.* Springer-Verlag.

Shamir, A. (1979). How to share a secret. *Communications ACM 1979, 22*(11), 612-613.

Sherman, T., & McGrew, A. (2003). Key establishment in large dynamic groups using one-way function trees. *IEEE Transactions on Software Engineering, 29*(5), 444-458.

Stallings, W. (2002). *Wireless communication and networks.* Pearson Education.

Steer, D., Strawczynski, L., Diffie, W., & Wiener, M. (1990). *A secure audio teleconference system.* Paper presented a the Advances in Cryptology – CRYPTO '88.

Steiner, M., Tsudik, G., & Waidner, M. (2000). Cliques: A new approach to group key agreement. In *Proceedings of the 18th International Conference on Distributed Computing Systems,* (pp. 380-387).

Tanenbaum, A. (2002). Network security. *Computer networks* (4th ed.). Prentice Hall PTR.

Tanenbaum, A. (2003). *Computer networks.* Prentice Hall PTR.

Wallner, D., Harder, E. J., & Agee, R. (1998). Key management for multicast: Issues and architectures, internet draft (work in progress). *Internet Eng. Task Force.* draft-wallner-key-arch-01.txt,.

Wong, C., Gouda, M., & Lam, S. (1998). Secure group communications using key graphs. In *Proceedings of the ACM SIGCOMM '98 conference on Applications, Technologies, Architectures, and Protocols for Computer Communication* (pp. 68-79).

Wong, T., Wang, C., & Wing, J. (2002). *Verifiable secret redistribution for threshold sharing schemes* (Tech. Rep. CMU-CS-02-114-R). School of Computer Science, Carnegie Mellon University.

Wu, B., Chen, J., Wu, J., & Cardei, M. (2006). A survey on attacks and countermeasures in mobile ad hoc networks. *Wireless/mobile network security* (Chapter 12). Springer.

Wu, B., & Wu, J., and Dong, Y. (in press). An efficient group key management scheme for mobile ad hoc networks. *International Journal of Security and Networks* (*IJSN*), *3*(4).

Wu, B., Wu, J., Fernandez, E., Ilyas, M., & Magliveras, S. (2005). Secure and efficient key management scheme in mobile ad hoc networks. *Journal of Network and Computer Applications* (*JCNA*), *30*(3), 937-954.

Wu, B., Wu, J., Fernandez, E., Magliveras, S., & Ilyas, M. (2005). Secure and efficient key management in mobile ad hoc networks. In *Proceedings of the 19th IEEE International Parallel & Distributed Processing Symposium.* Workshop 17, (pp. 288.1)

Yang, H., Luo, H., Ye, F., Lu, S., & Zhang, L. (2004). Security in mobile ad hoc networks: Challenges and solutions. *IEEE Wireless Communications, 11*(1), 38-47.

Yi, S., & Kravets, R. (2004). Composite key management for ad hoc networks. In *Proceedings of the 1st Annual International Conference on Mobile and Ubiquitous Systems: Networking and Services (MobiQuitous'04)* (pp. 52-61).

Yi, S., Naldurg, P., & Kravets, R. (2002). *Security aware ad hoc routing for wireless networks* (Report No. UIUCDCS-R-2002-2290, UIUC).

Zhou, L., & Haas, Z. (1999). Securing ad hoc networks. *IEEE Network Magazine, 13*(6), 24-30.

KEY TERMS

Certification Authority (CA): A trusted third party in an asymmetric cryptosystem that vouches for the binding of the public key with an identity.

Group Key: A common secret known by the group members

Key: A set of values that a cryptographic algorithm operates on

Key Distribution Center (KDC): A trusted third party in a symmetric cryptosystem that establishes a shared secret key between two parties

Key Management: The process of managing key materials in a cryptosystem which is related to key generation, storage, exchange, update, and replacement

Key Ring: A set of public or private keys used in PGP.

Mobile Ad Hoc Network (MANET): A collection of mobile hosts form a temporary network without centralized administration.

Chapter XXXI
Security Measures for Mobile Ad-Hoc Networks (MANETs)

Sasan Adibi
University of Waterloo, Canada

Gordon B. Agnew
University of Waterloo, Canada

ABSTRACT

Mobile ad hoc networks (MANETs) have gained popularity in the past decade with the creation of a variety of ad hoc protocols that specifically offer quality of service (QoS) for various multimedia traffic between mobile stations (MSs) and base stations (BSs). The lack of proper end-to-end security coverage, on the other hand, is a challenging issue as the nature of such networks with no specific infrastructure is prone to relatively more attacks, in a variety of forms. The focus of this chapter is to discuss a number of attack scenarios and their remedies in MANETs including the introduction of two entities; ad hoc key distribution center (AKDC) and decentralize key generation and distribution (DKGD), which serve as key management schemes.

INTRODUCTION

There are two classes of attacks on a network: *passive* and *active* attacks. In passive attacks, the intruder poses as an observer and only audits the information exchanged between communicating parties, without any intervention. Whereas in active attacks, the intruder actually takes part actively and performs actions such as additions, deletions, or delays.

The most basic requirements of a secure system should prevent common passive and active attacks, through the following functionalities:

- **Confidentiality:** Confidentiality or *privacy* is the ability to secure the content of the information communicated between authorized parties. When confidentiality is in place, the intruder should not be able to recover any information (part of the definition for passive attacks). In a broader sense, an intruder should not be able to determine the parties involved or whether a communication session occurred (anonymous routing). There are two levels of confidentiality:

- **Data confidentiality:** In which the unauthorized users are unaware of the existing protected data and their nature. This is further subcategorized as:
  - *Confidentiality of existing protected information*
  - *Confidentiality of protected data exposure*
- **Address confidentiality:** Which hides the identity of participating parties

- **Data integrity:** Integrity of data ensures the authorized recipient that data have not been altered in any sense, including addition, deletion, and undue delays. This requires data authentication. The following scenarios are associated with data integrity:
  - **Unauthorized modification protection:**Protecting against any illegitimate alteration.
  - **Detection of unauthorized protected data modification:** Detecting that a protected data has been modified in an unauthorized manner.
  - **Detection of a data deletion in a sequential order:** In a serial transmission (one bit at a time), it is important to detect if any part of the transmission has been deleted.
- **Authentication:** Authentication is a very important security requirement, which provides the facility to verify the identity of parties taking part in a communication. There are three types of authentication procedures (Kargl, 2006):
  - **Entity (user) authentication:** This type of authentication is used to authenticate an entity or a device to make sure entities wishing to communicate with other parties in the communication range are the ones they claim to be, such as people, clients, and servers.
  - **Geo-authentication:** In this type of authentication, the location of the nodes or any information about locations are to be verified and authenticated.
  - **Attribute authentication:** This is the process of establishing confidence in an attribute that applies to a specific device or entity.
  - **Data authentication:** Authentication of data is the ability of the authorized parties to ascertain the authenticity of data received from other authorized parties.
- **Nonrepudiation:** This is the ability to prevent an authorized user from denying the involvement in previous communications or activities. This is further subcategorized as follow:
  - **Protection against sender denial:** Protecting the receiver from the sender's denial that the data were sent by the sender.
  - **Protection against forward denial:** Protecting against the denial of forwarding entities on the path, disputing their forwarding actions.
  - **Protection against delivery denial:** Protecting against the delivery dispute of the data to the final destination.
  - **Protecting against receiving denial:** Protecting the sender from the recipient's denial of the fact that it has ever received the data.
- **Access control:** Access control is a mean for enabling the legitimate user to have access to the resources. Access control uses one or more of the other security mechanisms for granting access to the communications channel and/or applications. The following scenarios are categorized under access control:
  - **User identification:** Access control utilizes user-authentication to grant access for legitimate individuals.

- **Emergency access:** Accessing emergency procedures (i.e., disaster relief) is part of the access control scheme, where normal access procedures are interrupted with high-priority emergency access procedures.
- **Data encryption/decryption:** Privacy and data integrity procedure calls are used to perform user- and application-dependent encryption/decryption schemes.
- **Automatic logoff/logon:** Shutting down a part or parts of the network due to security breaches, is the task for access control. Granting access permission to those parts is also within acces control tasks.

- **Availability**: Availability is a probabilistic measure of entities being available for possible communication upon request. The higher the probability of communication entities being available, the stronger and more secure the communication channel is. Denial-of-service (DoS) causes less availability of the channel, therefore DoS and availability are opposite definitions.

LAYERED ATTACKS

Active attacks could be further categorized in a layered attack fashion (see Figure 1 and Table 1), since most security systems obey the open system interconnection (OSI) model, it is a common practice to discuss the types of attacks according

Figure 1. Active-layered attacks (Adapted from Manoj & Murthy, 2005)

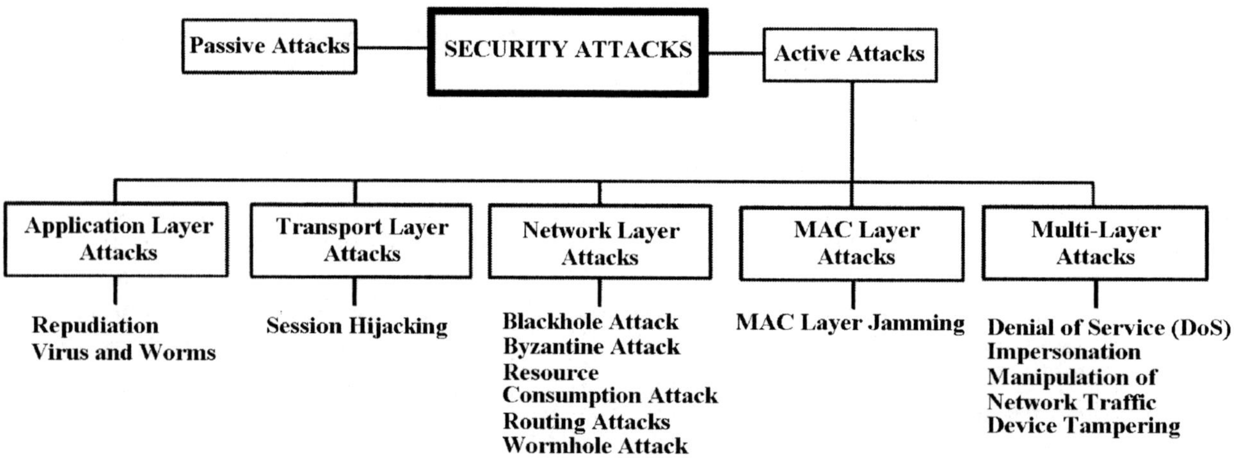

Table 1. Definition of a few active-layered attacks

DoS	Denial of service attack is a multilayer attack issue, which has the most impact on the physical layer, where communication signals are exposed to everyone and interfering (or interference) with the radio waves is the first step in DoS attack. Except for electromagnetic shielding (which can only limit the interferences) and physically securing the transmitters, there is not much to be done to prevent physical layer attacks. Fortunately for an effective physical attack, the intruder has to either have physical access to the network (tamper with the wired-infrastructures) or be physically close to the transmitters (in wireless-infrastructures). Therefore one can locate the infrastructures as distant as possible, or construct physical shielding around the transmitters.

continued on following page

Table 1. continued

Noise Signal	Increasing the noise level, which leads to the decrease of the signal to noise ratio (S/N), causes degradation of the bandwidth and roll-back of the transmission rates. In severe cases it can lead to DoS attack.
DoS	Denial of service attack can also impact the media access control (MAC) layer. For this, the attacker does not have to be physically tampering with the infrastructure, though the ability to inject frames directly into the channel is required. A MAC-layer-based DoS attack offers the following advantages to the attackers: - *Medium Independency:* Since many MAC-based communication protocols (i.e., 802.11) have similar MAC layer structures, a single MAC-layer attack can devastate many different infrastructures. - *Energy Efficiency:* A MAC layer attack does not necessarily and directly deal with the weakening of the communication signals, therefore these types of attacks require less amount of energy compared to the physical layer attacks
Jamming	Jamming happens when the communication channel is flooded with MAC layer queries. In this scenario, the MAC layer will not be able to service legitimate queries. Jamming can be considered as a DoS attack at MAC layer.
Blackhole Attack	In this type of attack, the attacker (or a malicious node) advertises a zero routing metric for all destinations. This causes all the neighbor nodes to route all their packets through the attacker (node). This can also be recognized as a DoS attack at the network layer.
Wormhole Attack	In this attack, the attacker records packets at one location in the network and tunnels them to another location in the network. This can cause an abrupt of service (DoS) due to the invalidity of routes for the packets, which are routed through this tunnel.
Byzantine Attack	This type of attack incorporates more than one attacker (malicious adversaries). A Byzantine attack involves the leaking of authentication/authorization secrets so that the malicious adversaries are indistinguishable from legitimate nodes. Therefore when adversaries are accepted in the communication schemes, they can cause various types of malicious activities, such as route changes, route loops, and nonoptimal routes. Byzantine attacks are very difficult to be identified.
Information Disclosure	In this scenario, a compromised node may leak confidential and vital information to unauthorized nodes in the network, such as, geographic location of nodes (sender, receiver, and intermediate nodes), network topology, and optimal routes.
Resource Consumption Attack	This type of attack can be discussed as a physical layer issue or a network layer issue. In the network layer, this type of attack directly deals with routing issues rather then energy related issues. Therefore, a malicious node tries to consume and waste the resources in the network through network layer-related activities, such as, unnecessary requests for routes, very frequent beacon packet creations, initiating a lot of route discoveries, and forwarding of staled packets to nodes.
Routing Attacks	These types of attacks deal with the routing algorithms and procedures, such as, routing table overflow and poisoning, packet replication, route cache poisoning, and rushing attack. These are further discussed more in the fifth section.
Others	Other types of network layer attacks include attacks on IP header/address (address sweep scan, timestamp attack, source route attack, record route attack, and fragment DoS attack) and internet control message protocol (ICMP) floods.
Attacks on TCP	Attacks on the transport control protocol (TCP) include acknowledgement (ACK) DoS, synchronization (SYN) flood, LAND attack (where spoofed TCP SYN is sent) "sending a spoofed TCP/SYN packet," session and tear-down attacks, session hijacking, and port-scan attack.
Attacks on UDP	Attacks on user datagram protocol (UDP) include port attack, (UDP flooding) and session hijacking (using a valid session ID).
Session, Presentation, Application	Higher layers (session, presentation, and application layers) are more specific and application oriented. Therefore these types of attacks vary in different networks and applications.

to the OSI layered model, namely, physical, MAC, network, transport, session, presentation, and application layers. Internet-based systems have adopted a more simplified five-layer approach based on transport control protocol (TCP)/IP protocol stack suite, in which the top three layers of the seven-layer model (session, presentation, and application layers) have been merged as a single layer: the TCP/IP application layer (see Figure 2) (Adibi, Erfani, & Harbi, 2006; Lu, 2002; Manoj & Murthy, 2005).

Wireless Routing Protocols in General

Ad hoc routing protocols are divided into the following categories:

Proactive (Table-driven)

In these types of routing protocols, nodes constantly search for routing information and storing them in tables, therefore when a route is needed, the route is already known. The major disadvantages of proactive routing protocols are:

- Consumption of relatively more bandwidth compared to identical amount of data transfer in other routing schemes.
- Increase of traffic overhead due to the constant updates.

The advantage is that there is no delay in route and destination determination. Examples of proactive routing protocols are (Lang, 2003):

- DSDV (destination-sequenced distance vector routing)
- OLSR (optimized link state routing)

Reactive (On-demand)

In reactive protocols, routes are determined as they are needed through "*route request (RREQ)*" and "*route reply (RREP)*" inquiries. The advantage of a reactive routing protocol is the fact that it requires relatively fewer traffic overhead. The disadvantage of reactive routing protocols, however, is relatively longer delays due to the sending and receiving RREQs and RREPs. Examples of reactive routing protocols are (Lang, 2003):

Figure 2. OSI Model vs. TCP/IP protocol stack

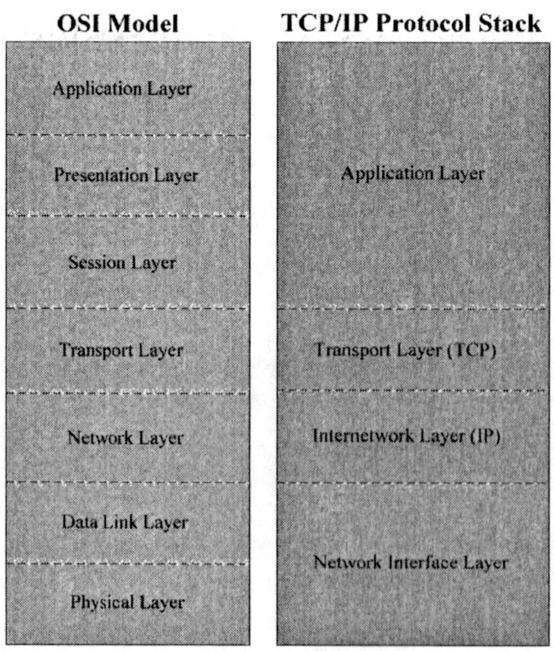

Figure 3. RREQ and RREP inquiries in reactive routing protocols

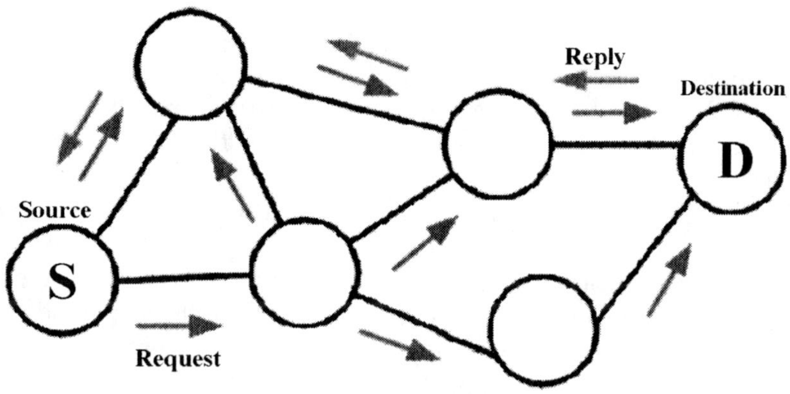

•DSR (dynamic source routing)

•AODV (ad hoc on-demand distance vector routing)

Besides reactive and proactive schemes, other types of routing protocols include (Lang, 2003):

- **Hybrid (pro-active/reactive):** A blend of reactive and proactive schemes, such as, *zone routing protocol (ZRP).*
- **Hierarchical:** Topology is divided into several local regions and local traffic is handled locally, such as, *hierarchical state routing protocol (HSR).*
- **Geographical:** These protocols use geographical coordinates in locating routing information, such as, *location-aided routing (LAR).*
- **Power aware:** In these protocols, power consumption is a serious factor, such as (Maleki, Dantu, & Pedram, 2002), *power-aware source routing (PSR).*
- **Multicast:** Multicasting is the transmission of data to groups of mobile-hosts identified by a single destination address, such as, *multicast ad hoc on-demand distance vector (MAODV).*

MANET SECURITY REQUIREMENTS

Ad hoc networks require relatively stronger security measures due to the nature of their topological weaknesses. The fact that there is no central infrastructure for ad hoc entities requires that every individual ad hoc element be part of the broader security scheme. Other issues also play roles, such as limited energy (relatively low battery life) and lack of physical security (i.e., the device could be stolen or tampered with). To remedy these limitations, it is necessary to establish cooperation enforcement between all entities and utilize secure routing schemes and efficient key management. The last two issues are discussed in this chapter in more details.

Secure Routing

The objective of secure routing is to provide a means for authenticating routing decisions and ensuring information integrity. The entity authentication includes authentication of source, destination, and all of the intermediate nodes. For specific ad hoc routing protocols, different measures, such as asymmetric or symmetric key cryptography, could be used.

ATTACKS ON AD HOC ROUTING PROTOCOLS

Attacks on ad hoc routing protocols are presented in Figure 4 and Table 2. Again these attacks are categorized into passive and active attacks. Each attack works in such a way as to paralyze a section of the routing protocol, therefore securing the routing protocols is very important.

In order to prevent attacks on routing protocols, security measures should be taken into consideration to prevent attacks and fortify the routing algorithms. These measures should provide the followings:

- **Availability:** Ultimately it should always be possible (with very high probability) to find an available route from any source to any destination within the wireless range. In ad hoc routing protocols, this feature should include preventing routing table overflow (an entry in the table to a nonexisting destination) and rushing attacks (an attacker disseminates RREQs quickly throughout the networks, suppressing any later legitimate RREQs when nodes drop them due to the duplicate suppression).
- **Isolation:** Ability to identify misbehaving nodes and disable them from interfering with the routing schemes. Preventing wormhole and black hole are examples of this category.
- **Lightweight computations:** Assigning heavy computing tasks to the least possible number of nodes (battery power protection) to prevent sleep deprivation.
- **Location privacy:** Protecting information about the location of nodes in a network and the network structure, to prevent location disclosure.
- **Self-Stabilization** – Automatically recover from any problem in a finite amount of time without human intervention.
- **Byzantine robustness:** This requires the function of the routing protocol to work correctly even if some of the nodes participating in routing are intentionally disrupting its operation. This is important in preventing impersonation attacks.

Figure 4. Active and passive attacks in ad hoc routing protocols (Adapted from Wang, Lu, & Bhargava, 2003)

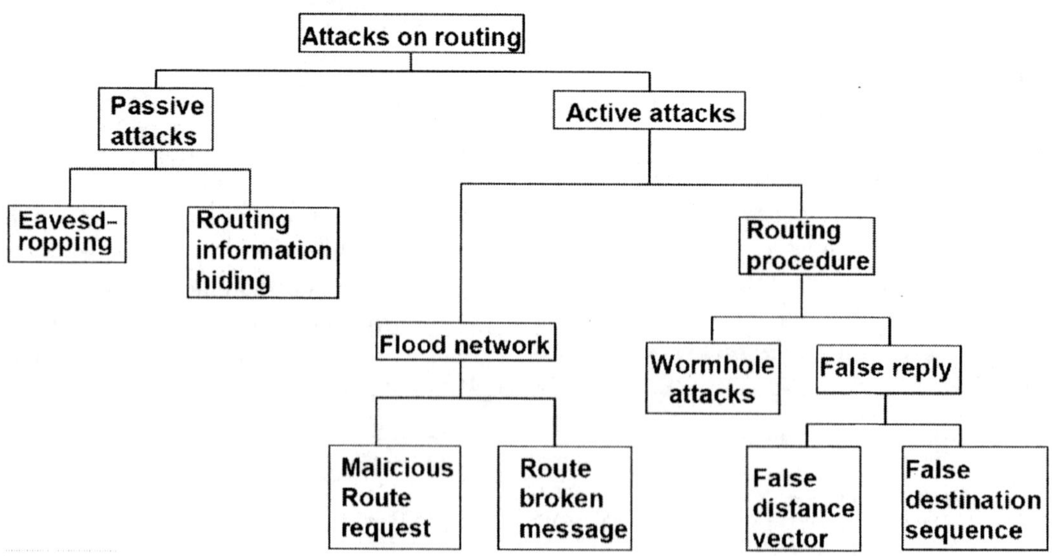

Table 2. Definition to a few of attacks for ad hoc routing protocols

Route Broken Message	Sets false route error to send a message back to the source (route discovery is reinitiated). This exhausts the limited bandwidth.
Malicious Route Request	Sends an invalid route request. This exhausts the limited bandwidth.
False Distance Vector	This involves replying "one hop to destination" to every request and selecting an enough large sequence number. This is an attack on the connectivity.
False Destination Sequence	This is to select a large number of hop to the destination, which is an attack on the connectivity.
Routing Table Overflow	A malicious node advertises routes to nonexisting nodes. Proactive routing protocols are more vulnerable.
Routing Table Poisoning	A malicious or compromised node sends fictitious routing updates or modifies genuine route updates, which causes suboptimal routing.
Packet Replication	A malicious or compromised node replicates stale packets causing excessive bandwidth consumption.
Route Cache Poisoning	An advisory can poison the route cache, which is a major issue for on-demand routing protocols, since they maintain a route cache to all known nodes.
Rushing Attack	An advisory that received a RREQ from a source floods the network quickly before any other legitimate nodes can react, causing other nodes to believe that they have received duplicates, thus discarding the legitimate responses. Therefore any route discovered by the source node would contain the advisory node information as one of the legitimate intermediate nodes.

Possible Solutions

To offer secure routing protocols, the following solutions are used:

- **Trusted route discovery:** To avoid internal attacks, the route discovery phase in ad hoc routing protocols should send packets via trusted routes.
- **Redundant paths and multipath routing:** Having redundant path increases route robustness by providing more route choices, such as in multipath ad hoc routing protocols.
- **Nondisclosure method and anonymous routing:** Anonymous routing avoids the location disclosure by using distributed independent security agents. This way outsiders could not identify the communicating parties.
- **Hierarchical structure or zone-based routing:** This type of routing protocol provides a foundation for authentication and local link-state routing.
- **Authentication among hosts:** This requires two-way authentication schemes for all parties to prevent impersonation (spoofing).
- **Preventing traffic pattern detection:** This is important in hiding the traffic patterns and frequency of transmitting information, as part of anonymous routing.
- **Intrusion detection:** Monitors the behavior of suspected hosts for anomaly detection and attack prevention.
- **Securing the medium:** To prevent physical-layer-based DoS attack, there has been a few methods introduced as security deterrence schemes, such as:
    - *Frequency inversion*
    - *Frequency hopping*

These two methods will be discussed in details in the next section.

CHALLENGES IN SECURE ROUTING FOR MANETS

As mentioned previously, securing routing protocols for wireless systems is more challenging than securing wired protocols, because not only do all of the possible wired-based attacks apply to ad hoc networks, but also mobility allows new attacks. The most important difference is the vulnerability of the medium. This is very important because everyone shares the same medium (open air) and if extra attention is not giving to its security, it could contribute to a DoS attack (lack of availability or jamming). Therefore the followings are the extra challenges for securing wireless routing protocols:

- **Intrusion detection:** Intrusion detection attempts to detect any malicious or unauthorized activity, either caused by an internal entity or an external source. There are a few types of intrusion detection systems:
  - *Anomaly-based* : Compares the activities in a network with a predefined normal activity map. In these systems, a sudden change in the activities would trigger anomaly alarm detection. Other types include: *network intrusion detection system (NIDS), host-based intrusion detection system (HIDS), application protocol-based intrusion detection system (APIDS),* and *protocol-based intrusion detection system (PIDS).*
- **Secure routing:** This shares a common ground with a layered approach, in a sense that security mechanisms have been integrated with the normal routing procedures. Routing is mostly covered in the network layer. Therefore secure routing is provisioned in the network layer. Secure routing will be discussed in details later in this chapter
- **Key management service:** Because of the difficulties in key exchange, the key management is a challenge in ad hoc networks. The following schemes are a few examples of existing key exchange methods ("Key Management," 2001):

  - Signature keys
  - Signature verification keys
  - Authentication (public, private, secret) keys
  - Data encryption (long-term, short-term) keys
  - Keys based on random number generation
  - Key encryption keys, which are further used for wrapping keys
  - Derivation keys used from master keys and master keys used from derivation keys
  - Key transport for public and private keys
  - Static key agreement used for public and private keys
  - Ephemeral key agreement for public and private keys

  Key management faces the following particular challenges:

  - Lack of a security infrastructure
  - Limited processing power
  - It should be fully distributed with minimal dependencies
  - Domain parameter and public key validations
  - Keys and related material compromise
  - Key recovery: consideration and policy
  - Audit and accountability issues

  Therefore the challenges are trust model, cryptosystems, key creation, key storage, and key distribution.
- **Securing the medium:** To prevent DoS, the following techniques are used as security deterrents:
  - *Frequency inversion*: The process of altering the signal's frequency spectrum in such a way that the signal could not be reconstructed and understandable without the knowledge of the inversion pattern.
  - *Frequency hopping*: Dividing the spectrum into various frequencies and using different frequencies in a predetermined fashion.
  - *Shared secret frequency key*: Sharing the secret of frequency-pattern between transmitters.

KEY MANAGEMENT APPROACHES

Due to the variable nature of ad hoc network topologies and the physical and resource limitations, key management is of great importance. There are many proposals for the key management for ad hoc protocols, however we introduce two methods, namely, *ad hoc key distribution center (AKDC)*, and *decentralized key generation and distribution (DKGD)* (Adibi et al., 2006):

- **Ad hoc key distribution center:** As shown in Figure 5, AKDC uses a centralized ad hoc scheme for key management, distribution, and access. In the AKDC, each device wishing to communicate with another device will have to undergo the following series of processes by the AKDC:
    - Identity and location determination
    - Authentication
    - Authorization
    - Key provision
    - Key delivery

  A lot of intelligence and power must be integrated into the design of an AKDC, however, there are a few downsides of having a central point. The fact that there is a known center for key distribution and its location is known to all, makes the AKDC prone to a variety of attacks, including DoS attack. This problem is remedied by the use of a decentralized and distributed scheme.
- **Decentralized key generation and distribution:** In a DKGD scheme (Figure 6), the key management scheme is distributed across the wireless range through DKGD agents. Every ad hoc element discovers the closest DKGD agent and binds with it. The fact that DKGDs are distributed across the network poses less of a security concern as the single point of failure is no longer an issue. No matter if AKDC or DKGD is used, all legitimate local ad hoc elements should register with the AKDC or the DKGD.
- **Ad hoc gateway access control (AGAC):** So far, the AKDC and DKGD schemes assume in-domain communications among ad hoc elements. However for inter-domain security measures when an outside element seeks communication to a local element, a new element, which is called the AGAC, is responsible for the security concerns. AGAC

Figure 5. AKDC scheme (Adapted from Adibi et al., 2006)

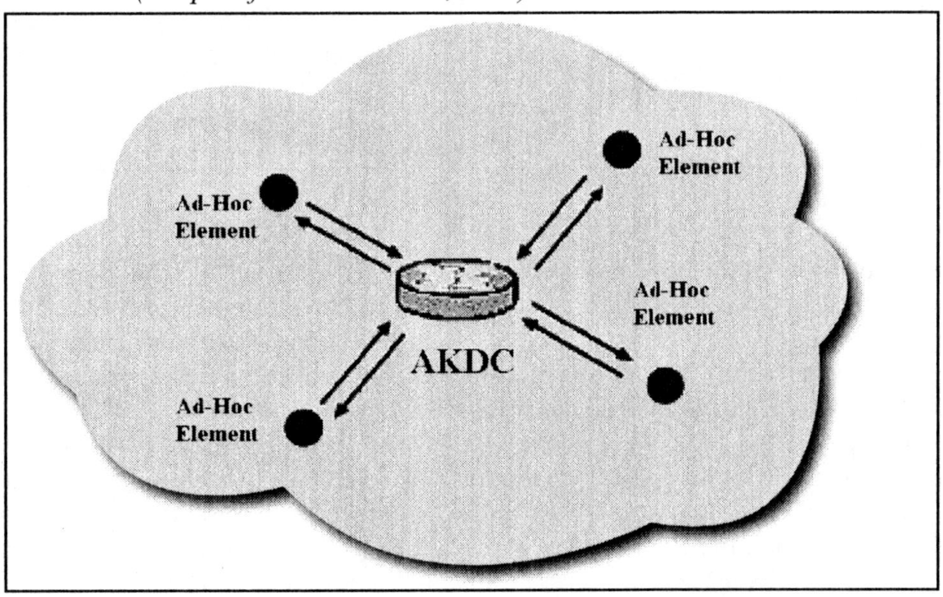

Figure 6. DKGD scheme (Adapted from Adibi, Erfani, & Harbi, 2006)

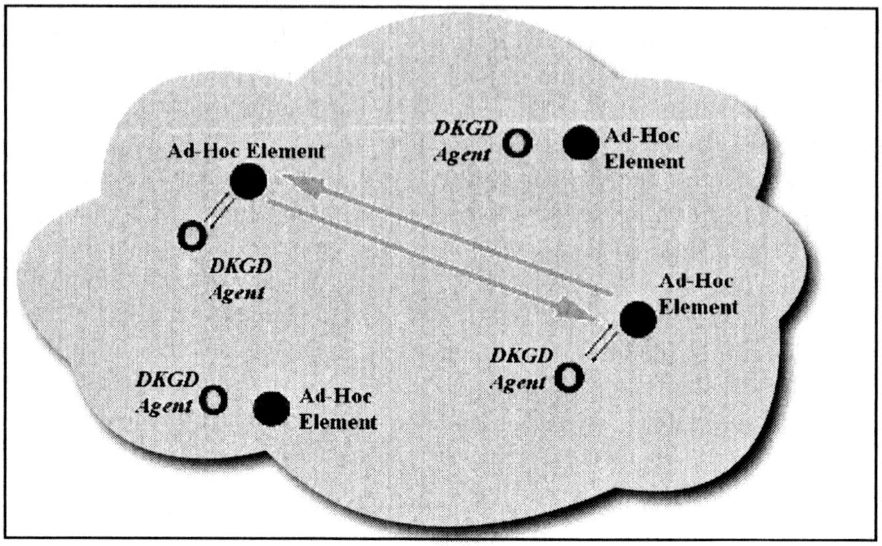

Figure 7: Self-organized certificate authorities (SOCA) (Adapted from Michiardi, 2004)

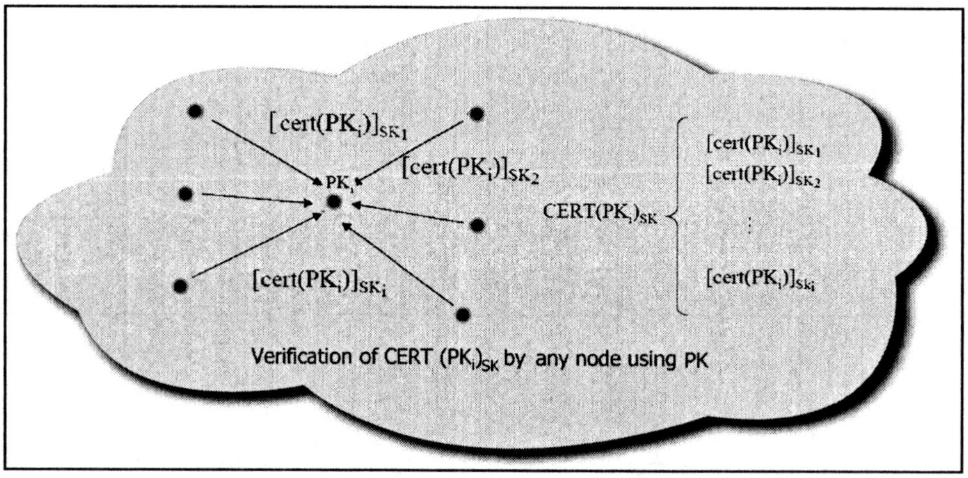

agents are located at the boundaries of radio domains, that is, where two or more local ad hoc domains intersect.

- **Secure and efficient key management (SEKM):** SEKM (Wu, Wu, Fernandez, Ilyas, & Magliveras, 2005) creates a public key infrastructure (PKI) using a secret shared key scheme and on top of an underlying multicast server groups. In SEKM, a view of the certificate authority (CA) is created by each server group. This provides and update for certificate services for all the participating nodes. For an efficient certificate delivery service, a ticket mechanism is introduced and used.
- **Self-organized CA (SOCA) (Michiardi, 2004):** In traditional cryptographic systems, there is one sender, one receiver, and an eavesdropper who is the opponent. However a SOCA is based on threshold cryptography.

Threshold cryptography allows one to share the power of a cryptosystem in which the power to regenerate a secret key is shared among several agents (Figure 7). The advantage of this is the distributed approach with self-organization. The downside is the network density.

SECURITY MECHANISMS IN MANETS

There are several mechanisms, which are embedded into the protocol schemes, which contribute to the robustness of security. Below is a list of a few of these mechanisms.

- **Multipath routing:** Multipath routing (Bonum & Othman, 2003) works by enhancing data confidentiality through the transmission of data via multiple paths. This is done to prevent any fixed unauthorized nodes from attaining useful data. This requires no encryption as data is already "split" among various paths.
- **Hierarchical routing:** Hierarchical routing (Rhee, Park, & Tsudik, 2004) is one of the categories of ad hoc routing protocols in which traffic handling is done through different layers and local activities are kept local. Therefore there is no need to broadcast all changes to the entire radio domain. Only global moves are reported across the entire network. In the security and key management cases, the architecture could use a two or more layered key management approach where groups of nodes are divided into *cell groups* consisting of ground nodes and *control groups* containing cell group managers.
- **Tunneling:** Tunneling (Choi, Song, Cao, & Porta, 2005) is widely used in many security schemes, such as virtual private networks (VPNs) and IP-Security (IPSec).
- **Other Measures in MANETs:** Other measures, which could be adopted in ad hoc scenarios are (Menezes, Oorschot, & Vanstone, 1996):
    - *Web of trust (PGP)*: Which is a Peer-based (one-to-one) system and requires no Certificate Authority. PGP symmetric and public-key cryptography schemes and includes a mechanism, which binds the public keys to the user identities.
    - *Crypto-based ID:* A crypto-based ID (CBID) requires no infrastructure and uses a binding between address and signature.
    - *ID-based crypto*: The ID-based crypto suggests that having an identity implies being authorized, therefore no certificates are needed.
    - *Context-dependent authentication*: Authentication is based on the content of the message.
    - *Password authenticated key exchange (PAKE):* In a PAKE, two or more communication parties, based on their knowledge of a password only, establish a cryptographic key through a message exchange, in such a way that an unauthorized entity cannot participate in the scheme and is kept from guessing the password. There are two forms of PAKE, which are balanced and augmented schemes.
    - *Cooperation Enforcement Mechanisms using Game Theoretical Approaches:* Game theory is a powerful tool that models interactions among participating entities. Each player tries to maximize some utility function in a distributed manner. Nash equilibrium is where the games settle, assuming the equilibrium exists, however, since nodes usually act selfishly, the equilibrium point might not be the optimal social point.

SECURE PROTOCOLS FOR MANETS

The main idea for these protocols is to offer extended security, therefore with only security in mind, the entire protocol functionalities have been designed for security in the the network layer. Four of these protocols are introduced as follow:

ARIADNE (A Secure On-Demand Routing Protocol for Ad Hoc Networks)

ARIADNE (Hu, Perrig, & Johnson, 2002) relies only on highly efficient symmetric cryptographic systems and does not require a trusted hardware or powerful processors. Routing messages can be authenticated using ARIADNE through one of the following three schemes: 1) Using shared secrets among each pair of nodes, 2) Using shared secrets among communicating nodes together with broadcast authentication, and 3) Using digital signatures. ARIADNE works well with timed efficient stream loss-tolerant authentication (TESLA) (Hu et al. 2002), which is an efficient broadcast authentication scheme that requires loose time synchronization, where a receiver knows an upper bound of difference between sender's local time and the receiver's local time.

SEAD (Secure Efficient Distance Vector Routing for Mobile Wireless Ad Hoc Networks)

SEAD (Hu, Johnson, & Perrig, 2002) is based on the design of the destination-sequenced distance-vector routing protocol (DSDV). To prevent DoS, SEAD uses efficient one-way hash functions and does not include the usage of asymmetric cryptographic operations. SEAD is robust against multiple uncoordinated attackers, which creates incorrect routing state for other nodes.

SADSR (Security-Aware Adaptive Dynamic Source Routing Protocol)

SADSR (Ghazizadeh, Ilghami, Sirin, & Yaman, 2002) includes an authentication scheme in which, the routing protocol messages are authenticated using asymmetric cryptographic-based digital signatures.The basic idea behind the functionality of SADSR is to have multiple routes to every destination and to store a local trust value related to each node throughout the network. A trust value is also assigned to each path based on nodes trust values. The paths with higher trust values are preferred and selected for routing.

SDSR (Secure Dynamic Source Routing)

SDSR (Kargl, Geiss, Schlott, & Weber, 2005) prevents various potential (active and passive) attacks to the ad-hoc-based networks. It also deals with selfish nodes in the following scenarios:

- *Motivation-based approaches*: Try to motivate network users to actively participate in the MANET.
- *Detect and exclude*: This scheme detects and excludes selfish nodes from the routing scheme
- *Mobile Intrusion Detection System (MobIDS)*: Focuses on integrating with other mechanisms for detecting selfish nodes.

CONCLUSION

Attacks can be categorized as per node behaviors, protocol schemes, or layered approaches. Security challenges in MANETs include securing the medium (preventing from DoS attack, etc.), securing the routing schemes, intrusion detection and prevention, key management, peer-to-peer security options, user and data authentication/authorization, data encryption, and digital signatures.

REFERENCES

Adibi, S., Erfani, S., & Harbi, H. (2006, May). *Security routing in MANETs: A comparative study.* Paper presented at the Electro/information Technology (EIT) Conference.

Bonum, S., & Othman, J. B. (2003). Data security in ad hoc networks using multipath routing. In *Proceedings of the 14th IEEE International Sym-*

posium on Personal, Indoor and Mobile Radio Communication (PIMRC 2003), vol. 2, (pp. 1331 -1335). Beijing, China.

Choi, H., Song, H., Cao, G., & Porta, T. L. (2005). *Mobile multi-layered IPsec.* Paper presented at the INFOCOM.

Ghazizadeh, S., Ilghami, O., Sirin, E., & Yaman, F. (2002). *Security-aware adaptive dynamic source routing protocol.* ILCN.

Hu, Y. C., Johnson, D. B., Perrig, A. (2002). *SEAD: Secure efficient distance vector routing for mobile wireless ad hoc networks.* MCSA.

Hu, Y. C., Perrig, A., & Johnson, D. B. (2002). *Ariadne: A secure on-demand routing protocol for ad hoc networks.* Paper presented at the MOBICOM.

Kargl, F. (2006, November). *Threats and security requirements for VANETs secure vehicle communication.* Paper presented at the C2C-CC Sec. Workshop.

Kargl, F., Geiss, A., Schlott, S., & Weber, M. (2005). *Secure dynamic source routing.* Paper presented at the HICSS.

Key Management, National Institute of Standards and Technology (NIST). (2001, November). Retrieved October 7, 2007, from http://csrc.nist.gov/encryption/kms/Key%20Mgmt%20Guideline%20Overview.ppt

Lang, D. (2003, March). A *comprehensive overview about selected ad hoc networking routing protocols* (Tech. Rep. No. TUM-I0311). Technische Universität München, Department of Computer Science.

Lu, Q. (2002, December). *Vulnerability of wireless routing protocols.* University of Massachusetts Amherst.

Maleki, M., Dantu, K., & Pedram, M. (2002, August). Power-aware source routing protocol for mobile ad hoc networks. In *Proceedings of the Symposium on Low Power Electronics and Design* (pp. 72-75).

Manoj, B. S., & Murthy, C. S. R. (2005, January). *Transport layer and security protocols for ad hoc wireless networks.* Retrieved October 7, 2007, from http://www.phptr.com/articles/article.asp?p=361984 &seqNum=10&rl=1

Menezes, A. J., Oorschot, P. C. V., & Vanstone, S. A. (1996). *CRC handbook of applied cryptography.* CRC Press.

Michiardi, P. (2004, March). *Security in wireless ad hoc networks.* Institut Eurecom.

Rhee, K. H., Park, Y. H., & Tsudik, G. (2004, June). An architecture for key management in hierarchical mobile ad-hoc networks. *Journal of Communications and Networks, 6*(2).

Wang, W., Lu, Y., & Bhargava, B. (2003, March). *On security study of two distance vector routing protocols for ad hoc networks.* Purdue University, CERIAS and Department of Computer Sciences.

Wu, B., Wu, J., Fernandez, E. B., Ilyas, M., & Magliveras, S. (2005). *Secure and efficient key management in mobile ad hoc networks.* Elsevier.

KEY TERMS

Access Control: This is a security mechanism to make sure that only legitimate parties have access to the data they are supposed to have access.

AKDC: Ad hoc key distribution center is a central component in an ad hoc network responsible for providing keys to ad hoc elements.

ARIADNE: A secure on-demand routing protocol for ad hoc networks.

Authentication: Authentication is required to make sure communicating parties are the ones who they claim to be.

Availability: A stochastic measure of predicting the availability of the communication channel and resources to the users

CA: Certificate authority is responsible for issuing digital certificates.

Confidentiality: This is a basic security requirement in which the address, location, and/or the data transferring between two communicating parties are to be kept as secrets.

DKGD: Decentralize key generation and distribution is another key distribution scheme for ad hoc networks in which the key distribution mechanism is done by distributed elements, not by a centralized entity.

Integrity: This is another basic security requirement. Integrity guarantees the correctness of the data transferring between two communicating parties, or their location information.

Nonrepudiation: This is a concept of ensuring that none of the communicating parties can deny the fact that they had sent or received certain data.

SADSR: Security-aware adaptive dynamic source routing protocol.

SEAD: Secure efficient distance vector routing for mobile wireless ad hoc networks.

SEKM: Secure and efficient key management is another key management scheme in which involves public key infrastructure.

SDSR: Secure dynamic source routing.

SOCA: Self-organized CA is a threshold-based cryptosystem in which the power is shared among different CAs.

Chapter XXXII
A Novel Secure Video Surveillance System Over Wireless Ad Hoc Networks

Hao Yin
Tsinghua University, China

Chuang Lin
Tsinghua University, China

Zhijia Chen
Tsinghua University, China

Geyong Min
University of Bradford, UK

ABSTRACT

The integration of wireless communication and embedded video systems is a demanding and interesting topic which has attracted significant research efforts from the community of telecommunication. This chapter discusses the challenging issues in wireless video surveillance and presents the detailed design for a novel highly-secure video surveillance system over ad hoc wireless networks. To this end, we explore the state-of-the-art cross domains of wireless communication, video processing, embedded systems, and security. Moreover, a new media-dependent video encryption scheme, including a reliable data embedding technique and real-time video encryption algorithm, is proposed and implemented to enable the system to work properly and efficiently in an open and insecure wireless environment. Extensive experiments are conducted to demonstrate the advantages of the new systems, including high security guarantee and robustness. The chapter would serve as a good reference for solving the challenging issues in wireless multimedia and bring new insights on the interaction of different technologies within the cross application domain.

INTRODUCTION

With the ever-increasing security demands of military and scientific applications, the development of a highly secure and reliable video surveillance system attracts significant interests from both academia and industry. The implementation and efficiency of such a system are greatly affected by the techniques in wireless communication, video processing, embedded systems, and security guarantee.

Recent advances in embedded system and wireless communications are enabling cost-effective digital wireless multimedia systems. The forthcoming integration of wireless communications and embedded video systems is a demanding and interesting research topic. Video surveillance has resorted to wireless transmission due to the several serious problems when the traditional coaxial or high-tech fiber-optic cables are adopted to transmit video images from the surveillance cameras to the stations at which the images are monitored and/or recorded. Compared with the traditional wire-line counterparts, wireless video surveillance systems do not require expensive and time-consuming system constructions and civil-engineering work. They can therefore be deployed rapidly with negligible environmental impact. Furthermore, wireless systems generally require lower costs of network maintenance, management, and operation.

However, some fundamental issues, such as framework design of wireless networks, video processing, video data transmission, video quality control, and system security should be resolved before wireless video surveillance systems can be successfully deployed (Garcia-Macias et al, 2003). Among these important issues, the system security is the most challenging problem that becomes the main concern of this chapter. Intel IXP425 network processor provides an ideal choice for implementing secure ad hoc video surveillance system, but the security issue is still a hot-spot that IXP425 cannot handle well. Therefore, an effective video encryption algorithm is necessary and meaningful in a wireless video surveillance system. At the same time, the secure routing protocol and system architecture should be carefully designed to avoid serious security flaws (Yin, Lin, Sebastien, & Chu, 2005).

This chapter explores the state-of-the-art cross domains of wireless communication, video processing, embedded systems and security, discusses the challenging issues in wireless video surveillance, and presents the detailed design of a novel highly-secure video surveillance system over ad hoc wireless networks. The rest of this chapter is organized as follows. Section 2 provides a review of wireless networks, ad hoc solution and security issues. Section 3 presents the design and implementation for the new video surveillance system and Section 4 evaluates its performance. Section 5 highlights the future trends in the relevant research areas. Finally, Section 6 concludes this chapter.

BACKGROUND

Wireless Networks

Wireless technologies, in the simplest sense, enable one or more devices to communicate without physical connections (without requiring peripheral cabling). Wireless networks serve as the transport mechanism among mobile devices or between these devices and the fixed wired networks (e.g., enterprise networks and the Internet). A wireless network has tremendous advantages in comparison with its wired counterpart: no network cable has to be installed through walls and floors, thus greatly reducing the cost and making the architecture more flexible.

The development of 802.11g (IEEE, 2003) based on the orthogonal frequency-division multiplexing (OFDM) technology allows high-load applications to be adapted in wireless environment. It is claimed that an optimal throughput of 54Mps and a range up to 100 feet indoors can be achieved. As the signal is modulated at 2.4 GHz, it is less affected by walls and physical obstacles than 802.11a (5 GHz). Thus our system is based on the 802.11g wireless infrastructure ad hoc networks.

Ad Hoc Solution

Ad hoc networks are a new wireless networking paradigm for mobile hosts. Unlike traditional mobile wireless networks, ad hoc networks do not rely on any fixed infrastructure. Instead, hosts rely on each other to keep the network connected. Ad hoc networks are designed to dynamically connect remote devices such as cell phones, laptops, and PDAs. These networks are termed "ad hoc" because of their shifting network topologies. Whereas wireless LANs use a fixed network infrastructure, ad hoc networks maintain random configurations, relying on a master-slave system connected by wireless medium to enable communication between mobile devices (Haas, 1999; Zhou, 1999).

The system we are designing is organized in an ad hoc manner. The nodes themselves (with camera) are carrying the flux towards the monitoring center, and all the routing tasks are performed by the camera nodes. A careful deployment can share the traffic load among all the camera nodes and effectively reduce the bottleneck effect as compared with an architectural network. It is also the cheapest solution as there is no need of extra networking hardware besides the cameras, network processors, and the monitoring center.

However, the design of ad hoc architecture is complex because of the routing and security issues. In a monitoring system, the node positions are static and predetermined by the topology of the building. The cameras are in a nonprotected environment, and they are susceptible to be damaged or even destroyed. Thus it would be preferable if every node has at least two direct neighbors on the way towards the monitoring center so that the system can still work properly in case some camera nodes are faulty.

Security Issues

Among the issues the wireless solution face, the system security is the most challenging problem. The NIST handbook *An Introduction to Computer Security* generically classifies security threats into nine categories ranging from errors and omissions to threats to personal privacy (Basgall, 1999). All of these represent potential threats in wireless networks as well. However, the more immediate concerns for wireless communications are device theft, denial-of-service, malicious hackers, malicious code, theft of service, and industrial and foreign espionage.

Data embedding techniques allow for a signal to be hidden without dramatically distorting the original content. Effective data embedding techniques should be able to invisibly embed data, allow for easy extraction, and achieve a high embedding rate. The most popular application of data embedding is digital watermark. Lots of research work has been done in this field over the past years. Although it is worthy noting that none of the existing schemes are capable of satisfying the demand for media-dependent access control in wireless video surveillance system, some ideas and framework of these digital watermark algorithms are valuable and may be extended to design the desired data embedding scheme (Yin, Lin, Qiu, Min, & Chu, in press).

The classical approach to watermark compressed video stream is to decompress the video, then use a spatial-domain or transform-domain watermarking technique to embed the watermark into the video signal, and finally recompress the watermarked video. Alattar, Lin, and Celik (2003) point out three major disadvantages of using the classical approach and further present a faster and more flexible approach to watermark compressed video named as compressed-domain watermarking. With this approach, the original compressed video is partially decoded to expose the syntactic elements of the compressed bitstream (such as encoded discrete cosine transform [DCT] coefficients) that is modified to insert the watermark and reassembled to form the compressed watermarked video.

Patchwork (Bender, Gruhl, Morimoto, & Lu, 1996) and quantization index modulation (QIM) (Chen & Wornell, 2001) are the two known techniques for the embedding algorithm. Patchwork (Bender et al., 1996) is a statistical scheme based on a pseudorandom and statistical process. Patchwork is host image independent and can invisibly embed a specific statistic pattern (composed of several pairs of specific pixels) in a host image with a Gaussian distribution. It shows reasonably high resistance to most nongeometric image modifications. But the major disadvantage is that only one bit can be embedded in one frame. Moreover, this algorithm operates specific pairs of points and the structure of video bitstream is changed by some adaptive processes such as transcoding. So during the detecting procedure these pairs of points at the same position are not the same as the original, or even out of borders due to the change of image size. As a result, the extracted data are likely to be wrong. Our proposed scheme is based on the statistical property of the luminance value, but differently we use image fields instead of pairs of points to overcome above mentioned problems.

Chen and Womell (2001) propose a QIM scheme for efficiently embedding and drawing out data. QIM method embeds information not simply by adding numbers to the host signal, but by first modulating an index of sequence of indices with the embedded information and then quantizing the host signal with the associated quantizer or sequence of quantizers. During the detecting procedure, the embedded information is determined by judging the minimum distance between the embedded signal and different quantized results. It is known that the QIM method is better than additive spread spectrum and generalized low-bits modulation (LBM) not only from the point of rate distortion-robustness tradeoffs, but also against bounded perturbation and fully informed attacks arising in several copyright applications. Since requantization is carried out in the transcoding procedure and the quantizers are different from the ones used in video encoding process, lots of computational errors are produced and the detection is likely failed. Our scheme improves the QIM by proposing an approach to alter the average luminance value of fields.

Routing Protocol

In recent years, a large number of ad hoc routing protocols have been proposed in the literature (Broch, Maltz, Johnson, Hu, & Jetcheva, 1998; Perkins & Royer, 1999; Per, 1999; Samir, Perkins, & Royer, 2000). In all these studies, two on-demand routing protocols show good performance: ad hoc distance vector (AODV) (Perkins & Royer, 1999) and DSR. In a scenario where a high volume of traffic goes through a static ad hoc network (by static we mean that the nodes configuration does not change or changes slowly), AODV performs better than DSR due to less additional load being imposed by source routes in data packets. Therefore our system is based on the AODV protocol.

AODV is an on-demand protocol. Each node maintains its routing table only for the routes they actually use to communicate with other nodes. If a node wants to initiate a new communication with another node that is not in its current routing table, a route request (RREQ) is broadcast. If a node receives such a request, it looks up its routing table to find whether there is a path to the destination nodes. If there exists a path, it replies a route reply (RREP); otherwise, it broadcasts the RREQ. If a node receives the same RREQ twice, it simply discards the message. Routes are maintained in the routing table as long as they send packets. If nothing is received after a predefined timeout value, the corresponding route entry is deleted. In case of nodes failure, neighbors on the active path send a special RREP to the source which can start a new path discovery phase. Neighbor's discovery is done either by local broadcasting of HELLO messages or by receiving a broadcast message from a neighbor given that the links between nodes are bidirectional.

Perkins and Royer (1999) try to avoid relying on the underlying MAC-layer protocol, but no solution has been proposed to avoid the overhead created by the HELLO message. In our system the routing protocols are coupled with the address resolution protocol (ARP) protocol as described by Desilva and Das (2000) so that we can avoid broadcasting HELLO messages. In addition, it is preferred to implement the routing protocol at link layer due to the following reasons (Johnson, Maltz, & Broch, 2001):

- Pragmatically, running the protocol at the link layer maximizes the number of mobile nodes that can participate in ad hoc networks.
- Historically, the protocol has grew from a multihop propagateing version of the Internet's address resolution protocol (Plummer, 1982), as well as from the routing mechanism used in IEEE 802 source routing bridges (Perlman, 1992).
- Technically, our design would expect the protocol to be simple enough so that it could be implemented directly in the firmware inside wireless network interface cards, well below the layer 3 software within mobile nodes.

Network Processor

The Intel IXP425 network processors are used as the basic processing unit to design our wireless video surveillance system. Intel® IXP425 network processor is a highly integrated, versatile single-chip processor used in a variety of products that need network connectivity and high performance to run their unique software applications. The Intel IXP425 network processor combines integration with support for multiple WAN and LAN technologies in a common architecture designed to meet requirements for high-end gateways, voice over IP (VoIP) applications, wireless access points, small-to-medium enterprise (SME) routers, switches, security devices, Mini-DSLAMs (digital subscriber line access multiplexers), xDSL line cards, industrial control, and networked imaging applications. The framework of Intel® IXP425 is shown in Figure 1 (Intel, 2006). Intel IXP425 network processor provides diverse functionalities, including data encryption, secure data transmission, and multimedia processing, thus making it an ideal choice for implementing secure ad hoc video surveillance systems.

Though Intel IXP425 network processor is an ideal choice for implementing secure ad hoc video surveillance systems, the security issue is still a hot-spot that IXP425 cannot handle well on its own. For example, video encryption is very important in wireless LAN environment since everyone can receive the video content and inject the faked video packets. Unfortunately, normal data encryption functions like AES provided by IXP425 is too computationally expensive to be applied to every single outgoing packet, especially in wireless ad hoc network environment which should consider power limitation of wireless devices (Allman, 2002; Borisov et al., 2003). Therefore, an effective video encryption algorithm is necessary. At the same time, the secure routing protocol and system architecture should all be carefully designed to avoid serious security flaws.

Figure 1. Framework of Intel® IXP425

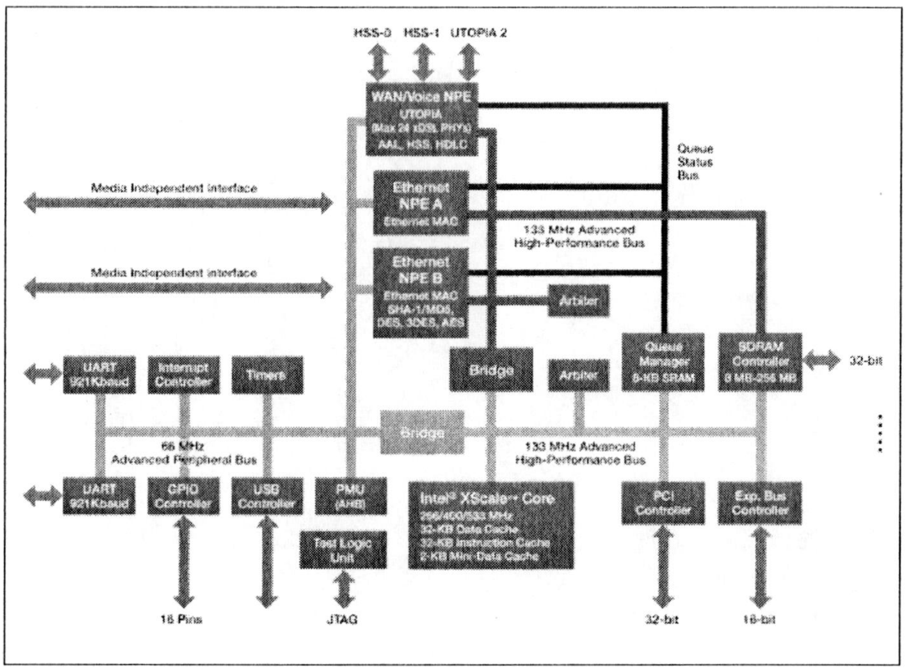

A NOVEL SECURE VIDEO SURVEILLANCE SYSTEM

Framework Design of Wireless Networks

The system is based on the 802.11g wireless ad hoc infrastructure. Intel IXP425 network processors are used as the basic processing unit. The 802.11g is a physical layer standard for WLANs with 2.4GHz and 5GHz radio bands. It specifies three available radio channels. The maximum link rate is 54Mbps per channel whereas 802.11b has 11Mbps. The 802.11g standard uses the OFDM modulation. However, for backward compatibility with 802.11b, 802.11g also supports complementary code-keying (CCK) modulation and, as an option for faster link rates, it allows packet binary convolutional coding (PBCC) modulation (Wentink, 2003).

Our wireless video system is composed of a set of camera nodes and a monitoring center, as shown in Figure 2. Each camera node is equipped with an IXP425 processor and 802.11g wireless card. The video source is captured by the camera, and then compressed by the processor locally. After that, a watermark containing the authentication information and encryption key of the video data is embedded into the video signal. The watermark is designed to be robust and difficult for the attacker to remove. The video data are then encrypted using our early proposed video selective encryption scheme (Yin, Lin, Qiu, Li, & Tan, 2005), which is implemented to be compatible with the hardware encryption engine supported by IXP425, so that the whole video processing can be performed in real-time.

The camera nodes are organized as a wireless ad hoc network in which every node also functions as a router to relay the video data from other nodes. Encrypted frames are finally routed to the monitoring center, traveling through a series of camera nodes. Usually the physical locations of all camera

Figure 2. Architecture of secure wireless video systems

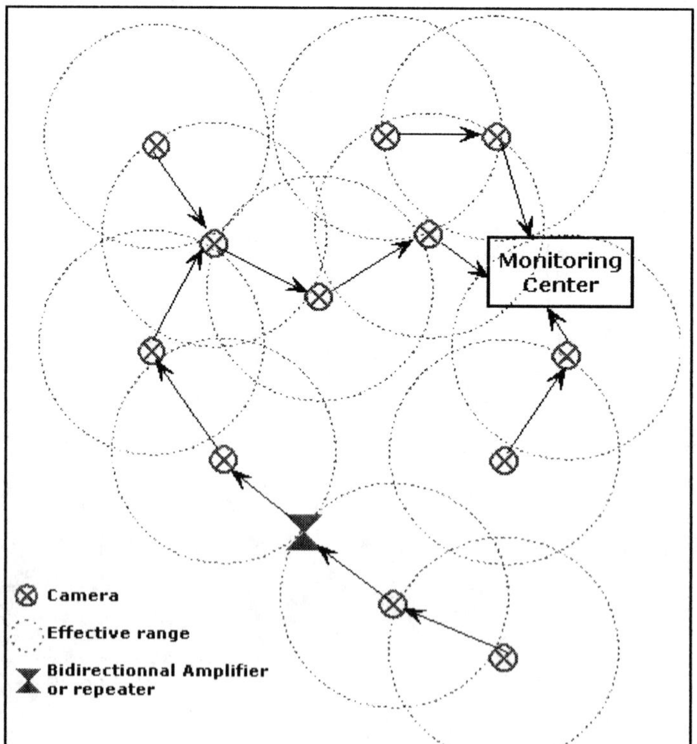

nodes are fixed. It is possible that some nodes are out of range from the main network. In this case, some bidirectional signal repeaters or amplifiers could be placed in strategic points to provide robust coverage for every node (Yin et al., 2005).

Video Processing

For the sake of security concerns, encryption keys are updated periodically. Thus, a core part of the system is how to embed encryption keys in the video data stream. Unlike normal watermarking techniques, in our system the camera nodes should not only detect the existence of a new encryption key but also extract it without losing any information, as shown in Figure 3.

Our key embedding algorithm focuses on the reliability and accuracy of embedded keys against the influences introduced by transmission errors and adaptive mechanisms. Real-time processing is also a requirement when designing the algorithm. The key embedding process can be divided into two parts: key embedding and key detecting. The first part is conducted by the video encoder, while the second part is conducted by the video decoder. New keys are embedded in the I-frames (Intra frame) of a group of pictures (GOP) and directly modulated into the direct current (DC) component of DCT coefficients of luminance blocks. In our algorithm, 200 bits of data can be embedded in one I-frame of size 640x480. Moreover, the key takes only 128 bits among the data.

In order to improve the error resilience capability, we employ Reed-Solomon (RS) code as the error control scheme. RS code is used to encode the key messages and an 8-bit flag and this takes the remainder 64 bits for RS code words. Then, all the GOPs are encrypted by the selective encryption algorithm, which contains some hash functions supported by IXP425 hardware, corresponding to the old key. Finally, the encrypted data are sent out to a neighbor node via the wireless network. After the data are received, the incoming packets are first decrypted and then decoded. When the embedded key messages are detected, the new key is used for future data decryption. If a GOP is badly damaged, the embedded key messages may not be extracted correctly. Therefore, we use more than one GOP to embed the rekey messages for redundant recovery (Yin et al., 2005). Figure 4 gives an example of three GOPs containing the redundant key messages.

Figure 3. Real-time key embedding and key detecting process, K_i is the 128-bit key information used to encrypt the video content

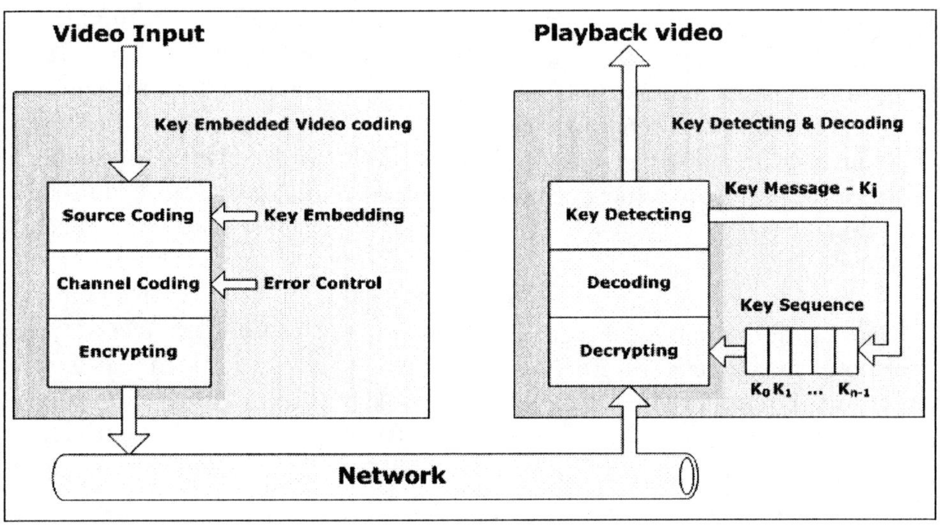

Figure 4. The redundant GOPs used in key updating process. From GOP_{i+1} to GOP_{i+3} there are three GOPs that contain the redundant key messages

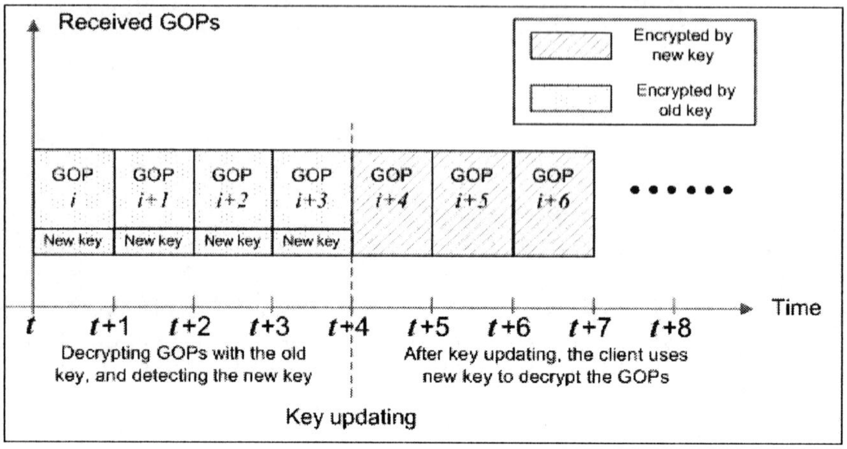

System Security Management

To develop a secure wireless video surveillance system, it is necessary to develop an effective video encryption algorithm, and meanwhile the secure routing protocol and system architecture should all be carefully designed to avoid serious security flaws.

Confidentiality

Data confidentiality is usually assured by encryption. However, encryption introduces large computational overhead. In stringent environment like real-time video transmission, encryption can become the system bottleneck and it is the common knowledge that full video stream encryption is not a good choice (Liu & Eskicioglu, 2003). Our video selective encryption algorithm takes advantage of the properties of monitored video to achieve secure, real-time encryption.

If the routing messages are not protected, eavesdroppers may discover the network topology by listening to the routing information and then attack the most active notes in the network. Topology information disclosure is not a threat by itself, but it can make other attacks more efficient. However, encrypting routing information could greatly increase the overhead. The basic AODV protocol has built in capabilities for extension headers. The secure ad hoc distance vector (SAODV) protocol is a proposal by Zapata (2005) for such extension headers. The extensions are used to send signatures and hash values that are later used for verification of the routing packets. The SAODV is not meant to yield any confidentiality since this is usually not needed or desired in general ad hoc networks. The protocol does provide means to get authentication, integrity, and nonrepudiation of the routing control packets. The protocol extensions use asymmetric cryptography to achieve authentication by signing the data packets with the private key. This allows for the destination node and all intermediate nodes to validate the request. Also, this allows for the nodes to be certain that no one has altered the packets. However, some fields of the packets must change and these are signed as if they were zeroed out. To allow for verification of the hop count field, a one-way hash chain is utilized. The initiator of the route request decides a max hop count, such as 10 or 15. It also generates a random value which is sent as the hash for the first hop count. The value is also hashed the max hop number of times producing a so-called top hash. Each node can verify the hop count by checking that the incoming hash value hashed max hop count minus the current hop count number of times is equal to the top hash. Since the top hash value is not changed, and thus signed, this provides the means to authenticate even the mutable hop count.

The SAODV extensions allow for two different ways for nodes to reply to a route request. The first way is to only allow the destination to reply. In this way the protocol works as described above. The destination node creates a route reply and signs it using its own secret key. The route reply is sent according to the usual AODV and each intermediate node can verify the reply and discard it if not valid. This approach does not consider the possibility of having intermediate nodes reply directly if they do have a valid route already. To add the ability for route discovery optimization a double signature scheme is devised. For each route request a second signature is added to the packet. This signature is stored in each intermediate node when they set up the reverse route. Later on, when a new route is needed because of node movement between the two peers an intermediate node that still has a route can reply directly by also including the second signature and the original signature (Yin et al., 2005). In addition to this, the actual life time is also sent in the reply which is signed by the intermediate node that sends the reply.

Authentication

The host-to-host authentication between the camera and the monitoring center is achieved by data encryption. But in ad hoc networks, we also have to consider the problem of neighbors' authentication, as nodes are "observing" the external world though the "eyes" of its neighbors. The neighbors must be authenticated before any other communication can be initiated. In a nonauthenticated environment, external nodes can insert themselves in the data path and then collect, disrupt, or corrupt the information using man-in-the-middle or black and grey holes attacks. To reduce the effect of computational power consumption attack, the authentication scheme is performed at link layer. Neighbors' authentication is assured by a certificated-based approach (Stallings, 1999) which provides practical solutions for data integrity, authentication, and nonrepudiation. The practical protocol is presented by Luo, Zerfos, Kong, Lu, and Zhang (2002).

Reactive Protection Scheme

The ad hoc environment is usually considered as physically insecure. For instance, cameras can easily be stolen or corrupted. A corrupted camera node can be used as a Byzantine enemy (Lamport, Shostak, & Pease, 1995) to attack the rest of the network. However, resources in the ad hoc network are limited due to the embedded nature of the nodes; especially computational power is the system bottleneck. In this situation, signing every packet between every node is not realistic for real-time multimedia streaming. Besides, if a malicious entity controls a node, it also controls the authentication keys, and systematic authentication is not useful against this type of attack.

In our system only routing protocol messages are systematically signed and time-stampeded to avoid basic attacks such as erroneous routing packet flooding. To prevent more subtle attacks like grey hole or session hijacking, we use the existing knowledge about the data stream (e.g., continuity, stability, fixed length, etc.) to detect misbehavior in the trusted network. Nodes which have detected misbehaving peers break the routing roads coming from the suspected nodes so that further traffic is ignored until a new (authenticated) road request is broadcasted. The level of intrusion detection capability depends on the computational power. The system would have a misbehaving threshold beyond which the system will cut itself from the rest of the network. The level of the threshold and the way to isolate the node from the network is worth further investigation.

Key Distribution

The key distribution solution proposed by Luo et al. (2002) has been chosen to safely distribute and refresh encryption keys and periodically check integrity of the camera nodes. This protocol is based on the threshold share secret revealed by Shamir (1979) and improves the shares refreshing proposed by Herzberg, Jarecki, Krawczyk, and Yung (1995). The system is based on RSA public key signatures. Each node gets a simple certificate in the form $< v_i, pk_i, T_{sign}, T_{expire} >$ where v_i is the identification number of the nodes, pk_i is the public key, T_{sign} is the time that the certificate is created, and T_{expire} is the certificate expiration time.

SYSTEM PERFORMANCE EVALUATION

This session will test the performance of the system we designed and meanwhile introduce one approach to evaluate such system, which may be applied to general wireless video surveillance systems.

Testing Environment

The testing procedure involves three steps. First we evaluate the video encoding and encryption algorithms, along with the basic network stack evaluation on a single link. In the second step, we measure the performance of a node for transmitting traffic to other nodes. The third step is a simulation study of a large scale network in order to analyze how the system evolves when the number of cameras increases.

- **Single node capability:** The testbed is composed of an IXP425 network processor and its evaluation board. The network interface of the camera node is a wireless 802.11g compatible network interface. A desktop computer equipped with the same network interface is used to stand for the monitoring center and to test the video decryption and playback.
- **Routing capability:** A set of low cost computers is equipped with wireless network interface and generates traffic towards the tested node. Different physical dispositions are set to test one-hop and multihops routing performance.
- **Scalability:** Large scale experiments are very challenging because they require too many hardware. We plan to use the results obtained from Steps 1 and 2 to build a realistic model of the node and simulate a large scale system.
- **Nodal processor:** Intel IXP425 network processor is chosen as the nodal processor of our system. Intel IXP425 is a member of Intel's IXP4XX product line of network processor, for small-to-medium enterprise, consumer, and other edge network applications. Like Intel's high-end network processors IXP2k series, IXP425 is also a multicore system that employs system-on-chip (SoC) techniques to support multiple WAN and LAN technologies in a highly integrated and versatile architecture. The Intel XScale core at up to 533 MHz provides headroom for customer-defined applications. It also supports a single-instruction stream multiple-data stream (SIMD) coprocessor for multimedia application acceleration. In our system, video encoder and watermark embedding are performed on XScale with optimization towards the SIMD coprocessor. Three network processor engines (NPEs), like a micro-engine of IXP1k, 2k network processors, are designed to complement the Intel XScale core for many computationally intensive data plane operations. These tasks include IP header inspection and modification, packet filtering, packet error checking, checksum computation, and flag insertion and removal. The NPE architecture includes an ALU, self-contained internal data memory, and an extensive list of I/O interfaces, together with hardware acceleration elements. The hardware acceleration elements associated with an NPE targets a set of networking applications. Each hardware acceleration element is designed to increase the speed of a specific networking task that would otherwise take many MIPS to complete by a standalone RISC processor. Among these functions, cryptographic hardware accelerators (SHA-1, MD5, DES, 3DES AES) in NPEB are used in our application for selected video encryption.

Experiments on Key Embedding Algorithm

This subsection focuses on the performance evaluation of the key embedding algorithm in a wireless environment. The algorithm is implemented on the platform of Intel IXP425 network processor.

We use two MPEG-2 test sequences, dinosaur and live-captured video, which are both encoded at 640x480 size and 20fps using 500 frames. The sequences are selected because of their different characteristic in motion and scene change. Dinosaur contains fast motion and scene change, while live-captured video contains slow motion and fixed scene. Besides, we should face the challenge derived from packet loss and bit error. We test the system in a real wireless network environment. The last module is a key detecting and decoding module, which contains selective encryption algorithm, MPEG-4 decoder, and the key embedding algorithm. They are used to decrypt the bitstream using old session key, and then detect the embedded key messages and decode the compressed video into playback video.

Based on this platform, we conduct a series of experiments to evaluate the system performance (Yin et al., 2005). The source-coding distortion introduced by our key embedding algorithm is illustrated in Figure 5. The video clip is MPEG-2 encoded with different modulation cycle. It is then transcoded and decoded by MPEG-4 decoder. All the four pictures are selected from the playback video. Obviously, the modulation cycle is the most important factor that affects the quality of the video sequence. When the modulation cycle is no larger than 4, the distortion derived from key embedding can be neglected. Figure 6 illustrates the PSNR of the dinosaur sequence at the receiver side. It is worth noting that that the larger modulation cycle can degrade not only the PSNR, but also introduce PSNR fluctuation, while modulation cycle less than 4 can provide a good quality of video.

Figure 7 illustrates the number of error bits found in the detection of all the 200 bits in a frame against the modulation cycle C. The downscaling in the transcoder reduces half of the width and height of the original video. This procedure reduces the blocks in each field, but does not have too much impact on the detection quality. However, it can be seen that the requantization greatly impacts the detection quality when the modulation cycle is less than 3. As shown in Figure 7, when the quantizer in the requantization (denoted by "new quantizer" in the figure) is higher and the quantizer in the source encoding (denoted by "old quantizer" in this figure) is closer to half of "new quantizer," more error bits appear in the detection procedure. When the modulation cycle is more than 4, errors have almost disappeared.

Figure 5. The effect of security management on video

Figure 6. The PSNR of frames and the probability of successfully detecting 200 bits in a frame changed with the modulation cycle at the receiver

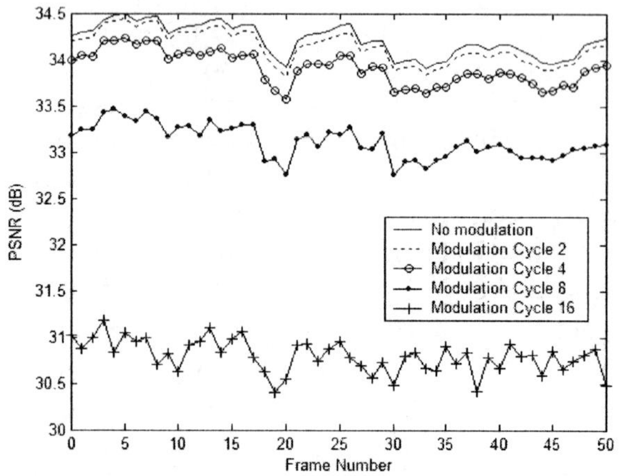

Figure 7. Average error bits in total 200bits embedded in an I-frame after transcoding with different modulation cycles and quantizers

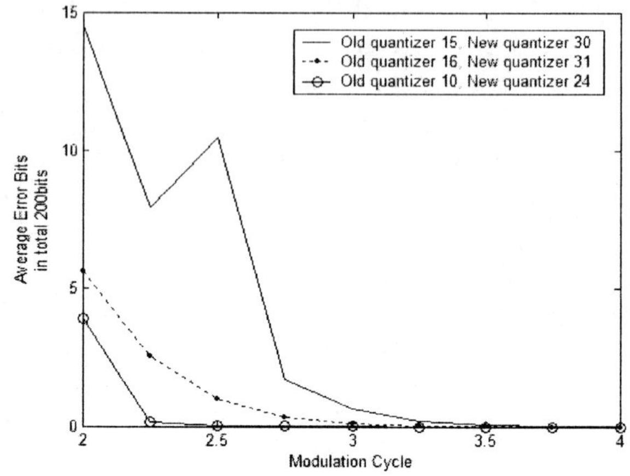

Figure 8. Average error bits in total 200bits of a GOP by using different packet loss rates, (a) RS code is not used, while (b) RS (25, 17) code is used

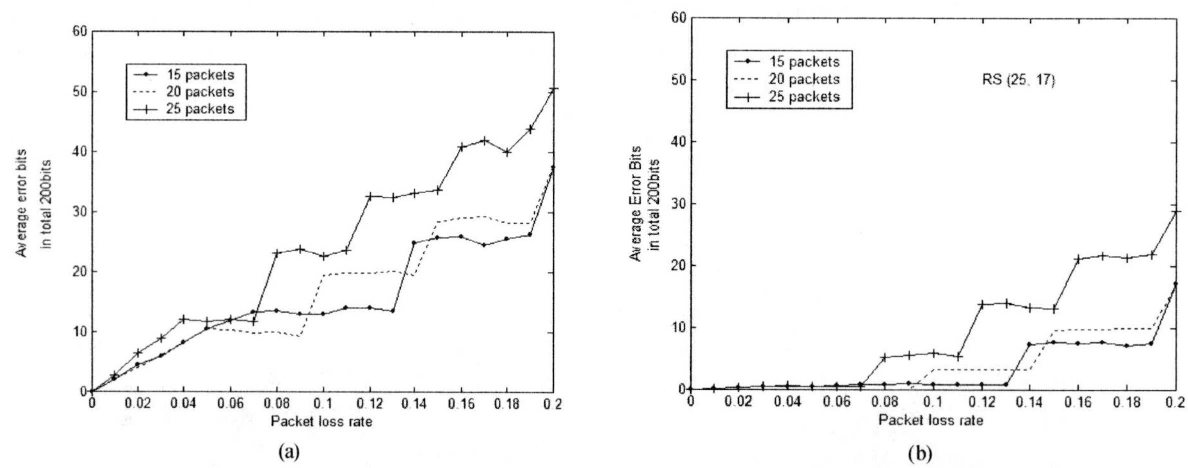

Figure 8 reveals the average error bits when receiving 200 bits data vs. the packet loss rate in the network. It can be seen that the extracted data error rate in a GOP rises as the packet loss rate increases. Usually the bitstream of an I-frame is divided into more than 10 packets for transmission in the network. As a result, key information is distributed into all the packets and the loss of packets leads to some error bits of the extracted key message.

As for the coding speed, Table 1 shows the coding time between the key embedded coding scheme and pure MPEG-2 encoder without data embedding. We can find that after the introduction of the key embedding algorithm, the processing time is only increased by around 6%.

Simulation of Routing Protocols

Preliminary simulations on AODV have been conducted in order to validate the choice of the routing protocol. The objective here is to have a qualitative evaluation of the routing protocol. Simulations have been conducted using NS2 simulator.

The arrows in Figure 9(a) represent the stream paths and we can see that the nodes are choosing the shortest paths to reach the monitoring center in order to reduce the number of hops per path in comparison with an architectural network.

Figure 9(b) reveals the volume of traffic received by each node in the scenario where a few cameras are placed at difference floors in a building and the distance between the nodes are greatly exaggerated (we do not consider the effect of reverberation against obstacles here). The figure shows that the monitoring center is the bottleneck of the architecture. This is inevitable in a monitoring system where all the streams are converging in one point. However, this phenomenon implies that the overall capacity is limited by the performance of the monitoring center.

Table 1. Complexity of the key embedding algorithm

Sequence	*Dinosaur*	*Live-captured*
Encoding speed without embedding (frame/sec)	35.45	37.27
Encoding speed with embedding (frame/sec)	33.50	35.00
Increased processing time (%)	5.8%	6.5%

Figure 9(a). Topology of a small monitoring system

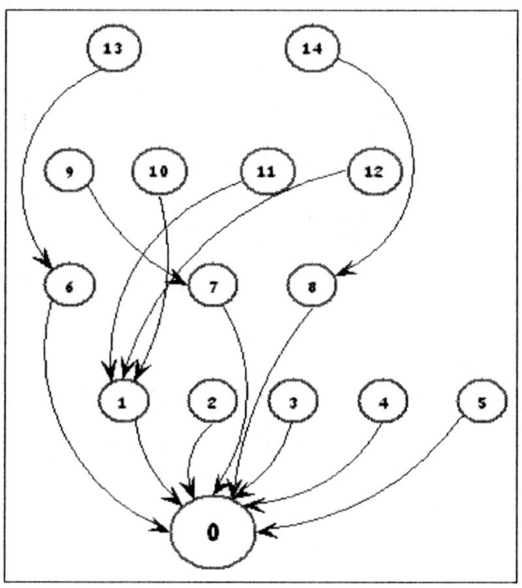

FUTURE TRENDS

With the continuing need for video surveillance in both fixed and remote locations, new advances in wireless networking would enable the development of a more secure, highly reliable wireless video network capable of supporting real-time high speed, high resolution video, and meanwhile maintaining the highest levels of data and network security without impacting the video stream. Technical trends and key issues in the wireless video system may include:

- **Load balanced routing protocols:** One problem of the routing protocol is that it is not reactive to the load in each node. Under the particular topology, if a node has a more critical location than others, a large portion of the traffic may converge toward the node and it may probably collapse under the heavy traffic. It would be more desirable for an ad hoc network that the routing protocol

Figure 9(b). Bandwidth of nodes in an ad hoc network under converging multimedia traffic

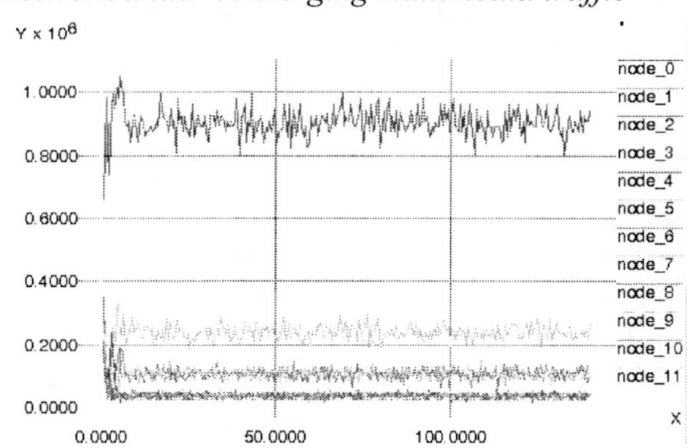

fairly distributes the traffic load among the nodes.

- **Local misbehavior detection system:** The system also needs to detect misbehaving neighbors. Only a few recent studies (e.g., Kargl, Klenk, Schlott, & Weber, 2005; Marti, Giuli, Lai, & Baker, 2000) have been conducted and reported in this field. Besides, in our case, misbehavior detection capability is limited by the computational power of the nodes. We hope to find an adaptive mechanism to suit our applications.
- **Scalability:** As demonstrated by the simulation results of our network layer, the monitoring center, as the only nondistributed component, is the bottleneck of the system. Some solutions must be found to scale the network size as far as possible.

CONCLUSION

A distributed video surveillance system typically consists of many video sources distributed over a wide area, transmitting live video streams to a central location for processing and monitoring. However, in the traditional wire-line solution, the deployment and maintenance of large-scale video surveillance system are often expensive and time-consuming. Thus there have been hot interests in wireless solution. But the practical implementation of wireless surveillance system still faces the challenges of framework design of wireless network, video processing, video data transmission, video quality control, and system security. Among them, the system security is the most challenging problem and also is the main concern in this chapter.

This chapter has presented the state-of-the-art cross domains of wireless communication, video processing, embedded systems, and security, through the design of a new secure video surveillance system. This system is based on the 802.11g ad hoc wireless infrastructure. Intel IXP425 network processors are used as the basic processing unit. A media-dependent video encryption scheme, including reliable data embedding technique and real-time video encryption algorithm, has been proposed and implemented to enable the system to work in an open and insecure wireless environment. The presented system offers several unique advantages: (1) it provides high security guarantee; (2) it does not require expensive access points/routers; (3) it can be readily deployed since it is built upon the existing wireless ad hoc infrastructure; and (4) it is robust in the presence of and adaptive mechanism and error-prone channel. This chapter would serve as a good reference for solving the issues of wireless multimedia and would bring new insights on the interaction of different technologies within the cross application domain.

ACKNOWLEDGMENT

This work was supported in part by grants from the National Natural Science Foundation of China (No.60673184, No. 60432030, No.60429202, No.90412012), national 863 program of China (No. 2007AA01Z419) and Microsoft Joint lab funding.

REFERENCES

Alattar, A.M., Lin, E.T., & Celik, M.U. (2003). Digital watermarking of low bit-rate advanced simple profile MPEG-4 compressed video. *IEEE Transaction on Circuits and Systems for Video Technology, 13*(8), 787-800.

Allman, S. (2002). Encryption and security: The advanced encryption standard. *EDN* (pp. 26-30). Retrieved October 7, 2007, from http://www.edn.com/article/CA253789.html?ref=nbsa

Basgall. (1999). *Experimental break-ins reveal vulnerability in Internet, Unix computer security.* Retrieved October 7, 2007, from http://www.cs.duke.edu/news/index.php?article=16

Bender, W., Gruhl, D., Morimoto, & Lu, A. (1996). Techniques for data hiding. *IBM System Journal, 38*(3-4), 313-316.

Borisov, N., et al. (2003). Intercepting mobile communications: The insecurity of 802.11. In *Proceedings of MOBICOM 2001* (pp. 180-189).

Broch, J.D., Maltz, A., Johnson, D.B., Hu, Y.-C., &. Jetcheva, J. (1998). A performance comparison of multi-hop wireless ad-hoc network routing protocols. *Mobile Computing and Networking*, 85-97.

Chen, B., & Wornell, G.W. (2001). Quantization index modulation: A class of provably good methods for digital watermarking and information embedding. *IEEE Transaction on Information Theory, 47*(4), 1423-1443.

Desilva, S., & Das, S.R. (2000). Experimental evaluation of a wireless ad hoc network. In *Proceedings of the 9th International Conference on Comp. Comm. & Networks* (pp. 528-534).

Garcia-Macias, J. A., et al. (2003). Quality of service and mobility for the wireless Internet. *ACM Wireless Networks, 9,* 341-352.

Haas, Z. J. (1999). The Performance of the zone routing protocol in reconfigurable wireless networks. *Special Issue on Wireless Ad Hoc Network, IEEE Journal on Selected Areas in Communications, 17*(8).

Herzberg, A., Jarecki, S., Krawczyk, H., & Yung, M. (1995). Proactive secret sharing or: How to cope with perpetual leakage. *Lecture Notes in Computer Science, 963,* 339.

IEEE. (2003). *802.11g IEEE Std 2003.* Retrieved October 7, 2007, from http://grouper.ieee.org/groups/802/11/

Intel. (2006). Intel® IXP425 network processor. *Intel product brief.* Retrieved October 7, 2007, from http://www.intel.com/design/network/products/npfamily/ixp425.htm

Johansson, P., Larsson, T., Hedman, N., Mielczarek, B., & Degermark, M. (1999). Scenario-based performance analysis of routing protocols for mobile ad-hoc networks. In *Proceedings of ACM Mobicom'99* (pp. 195-206).

Johnson, D.B., Maltz, D.A., & Broch, J. (2001). DSR the dynamic source routing protocol for multihop wireless ad hoc networks. *Ad hoc networking* (pp. 139-172). Addison-Wesley.

Kargl, F., Klenk, A., Schlott, S., & Weber, M. (2005). *Advanced detection of selfish or malicious nodes in ad hoc networks.* Paper presented at the First European Workshop on Security in Ad-hoc and Sensor Networks (LNCS 3313, pp. 152-165).

Lamport, L., Shostak, R., & Pease, M. (1982). The Byzantine generals problem. *ACM Transactions on Programming Languages and Systems, 4*(3), 382-401.

Liu, X., & Eskicioglu, A. (2003, November 17-19). *Selective encryption of multimedia content in distributed networks: Challenges and new directions.* Paper presented at the IASTED International Conference on Communications, Internet and Information Technology (CIIT 2003), Scottsdale, AZ.

Luo, H., Zerfos, P., Kong, J., Lu, S., & Zhang, L. (2002). Self-securing ad hoc wireless networks. In *Proceedings of the Seventh IEEE Symposium on Computers and Communications (ISCC '02)* (pp. 567-574).

Marti, S., Giuli, T.J., Lai, K., & Baker, M. (2000). Mitigating routing misbehavior in mobile ad hoc networks. In *Proceedings of International Conference on Mobile Computing and Networking* (pp. 255-265).

Perkins, C.E., & Royer, E.M. (1999). Ad-hoc on-demand distance vector routing. In *Proceedings of the 2nd IEEE Workshop on Mobile Computing Systems and Applications, New Orleans,* (pp. 90-100).

Perlman, R. (1992). *Interconnections: Bridges and routers.* Reading, MA: Addison-Wesley.

Plummer, D.C. (1982, November). *An Ethernet address resolution protocol: Or converting network protocol addresses to 48.bit Ethernet hardware* (RFC 826).

Samir, R.D., Perkins, C.E., & Royer, E.E. (2000). Performance comparison of two on-demand routing protocols for ad hoc networks. In *Proceedings of IEEE INFOCOM* (pp. 3-12).

Shamir, A. (1979). How to share a secret. *Communication of the ACM, 22*(11), 612-613.

Stallings, W. (1999). *Network security essentials: Applications and standards* (1st ed.). Prenctice Hall.

Wentink, M. (2003). Overcoming IEEE 802.11g's interoperability hurdles. *Communication Systems Design, 19-23.*

Yin, H., Lin, C., Sebastien, B., & Chu, X.W. (2005, October 13). A novel secure wireless video surveillance system based on Intel IXP425 network processor. In *Proceedings of the 1st ACM Workshop on Wireless Multimedia Networking and Performance Modeling (WMuNeP '05),* Montreal, Canada.

Yin, H., Lin, C., Qiu, F., Li, B., & Tan, Z. (2005). A media-dependent secure multicast protocol for adaptive video applications. In *Proceedings of the SIGCOMM Asia Workshop.*

Yin, H., Lin, C., Qiu,F., Min, G., & Chu, X. (in press). A novel key-embedded scheme for secure video multicast systems. *International Journal of Computers & Electrical Engineering* (No. 04-NM-904).

Zapata, M.G. (2005). *Ad hoc on-demand distance vector (SAODV) routing.* Retrieved October 7, 2007, from INTERNET-DRAFT draft-guerrero-manet-saodv-03.txt.

Zhou, L.D. (1999). Securing ad hoc networks. *IEEE Network, Special Issue on Network Security, 13*(6), 24-30.

KEY TERMS

Ad Hoc Network: A local area network created for a specific purpose and established for a single session and does not require a router or a wireless base station. Specially, a wireless ad hoc network is a self-organized computer network with wireless communication links.

Discrete Cosine Transform (DCT): A Fourier-related transform algorithm that is widely used for data compression. DCT converts data (pixels, waveforms, etc.) into sets of frequencies and expresses a function or a signal in terms of a sum of sinusoids with different frequencies and amplitudes. It is often used in signal and image processing, especially for lossy data compression.

GOP: The group of pictures (GOP) is a group of successive pictures within a MPEG-coded film or video stream. A GOP consists of all the pictures in successive two GOP headers.

Network Processor: An integrated circuit that is optimized for networking and communications functions, typically programmable CPU chip.

Scalability: A property of a system, a network, or a process that can be modified to fit the problem area, that is, scaled to perform well with large-scale users.

Surveillance System: A closed-circuit television system used to monitor something.

Watermarking (Digital Watermark): A technique used to add hidden copyright notices or other verification messages in a digital signal or video so that it cannot be detected by a standard playback device or viewer.

Wireless Network: A telecommunications network whose interconnections between nodes use standard protocol, but without the use of network cabling.

Chapter XXXIII
Cutting the Gordian Knot: Intrusion Detection Systems in Ad Hoc Networks

John Felix Charles Joseph
Nanyang Technological University, Singapore

Amitabha Das
Nanyang Technological University, Singapore

Boon-Chong Seet
Auckland Univerisity of Technology, New Zealand

Bu-Sung Lee
Nanyang Technological University, Singapore

ABSTRACT

Intrusion detection in ad hoc networks is a challenge because of the inherent characteristics of these networks, such as, the absence of centralized nodes, the lack of infrastructure, and so forth. Furthermore, in addition to application-based attacks, ad hoc networks are prone to attacks targeting routing protocols. Issues in intrusion detection in ad hoc networks are addressed by numerous research proposals in literature. In this chapter, we first enumerate the properties of ad hoc networks which hinder intrusion detection systems. After that, significant intrusion detection system (IDS) architectures and methodologies proposed in the literature are elucidated. Strengths and weaknesses of these works are studied and are explained. Finally, the future directions which will lead to the successful deployment of intrusion detection in ad hoc networks are discussed.

INTRODUCTION

Wireless ad hoc networks have attracted extensive attention among researchers in recent years. As the research activities matured, it has been widely realized that security in such networks is a major issue, and an extremely challenging one. The challenge arises mainly from the inherent characteristics of ad hoc networks. Chief among the characteristics, which affect the design of an effective security framework for such networks, are the highly distributed, decentralized, and dynamic natures of ad hoc networks. These properties, coupled with the lack of infrastructure in ad hoc networks, introduce some unprecedented issues, which are absent and never been explored in conventional networks.

A typical security system consists of two major components. The first is the intrusion prevention mechanism that aims to control access to the system and relies mainly on cryptographic techniques. The second one is the intrusion detection system that tries to detect if the prevention mechanism has been compromised by intruders, and if so, come up with an appropriate response to combat such intrusions. The intrusion detection system (IDS) thus forms the second line of defense (Nadkarni & Mishra, 2003).

Cryptographic techniques rely on secure key management and key distribution which require supporting infrastructure. The lack of infrastructure makes it extremely difficult to implement cryptographic access control mechanisms in ad hoc networks. This makes intrusion detection all the more important for such networks. However, it turns out that the inherent characteristics of ad hoc networks render conventional IDS unsuitable for such networks. This has spawned the research in ad hoc IDS design (Brutch & Ko, 2003).

This chapter illustrates the difficulties in providing an efficient intrusion detection system for ad hoc networks. In doing so, it discusses in detail interesting ad hoc IDS models proposed in literature. The strengths and weaknesses of these models are explained and promising future directions for cutting the Gordian knot of ad hoc IDS are discussed.

BACKGROUND

Although various analyses on intrusion detection mechanisms can be seen in the literature, only few qualify as significant. Mishra, Nadkarni, and Patcha (2004) give a detailed overview of various ad hoc IDS architectures and methodologies. They offer an extensive analysis and understanding of IDS in ad hoc networks. A comprehensive comparison between various proposed intrusion detection systems for ad hoc networks are discussed. Selected architectures and detection strategies explained by Mishra et al., which were found significant, are detailed in this writing.

Zhang, Huang, and Lee (2005) propose an evaluation environment for MANET (mobile ad hoc network) intrusion detection systems. They emulated routing attacks and evaluated application-based intrusion detection architectures over it. The work introduces a novel concept of evaluating ad hoc IDS models using known attacks. Routing attack libraries are used, which exhibit attack scenarios over the IDS model under-evaluation. The IDS models are evaluated for operational cost and effectiveness. Detection accuracy and false alarms are the primary evaluation parameters for assessing of the IDS model, in terms of detection effectiveness. The work is significant in providing a test-bed for ad hoc IDS models. Similarly, Little (2005) proposes a test-bed called TeaLab for ad hoc IDS design.

Concurrent to simulation-based ad hoc test-beds, Yang and Baras (2003) mathematically analyze vulnerabilities in ad hoc networks. The authors provide a great deal of understanding to the attack possibilities in ad hoc domain. Mathematical methods find attacks exhaustively. In this theoretical analysis all possible attacks are hypothesized. This comprehensive vulnerability analysis aids the design of an effective ad hoc IDS design.

CHARACTERISTICS OF AD HOC NETWORKS

Ad hoc networks differ from native wired/wireless networks in various aspects. These unique characteristics of ad hoc networks render typical security systems unsuitable (Awerbuch, Curtmola, Holmer, Rubens, & Nita-Rotaru, 2005; Papadimitratos & Haas, 2002). The fundamental concept of ad hoc networks is to have seamless connectivity without infrastructure or centralized control. The lack of infrastructure and a centralized control node makes it hard for security systems to be implemented. Furthermore, factors such as mobility, physical protection, and so forth affect the design of effective security models for ad hoc systems. These factors are enumerated below.

Lack of Infrastructure

Ad hoc networks do not have a fixed infrastructure. Typically, in conventional networks, the infrastructure provides a secure location for the implementation of critical security mechanisms (Debar, Dacier, & Wespi, 1999). Due to the absence of infrastructure, ad hoc networks do not provide a safe and efficient location to implement the security system. Additionally, operations such as control, maintenance, and other administrative functions have become hard in a distributed and infrastructure-less network. The only and apparent resort is to install these critical modules in end-user nodes. Implementing critical security systems in unreliable end-user nodes pose a real challenge.

Absence of a Central Authority

Conventional network have traffic concentration areas, otherwise called choke points, where security systems can be placed and implemented efficiently. Control nodes are placed in these choke points to monitor and control the network. Absence of centralized authority makes the network monitoring and control a challenging issue for ad hoc networks.

Every node in an ad hoc network has equal responsibility in network functions, such as routing, maintenance, and so forth. This unique characteristic will distribute the control authority to all nodes in the network. Nodes have to rely on other neighbor nodes for routing and data forwarding. In other words, nodes have to trust neighbor end-user nodes for critical functions. As neighbors can be potential attackers, trusting unknown neighbors is precarious to the integrity of security and other critical systems.

The above two issues are the crux of the security concern in the ad hoc network paradigm. The following are additional factors which also affect ad hoc network security design, but to a lesser degree.

Wireless Links

In respect to security, wireless links are the weakest. This is due the omnipresence of wireless channel and ease of physical access to the channel. Attacks such as eaves-dropping, active masquerading, and so forth are more possible in wireless networks than in a wired network. Furthermore, the most notorious of all attacks, the denial-of-service (DoS) attacks, can be achieved easily in wireless networks by jamming the wireless channel or by routing attacks.

Poor Physical Protection

Usually, the nodes in an ad hoc network are mobile and easily accessible physically. This raises concerns of physical protection of these devices. A single compromised node can bring down the entire network due to its prerogatives in the network.

Energy Constraints

Since ad hoc network nodes are mostly mobile and wireless, energy constraints are also a security issue. Typical symmetric encryption algorithms such as 3DES (triple data encryption standard), ADES (advanced encryption standard), and asymmetric encryption algorithms such as RSA (Rivest, Shamir, and Adleman) and its variants incurs high computation which may drain the battery of the mobile node. Additionally, rnergy-targeted attacks such as SDT (sleep deprivation torture), which aims to drain the mobile node's battery, also need consideration while designing ad hoc security system (Jacoby, Marchany, & Davis, 2004).

Unsuitability of Static Configurations

The obvious and immediate security solution for infrastructure-less and decentralized network is to provide static security systems installed in nodes. Ad hoc networks are mostly implemented over

Figure 1. Ad hoc routing insecurity: Route invasion

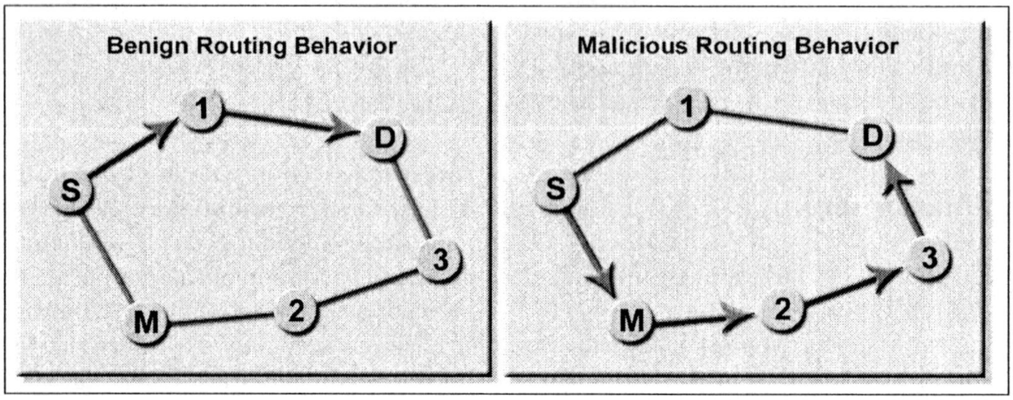

mobile nodes. Mobility introduces transient associations due to the dynamic nature of the network. Therefore, prediction of security configuration in a dynamic ad hoc network may not be possible. Also, static security systems have poor adaptability to new attacks, which renders them inefficient.

Delay Constraints

Security systems are delay-sensitive. Especially in highly dynamic environments, delay guarantees are necessary for the security system to function properly. This necessity arises from the transient associations in the ad hoc network, which is discussed in the succeeding section. However, delay guarantees are hard in dynamic networks because of wireless connectivity and mobility.

Transient Associations

The high mobility of nodes in ad hoc environment makes connections between the nodes transient. Therefore, a node will not be able to get security specific information from its neighbor node permanently. In other words, the time frame for particular information to be valid in the ad hoc network becomes very small because of transient associations.

Routing Security

Routing security is an issue that is unique to ad hoc networks (Papadimitratos & Haas, 2002). Conventional networks have security systems implemented in IP, transport, or application layer. Only ad hoc networks need security at the routing layer or protocol. The need arises from the nature of ad hoc network technology where every node can function as a router. Apparently, this has raised new challenges and issues, since securing a routing protocol has never been an issue for security system designers for legacy networks.

Any node in an ad hoc network can add/modify/delete routes. This functionality is the root cause of the vulnerability of ad hoc routing protocols. A malicious node can send malicious routing control messages to its neighbors. Since ad hoc networks are highly distributed, decentralized, and dynamic systems, preventing or detecting a malicious routing message becomes difficult. Moreover, semantically distinguishing between malicious and benign routing messages is infeasible. Routing insecurity introduces new attack possibilities. Active attacks such as route invasion, and route disruption, cause active damage to the network routing functions (Awerbuch et al., 2005). Route invasion and disruption attacks aim to modify, add, or delete benign routes by sending malicious routing information over the network. Passive attacks such as route monitoring and so forth try to eavesdrop for stealing sensitive information (Kong, Hong, & Gerla, 2003). To illustrate some of the difficulties and to familiarize routing insecurity in ad hoc networks, a trivial attack scenario is considered.

Let us examine route invasion, which is a trivial but destructive attack. In Figure 1(a), the benign route between S and D is through 1. In Figure 1(b), Node M sends a malicious routing control message, stating that it has a better route to D than through Node 1. This modifies the path for S → D from S → 1 → D to S → M → 2 → 3 → D. The modified path is not only inefficient; it includes the malicious Node M into the path. This extends the attack possibilities for the malicious node M on node A or B. To thwart intrusion detection, Node M can impersonate Node 1 and can provide falsified routing information which supports its cause.

Due to the absence of centralized authority and infrastructure, Node S has no trusted arbiter to get advice regarding whether the announced path is benign or otherwise. Malicious Node M has free access to the wireless channel and can exhibit anonymous routing attacks over S.

Static crypto systems fail here, due to poor physical protection, energy, and delay constraints. In the absence of centralized authority, dynamic crypto systems are not possible. Critical security systems such as key management, admission/access control, and authentication become hard to implement due to the lack of infrastructure. Analogous to IP spoofing, ad hoc routing protocols are prone to spoofing. However, unlike IP, spoofing in ad hoc networks is done at the routing protocol rather than the IP. Generically, ad hoc security needs to prevent or detect spoofing. However, the issue is more serious than in IP, since the target of the attack is the routing protocol itself.

Mobility and transient associations and dynamicity make the detection of malicious routing control messages impractical. In the above example, Node S will not be able to determine with its local knowledge whether Node M is on a shortest route to D or acting maliciously. Because, even if Node M is not on a shortest/optimal path to Node D now, due to changing topology, that may change at a point of time in the future. In other words, a malicious behavior highly resembles another benign behavior. Therefore, intrusion detection becomes very challenging.

INTRUSTION DETECTION TECHNIQUES

Intrusion detection systems are mechanisms which provide a "second wall of defense" (Nadkarni & Mishra, 2003) for the network system. In other words, IDS is a backup, in case the frontline security mechanisms fail. Therefore, IDS fundamentally assumes that cryptographic systems do not prevail or have failed. As mentioned earlier, IDS in ad hoc networks cannot trust information from other nodes. This limits the knowledge sharing between the nodes. Knowledge in IDS is the new benign/malicious behavior patterns. Typical systems use an arbiter (centralized) node to facilitate knowledge sharing. However, the absence of any centralized node in ad hoc networks renders knowledge sharing unreliable. Unreliable information in a security system is worth no information at all.

Conventional IDS are functional in application layer and monitor and detect malicious behavior exhibited by applications, such as, telnet, FTP, SMTP, and so forth. In rare cases, relatively simple IDS, such as firewalls are implemented in the IP layer. However, ad hoc networks' necessity for routing security has brought forth the need to implement IDS, which monitors and detects routing protocols, such as AODV, OLSR, DSR, and so forth. An IDS design for a routing protocol is an unexplored area of research. The requirements of IDS for a routing protocol differ vastly from the conventional IDS mechanisms.

Research in ad hoc IDS design is still in the rudimentary stages. Some research works (Hijazi & Nasser, 2005) on ad- hoc IDS, which try to cut the Gordian knot, follow strongly the IDS design methodologies of native IDS counterparts. In addition, most of the IDS models proposed in the literature focus on application-level IDS. The assumption that application level IDS for ad hoc network will suffice are the major weakness of these works. Therefore, though these IDS models consider ad hoc network characteristics and provide a decentralized and distributed IDS, they fail to address the routing insecurity.

Zhang, Lee, and Huang (2003) propose a distributed and decentralized IDS system at the routing layer but fail to describe the routing-level IDS model. Their work is similar to other research models on ad hoc IDS design, which provid application-level IDS. Eventually, Huang and Lee (2004) analyze AODV intensively and provid a strong understanding of AODV and a guide to design an AODV IDS at routing layer. However, they fail to state the statistical methodologies used in the IDS design.

In what follows, the existing IDS models are enumerated and its strengths and weaknesses are analyzed. Additionally, the feasibility of implementation of these methods is studied.

A SIMPLE IDS

Before venturing into the realm of ad hoc IDS designs, a short primer on a simple IDS model will be helpful. The fundamental working of an IDS is shown in Figure 2. A typical IDS model consists of three modules: detection module, response module, and audit trails (Athanasiades, Abler, Levine, Owen, & Riley, 2003). Audit trails is a database which stores known normal behavior or anomalous behavior patterns. The detection modules analyze the observed behavior by comparing them with the known behavior patterns in the audit trails' database. There are two types of pattern matching methodologies: misbehavior detection and anomaly detection. Misbehavior detection uses known malicious behavior patterns for comparison at the detection module. Anomaly detection uses known normal behavior patterns and measures the deviation of the node's behavior from the known normal behavior patterns.

The main strength of misbehavior detection is that the probability of false alarm is quite low. However, the probability of deduction is also low, as unknown attacks will skip detection. On the contrary, anomaly detection increases the probability of detection at the cost of increased false alarm rates. Typically, both mechanisms are used in concurrence to define a tradeoff point between probability of detection and false alarm rates.

AD HOC NETWORK IDS REQUIREMENTS

Having seen the fundamental operation of an IDS, this section explores the essentialities and challenges behind an efficient ad hoc IDS. Understanding the difference between an ad hoc IDS and the conventional IDS will help us to appreciate the requirements of an ad hoc IDS. Also, this will aid us to understand the strength and weakness of each proposed IDS models.

The notorious Mitnick IP (Shimomura & Markoff, 1996) spoofing attacks are classic examples which demonstrate the destructive capabilities of routing attacks. Legendary Mitnick used IP address spoofing to attack government and commercial networks by feigning IP address. In ad hoc networks,

Figure 2. A simple intrusion detection architecture

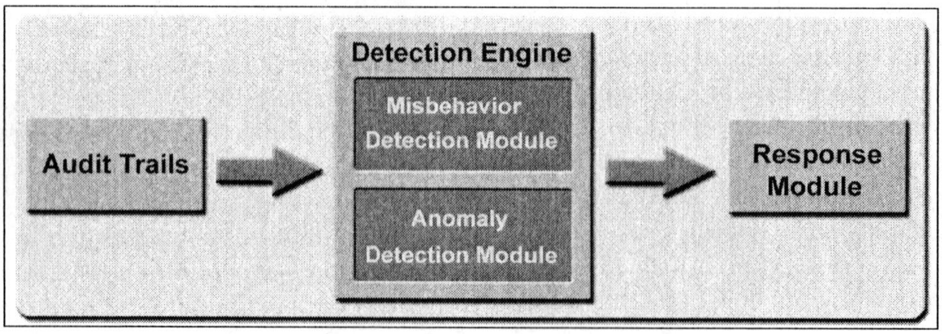

spoofing network address becomes naïve, since an attacker can get hands on the routing protocol itself. The gamut of attack possibilities is infinite (Awerbuch, Curtmola., Holmer., Nita-Rotaru., & Rubens., 2004). In the following sections, we will explore the factors that affect ad hoc IDS.

Knowledge Limitations in Audits

To assess whether a behavior is malicious or benign, a node needs knowledge about different behaviors. It is evident that with more knowledge, efficiency of distinguishing between malicious and benign behaviors increases. In conventional networks, knowledge is shared using a trusted arbitrary node. Absence of a centralized node in ad hoc networks limits the knowledge sharing. Knowledge sharing is precarious in a decentralized and distributed network. A malicious node can cast malicious information which may affect the integrity and security of IDS itself. When a node receives contradictory information from benign and malicious nodes, a decision dilemma occurs. IDS will not be able to decide which information is correct. To avoid this high risk scenario, the nodes can only resort to local knowledge. Local knowledge is information gained through the node's its own experience.

It can be argued that if the number of benign nodes is more than the malicious nodes, knowledge sharing will be reliable. It is true to some extent. According to Byzantine agreement (Lamport, Shostak, & Pease, 1982), for the distributed global knowledge to be reliable, benign nodes should be greater than two-thirds of the total number of nodes. However, ad hoc networks have an interesting attack scenario, which can thwart Byzantine agreement even in the presence of sufficient benign nodes. An attacker using address spoofing can create nonexistent neighbor nodes and can emulate malicious behavior for the nonexistent nodes. This gives the attacker the advantage of controlling the apparent number of malicious nodes in the network, thereby, invalidating Byzantine agreement.

It is important to consider that audit trails from routing protocols differ significantly from audit information from application layer protocols. Audits from application layer protocols record user behavior (Balajinath & Raghavan, 2001), such as login attempts and failures, access rate, and so forth. Whereas, audits from routing protocol record node behavior such as mobility, speed, connectivity, and so forth. The user behavior feature differs distinctly from routing protocol behavior features. Hence, the methodologies analyzing the audit trails have to be revised.

Detection Strategies

Detection methodologies can be classified as rule-based, statistical, and hybrid, which are explained below.

Rule-based detection use static rules to determine maliciousness in behavior. Rules are a set of logical conditions, and when these conditions are met, the behavior is categorized as malicious. Let us consider a simple rule to illustrate. Failure of three or more consecutive login attempts can reasonably be used to decide that the behavior is malicious. More complex rules are formed using typical logical reasoning mechanisms such as expert systems. Static rule-based approaches which are practical in conventional IDS fail due to the dynamic nature of ad hoc networks. The dynamic behavior creates transient connections which makes intrusion detection through static rules almost impossible. Furthermore, static security systems are known to perform inefficiently in dynamic and distributed systems.

Statistical approaches uses probability estimation theory (Duda, Hart, & Stork, 2000) to allow some flexibility to crisp logic and rule-based detection strategies. In statistical approach, probability of behavior being malicious is determined by statistically analyzing the known behavior patterns. However, statistical analysis of routing behavior in ad hoc networks is inconclusive and so is statistical IDS in ad hoc networks. Conventional IDS systems use statistical approach (Verwoerd & Hunt, 2002) after very intensive data analysis (Bykova, Ostermann, & Tjaden, 2001) and are derived using computational intelligence methodologies (Duda, Hart, & Stork, 2000). These analyses are done for audit trails from application layer protocols. The lack of similar research on routing behavior audits for ad hoc networks raises an interesting question on the suitability of statistical approaches for ad hoc IDS. This is another unexplored research area in ad hoc security.

Hybrid detection strategies combine the above two approaches. Hybrid mechanisms are expected to perform better than the two approaches, since they incorporate semantics (rule-based systems) and statistical intelligence. This is in fact supported by conventional IDS models where hybrid systems are usually superior.

In the ad hoc IDS paradigm, these detection methodologies face numerous shortcomings. A major impediment is the lack of features describing a routing behavior. Features are parameters or values describing a behavior. For example, the number of server logins is a feature describing a user behavior over server-client-based application layer protocol. Similarly, delay between two routing requests is an example of a feature describing a routing behavior. Typically, in a user behavior, the number of features can extend from 40-100 or more. On the contrary, a routing control message has very few independent features. The content of a routing message is kept as minimal as possible to increase the routing efficiency. This has decreased the features set describing a routing behavior. Different protocols have different features and the feature set is highly protocol dependent.

Inference

It can be inferred that an ad hoc IDS model requires a complete reconstruction of the current conventional IDS architecture. An IDS which functions with only local knowledge, without a centralized node, adapts to dynamic environments, and efficiently identifies malicious behavior will be a *magnum opus* in the field of ad hoc network security. Additionally, functions such as learning new attacks (part of adaptation) without corrupting the local knowledge base will be beneficial (Hossain, Bridges, & Vaughn, 2003; Pokrajac & Lazarevic, 2004). Learning is itself a dynamic process; therefore learning in a highly dynamic, distributed, decentralized, and insecure environment will be challenging.

Figure 3. LIDS architecture

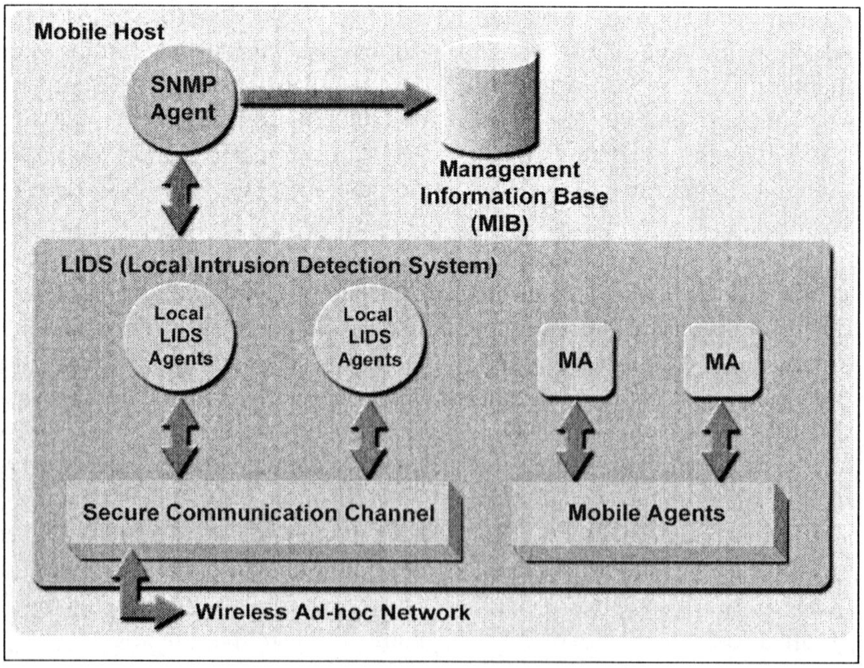

IDS MODELS

Intrusion detection systems have two major components: architectures and methodologies. IDS architectures are the design which depict the overall functioning of the IDS, like the system shown in Figure 2. On the other hand, IDS methodologies are models for detection strategies and their supplementary functions, which are the internal functions of the IDS. Efficiency of the IDS depends both on the architecture and methodologies. Mishra et al. (2004) analyzed various IDS architectures systematically. Succeeding sections discuss about these IDS architectures following Mishra et al.'s (2004) analysis.

Most of the architectures proposed in the literature assume that the methodologies used in conventional IDS model will suffice in an ad hoc environment. In the succeeding sections, the strength of this assumption is analyzed. Furthermore, it has been observed that research work which focuses on IDS architectures does not consider the limitations of IDS methodologies and vice versa.

Architectures

IDS Using Mobile Agents

IDS system is of two types: host-based and network-based. Host-based IDS, as the name implies, runs on individual nodes and functions partially or completely autonomously. On the other hand, in network-based IDS, a centralized node is used to monitor and detect intrusions on the network. It is obvious that network-based IDS is not possible in ad hoc environment because of the absence of a centralized authority.

Mobile agent is a software module, which aids in distributed host-based intrusion detection. The software module traverses through the nodes in the network to accomplish a particular task, such as collecting information, processing information, and so forth. Mobile agents try to emulate network-based intrusion detection by using a collective host-based IDS. The mobile agent provides a good framework to create distributed host-based intrusion detection system. However, mobile agents themselves pose a security threat to the ad hoc network. This is detailed in the following sections by analyzing the IDS architectures that uses mobile agents.

Local Intrusion Detection System Using Mobile Agents

LIDS (Patrick, Olivier, Jean-Marc, Bernard, Ludovic, & Ricardo, 2002) is an application-based IDS architecture for providing intrusion detection in ad hoc network. The IDS architecture is shown in Figure 3, which consists of agents. Agents are host-based intrusion detection modules running on all nodes. The architecture utilizes SNMP (simple network management protocol) to communicate with the neighbors.

A local LIDS agent is responsible for detecting the attacks locally. LIDS agents help neighbor nodes to decide on a suspected intrusion. Also, it receives updates of new attack patterns from the neighbor nodes. The attack patterns are stored in the information base. The MIB (management information base) agent is used to manage the information base. Between the neighbor nodes, SNMP is used to exchange information such as new attack patterns, decisions/responses, and so forth. The MIB agent is responsible for retrieving and sending information to/from neighbors using SNMP. The authors exploit the cooperative nature of ad hoc network by sharing the information about new attack patterns between the nodes.

Additionally, mobile agents are software modules which function autonomously for a dedicated task. For example, the LIDS may designate a mobile agent (MA) to determine the probability of a particular behavioral pattern to be malicious. The MA will autonomously travel between nodes and gather evidence from traversing nodes' MIB.

This approach is relatively naïve. First, the authors assume SNMP is secure in an ad hoc environment. In a network, where routing is insecure, SNMP is not as secure as in a conventional network. Second, as mentioned in the earlier section, knowledge sharing is highly insecure in an ad hoc network. This leads to the insecurity of the LIDS system itself. Compromised nodes can announce misleading intrusion detection information, which will eventually corrupt the information base of the entire network. Finally, in a network with transient associations, feasibility of mobile agents is questionable.

Stationary Secure Database IDS

Andrew (2001) proposes an IDS architecture which consists of a stationary secure database (SSD). Nodes post new information and decisions into this database. The architecture is simple, as shown in Figure 4. Only detection processing is done on the host and the information is stored in a secure stationary centralized point.

The other components of the IDS are typical, namely, misbehavior detection module (MDM), anomaly detection module (ADM), and communication port. These components form the mobile agent. A local intrusion database is also used to store node specific attack patterns and temporary information. The mobile agents will publish the newly found attack pattern to the SSD, only after a certain level of confidence is reached. The communication port is used to communicate with the other nodes' host-based intrusion detection system.

Apparently, it can be seen that stationary secure database (SSD) conflicts with the ad hoc characteristic of the absence of centralized authority. Even if a node is voted as the centralized node using trust mechanisms, there is no surety that the node will behave benignly. Furthermore, a malicious node can corrupt the SSD by sending incorrect intrusion detection information. SSD creates a hot spot, which is a single point of failure. Additionally, SSD assumes cryptographic mechanisms on the communication between the IDS and SSD. This violates the fundamental principle of IDS, which assumes *"no existence of cryptographic mechanisms."*

Modular Intrusion Detection Architecture

Kachirski and Guha (2002) propose an IDS where the intrusion-detection system is modularized into various submodules, as shown in Figure 5. The submodules are network monitoring, host monitoring, decision making, and response (action) modules. The modules are implemented in mobile agent framework. Network monitoring is packet monitoring over the network. Host monitor-

Figure 4. Secure stationary database architecture

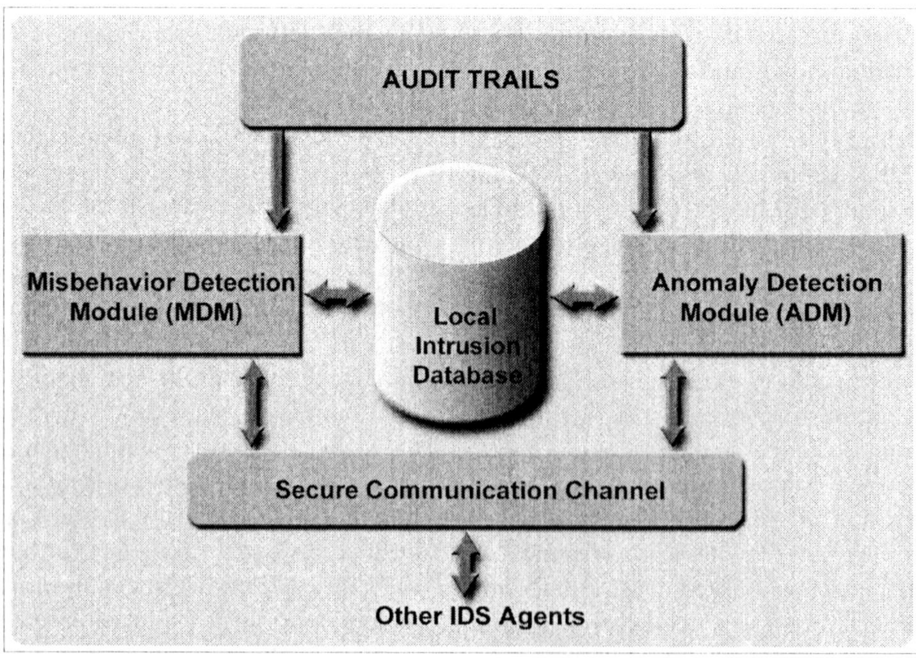

Figure 5. Modularized IDS architecture

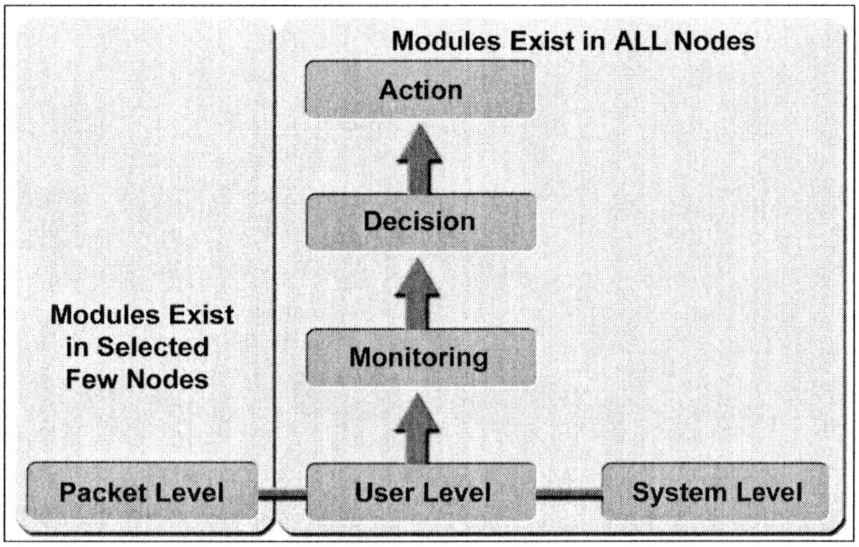

ing is monitoring of user behavior. A host-based monitor module exists in every node; however, network-based monitor exists only in a selected few. Decision making and response modules exist in every node.

The entire ad hoc network is segregated into clusters. Each cluster has a cluster-head, which runs the network-based monitoring. Therefore, packet-level monitoring is done by the cluster-head. Individual nodes use the packet-level audits from the cluster-head to improve the performance of the host-based intrusion detection system.

The strength of this IDS architecture is augmentation of network-based IDS with host-based IDS. The combination of these mechanisms has proved very efficient in conventional IDS. Furthermore, the authors have eliminated the single point of failure by distributing the cluster heads. This also distributes the management load between cluster-heads of the network. Also, host-level basis of decision making on an intrusion makes this approach robust against attacks on the IDS itself.

However, the architecture's trust on the cluster-head is its weak point. Malicious behavior of a cluster-head will lead to the compromise of all nodes under its control. In additions, similar to the other two mobile agent-based IDS, this architecture assumes secure routing, which may not be true.

Distributed IDS

Distributed IDS differs significantly from mobile agent-based IDS. Zhang, Lee, and Huang (2003) in their pioneering work propose a distributed IDS. The IDS architecture as shown in Figure 6 consists of local and cooperative intrusion detection engines. These detection engines are interfaced with their respective response modules.

A local intrusion detection system is a typical host-based IDS. The cooperative intrusion detection engine is used to decide globally about a particular behavior pattern. Collection of all cooperative detection engines on all nodes form a global intrusion detection engine. Semantically, Cooperative detection is analogous to network-based intrusion detection. However, global decision on behavior patterns will not dominate local decision. Nonetheless, global decision will aid local response. Few incorrect decisions about a behavioral pattern will not affect the global decision as more numbers of correct decisions will invalidate the incorrect decisions.

Local detection engine functions autonomously, independent of other nodes' detection engines. The cooperative engine will not aid the local-detection engine for identifying a malicious behavior pattern. This prevents propagation of malicious or

Figure 6. Distributed IDS architecture

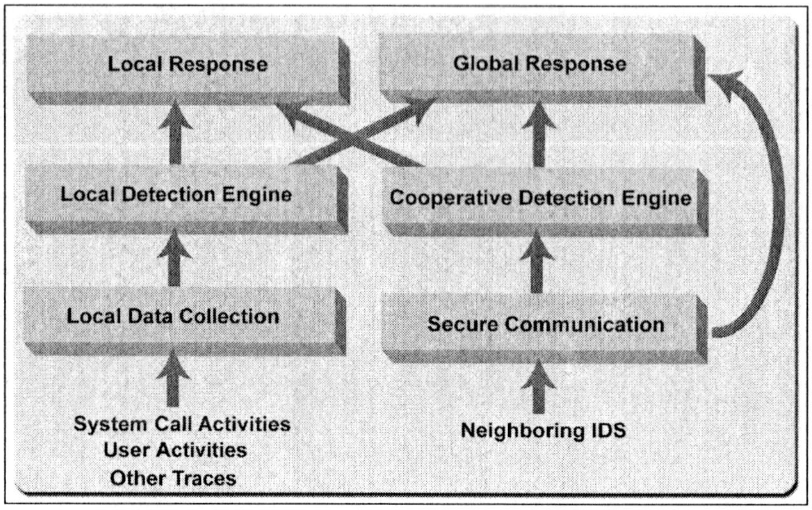

wrong detection to other nodes. However, the local response to an attack is aided by the cooperative detection engine. Furthermore, a global response is deduced by collecting information from various local intrusion detection engines of all nodes in the network. Eventually, this global response will be used for response action for that particular behavior pattern.

Although global decision sharing is secure comparing behavior-pattern sharing, the authors did not discuss how local intrusion detection relies on the global responses. Routing insecurity provides the ability to an attacker to create nonexistent nodes. Therefore, the attacker can emulate malicious behavior for these nonexistent nodes. Thus, the real majority of benign nodes will not help to guarantee security of the distributed IDS

Methodologies

TIARA

Techniques for intrusion-resistant ad hoc routing algorithms (TIARA) essentially an intrusion prevention model (Ramanujan, Ahamad, Bonney, Hagelstrom, & Thurber, 2000). TIARA is a conglomeration of innovative techniques which provides:

- Light-weight firewalls
- Traffic policing
- Intrusion tolerant routing
- Intrusion detection
- Flow monitoring
- Reconfiguration mechanisms
- Multipath routing
- Source initiated route switching

It aims to minimize the damage incurred on the ad hoc network by destructive attacks such as DoS, distributed denial of service (DDoS), and so forth. Routing and data traffic are protected by TIARA. TIARA is a distributed framework. TIARA is a highly efficient cross-layer intrusion prevention and detection mechanism. Exploring each of these techniques is beyond the scope of this chapter. Mishra has briefly discussed these techniques in his survey of ad hoc IDS.

However, it should be noted that intrusion detection is a module in the collection of techniques. The operational efficiency of the intrusion detection is unknown. Furthermore, tolerance to attacks is not the fundamental goal of an intrusion detection system. Unless the attackers are eliminated from the network or the attack is identified and segregated from benign traffic, the network is always under threat. Persistent attacks have high probability of success. Therefore, immediate response to attack is critical. TIARA has no response system for intrusions.

Threshold-Based Detection

A simplistic approach to ad hoc IDS is threshold-based detection. Bhargava and Agrawal (2001) propose an ad hoc IDS which prevents internal attacks (attacks within the network). Internal attacks are exhibited by nodes belonging to the network which behave maliciously, either by themselves or when compromised. Each node maintains a local variable called "MalCount" for every other node, which is increased for a particular node if its behavior is suspicious. Thus the MalCount array in a node tracks the level or state of suspicion that the host node has regarding the other nodes. Each node shares its local state of suspicion with respect to a particular node with other nodes in the network using a special packet REMAL. When a node receives REMAL, it increases its local MalCount for the particular node under suspicion.

The authors overlooked many aspects of ad hoc security. First, malicious knowledge sharing using REMAL will have cumulative malign effect on the network. Second, the security of the REMAL packet is unknown. Eventually, the entire network can be under threat by trusting unreliable REMAL packets. The crucial aspect of the security of the IDS is not considered in this methodology. Furthermore, routing security is not addressed.

Another interesting approach called watchdog-pathrater, which also uses threshold, is proposed by Sergio, Giuli, Kevin, and Mary (2000). Watchdog-pathrater, as the name implies, has a monitor and evaluator. Unlike Bhargava and Agrawal's (2001) approach, Watchdog-pathrater functions independently and does not share information with other nodes. When a packet is forwarded to a neighbor node, the forwarding node listens and monitors how the node behaves upon receiving a packet. A benign node will forward faithfully, which is overheard by the monitor. However, when the node does not forward the packet, the pathrater increases the failure rate for the path. The monitor does not distinguish between maliciousness and node faultiness. Upon the failure rate reaching the threshold, the node is discarded from any path.

This method is analogous to fault-tolerance in typical routing algorithms. This method effectively detects and responds to malicious packet dropping attacks (sinks). However, it fails to address attacks such as route invasion, route disruption, and so forth.

State-Based Anomaly Detection

One of the interesting approaches in conventional IDS models are state-based intrusion detection. Michael and Ghosh (2000) incorporate a state-based model in ad hoc intrusion detection. They propose two anomaly detection methodologies, which use finite-state machines (FSM). FSM have proved successful in conventional IDS because of their adaptability and dynamic learning capability of new attacks.

Anomaly detection methods proposed by Michael and Ghosh (2000) used protocol states. In the first method, the sequence and frequency of protocol states are monitored. Intrusion is affirmed when a particular sequence deviates significantly from normal behavior patterns or the frequency of states exceeds a threshold. To increase robustness, their second approach uses probabilistic state-based intrusion detection. Each occurrence of a suspicious protocol state increases the probability of the behavior being malicious.

These two approaches are well suited for transport and application layer protocols, which have many protocol states, and the protocol states are predictable. For example, attacks such as, TCP SYN flood attack can be detected using this approach.

However, this is not true in the case of routing protocols. State sequence or frequency of states does not distinguish a malicious behavior from a benign one. Traditionally, FSM were used to extract semantics from user behavior through application-layer protocols. In the case of ad hoc routing protocols, semantics is not represented by protocol states, but factors such as current topology, mobility, connectivity, and so forth are.

FUTURE TRENDS

Research in ad hoc network security is in its embryonic stages. Ad hoc network IDS is even more rudimentary, since the quest for an efficient intrusion prevention mechanism is not over yet (Hubaux, Buttyan, & Capkun, 2001). Intrusion prevention and detection mechanisms are mutually productive for ad hoc security. Clearly, a concrete and practical IDS model for ad hoc networks is yet to be evolved.

Historically, conventional IDS systems were subjected to intensive research and analysis before becoming practical. Analogous to conventional IDS, ad hoc intrusion detection needs more research. It is eminent that consideration of ad hoc network characteristics plays a vital role in the denouement of the IDS model. In literature, most of the research focus was on IDS architectures. However, IDS in ad hoc networks require innovative detection strategies to resolve the issue pertaining to IDS in ad hoc networks.

To summarize, we enumerate below the observations made from the study of ad hoc IDS models proposed in the literature.

First, routing security should be the crux of the IDS. Similar to conventional IDS, intensive statistical analysis and research is required on the feasibility of statistical and rule-based detection methodologies, in respect to routing behavioral data. Routing control messages produce a new kind of audit trails. New features linked to the properties of routing control message have to be derived. These derived parameters will aid in analyzing the feasibility of various detection methodologies.

Second, the absence of a centralized node necessitates innovative adaptation in the IDS. Adaptation is the process of learning new attacks, attack resolving techniques (responses), as well as changing statistical parameters with respect to the ad hoc network environment. Adaptation in a highly dynamic network is an interesting and new challenge. Efficiency of various computational intelligence methods, which are also used in conventional IDS, has to be analyzed. Learning new attacks through intelligence in ad hoc IDS paradigm is an unexplored research domain.

Finally, the most significant ad hoc network characteristics which affect the IDS model are the three Ds: distributed, decentralized and dynamic nature. An IDS architecture which considers these three factors will essentially be efficient. However, the IDS architecture should also consider the limitations of detection methodologies.

CONCLUSION

The implementation of intrusion detection systems in ad hoc networks is hindered by the inherent characteristics of these networks. These characteristics were examined and their significance was observed. The differences between conventional and ad hoc intrusion detection systems are detailed. Requirements of an effective ad hoc IDS are studied. Various proposed IDS architectures and methodologies are explored and their strengths and weakness are discussed. The future of ad hoc IDS depends mostly on the statistical properties of ad hoc network's routing behaviors. Therefore, considerable research and development is required in this domain.

REFERENCES

Andrew, B., Smith. (May 2001). *An examination of an intrusion detection architecture for wireless ad hoc networks.* Paper presented at the 5th National Colloquium for Information System Security Education.

Athanasiades, N., Abler, R., Levine, J., Owen, H., & Riley, G. (2003). *Intrusion detection testing and benchmarking methodologies.* Paper presented at the First IEEE International Workshop on Information Assurance, IWIAS 2003.

Awerbuch, B., Curtmola, R., Holmer, D., Nita-Rotaru, C., & Rubens, H. (2004). *Mitigating Byzantine attacks in ad hoc wireless networks.* John Hopkins University, Department of Computer Science.

Awerbuch, B., Curtmola, R., Holmer, D., Rubens, H., & Nita-Rotaru, C. (2005). *On the survivability of routing protocols in ad hoc wireless networks.*

Paper presented at the Security and Privacy for Emerging Areas in Communications Networks, SecureComm 2005.

Balajinath, B., & Raghavan, S. V. (2001). Intrusion detection through learning behavior model. *Computer Communications, 24*(12), 1202-1212.

Bhargava, S., & Agrawal, D. P. (2001, Fall). *Security enhancements in AODV protocol for wireless ad hoc networks.* Paper presented at the IEEE 54th Vehicular Technology Conference, VTC 2001.

Brutch, P., & Ko, C. (2003). *Challenges in intrusion detection for wireless ad-hoc networks.* Paper presented at the Applications and the Internet Workshops, 2003.

Bykova, M., Ostermann, S., & Tjaden, B. (2001). Detecting network intrusions via a statistical analysis of network packet characteristics. In *Proceedings of the 33rd Southeastern Symposium on System Theory, 2001.*

Debar, H., Dacier, M., & Wespi, A. (1999). Towards a taxonomy of intrusion-detection systems. *Computer Networks-the International Journal of Computer and Telecommunications Networking, 31*(8), 805-822.

Duda, R. O., Hart, P. E., & Stork, D. G. (2000). *Pattern classification* (2nd ed.). Wiley Inter-Science Publication.

Hijazi, A., & Nasser, N. (2005). *Using mobile agents for intrusion detection in wireless ad hoc networks.* Paper presented at the Second IFIP International Conference on Wireless and Optical Communications Networks, WOCN 2005

Hossain, M., Bridges, S. M., & Vaughn, R. B., Jr. (2003). *Adaptive intrusion detection with data mining.* Paper presented at the IEEE International Conference on Systems, Man and Cybernetics, 2003.

Huang, Y. A., & Lee, W. (2004). Attack analysis and detection for ad hoc routing protocols. *Recent advances in intrusion detection, proceedings* (Vol. 3224, pp. 125-145). Berlin: Springer-Verlag Berlin.

Hubaux, J.-P., Buttyan, L., & Capkun, S. (2001). *The quest for security in mobile ad hoc networks.* Paper presented at the 2nd ACM international Symposium on Mobile Ad hoc Networking & Computing, Long Beach, CA.

Jacoby, G. A., Marchany, R., & Davis, N. J., IV. (2004). *Battery-based intrusion detection a first line of defense.* Paper presented at the Information Assurance Workshop, 2004/Proceedings from the Fifth Annual IEEE SMC.

Kachirski, O., & Guha, R. (2002). *Intrusion detection using mobile agents in wireless ad hoc networks.* Paper presented at the IEEE Workshop on Knowledge Media Networking, 2002.

Kong, J., Hong, X., & Gerla, M. (2003). *A new set of passive routing attacks in mobile ad hoc networks.* Paper presented at the Military Communications Conference, MILCOM 2003. IEEE.

Lamport, L., Shostak, R., & Pease, M. (1982). The Byzantine generalsproblem. *ACM Transactions on Programming Languages and Systems, 4*(3), 382-401.

Little, M. (2005). *TEALab: A testbed for ad hoc networking security research.* Paper presented at the Military Communications Conference, MILCOM 2005. IEEE.

Michael, C. C., & Ghosh, A. (2000). *Two state-based approaches to program-based anomaly detection.* Paper presented at the 16th Annual Conference Computer Security Applications, ACSAC '00.

Mishra, A., Nadkarni, K., & Patcha, A. (2004). Intrusion detection in wireless ad hoc networks. *IEEE Wireless Communications, 11*(1), 48-60.

Nadkarni, K., & Mishra, A. (2003). *Intrusion detection in MANETS: The second wall of defense.* Paper presented at the 29th Annual Conference of the IEEE Industrial Electronics Society, IECON 2003.

Papadimitratos, P., & Haas, Z. (2002, January 27-31). *Secure routing for mobile ad hoc networks.* Paper presented at the SCS Communication Networks and Distributed Systems Modeling and Simulation Conference (CNDS 2002), San Antonio.

Patrick, A., Olivier, C., Jean-Marc, P., Bernard, J., Ludovic, M., & Ricardo, P. (2002). *Security in ad hoc networks: A general intrusion detection architecture enchancing trust based approaches.* Paper presented at the 1st International Workshop on Wireless Info. Sys., Cicudad Real, Spain.

Pokrajac, D., & Lazarevic, A. (2004). *Applications of unsupervised neural networks in data mining.* Paper presented at the 7th Seminar on Neural Network Applications in Electrical Engineering, NEUREL 2004.

Ramanujan, R., Ahamad, A., Bonney, J., Hagelstrom, R., & Thurber, K. (2000). *Techniques for intrusion-resistant ad hoc routing algorithms (TIARA).*

Sergio, M., Giuli, T. J., Kevin, L., & Mary, B. (2000). *Mitigating routing misbehavior in mobile ad hoc networks.* Paper presented at the Conference Name|. Retrieved Access Date|. from URL|.

Shimomura, T., & Markoff, J. (1996). *Take down: The pursuit and capture of Kevin Mitnick, America's most notorious cyber-criminal; by the man who did it.* London: Secker & Warburg.

Verwoerd, T., & Hunt, R. (2002). Intrusion detection techniques and approaches. *Computer Communications, 25*(15), 1356-1365.

Yang, S., & Baras, J. S. (2003). *Modeling vulnerabilities of ad hoc routing protocols.* Paper presented at the 1st ACM Workshop on Security of Ad Hoc and Sensor Networks, Fairfax, Virginia.

Zhang, Y., Huang, Y.-A., & Lee, W. (2005). *An extensible environment for evaluating secure MANET.* Paper presented at the First International Conference on Security and Privacy for Emerging Areas in Communications Networks, SecureComm 2005.

Zhang, Y. G., Lee, W. K., & Huang, Y. A. (2003). Intrusion detection techniques for mobile wireless networks. *Wireless Networks, 9*(5), 545-556.

KEY TERMS

Ad Hoc Networks: Ad hoc networks are loosely organized and configured network. There are no centralized nodes, such as routers, gateways, and so forth. All network functions are done by every node and thereby every node supports the network's functioning.

Anomaly Detection: Anomaly detection is a type of intrusion detection in which historical normal behavior of the network is used. Any deviation of a behavior from the normal will raise an alarm.

Audit Trails: Audit trails describe a network or node behavior. It contains values for a set of parameters, which is recorded in periodic intervals of time. The parameter set is called as the feature set and usually differs between different network environments, protocols, and systems.

Intrusion/Attack: Intrusion is a behavior of an external or internal node(s) with malign intent, which aims to affect other benign nodes in the network.

Intrusion Detection: Intrusion detection is the process of identifying and distinguishing malicious behavior from the normal network traffic.

Misbehavior Detection: Misbehavior detection is a complement to anomaly detection. In this type of intrusion detection, known intrusion behavior patterns are used. Any resemblance of a behavior with these patterns will result in an alarm.

Mobile Agents: Mobile agents are specialized software which move between nodes to accomplish their assigned tasks, such as data collection and so forth.

Chapter XXXIV
Security in Wireless Sensor Networks

Luis E. Palafox
CICESE Research Center, Mexico

J. Antonio Garcia-Macias
CICESE Research Center, Mexico

ABSTRACT

In this chapter we present the growing challenges related to security in wireless sensor networks. We show possible attack scenarios and evidence the easiness of perpetrating several types of attacks due to the extreme resource limitations that wireless sensor networks are subjected to. Nevertheless, we show that security is a feasible goal in this resource-limited environment; to prove that security is possible we survey several proposed sensor network security protocols targeted to different layers in the protocol stack. The work surveyed in this chapter enable several protection mechanisms vs. well documented network attacks. Finally, we summarize the work that has been done in the area and present a series of ongoing challenges for future work.

INTRODUCTION

Recently, wireless sensor networks (WSN) have gained great popularity, mainly because they provide a low cost alternative to solving a great variety of real-world problems (Akyildiz, Su, & Sankarasubramaniam, 2003). Their low cost enableds the deployment of large amounts of sensor nodes (in the order of thousands, and in the future perhaps millions), which most of the time operate under harsh environments. WSN present extreme resource limitations, mainly in available memory space and energy source. Both limitations represent great obstacles for the integration of traditional security techniques. The highly unreliable communication channels that are used in WSN and the fact that they operate unattended make the integration of security techniques even harder.

Wireless sensor networks today offer the processing capabilities of computers of a few decades ago and the industry's trend is to reduce the cost of wireless sensing nodes while maintaining the same processing power. Based on this idea, many researchers have started to face the challenge of maximizing processing capabilities and reducing energy consumption while protecting sensor networks from possible attacks.

BACKGROUND

WSN have many more limitations than other traditional computer networks. Due to these limitations, it is unfeasible to use the traditional security approaches in these resource-constrained networks. Thus, to develop efficient security techniques, it is imperative to consider the limitations involved.

Extremely Limited Resources

Every security mechanism requires a certain amount of resources for its implementation, these resources include data memory, program memory, and energy source to power the sensor node; however, these resources are very scarce in sensor nodes.

- Memory limitations. In order to implement an efficient security mechanism, the algorithm used for such implementation must have a small footprint.
- Energy limitations. When including security mechanisms, careful attention should be paid to energy-depleting factors including the consumed energy in computation of the security functions (i.e., encrypt, decrypt, data signatures, signature verification), the consumed energy of additional security related data transmissions or overhead (i.e., initialization vectors required for encrypt/decrypt), and the energy spent in storing the security related parameters (i.e., cryptographic keys).

Highly Unreliable Communication Medium

Unreliable communication is another threat to WSN. The security relies heavily on a defined protocol, which depends on communication.

- Unreliable transfers. The packets can be corrupted or even discarded due to errors in the communication channel or to congested nodes which results in packet loss; as a consequence, application developers are forced to allocate extra resources for error handling. Most importantly is the fact that if a protocol does not have the appropriate mechanisms for error handling, packets including critical security information could be lost (e.g., a cryptographic key).
- Conflicts. Even if we had a reliable communication channel, the communication still could be unreliable due to the broadcast nature of sensor networks. If a collision occurs in the middle of a transfer, there would be conflicts and the transfer itself would fail. On a highly populated network this can be a big problem, as has already been pointed out (Akyildiz, Su, Sankarasubramaniam, & Cayirci, 2002).
- Latency. Multihop routing, network congestion, and in-network processing can introduce latency to the network, making synchronization difficult between nodes. Synchronization problems can be critical for network security mechanisms that rely on error reporting and cryptographic key distribution. Some real/time communications techniques could be used in WSN (Stankovic, Abdelzaher, Lu, Sha, & Hou, 2003).

Unattended Operation

On most wireless sensor network applications, nodes are left unattended for long time periods. The three main disadvantages of leaving the network unattended are:

- Exposure to physical attacks. The network can be deployed in an environment open to adversaries, in undesirable climatologic conditions, and so forth. Thus, the probability of a node suffering a physical attack is much higher than in typical computers on traditional networks, which normally are placed on a secure location and only face attacks through the network.

- Remote management. Remote network management makes practically impossible the detection of physical attacks or network maintenance problems. The most extreme example is perhaps when a node is being used on a military battlefield reconnaissance application; in that case the node would no longer have physical contact with the user once deployed.
- No central point of administration. A sensor network must be distributed, with no central point of administration. However, if its design is not adequate, network organization would be hard, inefficient, and fragile.

Summarizing, the time the sensor network spends unattended is directly proportional to the probability of an adversary performing an attack on any of its nodes.

Security Requirements

Wireless sensor networks share many characteristics with traditional networks, including their security requirements; however, they also introduce several requirements that are exclusive to them.

Data Confidentiality

Data confidentiality is the biggest problem in network security. Every network with any security approach would probably address this issue before any other. In sensor networks, confidentiality relates to the following (Carman, Kruus, & Matt, 2000; Perrig, Szewczyk, Tygar, Wen, & Culler, 2002):

- A sensor node must not filter sensor readings to its neighbors; particularly on military applications where the stored data in a node can be highly confidential.
- On many applications, the nodes need to communicate highly confidential data (i.e., key distribution), thus, it is very important to build a secure communication channel in WSN.
- The nodes' public information, such as their identity and their public keys, can be encrypted to a certain extent for protecting against traffic analysis attacks.

The traditional approach for keeping confidential information secret is to encrypt it using a secret key that only the destination node knows, thus, resulting in confidentiality.

Data Integrity

With the implementation of confidentiality an adversary may be unable to steal any data from the sensor network. However, this does not imply that the data are secure. The adversary could still be able to modify the data to the degree of affecting the overall operation of the network. For instance, a malicious user may add or remove certain fragments to a packet. Then, this packet could be sent to its original destination. The data loss or corruption can occur even without the presence of a malicious user due to harsh environmental conditions. Thus, data integrity helps to assure that the received data have not been modified in transit.

Data Freshness

Even though data confidentiality and integrity has been achieved, we must assure that each message is fresh. Data freshness suggests that the data are recent, and assures that no old message has been resent. This requirement is especially important when shared keys strategies are being used. Typically, shared keys need to be renewed over time. However, it takes time to propagate the new keys through the entire network. Under this scheme, it would be easy for an adversary to perpetrate a packet replay attack. Furthermore, it would be easy to corrupt the operation of the network if the nodes are not well informed of the time at which the key will change. To solve this problem, a time dependent counter may be added to the packet for assuring data freshness.

Authentication

Besides modifying packets, an adversary can also potentially alter the flow of the packets through the addition of fake packets to the network. Consequently, the adversary can make receiving node believe that the data comes from an authentic source. Additionally, authentication is needed for several administrative tasks (i.e., dynamic network reprogramming, controlling node duty cycle). Thus, we can determine that message authentication is important for many sensor network applications.

Availability

Adjusting current traditional encryption algorithms to sensor network implies an additional cost. Some approaches suggest modifying code to favor code reutilization as much as possible. Other approaches tend to use additional communication to achieve the same goal. Other more radical approaches impose restrictions to the data or propose less robust schemes (like centralized schemes) to simplify algorithms. But all of these approaches decrease the level of availability of the nodes and consequently, the availability of the entire network for the following reasons:

- The introduction of additional processing results in additional power consumption. If we exhaust the available energy of a node, its data would no longer be available.
- Introducing additional communication operations also consumes more energy. Furthermore, adding more communication considerably increases the probability of generating a collision.
- If we introduce a centralized scheme, it would only have a single point, which can be a constant threat to the availability of the entire network.

The implementation of security mechanisms not only interferes with network operation, it also can considerably affect availability of the entire network.

Autoconfiguration

WSN are an extreme case of ad hoc networks, which require that each node be independent and flexible for configuring itself according to several situations. There is no fixed infrastructure to administer a sensor network. This also brings a great challenge for security in this type of networks. For instance, the dynamic nature of the network suggest of preinstalling a key shared between the base station and the rest of the nodes (Eschenauer & Gligor, 2002). Several schemes of random key distribution have been proposed in the context of symmetric encryption techniques (Chan, Perrig, & Song, 2003; Eschenauer & Gligor, 2002; Hwang & Kim, 2004; Liu, Li, & Ning, 2005). In the area of public key cryptography on wireless sensor networks, this same dynamicity requires efficient mechanisms for key distribution. WSN must autoconfigure for key management and for establishing trust relationships among nodes, in a similar way as they autoconfigure to perform multihop routing.

If a sensor network lacks of autoconfiguration, the damage done by an adversary or even by the hostile environment could be fatal.

Security Attacks on Wireless Sensor Networks

The nature of the WSN makes them vulnerable to several types of attacks. Such attacks can be perpetrated in a variety of ways, most notably are the denial or service attacks (DoS), but there are also traffic analysis attacks, eavesdropping, physical attacks, and others. DoS attacks in wireless sensor networks go from simple communication channel saturation techniques to more sophisticated designed to tamper with the message authentication code (MAC) layer protocol (Perrig, Stankovic, & Wagner, 2004).

Due to the great differences in available energy and computational power, protecting against a well designed denial-of-service attack is practically impossible. A more powerful node could easily block any other normal node, and consequently, prevent the sensor network from performing its function.

We can observe that attacks on sensor networks are not exclusively restricted to denial-of-service attacks; among these other types of attacks we can include compromised nodes, attacks to routing protocols, and physical attacks.

Attack Scenario

To propose and develop efficient prevention and recuperation mechanisms for attacks on wireless sensor networks it is important to know and understand the nature of the potential adversaries; these can be classified in two groups (Karlof &Wagner, 2003): mote class adversaries and laptop class adversaries. In the first case, the adversary has access to sensor nodes. In contrast, the laptop class adversary has access to more powerful devices such as personal computers, PDAs, and so forth. Thus, in this case, the devices have many advantages over legit nodes: larger energy source, more powerful processors, and they could also have high-power transmitters or a highly sensitive antenna to eavesdrop on traffic.

A laptop class adversary can produce more damage as opposed to an adversary that only has access to a few sensor nodes. For instance, a sensor node can only block radio links in a small neighborhood while an adversary with a laptop computer could block the entire sensor network with the help of a more powerful transmitter. Furthermore, a laptop class adversary could potentially eavesdrop on the traffic of the entire network, while a mote class adversary could only eavesdrop on the traffic in a very limited area.

Another commonly used adversary classification considers external and internal adversaries. Previously, we discussed external attacks, where the adversaries do not have any access to the sensor network. Conversely, internal attacks are those perpetrated by an authorized participant in the network that has turned malicious. Internal attacks can be mounted from compromised nodes that are executing malicious codes or from laptop computers that have access to cryptographic materials, data, and codes from authorized nodes.

Attacks to Routing Protocols

Most routing protocols for WSN are very simple; due to this simplicity, they are generally more vulnerable to attacks than their counterparts in ad hoc networks. Most attacks on network layer protocols fall into one of the following categories:

- Spoofed, altered, or replayed routing information. This attack is directed toward the routing information that is exchanged between nodes. By spoofing, altering, or replaying routing information, the adversaries could potentially create routing loops, attract or repel network traffic, lengthen or shorten routes, generate fake error messages, partition the network, increase node to node latency, and so forth.
- Selective forwarding. Multihop networks often operate assuming faithfully that messages will be received by their destination. On a selective forwarding attack, malicious nodes could prevent forwarding certain messages or even discard them; consequently, these messages would not propagate through the network. A simple form of this attack is very easy to be detected because the neighbor nodes could easily infer that the route is no longer valid and use an alternate one. A more subtle form of this attack is when and adversary selectively forwards packets. Therefore, if an adversary is interested in suppressing or modifying packets that come from certain source, the adversary could selectively forward the rest of the traffic, thus, the adversary would not raise any suspicion of the attack.
- Sinkhole attacks. In a sinkhole attack, the goal of the adversary is to attract all the traffic to a certain area or the network through a compromised node, creating a sinkhole (metaphorically speaking). Due to the fact that the nodes that are located across the route have the ability to alter application data, the sinkhole attacks could facilitate other types of attacks (like selective forwarding for instance).
- Sybil attacks. In a Sybil attack (Douceur, 2002), a node presents multiple identities to the rest of the nodes. Sybil attacks are a threat to geographical routing protocols, since they require the exchange of coordinates for effi-

cient packet routing. Ideally, we would expect that a node only sends a set of coordinates, but under a Sybil attack, an adversary could pretend to be in many places at once.

- Wormhole attacks. In a wormhole attack (Hu, Perrig, & Johnson, 2002) an adversary builds a virtual tunnel through a low latency link that takes the messages from one part of the network and forwards them to another. The simplest case of this attack is when one node is located between two other nodes that are forwarding. However, wormhole attacks commonly involve two distant nodes that are colluded to underestimate the distance between them and forward packets through an external communication channel that is only available to the adversary.
- HELLO flood attacks. Some protocols require nodes to send HELLO packets to advertise themselves to their neighbors. If a node receives such packet, it would assume that it is inside the RF range of the node that sent that packet. However, this assumption could be false because a laptop class adversary could easily send these packets with enough power to convince all the network nodes that the adversary is their neighbor. Consequently, nodes close to the adversary may try to use the adversary as a route to the base station, while nodes further away would send packets directly to the adversary. But the transmission power of those nodes is much less that the adversary's, thus, the packets would get lost, and that would create a state of confusion in the sensor network.
- Acknowledgement spoofing. Some routing algorithms require the use of acknowledgement signals (ACK). In this case, an adversary could spoof this signal in response to the packets that the adversary listens to. This results in convincing the transmitting node that a weak link is strong. Thus, an adversary could perform a selective forwarding attack after spoofing ACK signals to the node that the adversary intends to attack.

Attacks to Data Aggregation Techniques

Data aggregation in wireless sensor networks can significantly reduce communication overhead compared to all the nodes sending their data to the base station. However, data aggregation complicates even more network security. This is due to the fact that every intermediate node could potentially modify, forge, or discard messages. Therefore, a single compromised node could be able to alter the final aggregation value. Intruder node and compromised node attacks are two major threats to security in sensor networks that use data aggregation techniques.

Physical Attacks

Sensor networks often operate in hostile environments. In those environments, the size of the nodes plus the unattended operation mode contributes to make them very vulnerable to physical attacks (i.e., node destruction) (Wang, Gu, Schosek, Chellappan, & Xuan, 2005c). In contrast to other types of attacks, physical attacks destroy the nodes permanently, thus, their loss is irreversible. For instance, an adversary could extract cryptographic keys, alter the node's circuitry, and reprogram it or replace it with malicious nodes (Wang, Gu, Chellappan, Xuan, & Lai, 2005b). Previous work shows that a Berkeley MICA2 mote (one of the most commonly used in the research community) can be compromised in less than a minute. Even though these results are not surprising, because MICA2 motes do not have any physical protection mechanism, they give us a good idea of what a well-trained adversary can do.

Defense Countermeasures

In this section we will present some security mechanisms that have been proposed in the literature and that help in meeting the security requirements discussed earlier. For this purpose, we will begin by discussing the key establishment process in WSN which is the base for security in this type of networks. We will follow that with a description of security mechanisms for preventing denial-of-service attacks, defense against routing protocol attacks, how to protect from traffic analysis attacks, defending against sensor node privacy attacks, and protection against physical and data aggregation attacks.

Key Establishment Process

One important aspect of security that has received a great deal of attention from the research community is the key establishment process in WSN. Due to the fact that encryption and key establishment are crucial elements in security defense mechanisms, and most security mechanisms rely on pure encryption, we will give a general overview on encryption before going into any details about specific security defense mechanisms.

Overview

The key establishment and key management problems are not exclusive to sensor networks. In fact, this type of problems has been thoroughly studied in the wireless network community. Traditionally, key establishment is performed through some public key protocol. The most commonly used is Diffie-Hellman (Diffie & Hellman, 1976), but there are many more.

However, most of the traditional techniques are not suitable for low-power devices such as sensor nodes. This is due to the fact that these techniques use asymmetric cryptography, which is also known as public key cryptography. In this case it is required to maintain two mathematically related keys, one of which is public while keeping the other private. The problem with public key cryptography in WSN

Table 1. A summary of the analysis for cipher performance (Law et al. 2004)

By key setup						
Rank	***Size Optimized***			***Speed Optimized***		
	Code mem.	***Data mem.***	***Speed***	***Code mem.***	***Data mem.***	***Speed***
1	RC5-32	MISTY1	MISTY1	RC6-32	MISTY1	MISTY1
2	KASUMI	Rijndael	Rijndael	KASUMI	Rijndael	Rijndael
3	RC6-32	KASUMI	KASUMI	RC5-32	KASUMI	KASUMI
4	MISTY1	RC6-32	Camellia	MISTY1	RC6-32	Camellia
5	Rijndael	RC5-32	RC5-32	Rijndael	Camellia	RC5-32
6	Camellia	Camellia	RC6-32	Camellia	RC5-32	RC6-32
By encryption mode						
Rank	***Size Optimized***			***Speed Optimized***		
	Code mem.	***Data mem.***	***Speed***	***Code mem.***	***Data mem.***	***Speed***
1	RC5-32	RC5-32	Rijndael	RC6-32	RC5-32	Rijndael
2	RC6-32	MISTY1	MISTY1	RC5-32	MISTY1	Camellia
3	MISTY1	KASUMI	KASUMI	MISTY1	KASUMI	MISTY1
4	KASUMI	RC6-32	Camellia	KASUMI	RC6-32	RC5-32
5	Rijndael	Rijndael	RC6-32	Rijndael	Rijndael	KASUMI
6	Camellia	Camellia	RC5-32	Camellia	Camellia	RC6-32

is that, computationally speaking, it is very heavy for the sensor nodes. However, there has been work that shows that implementation is viable if a proper selection of algorithms is made (Gaubatz, Kaps, & Sunar 2004; Gura, Patel, Wander, Eberle, & Shantz, 2004; Malan, Welsh, & Smith, 2004; Watro, Kong, Fen Cuti, Gardiner, Lynn, & Kruus, 2004).

For these reasons, symmetric encryption is the more widely selected technique for applications that cannot handle the computational complexity of asymmetric encryption. Symmetric techniques use a single key that is shared by the two communicating parties. This key is used for data encryption and decryption. The traditional example of symmetric encryption is the DES (data encryption standard) algorithm. However, the use of DES has decreased significantly because it can be easily broken. Currently, other algorithms such as 3DES (triple DES), RC5, AES, and others (Schneier, 1996).

An analysis of several cipher algorithms (Law, Doumen, & Hartel, 2004) is summarized in Table I, where two classifications are made: one by key setup and the other by encryption mode. In both classifications the algorithms were optimized for code size and speed and aspects such as speed, code size, and required data memory were evaluated.

A great challenge for symmetric encryption is the problem of key management. The problem resides in the fact that both parties need to know the key prior to starting secure communication. Thus, the problem can be summarized as follows: how can we assure that only the two communicating parties know the key and no one else does? Distributing secret keys is not an easy problem to solve because preinstalling the key in the sensor node is not always an option.

Key Establishment Protocols

There are several random key predistribution techniques that have been proposed. Eschenauer and Gligor (2002) propose a scheme based on probabilistic key sharing among sensor nodes. This scheme operates first by distributing a key chain to all participant nodes before their deployment. Each key chain consists of a set of keys that has been randomly selected from a larger offline-generated key set. To use the random key predistribution technique it is not necessary that each pair of nodes share a key. However, every pair of nodes that does share a key may use that key to establish a direct secure connection between them. Eschenauer and Gligor (2002) show that under this scheme it is highly probable that sensor nodes can operate with shared keys.

The LEAP protocol (Zhu, Setia, & Jajodia, 2003) adopts the approach of using multiple techniques for key establishment. Here, the authors make the observation than any mechanism by itself provides security for every type of connection in wireless networks. Thus, in this work they present four different types of keys that are used depending on the communication type to be established.

In *PIKE* (Chan & Perrig, 2005), the authors describe a mechanism for establishing a key between two nodes based on the trust that both nodes have toward a third node in the same network. The shared keys of each node are propagated throughout the network in such a way that for every node A and B a node C exists that shares a key with A and B. Thus, the key establishment protocol between A and B can be securely routed through C.

Perrig et al. (2002) propose a key distribution scheme for secure broadcast authentification named μTESLA. The main idea of μTESLA is to achieve asymmetric cryptography through the delayed disclosure of symmetric keys.

It is important to point out that the most significant advances in the integration of public cryptography to WSN (which will be discussed next) have been made recently. This makes random key predistribution a less interesting topic.

Public Key Cryptography

Two of the more commonly used public key cryptography algorithms are RSA and ECC (Schneier, 1996). Traditionally, it was thought that these techniques were way too complex for applying them to WSN. However, successful implementations of public key cryptographic systems in WSN have been published recently.

Gura et al. (2004) report that it is possible to implement RSA and ECC in 8-bit microprocessors, demonstrating a performance advantage of ECC over RSA. Another advantage is that the 160-bit key in ECC generates shorter messages during transmission compared to the 1024-bit key of RSA. Particularly, this work demonstrates that the dot product operations used in ECC execute faster than the operations in RSA.

Watro et al. (2004) show that certain parts of the RSA cipher can be implemented on current sensor network platforms, particularly in the MICA2 Berkeley motes (Hill, Szewczyk, Woo, Hollar, Culler, & Pister, 2000). They implemented the public key operations in the sensor nodes while the private ones were performed in more powerful devices. In this case they used a laptop computer.

Malan et al. (2004) propose a scheme based on ECC and show an implementation of the Diffie-Hellman algorithm based on the elliptic curve discrete logarithm problem. While key generation is by no means fast (around 34 seconds for generating the pair of keys and another 34 seconds for generating the secret key), this probably would suffice for applications that do not require frequent key renewal.

Preventing Against Denial-of-Service

In Table 2 we show the most common denial of service attacks and their corresponding countermeasures classified by layers. Due to the fact that DoS attacks are very common, efficient countermeasures mechanisms are required. One approach to defend against the classic channel jamming attack is to identify the part of the network that is jammed and route traffic around that area. Wood and Stankovic (2002) describe a two phase approach where nodes along the perimeter of the jammed area report their status to their neighbors who then collaboratively define the jammed region and simply route around it.

To protect against jamming at the MAC layer, nodes could use an admission control mechanism that limits their transmission rate. This would allow the network to ignore the requests designed to exhaust the node's energy source. However, this is not an optimal solution because the network must be able to handle large volumes of traffic.

To protect against malicious nodes that intentionally misroute traffic could be done at the cost of redundancy. In this case, a node can send the message through multiple routes, thus increasing the probability that the message will arrive to its final destination simply because the message does not rely on a single route to get there.

Defending Against Routing Protocol Attacks

Routing protocols for WSN has been a well studied topic to a certain extent. However, most of the research efforts focus mainly in providing energy

Table 2. Wireless sensor network DoS attacks/defenses

Layer	Attacks	Defenses
Physical	Jamming	Spread-spectrum, priority messaging, lower duty cycle, region mapping, mode change
	Tampering	Tamper-proof, hiding
Link	Collision	Error correcting code
	Exhaustion	Rate limitation
	Unfairness	Small frames
Network (routing)	Neglect and greed	Redundancy, probing
	Homing	Encryption
	Misdirection	Egress filtering, authorization monitoring
	Black holes	Authorization, monitoring, redundancy
Transport	Flooding	Client puzzles
	Desynchronization	Authentication

efficient routing mechanisms. There is a large demand of routing protocols that besides offering energy efficiency they also offer security against certain network attacks such as sinkhole attacks, wormholes attacks, and the Sybil attack. As the WSN range of applications is increasing as well as its network densities, secure routing will be a design factor that must be considered for future applications.

Security Techniques for Routing Protocols

Deng, Han, and Mishra (2002) introduce an INtrusion tolerant routing protocol for sensor networks (INSENS). This protocol is based on minimizing the damage caused by an intruder and keep routing despite its presence, without having to identify the intruder. In this work, the authors state that an intruder does not have to be a malicious node necessarily, it very well could be a node that is just malfunctioning for physical reasons. Identifying a malicious node from a malfunctioning one could be extremely difficult. For this reason they make no distinction between them. The first technique that they propose is to mitigate the damage caused by a potential intruder by applying redundancy. This is, as we previously mentioned, sending a packet through multiple routes.

They also assume that there are large differences in available resources between the base station and the sensor nodes, thus, they propose that routing table computation is to be performed at the base station. This is done in three phases. In the first phase the base station broadcast a request that propagates through the entire network. On the next phase, the base station collects information about node connectivity. Finally, the base station computes a series of routing tables for each node. These tables include redundant routing information used for the redundant message transmission we discussed earlier.

There are several attacks that could be launched to the routing protocol during each one of the three phases. On the first phase, a node could spoof a request done by the base station. A malicious node could forward the request through a fake route or simply not forward the request done by the base station.

To avoid this, Deng et al. (2002) use a technique similar to μTESLA where one-way key chains are used to authenticate the message from the base station.

Tanachaiwiwat, Dave, Bhindwale, and Helmy (2003) introduce a novel technique that they called TRANS (trust routing for location aware networked sensors). This routing protocol was proposed for data-centric networks. It also uses delayed key disclosure to achieve asymmetric cryptography. In their implementation, they use μTESLA for message authentication and confidentiality. By using μTESLA, TRANS can be sure that a message follows a trusted route through location-based routing. The approach consists of the base station sending an encrypted broadcast message to its neighbors. Only those trusted neighbors would have the key required to decrypt that message. The trusted neighbors would add their location to the route (for returning messages), and would encrypt the message now with their own keys and send the message to the neighbor closest to the destination. When the message arrives to its destination, the receiving node must authenticate the source (in this case the base station) using a MAC that belongs to the base station. Afterwards, the node can simply send a message to the base station through the trusted route that the original message followed.

An important challenge in the area of secure routing for wireless networks is that it is very easy to disrupt the routing protocol by simply disrupting the route discovery process. Papadimitratos and Haas (2002) propose a secure route discovery protocol that guarantees, under certain conditions, that the correct network topology would be obtained. This protocol is very similar to TRANS. The security relies on the MAC layer and in an accumulation on the node identities that are included in the route. By doing this, a source node can discover the network topology because each node from the source to the destination appends its identity to the message. In order to ensure that the message has not been tampered with, a MAC code is also appended to the message, which can be authenticated by either the destination or by the source (for returning messages).

How to Protect from Traffic Analysis Attacks

There are some strategies to protect from traffic analysis attacks. Deng, Han, and Mishra (2004) propose a technique based on a random walk through the network. This technique also send packets randomly to nodes different from the parent node in the routing tree. The main goal of this technique is to make it harder to a potential adversary to infer the route from a given node to the base station and also to prevent against a possible rate monitoring attack, but it would not protect against a time correlation attack. To protect against a time correlation attack, they propose a fractal strategy. With this technique a node would generate a fake packet (with certain probability) while one of its neighbors is sending a packet to the base station. The fake packet would be sent to another neighbor that consequently may send another fake packet, thus, deceiving the potential adversary. These fake packets would use the time-to-live (TTL) parameter to decide for how long they would be circulating throughout the network.

Defending Against Sensor Node Privacy Attacks

To protect against privacy attacks, several proposals have been made that reduce the effects of those attacks, we will discuss some of those proposals in this section (Gruteser, Schelle, Jain, Han, & Grunwald, 2003).

Anonymity Mechanisms

When very precise location information is being used it is easy to identify the user and monitor the user's activity, thus, this opens the door for a privacy attack. The anonymity mechanisms depersonalize the data before releasing them; these techniques are an alternative approach to policy-based access control.

Some researchers have proposed certain techniques that make use of anonymity mechanisms. For instance, Gruteser and Grunwald (2003a) analyze the feasibility of anonymizing location information for location-based services in an automotive telematic environment. Beresford and Stajano (2003) evaluate anonymity techniques for an indoor location-based system based on the active nat.

Producing total anonymity is a difficult problem given the lack of knowledge about the concerning node's location. Therefore, for the privacy problem, there is a tradeoff between the required anonymity level and the need for public information. Three approaches have been proposed to address this problem (Gruteser & Grunwald, 2003b; Gruteser et al., 2003; Priyantha, Chakraborty, & Balakrishnan, 2000; Smailagic & Kogan, 2002):

- Decentralize sensitive data. The main idea in this approach is to distribute the sensed location data through a spanning tree. By doing so, no single node will contain the original data.
- Secure the communication channel. By using secure communication protocols such as SPINS (Perrig et al., 2002), eavesdropping and active attacks can be prevented.
- Node mobility. Making the nodes move can be an effective defense mechanism against privacy attacks, particularly due to the fact that location information would be changing constantly. For instance, the Cricket system (Priyantha et al., 2000) is a system with location support for mobile object inside buildings.

Policy-Based Approach

The policy-based approach is a topic that is currently receiving a great deal of attention from the research community. Access control decisions and authentication are based on the specifications provided by the privacy policies. Molnar and Wagner (2004) introduce the concept of private authentication in RFID applications, which can be considered passive nodes. In the automotive telematics domain, Duri, Gruteser, Liu, Moskowitz, Perez, Singh et al. (2002) propose a policy-based framework to protect data from the sensors, where an on-board computer can act as a trusted agent. Snekkenes (2001) presents advanced concepts for policy specification on cell phone networks. These concepts allow access control based on criteria such as request time, location, object speed, and identity. Myles, Friday and Davies (2003) describe an architecture for a centralized server that controls the access of client applications through the use of validation modules that verify the XML-formatted application policies. Hengartner and Steenkiste (2003) point out that access control policies must be governed by room or user policies. The room policies specify who is authorized to find out about the people currently in the room, while user policies state who is permitted to access location information about another user.

Langheinrich (2005) proposed a framework called PawS (privacy awareness system). This framework is based on privacy policy advertisements through special packets called privacy beacons. Those policies are maintained with privacy proxies, which keep databases that store those policies.

Information Flooding

Ozturk, Zhang, Trappe, and Ott (2004) propose antitraffic analysis mechanisms to prevent an external adversary from obtaining the location of a data source. Random data routing and phantom traffic are used to hide real traffic, so that it is difficult for an adversary to track the data source through traffic analysis. Ozturk et al. have developed comparable methods that rely on flooding-based routing protocols.

Some similar mechanisms can be used to prevent an adversary to track the base station through traffic analysis (Gura et al., 2004). A key problem with these techniques is that they involve an energy cost in order to provide information anonymity.

Protecting from Physical Attacks

Physical attacks, as we pointed out earlier, represent an important threat to sensor networks because of their unattended operation mode and their extremely limited resources. Nodes may be equipped with tamper-proof physical protection. For instance, an alternative to this is tamper-proof packaging (Wood & Stankovic, 2002). Related research work focuses in the design of hardware that make their memory content inaccessible to adversaries. Another alternative is to use special software and hardware to detect physical tampering.

As the hardware costs decrease, integrating tamper-proof hardware would be a feasible solution for sensor network applications. However, the research community has agreed by consensus that the trend should be making cheaper sensor nodes without adding extra functionalities; thus, integrating physical protection is not a solution that would be commonly accepted in the near future. One possible approach for protecting against physical attacks is self-destruction. The main idea behind this approach is that whenever a node detects a possible attack it self-destructs. This is particularly feasible on networks where there are redundant nodes and when the cost per node is low. Obviously, the key to this approach is detecting a possible attack. One possible solution is to statically verify the status of their neighbors, but in mobile networks this still is an open problem.

Regarding the deployment of security components outside the nodes, several proposals have been made (Bulusu & Jha, 2005). Sastry, Shankar, and Wagner (2003) introduce the concept of secure location verification and propose a secure localization scheme called ECHO that assures node location legitimacy. In this scheme, the security relies over physical sound properties and RF. The adversary cannot claim to have a shorter distance by starting the ultrasound response early because it will not have the nonce.

Hu and Evans (2004) use directional antennas to defend against wormhole attacks. In the work presented by Wang et al. (2005b) the authors study the modeling and defense of sensor networks against search-based physical attacks. They define a physical attack-based model, where an adversary walks the network using signal detecting equipment to locate active nodes and destroy them. In prior work, the authors identified and modeled blind physical attacks (Wang, Gu, Chellappan,

Schosek, & Xuan, 2005a). The defense algorithm is executed by individual nodes in two phases: in the first phase, the nodes detect the attacker and notify other nodes; in the second phase, the nodes receive the notification and change their state to safe mode.

Seshadri, Perrig, Van Doorn, and Khosla (2004) introduce a mechanism called SWATT to verify and detect when memory content is altered. This mechanism can be use as defense against a physical attack by modifying code in the nodes.

Secure Data Aggregation

As sensor networks increase in size, the amount of data that they collectively sense also increases. However, due to the computational limitations of each node, a small sensor is only responsible for a very small portion of the entire data. Due do this, a network search would probably return a large amount of raw data, most of which would not be of the user's interest.

For this reason, raw data preprocessing is recommended to produce more meaningful results to the user. This is typically done by a series of aggregators. An aggregator is responsible for collecting raw data from a subset of nodes and processing that raw data into more usable data.

However, aggregation techniques are particularly vulnerable to attacks because a single aggregator node is responsible for processing the data from multiple nodes. Due to this fact, secure data aggregation techniques are required by sensor network that consider the possibility of one or more malicious nodes.

Overview

If an aggregator node is compromised, then all the transmitted data in the network to the base station may be forged. To detect this, Ye, Luo, Lu, and Zhang (2005) define a mechanism based on statistical filters. This uses multiple MAC codes across the entire route from the aggregator node to the base station. Any packet that does not pass verification would be discarded.

Wagner (2004) analyzes the resiliency of aggregation techniques, and argues that current aggregation techniques were proposed without security in mind, and thus, are vulnerable to attacks. A mathematical framework is proposed to formally evaluate security for aggregation. This theory allows quantifying the robustness of an aggregation operation against a malicious attack. By using the framework, it is argued that the aggregation functionalities that can be securely computed under the presence of k compromised nodes are exactly the functions that are (k, α)-resilient for some α that is not too large. This work opened the door for secure data aggregation in sensor networks. However, the presented level of aggregation model is fairly simple compared to real sensor network implementations. Extending this technique to multilevel aggregation scenarios with heterogeneous devices is an interesting challenge.

Secure Data Aggregation Techniques

As we pointed out earlier, data aggregation has been studied in reasonable depth. The problem with classical data aggregation is that they all assume trusted nodes. Of course, in practice this may not be the case, and for this reason, secure data aggregation techniques are required.

Przydatek, Song, and Perrig (2003) describe a secure information aggregation (SIA). They point out that aggregation techniques and sensor networks are vulnerable to a variety of attacks including denial-of-service attacks. However, this work focuses on protecting against a specific type of attack called stealthy attack. The goal of SIA is to ensure that if a user accepts the result of an aggregation as correct, then there is a high probability that the value is close to the true aggregation value. In case that the aggregated value has been tampered with, the user must reject the forged value with a high probability.

Hu and Evans (2003) propose a secure aggregation technique that uses the μTESLA protocol to provide security. In this case, the nodes organize into a hierarchy tree where intermediate nodes play the aggregator role. Recall that the μTESLA achieves asymmetry through delayed disclosure of symmetric keys. For this, a child cannot verify the data authenticity immediately because the key used to generate the MAC code has not been disclosed.

However, this technique does not guarantee that the data being reported by the nodes and the aggregator are correct. To address this problem, the base station is responsible for distributing temporary keys to the network as well as the μTESLA key used for validating the MAC. By using this key, the node can verify their children's MAC codes.

We can note that secure data aggregation techniques play an important role in adopting WSN technology due to the large amount of raw data and the localized in-network processing required in these networks. Research efforts in this area have been limited, thus, much more investigation is needed in this particular topic.

CONCLUSION

Certainly, incorporating efficient security mechanisms to WSN is a huge challenge, mainly because of the differences they have compared to traditional networks. Their resource constraints, their large scale deployments, along with their operating environments, represent great obstacles to achieve security. Nevertheless, efficient mechanisms have been proposed to deal with a great variety of attacks to which WSN presumably are subjected to. These security techniques confront specific attacks that operate across different layers of the protocol stack. Attacks like signal jamming (physical layer), induced collisions (MAC sublayer), packet redirection (routing layer), and many others have been the addressed through many security mechanisms, many of which we described in this chapter.

However, most of the security techniques rely heavily on a key distribution protocol and assume that secret keys have already been placed on the distributed nodes. However as we showed in this chapter, efficient key distribution in WSN is no easy task. In fact, most of the research efforts in WSN security are directed to proposing efficient key distribution techniques; in this chapter we discussed research work in the area of WSN key distribution. As of now, we still believe that there is much room for improvement in efficient key distribution in wireless sensor networks. As more efficient key distribution key mechanisms continue to appear, more efficient application-specific security techniques will also emerge.

But overall, perhaps the biggest challenge of all is proving that the proposed security techniques work well in real-world sensor network applications. Currently, there is a huge gap between real-world WSN development and WSN security research. Thus, we consider that integrating the proposed security techniques to real-world applications is a challenge that should be faced in the near future, as opposed to proposing new techniques that most of the time does not go beyond lab implementations.

REFERENCES

Akyildiz, I., Su, W., Sankarasubramaniam, Y., & Cayirci, E. (2002). A survey on sensor networks. *IEEE Communications Magazine, 40*(8), 102-114.

Anderson, R. J., & Kuhn, M. G. (1996, November). *Tamper resistance: A cautionary note.* Paper presented at the Second USENIX Workshop on Electronic Commerce, Oakland, CA.

Anderson, R. J., & Kuhn, M. G. (1997). Low cost attacks on tamper resistant devices. In B. Christianson, B. Crispo, T. M. A. Lomas, & M. Roe (Eds.), *Security Protocols Workshop* (LNCS 1361, pp. 125-136). Springer.

Beresford, A., & Stajano, F. (2003). Location privacy in pervasive computing. *IEEE Pervasive Computing, 2*(1), 46-55.

Bulusu, N., & Jha, S. (2005). *Wireless sensor networks: a system perspective.* Artech House.

Carman, D. W., Kruus, P. S., & Matt, B. J. (2000). *Constraints and approaches for distributed sensor network security* (Tech. Rep. No. 00-010). NAI Labs, The Security Research Division.

Chan, H., & Perrig, A. (2005, March). *PIKE: Peer intermediaries for key establishment in sensor networks.* Paper presented at IEEE INFOCOM, Miami.

Chan, H., Perrig, A., & Song, D.X. (2003, May). *Random key predistribution schemes for sensor networks*. Paper presented at the IEEE Symposium on Security and Privacy, Oakland, CA.

Deng, J., Han, R., & Mishra, S. (2002). *INSENS: Intrusion-tolerant routing in wireless sensor networks* (Tech. Rep. No. CU-CS-939-02). University of Colorado, Department of Computer Science.

Deng, J., Han, R., & Mishra, S. (2004). *Countermeasures against traffic analysis in wireless sensor networks* (Tech. Rep. No. CU-CS-987-04). University of Colorado, Department of Computer Science.

Diffie, W., & Hellman, M. E. (1976). New directions in cryptography. *IEEE Transactions on Information Theory, 22*(6), 644-654.

Douceur, J. R. (2002). The sybil attack. In P. Druschel, M. F. Kaashoek, & A. I. T. Rowstron (Eds.), *IPTPS* (pp. LNCS 2429, pp. 251-260). Springer.

Du, W., Deng, J., Han, Y.S., & Varshney, P. K. (2003). *A pairwise key pre-distribution scheme for wireless sensor networks.* In Jajodia (pp. 42-51).

Duri, S., Gruteser, M., Liu, X., Moskowitz, P., Perez, R., Singh, M., et al. (2002). Framework for security and privacy in automotive telematics. In *Proceedings of the 2nd International Workshop on Mobile Commerce (WMC '02)*, New York, (pp. 25-32). ACM Press.

Eschenauer, L., & Gligor, V. D. (2002). A key-management scheme for distributed sensor networks. In V. Atluri (Ed.), *ACM Conference on Computer and Communications Security* (pp. 41-47).

Estrin, D., Govindan, R., Heidemann, J. S., & Kumar, S. (1999). Next century challenges: Scalable coordination in sensor networks. In *Proceedings of the MOBICOM* (pp. 263-270).

Gaubatz, G., Kaps, J.-P., & Sunar, B. (2004). Public key cryptography in sensor networks: Revisited. In C. Castelluccia, H. Hartenstein, C. Paar, & D. Westhoff (Eds.), *ESAS* (LNCS 3313, pp. 2-18). Springer.

Gruteser, M., & Grunwald, D. (2003a). Anonymous usage of location-based services through spatial and temporal cloaking. In *Proceedings of the USENIX MobiSys.*

Gruteser, M., Schelle, G., Jain, A., Han, R., & Grunwald, D. (2003). Privacy-aware location sensor networks. In M. B. Jones (Ed.), *USENIX HotOS* (pp. 163-168).

Gura, N., Patel, A., Wander, A., Eberle, H., & Shantz, S. C. (2004). Comparing elliptic curve cryptography and RSA on 8-bit CPUs. In M. Joye & J.-J. Quisquater (Eds.), *CHES* (LNCS 3156, pp. 119-132). Springer.

Hartung, C., Balasalle, J., & Han, R. (2005). *Node compromise in sensor networks: The need for secure systems* (Tech. Rep. No. CU-CS-990-05). University of Colorado, Department of Computer Science.

Hengartner, U., & Steenkiste, P. (2003). Protecting access to people location information. In Hutter (pp. 25-38).

Hill, J., Szewczyk, R., Woo, A., Hollar, S., Culler, D. E., & Pister, K. S. J. (2000). System architecture directions for networked sensors. In *Proceedings of the 9th International Conference on Architectural Support for Programming Languages and Operating Systems* (pp. 93-104).

Hu, L., & Evans, D. (2003). *Secure aggregation for wireless network.* Paper presented at the SAINT Workshops IEEE Computer Society (pp. 384-394).

Hu, L., & Evans, D. (2004). *Using directional antennas to prevent wormhole attacks.* Paper presented at the NDSS. The Internet Society.

Hu, Y.-C., Perrig, A., & Johnson, D. B. (2002). *Wormhole detection in wireless ad hoc networks* (Tech. Rep. No. TR01-384). Rice University, Department of Computer Science.

Karlof, C., & Wagner, D. (2003). Secure routing in wireless sensor networks: Attacks and countermeasures. *Ad Hoc Networks, 1*(2-3), 293-315.

Karp, B., & Kung, H. T. (2000). *GPSR: Greedy perimeter stateless routing for wireless networks*. Paper presented at the MOBICOM (pp. 243-254).

Langheinrich, M. (2005). *Personal privacy in ubiquitous computing: Tools and system support*. Unpublished doctoral dissertation, Swiss Federal Institute of Technology Zurich.

Law, Y. W., Doumen, J., & Hartel, P. (2004). *Survey and benchmark of block ciphers for wireless sensor networks* (Tech. Rep. No. TR-CTIT-04-07). Mathematics and Computer Science University of Twente, Faculty of Electrical Engineering, The Netherlands.

Madden, S., Franklin, M. J., Hellerstein, J. M., & Hong, W. (2002). TAG: A tiny aggregation service for ad-hoc sensor networks. *SIGOPS Oper. Syst. Rev., 36*(SI), 131-146.

Malan, D. J., Welsh, M., & Smith, M. D. (2004). *A public-key infrastructure for key distribution in TinyOS based on elliptic curve cryptography*. Paper presented at the SECON (pp. 71-80).

Molnar, D., & Wagner, D. (2004). Privacy and security in library RFID: Issues, practices, and architectures. In V. Atluri, B. Pfitzmann, & P. D. McDaniel (Eds.), *ACM Conference on Computer and Communications Security* (pp. 210-219).

Myles, G., Friday, A., & Davies, N. (2003). Preserving privacy in environments with location-based applications. *IEEE Pervasive Computing, 2*(1), 56-64.

Ozturk, C., Zhang, Y., Trappe, W., & Ott, M. (2004). *Source-location privacy for networks of energy-constrained sensors*. Paper presented at the WSTFEUS (pp. 68-81). IEEE Computer Society.

Papadimitratos, P., & Haas, Z. (2002). Secure routing for mobile ad hoc networks. In *Proceedings of SCS Communication Networks and Distributed System Modeling and Simulation Conference, CNDS '04.*

Perrig, A., Stankovic, J. A., & Wagner, D. (2004). Security in wireless sensor networks. *Communications of the ACM, 47*(6), 53-57.

Perrig, A., Szewczyk, R., Tygar, J. D., Wen, V., & Culler, D. E. (2002). SPINS: Security protocols for sensor networks. *Wireless Networks, 8*(5), 521-534.

Pietro, R. D., Mancini, L. V., Law, Y. W., Etalle, S., & Havinga, P. J. M. (2003). *LKHW: A directed diffusion-based secure multicast scheme for wireless sensor networks*. Paper presented at the ICPP Workshops. IEEE Computer Society.

Priyantha, N. B., Chakraborty, A., & Balakrishnan, H. (2000). *The Cricket location support system*. Paper presented at the MOBICOM (pp. 32-43).

Przydatek, B., Song, D. X., & Perrig, A. (2003). SIA: Secure information aggregation in sensor networks. In I. F. Akyildiz, D. Estrin, D. E. Culler, & M. B. Srivastava (Eds.), *SenSys. ACM* (pp. 255-265).

Sastry, N., Shankar, U., & Wagner, D. (2003). Secure verification of location claims. In *Proceedings of the 2003 ACM Workshop on Wireless Security, WiSe '03*, New York, (pp 1–10). ACM Press.

Schneier, B. (1996) *Applied cryptography: Protocols, algorithms, and source code in C* (2nd ed.). John Wiley.

Seshadri, A., Perrig, A., Van Doorn, L., & Khosla, P. K. (2004). *SWATT: Software-based attestation for embedded devices*. Paper presented at the IEEE Symposium on Security and Privacy. IEEE Computer Society.

Shrivastava, N., Buragohain, C., Agrawal, D., & Suri, S. (2004). Medians and beyond: New aggregation techniques for sensor networks. In *Proceedings of the 2nd International Conference on Embedded Networked Sensor Systems, SenSys '04*, New York, (pp. 239-249). ACM Press.

Smailagic, A., & Kogan, D. (2002). Location privacy in pervasive computing. *IEEE Wireless Communications, 9*(5), 10-17.

Snekkenes, E. (2001). *Concepts for personal location privacy policies*. Paper presented at the ACM Conference on Electronic Commerce (pp. 48-57).

Stankovic, J. A., Abdelzaher, T. F., Lu, C., Sha, L., & Hou, J. C. (2003). Real-time communication and coordination in embedded sensor networks. *Proceedings of the IEEE, 91*(7), 1002-1022.

Tanachaiwiwat, S., Dave, P., Bhindwale, R., & Helmy, A. (2003). Secure locations: Routing on trust and isolating compromised sensors in location-aware sensor networks. In *Proceedings of the 1st International Conference on Embedded Networked Sensor Systems, SenSys '03*, New York, (pp. 324-325). ACM Press.

Wagner, D. (2004). Resilient aggregation in sensor networks. In *Proceedings of the 2nd ACM Workshop on Security of Ad Hoc and Sensor Networks, SASN '04*, New York, (pp. 78-87). ACM Press.

Wang, X., Gu, W., Chellappan, S., Schosek, K., & Xuan, D. (2005a). *Lifetime optimization of sensor networks under physical attacks.* Paper presented at the IEEE International Conference on Communications, ICC '05 (Vol. 5, pp. 3295-3301).

Wang, X., Gu, W., Chellappan, S., Xuan, D., & Lai, T. H. (2005b). *Sacrificial node-assisted defense against search-based physical attacks in sensor networks* (Tech. Rep.). Ohio State University, Department of Computer Science and Engineering.

Wang, X., Gu, W., Schosek, K., Chellappan, S., & Xuan, D. (2005c). *Sensor network configuration under physical attacks.* In X. Lu & W. Zhao (Eds.), *ICCNMC* (LNCS 3619, pp. 23-32). Springer.

Watro, R. J., Kong, D., Fen Cuti, S., Gardiner, C., Lynn, C., & Kruus, P. (2004). *TinyPK: Securing sensor networks with public key technology.* In Setia & Swarup (pp. 59-64).

Wood, A. D., & Stankovic, J. A. (2002). Denial of service in sensor networks. *IEEE Computer, 35*(10), 54-62.

Ye, F., Luo, H., Lu, S., & Zhang, L. (2005). Statistical en-route filtering of injected false data in sensor networks. *IEEE Journal on Selected Areas in Communications, 23*(4), 839-850.

Zhu, S., Setia, S., & Jajodia, S. (2003). *LEAP: Efficient security mechanisms for large-scale distributed sensor networks.* In Jajodia (pp. 62-72).

KEY TERMS

Compromised Node: A node on which an attacker has gained control after network deployment. Generally compromise occurs once an attacker has found a node, and then directly connects the node to their computer via a wired connection of some sort. Once connected the attacker controls the node by extracting the data and/or putting new data or controls on that node.

Data Aggregation: Process of reducing large amounts of sensor generated data to smaller and more representative data sets that synthesize the state of the phenomena that the network is monitoring.

Data Freshness: Implies that the sensed data are recent, and it ensures that no adversary replayed old messages.

Insider Attacks: These types of attacks are those launched by adversaries that have access to one or more compromised nodes in a network. Insider attacks are the most challenging ones because the adversary has access to the network's cryptographic materials (i.e., keys, ciphers, and data).

Key Distribution: Process of efficiently distributing cryptographic keys to the nodes that belong to a network. These keys could either be pairwise keys (for two party communications), group keys (for cluster-wide communication), or network keys (for secure broadcast communication).

Mote: A wireless receiver/transmitter that is typically combined with a sensor of some type to create a remote sensor. Some motes are designed to be incredibly small so that they can be deployed by the hundreds or even thousands for various applications

Node Authentication: Process of ensuring that a given node and its data are legit.

Outsider Attacks: Attacks perpetrated by adversaries that do not have access to direct access to any of the authorized nodes in the network. However, the adversary may have access to the physical medium, particularly if we are dealing with wireless networks. Therefore, attacks such as replay messages and eavesdropping fall into this classification. However, coping with this attack is fairly easy by using traditional security techniques such as encryption and digital signatures.

Chapter XXXV
Security and Privacy in Wireless Sensor Networks:
Challenges and Solutions

Mohamed Hamdi
University of November 7th at Carthage, Tunisia

Noreddine Boudriga
University of November 7th at Carthage, Tunisia

ABSTRACT

The applications of wireless sensor networks (WSNs) are continuously expanding. Recently, consistent research and development activities have been associated to this field. Security ranks at the top of the issues that should be discussed when deploying a WSN. This is basically due to the fact that WSNs are, by nature, mission-critical. Their applications mainly include battlefield control, emergency response (when a natural disaster occurs), and healthcare. This chapter reviews recent research results in the field of WSN security.

INTRODUCTION

The applications of wireless sensor networks (WSNs), which cover both the civil and military contexts, are continuously expanding. The ability to develop miniaturized, battery powered motes that combine sensing, correlation, fusion, and wireless communication capabilities makes the WSN technology cost-effective for being used in future. In fact, WSNs can be used to gather and analyze information about vehicular movement, humidity, temperature, pressure, as well as many other parameters.

However, the enormous potential of WSNs can be unlocked only if the corresponding infrastructures are adequately safeguarded. In fact, violating one or more security properties would lead to wrong decisions, and consequently wrong reactions. Hence, security should rank at the top of the issues that should be discussed when designing a WSN. Another motivation is that WSNs are, by nature, mission-critical, meaning that they are developed for sensitive tasks where error-tolerance is very small. The importance of security in the WSN context is exacerbated by certain factors including the following:

- Sensor nodes have limited storage, computation, and power resources. For this reason, security mechanisms should be adapted to the WSN capabilities.
- The network does not have a static infrastructure. WSN architectures can be only timely defined. This renders the application of existing robust cryptographic mechanisms (e.g., public key infrastructure [PKI], digital signature) more difficult than in customary networks.
- The sensing and communication tasks are often performed in a hostile environment where the gathered events are subjected to numerous threats that might affect the final decision.
- The detected events are forwarded through the sensor nodes themselves, preventing the application of strong communication security mechanisms.

This chapter surveys recent research activities in the area of WSN security. More accurately, the following aspects will be discussed:

1. **Wireless sensor networks:** This section addresses several WSN basic issues to highlight the related scientific challenges. Components, architecture, topology, routing, mobile target tracking, and alert management will be, among others, discussed.
2. **WSN security objectives:** Traditional security goals (i.e., confidentiality, authenticity, integrity, and availability) should be extended to fit the requirements of WSNs. Several particular concepts are introduced at this level. For instance, confidentiality, authenticity, and integrity, which have been customarily associated to data and node identity, should be extended to cover node location. This poses several new security challenges in the WSN context.
3. **Attacks against WSNs:** This section describes the most important attacks techniques concerning WSNs. Attacks are classified according to the basic security properties they violate. A taxonomy of these attacks will also be proposed. This taxonomy is based on three major attack activities: (1) attacks on transmitted information, (2) attacks on architecture, structure, protocols, and (3) attacks on the localization framework.
4. **Countermeasures:** Potential security solutions that allow countering the aforementioned threats will be proposed. They will be classified according to the level at which they act (e.g., link level, routing, and application). Countermeasures will be also categorized into preventive and reactive solutions. For example, robust localization (resp. fault-tolerance) schemes belong to the first (resp. second) category.
5. **Building security policies for WSNs:** Several key security processes, such as monitoring and incident response, can not be directly applied in the WSN field. They should therefore be heavily adapted in order to support WSN specific constraints.

WIRELESS SENSOR NETWORKS

Due to advances in wireless communications and electronics over the last few years, the development of networks of low-cost, low-power, multifunctional sensors has received increasing attention. These sensors are small in size and able to sense, process data, and communicate with each other, typically over an radio frequency (RF) channel. A sensor network is designed to detect events or phenomena, collect and process data, and transmit sensed information to interested users. Basic features of sensor networks are:

- Self-organizing capabilities
- Short-range broadcast communication and multihop routing
- Dense deployment and cooperative effort of sensor nodes
- Frequently changing topology due to fading and node failures
- Limitations in energy, transmit power, memory, and computing power

These characteristics, particularly the last three, make sensor networks different from other wireless ad hoc or mesh networks. Clearly, the idea of mesh networking is not new; it has been suggested for some time for wireless Internet access or voice communication. Similarly, small computers and sensors are not innovative per se. However, combining small sensors, low-power computers, and radios makes for a new technological platform that has numerous important uses and applications.

Wireless sensor networks are interesting from an engineering perspective, because they present a number of serious challenges that cannot be adequately addressed by existing technologies:

- **Extended lifetime:** As mentioned above, WSN nodes will generally be severely energy constrained due to the limitations of batteries. A typical alkaline battery, for example, provides about 50 watt-hours of energy; this may translate to less than a month of continuous operation for each node in full active mode. Given the expense and potential infeasibility of monitoring and replacing batteries for a large network, much longer lifetimes are desired. In practice, it will be necessary in many applications to provide guarantees that a network of unattended wireless sensors can remain operational without any replacements for several years.
- **Responsiveness:** A simple solution to extending network lifetime is to operate the nodes in a duty-cycled manner with periodic switching between sleep and wake-up modes. While synchronization of such sleep schedules is challenging in itself, a larger concern is that arbitrarily long sleep periods can reduce the responsiveness and effectiveness of the sensors. In applications where it is critical that certain events in the environment be detected and reported rapidly, the latency induced by sleep schedules must be kept within strict bounds, even in the presence of network congestion.
- **Robustness:** The vision of wireless sensor networks is to provide large scale, yet fine-grained coverage. This motivates the use of large numbers of inexpensive devices. However, inexpensive devices can often be unreliable and prone to failures. Rates of device failure will also be high whenever the sensor devices are deployed in harsh or hostile environments. Protocol designs must therefore have built-in mechanisms to provide robustness. It is important to ensure that the global performance of the system is not sensitive to individual device failures. Further, it is often desirable that the performance of the system degrade as gracefully as possible with respect to component failure.
- **Synergy:** Moore's law-type advances in technology have ensured that device capabilities in terms of processing power, memory, storage, radio transceiver performance, and even accuracy of sensing improve rapidly (given a fixed cost). However, if economic considerations dictate that the cost per node be reduced drastically from hundreds of dollars to less than a few cents, it is possible that the capabilities of individual nodes will remain constrained to some extent. The challenge is therefore to design synergistic protocols, which ensure that the system as a whole is more capable than the sum of the capabilities of its individual components. The protocols must provide an efficient collaborative use of storage, computation, and communication resources.
- **Scalability:** For many envisioned applications, the combination of fine granularity sensing and large coverage area implies that wireless sensor networks have the potential to be extremely large scale (tens of thousands, perhaps even millions of nodes in the long term). Protocols will have to be inherently distributed, involving localized communication, and sensor networks must utilize hierarchical architectures in order to provide such scalability. However, visions of large numbers of nodes will remain unrealized in practice until some fundamental problems, such as failure handling and in-situ reprogramming, are addressed even in small settings involving tens to hundreds of nodes.

There are also some fundamental limits on the throughput and capacity that impact the scalability of network performance.

- **Heterogeneity:** There will be a heterogeneity of device capabilities (with respect to computation, communication, and sensing) in realistic settings. This heterogeneity can have a number of important design consequences. For instance, the presence of a small number of devices of higher computational capability along with a large number of low-capability devices can dictate a two-tier, cluster-based network architecture, and the presence of multiple sensing modalities requires pertinent sensor fusion techniques. A key challenge is often to determine the right combination of heterogeneous device capabilities for a given application.
- **Self-configuration:** Because of their scale and the nature of their applications, wireless sensor networks are inherently *unattended* distributed systems. Autonomous operation of the network is therefore a key design challenge. From the very start, nodes in a wireless sensor network have to be able to configure their own network topology: localize, synchronize, and calibrate themselves, coordinate inter-node communication, and determine other important operating parameters.
- **Privacy and security:** The large scale, prevalence, and sensitivity of the information collected by wireless sensor networks (as well as their potential deployment in hostile locations) give rise to the final key challenge of ensuring both privacy and security.

WSN SECURITY OBJECTIVES

WSN Security Challenges

WSNs are characterized by many constraints compared to traditional communication networks. Due to these particular constraints, the application of existing network security approaches does not allow to fulfill the required security properties. Hence, appropriate security needs and techniques should be defined for WSN environments while borrowing concepts from the currently used security mechanisms. In the following, we highlight the most relevant, from security point of view, WSN intrinsic features.

Resource Limitations

Security mechanisms and processes necessarily require a certain amount of processing, power, storage, and memory resources. However, sensor nodes are often resource-impoverished. In the following, we detail the basic resource limitations characterizing WSNs.

- **Processing limitations:** A custom processor for sensor nodes should essentially have a low-power sleep mode, allowing reducing energy consumption, and a low-overhead wakeup mechanism, preventing the occurrence of network congestion due to signalling messages. Ekanayake (2004) shows that the processing speed offered by most of the available microcontrollers ranges between 4 and 400 MIPS. Even though this is a performance to implement the communication functions, it turns out to be not sufficient to support advanced security mechanisms, especially when a heavy traffic is exchanged across the WSN. For instance, it has been shown by Blaß (2005) that a traditional Diffie-Hellman key exchange operation would last 48.04 seconds on the AmtelMega processor. As a result, novel security algorithms should be considered to keep up with the sensor node processing limitations.
- **Limited memory and storage space:** A sensor is a tiny device with only a small amount of memory and storage space for the code. In order to build an effective security mechanism, it is necessary to limit the code size of the security algorithm. For example, one common sensor type (TelosB) has an 16-bit, 8 MHz RISC CPU with only 10K RAM, 48K program memory, and 1024K flash storage. With such a limitation, the software

built for the sensor must also be quite small. The total code space of TinyOS, the de-facto standard operating system for wireless sensors, is approximately 4 K (Hill 2000), and the core scheduler occupies only 178 bytes. Therefore, the code size for the all security related code must also be reduced.

- **Power limitation:** Energy is the biggest constraint to wireless sensor capabilities. We assume that once sensor nodes are deployed in a sensor network, they cannot be easily replaced (high operating cost) or recharged (high cost of sensors). Therefore, the battery charge taken with them to the field must be conserved to extend the life of the individual sensor node and the entire sensor network. When implementing a cryptographic function or protocol within a sensor node, the energy impact of the added security code must be considered. When adding security to a sensor node, we are interested in the impact that security has on the lifespan of a sensor (i.e., its battery life). The extra power consumed by sensor nodes due to security is related to the processing required for security functions (e.g., encryption, decryption, signing data, and verifying signatures), the energy required to transmit the security related data or overhead (e.g., initialization vectors needed for encryption/decryption), and the energy required to store security parameters in a secure manner (e.g., cryptographic key storage).

Data Loss

Certainly, unreliable communication is another threat to sensor security. The security of the network relies heavily on a defined protocol, which in turn depends on communication.

- **Unreliable transfer:** Normally the packet-based routing of the sensor network is connectionless and thus inherently unreliable. Packets may get damaged due to channel errors or dropped at highly congested nodes. The result is lost or missing packets. Furthermore, the unreliable wireless communication channel also results in damaged packets. A higher channel error rate also forces the software developer to devote resources to error handling. More importantly, if the protocol lacks the appropriate error handling it is possible to lose critical security packets. This may include, for example, a cryptographic key.
- **Collisions:** WSNs impose strict requirements on a medium access protocol. This is basically due to the ad hoc architecture characterizing WSNs as well as the long network lifetime needs. Moreover, as data are broadcasted over the radio link, packets may collide resulting in decreasing of the channel throughput. Depending on the medium access and transport layer protocols, the information loss can reach a certain degree such that the analysis center becomes no longer able to identify the events corresponding to the gathered data.
- **Latency:** Multihop routing, network congestion, and node processing can lead to greater latency in the network, thus making it difficult to achieve synchronization among sensor nodes. The synchronization issues can be critical to sensor security where the security mechanism relies on critical event reports and cryptographic key distribution. Interested readers please refer to Stankovic (2003) on real-time communications in wireless sensor networks.

Uncontrollable Behavior

Depending on the function of the particular sensor network, the sensor nodes may be left unattended for long periods of time. There are three main caveats to unattended sensor nodes:

- **Exposure to physical attacks:** The sensor may be deployed in an environment open to adversaries, bad weather, and so on. The likelihood that a sensor suffers a physical attack in such an environment is therefore much higher than the typical PCs, which is located in a secure place and mainly faces attacks from a network.

- **Managed remotely:** Remote management of a sensor network makes it virtually impossible to detect physical tampering (i.e., through tamperproof seals) and physical maintenance issues (e.g., battery replacement). Perhaps the most extreme example of this is a sensor node used for remote reconnaissance missions behind enemy lines. In such a case, the node may not have any physical contact with friendly forces once deployed.
- **No central management point:** A sensor network should be a distributed network without a central management point. This will increase the vitality of the sensor network. However, if designed incorrectly, it will make the network organization difficult, inefficient, and fragile.

Security Requirements

A sensor network is a special type of network. It shares some commonalities with a typical computer network, but also poses unique requirements of its own as discussed in Section 3. Therefore, we can think of the requirements of a wireless sensor network as encompassing both the typical network requirements and the unique requirements suited solely to wireless sensor networks.

Data Confidentiality

Data confidentiality is the most important issue in network security. Every network with any security focus will typically address this problem first. In sensor networks, the confidentiality relates to the following (Carman 2000; Perrig 2002):

- A sensor network should not leak sensor readings to its neighbors. Especially in a military application, the data stored in the sensor node may be highly sensitive.
- In many applications nodes communicate highly sensitive data, for example, key distribution; therefore it is extremely important to build a secure channel in a wireless sensor network.
- Public sensor information, such as sensor identities and public keys, should also be encrypted to some extent to protect against traffic analysis attacks.

The standard approach for keeping sensitive data secret is to encrypt the data with a secret key that only intended receivers possess, thus achieving confidentiality.

Data Integrity

With the implementation of confidentiality, an adversary may be unable to steal information. However, this does not mean the data are safe. The adversary can change the data, so as to send the sensor network into disarray. For example, a malicious node may add some fragments or manipulate the data within a packet. This new packet can then be sent to the original receiver. Data loss or damage can even occur without the presence of a malicious node due to the harsh communication environment. Thus, data integrity ensures that any received data have not been altered in transit.

Data Freshness

Even if confidentiality and data integrity are assured, we also need to ensure the freshness of each message. Informally, data freshness suggests that the data are recent, and it ensures that no old messages have been replayed. This requirement is especially important when there are shared-key strategies employed in the design. Typically shared keys need to be changed over time. However, it takes time for new shared keys to be propagated to the entire network. In this case, it is easy for the adversary to use a replay attack. Also, it is easy to disrupt the normal work of the sensor, if the sensor is unaware of the new key change time. To solve this problem a nonce, or another time-related counter, can be added into the packet to ensure data freshness.

Availability

Adjusting the traditional encryption algorithms to fit within the wireless sensor network is not free, and will introduce some extra costs. Some approaches choose to modify the code to reuse as much code as possible. Some approaches try to make use of additional communication to achieve the same goal. What is more, some approaches force strict limitations on the data access, or propose an unsuitable scheme (such as a central point scheme) in order to simplify the algorithm. But all these approaches weaken the availability of a sensor and sensor network for the following reasons:

- Additional computation consumes additional energy. If no more energy exists, the data will no longer be available.
- Additional communication also consumes more energy. What is more, as communication increases so too does the chance of incurring a communication conflict.
- A single point failure will be introduced if using the central point scheme. This greatly threatens the availability of the network.
- The requirement of security not only affects the operation of the network, but also is highly important in maintaining the availability of the whole network.

Self-Organization

A wireless sensor network is typically an ad hoc network, which requires every sensor node be independent and flexible enough to be self-organizing and self-healing according to different situations. There is no fixed infrastructure available for the purpose of network management in a sensor network. This inherent feature brings a great challenge to wireless sensor network security as well. For example, the dynamics of the whole network inhibits the idea of preinstallation of a shared key between the base station and all sensors (Eschenauer 2002). Several random key predistribution schemes have been proposed in the context of symmetric encryption techniques (Chan 2003; Eschenauer 2002; Hwang 2004; Liu 2005). In the context of applying public-key cryptography techniques in sensor networks, an efficient mechanism for public-key distribution is necessary as well. In the same way that distributed sensor networks must self-organize to support multihop routing, they must also self-organize to conduct key management and building trust relation among sensors. If self-organization is lacking in a sensor network, the damage resulting from an attack or even the hazardous environment may be devastating.

Time Synchronization

Most sensor network applications rely on some form of time synchronization. In order to conserve power, an individual sensor's radio may be turned off for periods of time. Furthermore, sensors may wish to compute the end-to-end delay of a packet as it travels between two pair-wise sensors. A more collaborative sensor network may require group synchronization for tracking applications and so forth. Ganeriwal (2005), proposes a set of secure synchronization protocols for sender-receiver (pair-wise), multihop sender-receiver (for use when the pair of nodes are not within single-hop range), and group synchronization.

Secure Localization

Often, the utility of a sensor network will rely on its ability to accurately and automatically locate each sensor in the network. A sensor network designed to locate faults will need accurate location information in order to pinpoint the location of a fault. Unfortunately, an attacker can easily manipulate nonsecured location information by reporting false signal strengths, replaying signals, and so forth.

A technique called verifiable multilateration (VM) is described by Capkun (2006). In multilateration, a device's position is accurately computed from a series of known reference points. Capkun (2006) uses authenticated ranging and distance bounding to ensure accurate location of a node. Because of distance bounding, an attacking node can only increase its claimed distance from a reference point. However, to ensure location consistency, an attacking node would also have to prove that its distance from another reference point is shorter. Since it cannot do this, a node manipulating the localization protocol can be found. For large sensor networks, the secure positioning for sensor networks (SPINE) algorithm is used. It is a three phase algorithm based upon verifiable multilateration.

Lazos (2005) describes secure range-independent localization (SeRLoc). Its novelty is its decentralized, range-independent nature. SeRLoc uses locators that transmit beacon information. It is assumed that the locators are trusted and cannot be compromised. Furthermore, each locator is assumed to know its own location. A sensor computes its location by listening for the beacon information sent by each locator. The beacons include the locator's location. Using all of the beacons that a sensor node detects, a node computes an approximate location based on the coordinates of the locators. Using a majority vote scheme, the sensor then computes an overlapping antenna region. The final computed location is the centroid of the overlapping antenna region. All beacons transmitted by the locators are encrypted with a shared global symmetric key that is preloaded to the sensor prior to deployment. Each sensor also shares a unique symmetric key with each locator. This key is also preloaded on each sensor.

Authentication

An adversary is not just limited to modifying the data packet. It can change the whole packet stream by injecting additional packets. So the receiver needs to ensure that the data used in any decision-making process originate from the correct source. On the other hand, when constructing the sensor network, authentication is necessary for many administrative tasks (e.g., network reprogramming or controlling sensor node duty cycle). From the above, we can see that message authentication is important for many applications in sensor networks. Informally, data authentication allows a receiver to verify that the data really are sent by the claimed sender. In the case of two-party communication, data authentication can be achieved through a purely symmetric mechanism: the sender and the receiver share a secret key to compute the message authentication code (MAC) of all communicated data.

Adrian Perrig et al. (2002) propose a key-chain distribution system for their μTESLA secure broadcast protocol. The basic idea of the μTESLA system is to achieve asymmetric cryptography by delaying the disclosure of the symmetric keys. In this case a sender will broadcast a message generated with a secret key. After a certain period of time, the sender will disclose the secret key. The receiver is responsible for buffering the packet until the secret key has been disclosed. After disclosure the receiver can authenticate the packet, provided that the packet was received before the key was disclosed. One limitation of μTESLA is that some initial information must be unicast to each sensor node before authentication of broadcast messages can begin. Liu and Ning (2003, 2004) propose an enhancement to the μTESLA system that uses broadcasting of the key chain commitments rather than μTESLA's unicasting technique. They present a series of schemes starting with a simple predetermination of key chains and finally settling on a multilevel key chain technique. The multilevel key chain scheme uses predetermination and broadcasting to achieve a scalable key distribution technique that is designed to be resistant to denial-of-service (DoS) attacks, including jamming.

Attacks against WSNs

Sensor networks are particularly vulnerable to several key types of attacks. Attacks can be performed in a variety of ways, most notably as denial-of-service attacks, but also through traffic analysis, privacy violation, physical attacks, and so on. Denial-of-service attacks on wireless sensor networks can range from simply jamming the sensor's communication channel to more sophisticated attacks designed to violate the 802.11 MAC protocol (Perrig 2004) or any other layer of the wireless sensor network.

Due to the potential asymmetry in power and computational constraints, guarding against a well orchestrated denial-of-service attack on a wireless sensor network can be nearly impossible. A more powerful node can easily jam a sensor node and effectively prevent the sensor network from performing its intended duty. We note that attacks on wireless sensor networks are not limited to simply denial-of-service attacks, but rather encompass a variety of techniques including node takeovers, attacks on the routing protocols, and attacks on a node's physical security. In this section, we first address some common denial-of-service attacks and then describe additional attacking, including those on the routing protocols as well as an identity based attack known as the Sybil attack.

Denial-of-Service Attacks

A standard attack on wireless sensor networks is simply to jam a node or set of nodes. Jamming, in this case, is simply the transmission of a radio signal that interferes with the radio frequencies being used by the sensor network (Wood 2002). The jamming of a network can come in two forms: constant jamming and intermittent jamming. Constant jamming involves the complete jamming of the entire network. No messages are able to be sent or received. If the jamming is only intermittent, then nodes are able to exchange messages periodically, but not consistently. This too can have a detrimental impact on the sensor network as the messages being exchanged between nodes may be time sensitive. Attacks can also be made on the link layer itself. One possibility is that an attacker may simply intentionally violate the communication protocol, for example, ZigBee or IEEE 801.11b (Wi-Fi) protocol, and continually transmit messages in an attempt to generate collisions. Such collisions would require the retransmission of any packet affected by the collision. Using this technique it would be possible for an attacker to simply deplete a sensor node's power supply by forcing too many retransmissions. At the routing layer, a node may take advantage of a multihop network by simply refusing to route messages. This could be done intermittently or constantly with the net result being that any neighbor who routes through the malicious node will be unable to exchange messages with, at least, part of the network. The transport layer is also susceptible to attack, as in the case of flooding. Flooding can be as simple as sending many connection requests to a susceptible node. In this case, resources must be allocated to handle the connection request. Eventually, a node's resources will be exhausted, thus rendering the node useless.

Traffic Analysis Attacks

Wireless sensor networks are typically composed of many low-power sensors communicating with a few relatively robust and powerful base stations. It is not unusual, therefore, for data to be gathered by the individual nodes where they are ultimately routed to the base station. Often, for an adversary to effectively render the network useless, the attacker can simply disable the base station. To make matters worse, Deng et al. (2005) demonstrate two attacks that can identify the base station in a network (with high probability) without even understanding the contents of the packets (if the packets are themselves encrypted).

A rate monitoring attack simply makes use of the idea that nodes closest to the base station tend to forward more packets than those farther away from the base station. An attacker needs only to monitor which nodes are sending packets and follow those nodes that are sending the most packets. In a time correlation attack, an adversary simply generates events and monitors to whom a node sends its packets. To generate an event, the adversary could simply generate a physical event that would be monitored by the sensor(s) in the area (turning on a light, for instance).

Wormhole Attacks

In a wormhole attack, an attacker receives packets at one point in the network, "tunnels" them to another point in the network, and then replays them into the network from that point. For tunnelled distances longer than the normal wireless transmission range of a single hop, it is simple for the attacker to make the tunneled packet arrive with better metric than a normal multihop route, for example, through use of a single long-range directional wireless link or through a direct wired link to a colluding attacker. It is also possible for the attacker to forward each bit over the wormhole directly, without waiting for an entire packet to be received before beginning to tunnel the bits of the packet, in order to minimize delay introduced by the wormhole. Due to the nature of wireless transmission, the attacker can create a wormhole even for packets not addressed to it, since it can overhear them in wireless transmission and tunnel them to the colluding attacker at the opposite end of the wormhole. If the attacker performs this tunneling honestly and reliably, no harm is done; the attacker actually provides a useful service in connecting the network more efficiently. However, the wormhole puts the attacker in a very powerful position relative to other nodes in the network, and the attacker could exploit this position in a variety of ways. The attack can also still be performed even if the network communication provides confidentiality and authenticity, and even if the attacker has no cryptographic keys. Furthermore, the attacker is invisible at higher layers; unlike a malicious node in a routing protocol, which can often easily be named, the presence of the wormhole and the two colluding attackers at either endpoint of the wormhole are not visible in the route. The wormhole attack is particularly dangerous against many ad hoc network routing protocols in which the nodes that hear a packet transmission directly from some node consider themselves to be in range of (and, thus a neighbor of) that node.

Attacks against Privacy

Sensor network technology promises a vast increase in automatic data collection capabilities through efficient deployment of tiny sensor devices. While these technologies offer great benefits to users, they also exhibit significant potential for abuse. Particularly relevant concerns are privacy problems, since sensor networks provide increased data collection capabilities (Gruteser 2003). Adversaries can use even seemingly innocuous data to derive sensitive information if they know how to correlate multiple sensor inputs. For example, in the famous "panda-hunter problem" (Ozturk 2004), the hunter can imply the position of pandas by monitoring the traffic. The main privacy problem, however, is not that sensor networks enable the collection of information. In fact, much information from sensor networks could probably be collected through direct site surveillance. Rather, sensor networks aggravate the privacy problem because they make large volumes of information easily available through remote access. Hence, adversaries need not be physically present to maintain surveillance. They can gather information in a low-risk, anonymous manner. Remote access also allows a single adversary to monitor multiple sites simultaneously (Chan 2003). Some of the more common attacks (Chan 2003; Gruteser 2003) against sensor privacy are:

- **Monitor and eavesdropping:** This is the most obvious attack to privacy. By listening to the data, the adversary could easily discover the communication contents. When the traffic conveys the control information about the sensor network configuration, which contains potentially more detailed information than accessible through the location server, the eavesdropping can act effectively against the privacy protection.
- **Traffic analysis:** Traffic analysis typically combines with monitoring and eavesdropping. An increase in the number of transmitted packets between certain nodes could signal that a specific sensor has registered activity. Through the analysis on the traffic, some sensors with special roles or activities can be effectively identified.
- **Camouflage:** Adversaries can insert their node or compromise the nodes to hide in the sensor network. After that these nodes can masquerade as a normal node to attract the packets, then misroute the packets, for example, forward the packets to the nodes conducting the privacy analysis.

Physical Attacks

Sensor networks typically operate in hostile outdoor environments. In such environments, the small form factor of the sensors, coupled with the unattended and distributed nature of their deployment, make them highly susceptible to physical attacks, that is, threats due to physical node destructions (Wang 2004).

Unlike many other attacks mentioned above, physical attacks destroy sensors permanently, so the losses are irreversible. For instance, attackers can extract cryptographic secrets, tamper with the associated circuitry, modify programming in the sensors, or replace them with malicious sensors under the control of the attacker (Wang 2004). Recent work has shown that standard sensor nodes, such as the MICA2 motes, can be compromised in less than one minute (Hartung 2004). While these results are not surprising given that the MICA2 lacks tamper resistant hardware protection, they provide a cautionary note about the speed of a well-trained attacker. If an adversary compromises a sensor node, then the code inside the physical node may be modified.

Countermeasures

Now we are in a position to describe the measures for satisfying security requirements and protecting the sensor network from attacks. We start with key establishment in wireless sensor networks, which lays the foundation for the security in a wireless sensor network, followed by defending against DoS attacks, secure broadcasting and multicasting, defending against attacks on routing protocols, combating traffic analysis attacks, defending against attacks on sensor privacy, intrusion detection, secure data aggregation, defending against physical attacks, and trust management.

Key Management Fundamentals

Key management issues in wireless networks are not unique to wireless sensor networks. Indeed, key establishment and management issues have been studied in depth outside of the wireless networking arena. Traditionally, key establishment is done using one of many public-key protocols. One of the more common is the Diffie-Hellman public key protocol, but there are many others. Most of the traditional techniques, however, are unsuitable in low power devices such as wireless sensor networks. This is due largely to the fact that typical key exchange techniques use asymmetric cryptography, also called public key cryptography. In this case, it is necessary to maintain two mathematically related keys, one of which is made public while the other is kept private. This allows data to be encrypted with the public key and decrypted only with the private key. The problem with asymmetric cryptography, in a wireless sensor network, is that it is typically too computationally intensive for the individual nodes in a sensor network. This is true in the general case, however, Gaubatz (2004), Gura (2004), Malan (2004), and Watro (2004) show that it is feasible with the right selection of algorithms.

Symmetric cryptography is therefore the typical choice for applications that cannot afford the computational complexity of asymmetric cryptography. Symmetric schemes utilize a single shared key known only between the two communicating hosts. This shared key is used for both encrypting and decrypting data. The traditional example of symmetric cryptography is data encryption standard (DES). The use of DES, however, is quite limited due to the fact that it can be broken relatively easily. In light of the shortcomings of DES, other symmetric cryptography systems have been proposed including triple DES (3DES), RC5, AES, and so on.

One major shortcoming of symmetric cryptography is the key exchange problem. Simply put, the key exchange problem derives from the fact that two communicating hosts must somehow know the shared key before they can communicate securely. So the problem that arises is how to ensure that the shared key is indeed shared between the two hosts who wish to communicate and no other rogue hosts who may wish to eavesdrop. How to distribute a shared key securely to communicating hosts is a nontrivial problem since predistributing the keys is not always feasible.

Key Establishment

One security aspect that receives a great deal of attention in wireless sensor networks is the area of key management. Wireless sensor networks are unique (among other embedded wireless networks) in this aspect due to their size, mobility, and computational/power constraints. Indeed, researchers envision wireless sensor networks to be orders of magnitude larger than their traditional embedded counterparts. This, coupled with the operational constraints described previously, makes secure key management an absolute necessity in most wireless sensor network designs. Because encryption and key management/establishment are so crucial to the defense of a wireless sensor network, with nearly all aspects of wireless sensor network defenses relying on solid encryption, we first begin with an overview of the unique key and encryption issues surrounding wireless sensor networks before discussing more specific sensor network defenses.

WSN Key Management Protocols

Random key predistribution schemes have several variants. Eschenauer and Gligor (2002) propose a key predistribution scheme that relies on probabilistic key sharing among nodes within the sensor network. Their system works by distributing a key ring to each participating node in the sensor network before deployment. Each key ring should consist of a number randomly chosen keys from a much larger pool of keys generated offline. An enhancement to this technique utilizing multiple keys is described by Chan (2003). Further enhancements are proposed by Deng (2005) and (Liu 2005) with additional analysis and enhancements provided by Hwang (2004). Using this technique, it is not necessary that each pair of nodes share a key. However, any two nodes that do share a key may use the shared key to establish a direct link to one another. Eschenauer and Gligor show that, while not perfect, it is probabilistically likely that large sensor networks will enjoy shared-key connectivity. Further, they demonstrate that such a technique can be extended to key revocation, rekeying, and the addition/deletion of nodes. The LEAP protocol described by Zhu et al. (2003) takes an approach that utilizes multiple keying mechanisms. Their observation is that no single security requirement accurately suites all types of communication in a wireless sensor network. Therefore, four different keys are used depending on whom the sensor node is communicating with. Sensors are preloaded with an initial key from which further keys can be established. As a security precaution, the initial key can be deleted after its use in order to ensure that a compromised sensor cannot add additional compromised nodes to the network.

In *PIKE* (Chan 2005), Chan and Perrig describe a mechanism for establishing a key between two sensor nodes that is based on the common trust of a third node somewhere within the sensor network. The nodes and their shared keys are spread over the network such that for any two nodes A and B, there is a node C that shares a key with both A and B. Therefore, the key establishment protocol between A and B can be securely routed through C.

Huang et al. (2003) propose a hybrid key establishment scheme that makes use of the difference in computational and energy constraints between a sensor node and the base station. They posit that an individual sensor node possesses far less computational power and energy than a base station.

In light of this, they propose placing the major cryptographic burden on the base station where the resources tend to be greater. On the sensor side, symmetric-key operations are used in place of their asymmetric alternatives. The sensor and the base station authenticate based on elliptic curve cryptography. Elliptic curve cryptography is often used in sensors due to the fact that relatively small key lengths are required to achieve a given level of security.

Huang et al. also use certificates to establish the legitimacy of a public key. The certificates are based on an elliptic curve implicit certificate scheme (Huang et al., 2003). Such certificates are useful to ensure both that the key belongs to a device and that the device is a legitimate member of the sensor network.

Each node obtains a certificate before joining the network using an out-of-band interface.

WSN and Public Key Cryptography

Two of the major techniques used to implement public-key cryptosystems are RSA and elliptic curve cryptography (ECC). Traditionally, these have been thought to be far too heavy in weight for use in wireless sensor networks.

Recently, however, several groups have successfully implemented public-key cryptography (to varying degrees) in wireless sensor networks. Gura et al. (2004) report that both RSA and elliptic curve cryptography are possible using 8-bit CPUs with ECC, demonstrating a performance advantage over RSA. Another advantage is that ECC's 160 bit keys result in shorter messages during transmission compared the 1024 bit RSA keys. In particular Gura et al. demonstrate that the point multiplication operations in ECC are an order of magnitude faster than private-key operations within RSA, and are comparable (though somewhat slower) to the RSA public-key operation.

Watro et al. (2004) show that portions of the RSA cryptosystem can be successfully applied to actual wireless sensors, specifically the UC Berkeley MICA2 motes (Hill et al., 2000). In particular, they implemented the public operations on the sensors themselves while offloading the private operations to devices better suited for the larger computational tasks. In this case, a laptop that was the TinyPK system described by Watro (2004) is designed specifically to allow authentication and key agreement between resource constrained sensors. The agreed upon keys may then be used in conjunction with the existing cryptosystem, TinySec (Karlof 2003). To do this, they implement the Diffie-Hellman key exchange algorithm and perform the public-key operations on the Berkeley motes.

The Diffie-Hellman key exchange algorithm used by Malan et al. (2004) is detailed in the following. In this case, a point G is selected from an elliptic curve E, both of which are public. A random integer KA is selected, which will act as the private key. The public key (TA in the case of the sender) is then TA = KA * G. The receiver performs a similar set of operations to compute TB = KB * G. Both peers can now easily compute the shared-secret using their own private keys and the public keys that have been exchanged. In this case, the sender computes KA * TB = KA * KB * G while the receiver computes KB * TA = KB * KA * G. Because KA * TB = KB * TA, the sender and the receiver now share a secret key.

DoS Countermeasures

Since denial-of-service attacks are so common, effective defenses must be available to combat them. One strategy in defending against the classic jamming attack is to identify the jammed part of the sensor network and effectively route around the unavailable portion. Wood and Stankovic (2002) describe a two phase approach where the nodes along the perimeter of the jammed region report their status to their neighbors who then collaboratively define the jammed region and simply route around it. To handle jamming at the MAC layer, nodes might utilize a MAC admission control that is rate limiting. This would allow the network to ignore those requests designed to exhaust the power reserves of a node. This, however, is not fool-proof as the network must be able to handle any legitimately large traffic volumes.

Overcoming rogue sensors that intentionally misroute messages can be done at the cost of redundancy. In this case, a sending node can send the message along multiple paths in an effort to increase the likelihood that the message will ultimately arrive at its destination. This has the advantage of effectively dealing with nodes that may not be malicious, but rather may have simply failed as it does not rely on a single node to route its messages. To overcome the transport layer flooding denial-of-service attack, Aura, Nikander and Leiwo (2001) suggest using the client puzzles posed by Juels and Brainard in an effort to discern a node's commitment to making the connection by utilizing some of their own resources. Aura et al. advocate that a server should force a client to commit its own resources first. Further, they suggest that a server should always force a client to commit more resources up front than the server. This strategy would likely be effective as long as the client has computational resources comparable to those of the server.

Detecting Node Replication Attacks

Parno et al. (2005) describe two algorithms: randomized multicast and line-selected multicast. Randomized multicast is an evolution of a node broadcasting strategy. In the simple node broadcasting strategy each sensor propagates an authenticated broadcast message throughout the entire sensor network. Any node that receives a conflicting or duplicated claim revokes the conflicting nodes. This strategy will work, but the communication cost is far too expensive. In order to reduce the communication cost, a deterministic multicast could be employed where nodes would share their locations with a set of witness nodes. In this case, witnesses are computed based on a node's ID. In the event that a node has been replicated on the network, two conflicting locations will be forwarded to the same witness who can then revoke the offending nodes. But since a witness is based on a node's ID, it can easily be computed by an attacker who can then compromise the witness nodes. Thus, securely utilizing a deterministic multicast strategy would require too many witnesses and the communication cost would be too high.

FUTURE TRENDS

Research on WSN security is still in infancy. Many key issues have not been sufficiently detailed or have even remained unexplored. In the near future, advanced security features may be built into the sensor nodes available in the market. While their prospects look shiny, these security functionalities have surprisingly received little attention from the research community. In the following, we describe the most interesting (in our sense) WSN-related research aspects.

1. **Building security policies for WSNs:** Due to their ad hoc topology, WSNs can not conform to traditional rigid security policies. WSN-oriented security policies should be flask enough to support the continuously modified network constituency and structure. The WSN architecture should therefore be flexible in their support of security policies, providing sufficient mechanisms for supporting the wide variety of real-world security policies. Appropriate formalisms to build, model, validate, verify, and test such architectures should be evolved.
2. **Developing scalable security mechanisms:** A common practice is to use exaggerated tools of information security, which decrease efficiency and system availability and introduce redundancy. Another effect of exaggeration of the security mechanisms is increasing the system complexity, which later influences implementation of a given project in practice, especially increasing expenses and decreasing efficiency. The solution of this inconsistency seems to be the introduction of scalable security model, which can change the security level depending on particular conditions of a given case. In this chapter a mechanism, which can modify the level of information security for each phase of a protocol, is presented. Parameters, which influence modification of the security level, are the risk of successful attack, probability of successful attack, and some measures of independence (leading to completeness) of security elements. The used security elements, which take care of the protection of information, are based mainly on PKI services and cryptographic modules.
3. **Securing hybrid broadband wireless sensor networks (HBWSNs):** High-speed WSNs begin to be widely used in different applications. Securing the corresponding flows encompasses the development of novel concepts that do not rely on thorough inspection of the transmitted packets but rather on the control of a set of relevant samples that are representative with respect to the total flow.
4. **Defining secure correlation functions: Two novels aspects are being investigated in the field of WSN security:** blind correlation and recursive signature. The first consists in correlating encrypted events without revealing their content in order to optimize the use of networking and processing resources. The second is applicable when, within a transmission chain, a set of nodes recursively sign the event. This is a particularly challenging problem in the WSN context because the intermediary nodes are resource-impoverished.

REFERENCES

Aura, T. N., P., & Leiwo, J. (2001). *DoS-resistant authentication with client puzzles.* Paper presented at the 8th International Workshop on Security Protocols (Vol. 2133, pp. 170-183). Springer-Verlag.

Blaß, E.-O. Z., & M. (2005). *Towards acceptable public-key encryption in sensor networks.* Paper presented at the ACM 2nd International Workshop on Ubiquitous Computing, Miami. INSTICC Press.

Capkun, S. H., & J.-P. (2006). Secure positioning in wireless networks. *IEEE Journal on Selected Areas in Communications, 24*(2), 221-232.

Carman, D. W. K., P. S., & Matt, B. J. (2000). Constraints and approaches for distributed sensor network security. *Glenwood, NAI Labs, Network Associates.* Retreived October 9, 2007, from www.cs.umbc.edu/courses/graduate/CMSC691A/Spring04/papers/nailabs_report_00-010_final.pdf

Chan, H. P., & A. (2003). Security and privacy in sensor networks. *IEEE Communications Magazine,* 103-105.

Chan, H. P., & A. (2005). *PIKE: Peer Intermediaries for key establishment in sensor networks.* Paper presented at the IEEE INFOCOM, Miami.

Chan, H. P., A., & Song, D. (2003). *Random key predistribution schemes for sensor networks.* Paper presented at the IEEE Symposium on Security and Privacy.

Deng, J. H., R., & Mishra, S. (2005). *Security, privacy, and fault-tolerance in wireless sensor networks.* Artech House.

Ekanayake, V. K., C., & Manohar, R. (2004). *An ultra low-power processor for sensor networks.* Paper presented at the ACM ASPLOS Conference, Boston.

Eschenauer, L. G., & V. D. (2002). *A key-management scheme for distributed sensor networks.* Paper presented at the 9th ACM Conference on Computer and Communications Security. Washington D.C: ACM Press.

Ganeriwal, S. C., S., & Han, C.-C., & Srivastava, M. B. (2005). *Secure time synchronization service for sensor networks.* Paper presented at the 4th ACM Workshop on Wireless Security, New York.

Gaubatz, G. K., J. P., & Sunar B. (2004). *Public key cryptography in sensor networks: Revisited.* Paper presented at the 1st European Workshop on Security in Ad hoc and Sensor Networks, Heidelberg, Germany.

Gruteser, M. S., G., Jain, A., Han, R., & Grunwald, D. (2003). *Privacy-aware location sensor networks.* Paper presented at the 9th Usenix Workshop on Hot topics in Operating Systems, Hawaii.

Gura, N. P., A., Wander, A. Eberle, A., & Shantz, S. (2004). *Comparing elliptic-curve cryptography and RSA on 8-bit CPUs.* Paper presented at the Workshop on Cryptographic hardware and Embedded Systems, San Francisco.

Hartung, C. B., J., & Han, R. (2004). *Node compromise in sensor networks: The need for secure systems.* University of Colorado at Boulder, Department of Computer Science. Retrieved October 9, 2007, from www.cs.colorado.edu/department/publications/reports/docs/CU-CS-990-05.pdf

Hill, J. S., R. Woo, A., Hollar, S., Culler, D. E., & Psiter, K. (2000). System architecture directions for networked sensors. *Architectural Support for Programming Languages and Operating Systems,* 93-104. Cambridge, MA.

Huang, Q. C., J. Kobayashi, H., Liu, B., & Zhang, J. (2003). *Fast authenticated key establishment protocols for self-organizing sensor networks.* Paper presented at the 2nd ACM Conference on Wireless Sensor Networks and Applications. San Diego: ACM Press.

Hwang, J. K., & Y. (2004). *Revisiting random key pre-distribution schemes for wireless sensor networks.* Paper presented at the 2nd ACM Workshop on Security of Ad Hoc and Sensor Networks, New York.

Karlof, C. S., N., & Wagner, D. (2003). Secure routing in wireless sensor networks: Attacks and countermeasures. *Elsevier's AdHoc Networks*

Journal, Special Issue on Sensor Network Applications and Protocols, 1(2-3), 293-315.

Lazos, L. P., & R. (2005). SERLOC: Robust localization for wireless sensor networks. *ACM Transactions on Sensor Networks, 1*(1), 73-100.

Liu, D. N., & P. (2003). *Efficient distribution of key chain commitments for broadcast authentication.* Paper presented at the10th Annual Network and Distributed System Security Symposium, San Diego.

Liu, D. N., & P. (2004). Multilevel μTesla: Brodcast authentication for distributed sensor networks. *Transactions on Embedded Computing Systems, 3*(4), 800-836.

Liu, D. N., P., & Li, R. (2005). Establishing pairwise keys in distributed sensor networks. *ACM Transactions on Information Systems Security, 8*(1), 41-47.

Malan, D. J. W., M., & Smith, M. D. (2004). *A public-key infrastructure for key distribution in TinyOS based on elliptic-curve cryptography.* Paper presented at the 1st Annual IEEE Communications Society Conference on Sensor and Ad Hoc Communications and Networks, Santa Clara, CA.

Ozturk, C. Z., Y., & Trappe, W. (2004). *Source-location privacy in energy-constrained sensor network routing.* Paper presented at the 2nd ACM Workshop on Security of Ad Hoc and Sensor Networks, New York.

Parno, B. P., A., & Gligor, V. (2005). *Distributed detection of node replication attacks in sensor networks.* Paper presented at the IEEE Symposium on Security and Privacy, Oakland, CA

Perrig, A. S., R. Tygar, J. D., Wen, V., & Culler, D. E. (2002). SPINS: Security protocols for sensor networks. *Wireless Networking, 8*(5), 521-534.

Perrig, A. S., J., & Wagner, D. (2004). Security in wireless sensor networks. *Communications ACM, 47*(6), 53-57.

Stankovic, J. A. (2003). *Real-time communication and coordination in embedded sensor networks. Proceedings of the IEEE, 91*(7).

Wang, X. G., W. Schosek, K., Chellappan, S., & Xuan, D. (2004). Sensor network configuration under physical attacks. *D. o. C. S. a. engineering.* Ohio-State University. Retrieved October 9, 2007, from www.springerlink.com/index/E5T6KWNK-MABWR672.pdf

Watro, R. K., D. Cuti, S., Gardiner, C., Lynn, C., & Kruus, P. (2004). *TinyPK: Securing sensor networks with public key technology.* Paper presented at the 2nd ACM Workshop on Security of Ad hoc and Sensor Networks, New York. ACM Press.

Wood, A. D. S., & J. A. (2002). Denial of service in sensor networks. *Computer, 35*(10), 54-62.

Zhu, S. S., S., & Jajodia, S. (2003). *LEAP: Efficient security mechanisms for large-scale distributed sensor networks.* Paper presented at the 10th ACM Conference on Computer and Communications Security, New York. ACM Press.

KEY TERMS

Camouflage: Adversaries can insert their node or compromise the nodes to hide in the sensor network. After that these nodes can masquerade as a normal node to attract the packets, then misroute the packets.

Denial-of-Service Attack: An attack aiming at disrupting the acquisition of information within a geographical zone or preventing the communication of alert and signalling messages between sensor nodes.

Key Management: Process of generating, validating, exchanging, and renewing asymmetric and symmetric keys.

Rate Monitoring Attack: A rate monitoring attack simply makes use of the idea that nodes closest to the base station tend to forward more packets than those farther away from the base station.

Wireless Sensor Network (WSN): Dense collection of tiny sensor motes deployed in a region of interest to gather information about a specific phenomenon for later analysis. WSNs allow efficient, distributed, and collaborative control of various natural and human events.

Wormhole Attack: In a wormhole attack, an attacker receives packets at one point in the network, "tunnels" them to another point in the network, and then replays them into the network from that point.

Chapter XXXVI
Routing Security in Wireless Sensor Networks

A.R. Naseer
King Fahd University of Petroleum & Minerals, Dhahran

Ismat K. Maarouf
King Fahd University of Petroleum & Minerals, Dhahran

Ashraf S. Hasan
King Fahd University of Petroleum & Minerals, Dhahran

ABSTRACT

Since routing is a fundamental operation in all types of networks, ensuring routing security is a necessary requirement to guarantee the success of routing operation. Securing routing task gets more challenging as the target network lacks an infrastructure-based routing operation. This infrastructure-less nature that invites a multihop routing operation is one of the main features of wireless sensor networks that raises the importance of secure routing problem in these networks. Moreover, the risky environment, application criticality, and resources limitations and scarcity exhibited by wireless sensor networks make the task of secure routing much more challenging. All these factors motivate researchers to find novel solutions and approaches that would be different from the usual approaches adopted in other types of networks. The purpose of this chapter is to provide a comprehensive treatment of the routing security problem in wireless sensor networks. The discussion flow of the problem in this chapter begins with an overview on wireless sensor networks that focuses on routing aspects to indicate the special characteristics of wireless sensor networks from routing perspective. The chapter then introduces the problem of secure routing in wireless sensor networks and illustrates how crucial the problem is to different networking aspects. This is followed by a detailed analysis of routing threats and attacks that are more specific to routing operation in wireless sensor networks. A research-guiding approach is then presented to the reader that analyzes and criticizes different techniques and solution directions for the secure routing problem in wireless sensor network. This is supported by state-of-the-art and familiar examples from the literature. The chapter finally concludes with a summary and future research directions in this field.

INTRODUCTION

Wireless sensor networks (WSNs) are gaining popularity due to the fact that they provide feasible and economical solution to many of the most challenging problems in a wide variety of applications such as military applications, healthcare, traffic monitoring, pollution/weather monitoring, wildlife tracking, remote sensing, and so forth. This has fuelled extensive research to address the critical issues of providing security, intrusion detection/tolerance, high availability, and survivability of the sensor network.

The issue of secure routing in wireless and mobile computing is a major challenging design factor in different networking aspects. However, the problem gets more complicated when considering infrastructure-less networks that exhibit even more constraints and new types of attacks. In the continuously and rapidly evolving area of wireless communication, the field of wireless sensor networks comes into the picture as a very hot area of research in all its aspects. WSN is a multihop network that is actually one type of ad hoc networks. However, WSN draws the special attention of researchers due to the fact that it exhibits more constraints and critical conditions than normal ad hoc networks in terms of power sources, computing capabilities, memory capacity, and other factors. This requires different approaches and protocol engineering directions from those applied to normal ad hoc networks.

WSNs are susceptible to several types of attacks at different layers of the network since they are normally deployed in open and unprotected environments and are constituted of cheap small devices with limited computational power, limited memory, and limited battery life. Nodes of a sensor network cannot be trusted for the correct execution of the critical network functions. Node misbehavior may range from simple selfishness or lack of collaboration due to the need for power saving, to active attacks aiming at denial-of-service and subversion of traffic. A sensor network without sufficient protection from these attacks may not be deployable in many areas. Intrusion preventive mechanisms such as encryption and authentication can be applied to protect WSNs against some types of attacks. Key management is the cornerstone of security services such as encryption and authentication in wireless sensor network. Research seeking low-cost key management techniques that can survive node compromises in sensor networks has been a very active area, yielding several novel key predistribution schemes. However, there are some attacks for which there is no known prevention method, such as wormhole attack. Moreover, there are no guarantees that the preventive methods will be able to hold the intruders. Hence it is necessary to use some mechanisms of intrusion detection. Besides preventing the intruder from causing damages to the network, the intrusion detection system can acquire information related to the attack techniques, helping in the development of better prevention systems.

One special aspect in WSN is the provision of secure routing. As mentioned previously, the nature of WSN complicates the security requirements and adds difficulties in solving security problems. In fact, secure routing in WSN is actually still not captured well in the research field. One main reason is that the design of a routing protocol is biased towards solving the problem of power limitations and reducing communication overhead while keeping security concerns at a later phase to be integrated with the current routing solutions.

Among different approaches in solving the problem of secure routing in WSN, reputation system-based solution is one technique that has generated enough interest among WSN researching community. Reputation systems attempt to provide security by allowing different nodes rate each other based on their routing activities and behavior analysis. When a node has an experience profile about its neighbors, it may select the node that it trusts more, and, hence, achieve a secure routing operation.

The rest of the chapter is organized as follows. Section 2 of the chapter provides the relevant background material covering an overview of WSN that includes WSN definition, sensor node structure, applications, and so forth. As WSN is a class of mobile ad hoc networks (MANET), the main differences between WSN and MANET will be presented. These differences are explained in a way that emphasizes to the reader how they make WSN an independent research target as compared with MANET.

Section 3, being the routing security section, defines precisely the problem of secure routing in general. This section will discuss the requirements for secure routing in WSN. This will be followed by the challenges and constraints in WSN to achieve secure routing. After the reader understands the routing security problem in WSN, the reader will be given a critical discussion about the importance of this problem. This will also include an explanation of the relationship between routing security and different network aspects like survivability, connectivity and network partitioning, throughput, packet delay, and so forth.

Section 4 on routing attacks and threats presents in brief the different possible communication models and trust relationships between WSN nodes that a threat will be based on. It will clearly show how researcher assumptions on nodes communication models and nodes relationships will impact the security analysis. In this section, the reader will be provided with a global picture of the approaches and techniques that are used by the attackers. This will also include a discussion of the holes and weakness points that are exploited to achieve such attacks. Some examples of famous attacks will be given with explanation. The explanation will focus on how the attack works by exploiting the routing protocol aspects. Thus, the section will also show the robustness level that is provided by different routing protocols. How we can think secure and provide robust solutions against routing attacks and threats will be the subject of this section. The section gives examples of how an attack can be prevented or detected as a tip for a more general approach.

Section 5, "Routing Security Solutions and Techniques," explains the objectives to be met when developing a routing security solution. These security objectives are explained under the lights of WSN constraints. Thus, the reader will be aware of the tradeoffs that should be considered in the design. A first step in the solution design is to decide whether the solution will prevent the attack or avoid it after detection. This section gives a comparison between these approaches based on the severity of the threats and WSN conditions and resources availability. In this section, cryptographic-based and noncryptographic-based approaches will be discussed and the tradeoffs with resources will also be analyzed. Examples of such solutions will be provided with a focus on how these solutions meet secure routing goals and what drawbacks they exhibit. Reputation-based solution will be discussed as a detection approach by presenting the general concept of reputation systems, followed by suggestions and approaches in reputation system solutions that can fit WSN secure routing requirements.

BACKGROUND

Wireless Sensor Network Overview

WSN is an ad hoc-like deployment of a large number of sensor nodes that are intended to monitor and communicate information pertaining to a phenomenon or an event of interest. The deployment is either random or utilizes predetermined locations near or inside the phenomenon. The typical deployment scenario of WSN is depicted in Figure 1, where a number of sensor nodes are scattered in the sensor field. The sensor nodes collect data from the field and route the data through the multihop structure of the network to a specialized node referred to as the sink or base station. Finally, the sink may communicate the raw data or a processed version to the end-user utilizing an infrastructure network such as the Internet.

Applications of WSN

Due to the versatility and flexibility of WSN, it has found many applications especially in situations where direct probing or measurement of the event of interest is either costly or risky. WSNs facilitate many applications including:

Figure 1. Sensor nodes deployment in a sensor field

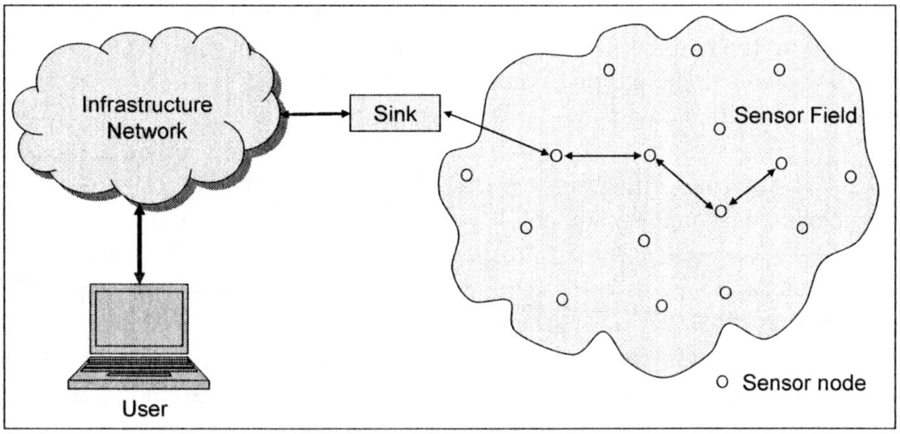

- **Military applications:**
  - Monitoring friendly forces, equipment, and ammunition
  - Battlefield surveillance
  - Reconnaissance of opposing forces and terrain
  - Targeting guidance
  - Battle damage assessment
  - Nuclear, biological ,and chemical (NBC) attack detection and reconnaissance
- **Environmental and precision agriculture Applications:**
  - Tracking the movements of birds, small animals, and insects
  - Monitoring environmental conditions that affect irrigation
  - Earth, and environmental monitoring in marine, soil, and atmospheric contexts
  - Forest fire detection
  - Meteorological or geophysical research
  - Flood detection
  - Pollution study
  - Fertilizer and humidity sensing for farms
- **Health applications:**
  - Providing interfaces for the disabled
  - Integrated patient monitoring
  - Administration in hospitals
  - Telemonitoring of human physiological data
  - Tracking and monitoring doctors and patients inside a hospital
- **Facility management and commercial aApplications:**
  - Managing inventory
  - Monitoring product quality
  - Robot control and guidance in automatic manufacturing environments
  - Interactive museums
  - Smart structures with sensor nodes embedded inside
  - Vehicle tracking and detection
  - Machine surveillance and preventive maintenance
  - Intelligent building
- **Telematics:**
  - For roads and traffic management

Sensor Node Structure

The basic structure of the sensor node is shown in Figure 2(a), while the protocol stack for the node is shown in Figure 2(b). The node contains an embedded system that performs the following main functions:

- **Sensing:** Every node should have the ability to observe and/or control the physical environment.
- **Computing:** The collected data from physical environment through sensing function are

processed to produce beneficial information.

- **Communication:** Every node should be able to communicate and exchange raw data or processed information among them.

To accomplish the above tasks, the sensor node comprises of four main components: the controller/memory module, the power supply module, the RF transceiver module, and the sensors/actuators module. In addition, the sensor unit may optionally contain two other modules, namely, the position finding module and the mobilizer module. While the former module is sometimes needed to determine the location of the node, the latter allows mobilizing the node to carry out certain tasks in the field of interest. The brief description of these modules of a sensor node is presented next.

Controller/memory module: The controller consists of a processor and a memory system. The processor manages the procedures that make the sensor node collaborate with the other nodes to carry out the assigned sensing tasks. The memory system stores data, software, and application programs required to run the node. Though the higher computational powers are being made available in smaller and smaller processors and controllers, processing and memory units of sensor nodes are still scarce resources. For instance, the processing unit of a smart dust mote prototype is a 4 MHz Atmel AVR8535 micro-controller with 8 KB instruction flash memory, 512 bytes RAM, and 512 bytes EEPROM (Perrig, Szewezyk, Wen, Culler, & Tygar, 2001). TinyOS operating system is used on this processor, which has 3500 bytes OS code space and 4500 bytes available code space.

Power supply module: One of the most important components of a sensor node is the power unit. Since the sensor nodes are often inaccessible, power is considered a scarce resource and the lifetime of a sensor network depends on the lifetime of the power resources of the nodes. Power is also a scarce resource due to the size limitations. For instance, the total stored energy in a smart dust mote is of the order of 1 J (Pottie & Kaiser, 2000). It is possible to extend the lifetime of the sensor networks by energy scavenging, which means extracting energy from the environment. Solar cells are an example for the techniques used for energy scavenging.

Radio transceiver module: The radio transceiver unit is responsible for connecting the node to the network.

Figure 2. Wireless sensor network: (a) node structure, and (b) protocol stack

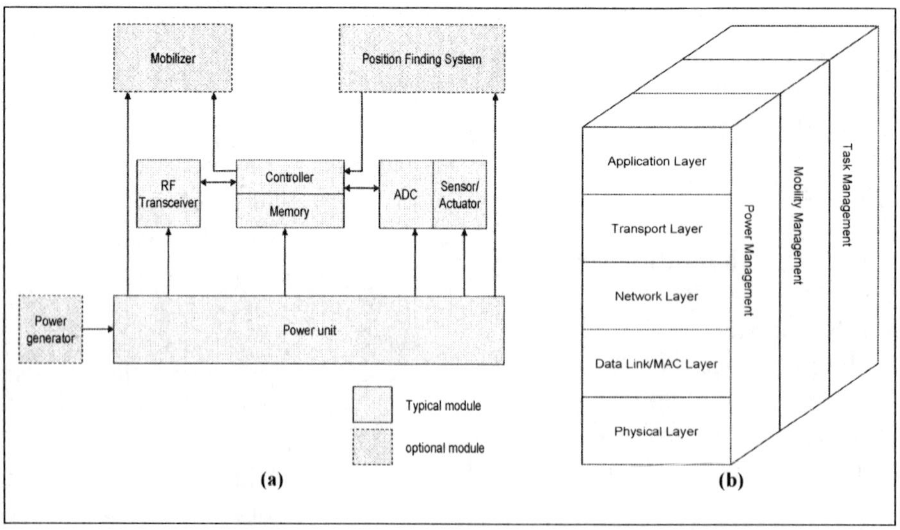

Sensors/actuators module: Sensing and actuator units are usually composed of sensors, actuators, and analog to digital (for sensing) and digital to analog (for actuating) converters (ADC/DAC). The analog signals produced by the sensors based on the observed phenomenon are converted to digital signals by the ADC, and then fed into the processing unit or the controller. On the other hand, the digital signals produced by the controller are converted to analog signals by the DAC to feed the actuators.

Position finding module: In some instances, the operation of the sensor node, as in some of the routing techniques, requires knowledge of location with high accuracy. These nodes will be equipped with a module that is used to determine either the relative or absolute location of the node. The determination of the absolute location can be obtained using global positioning system (GPS), while the relative location can be calculated using the less expensive signal triangulation or multilateration techniques (Savvides, Han, & Srivastava, 2001).

Mobilizer module: For particular application of WSN, it may be required to move a subset of the deployed sensor node into specific positions in the field of interest. For these nodes, the mobilizer module allows the node to move or change its location to perform the required task.

WSN vs. MANET

While WSN and mobile ad hoc networks share a lot of commonalities, there are distinct differences that exist between the two technologies. These differences include the following characteristics.

Node population: Typically the number of nodes deployed in a WSN is orders of magnitude greater than the number of nodes in a MANET. This is of course a function of the application and the sensor field. In addition, wireless sensor nodes usually have shorter communication range compared to their counterparts in MANET. This implies that the deployment density for sensor nodes may be significantly higher than that for the MANET.

Resource constraints: An obvious difference between MANET and WSN is resource constraints. Resources include power, memory, and processing capabilities. Although both networks suffer from resource deficiency, WSNs are more constrained and limited by such resources, especially in power. MANET nodes are typically laptops or handheld devices that have greater provisions in terms of power and processing capability which is not the case for little sized sensor nodes. Any protocol design and implementation targeting WSN from the physical to the application layer must consider resource usage optimization not as an additional feature in the system but as a main design goal.

Communication and topology models: The most used communication model adopted for MANET is the point-to-point model. While this model is also applicable in WSN, other communication models such as broadcasting and multicasting are more realistic and representative of the intended applications. In addition, the topology for WSN is highly variable compared to that for MANETs. The loss of a sensor node due to a battery running out or destruction results in a topological change in the network for which the WSN has to respond and self-organize.

Application characteristics: WSN targets a great range of applications as mentioned in section 2.2. Therefore, WSN is expected to have different requirements in terms of node density, sensing functions, routing activity, and so forth compared to those for MANET. This variation also exists in MANET but to a lesser degree as the number of applications for ad hoc networks is not as great as that for WSN.

Addressing and identification: WSN nodes usually do not possess a unique identification ID as opposed to ad hoc node in a MANET where every node is identified by its media access control (MAC) address or the Internet address. Nodes within a sensor field organize and establish a mechanism to identify adjacent nodes and perform the required functionality.

ROUTING SECURITY

What is Routing Security

Routing is a fundamental operation in almost all types of networks because of the introduction of interdomain communication. Ensuring routing security is a necessary requirement to guarantee the success of routing operation. When we talk about secure routing, we are concerned with security problems that may occur due to improper actions from an assumed router. These undesired actions can be related either to the router identity or the router behavior. If the router has an undesirable identity or authorization, the router is considered as an intruder who might perform serious attacks. Such attacks can be avoided by providing security services that validate the routers' identities. On the other hand, a router that misbehaves in the network by performing undesirable routing operations also contributes to the routing security problem. However, the attacks caused by misbehaving routers can be avoided by mechanisms that validate and evaluate the router behavior in the network.

Secure routing tasks get more challenging as the target network lacks an infrastructure-based routing operation. This infrastructure-less nature that invites a multihop routing operation is one of the main features of WSN networks that raise the importance of secure routing problem in these networks. Moreover, the risky environment, application criticality, and resources limitations and scarcity exhibited by WSN networks make the tasks of secure routing much more challenging.

Why Routing Security is Important in WSN

Secure routing in WSN is important for both securing obtained information as well as protecting the network performance from degradation. Most WSN applications carry and deliver very critical and secret information like in military and health applications. A WSN network infected by malicious nodes can alter or inject incorrect information, misroute packets, analyze data, or do not forward packets to their destination. Thus, having a secure routing protocol or framework can protect data exchange, secure information delivery, and maintain and protect the value of the communicated information.

Insecure routing can cause performance degradation as well. For example, nonforwarding attacks decrease the system throughput since packets will be retransmitted many times and they are not delivered. Denial-of-service attacks can increase the packet delay since some nodes acting as routers will be busy in responding to the attack and enforced to delay the processing of other packets. An infected WSN network can be partitioned into different parts that cannot communicate among each other due to nonforwarding attacks. This leads to the demand of increasing the number of sensors or changing the node deployment to return the network connectivity. This is very expensive; however, it can be avoided if a good secure routing solution is adopted.

Network resources are also affected by insecure routing. For example, denial-of-service attacks effect resource availability, whether we consider an offended node as a resource for routing or we consider the availability of data itself. Also, this attack forces offended nodes to consume unnecessary energy on packet reception and processing.

As we can see, the information value and the network performance are directly affected by the security level of the routing operation in WSN. A secure routing solution, thus, should provide information protection and performance maintenance. However, any proposed solution should account for the overhead impact on the network performance. For example, a secure routing solution may introduce an overhead that decreases the network throughput. If this degradation is more than the one resulted from the attack; it is better not to implement the solution in this regard.

ROUTING ATTACKS AND THREATS IN WSN

The designing of a secure routing protocol becomes very essential as the weaker defender (i.e., sensor nodes) in the sensor network has the greatest inherent disadvantage of insecure wireless communication, limited node capabilities, possible insider threats, and the stronger attacker has the all-time advantage of possessing powerful laptops with high energy and long range communication to launch severe attack to the network. Most of the routing protocols have not been designed with security as a goal. All of the proposed network routing protocols in the literature are more prone to attacks. Attackers can attract or repel traffic flows, increase latency, or disable the entire network, sometimes with little effort.

Threat Models

In order to define a robust security model, specification of both the security requirements and the threat model are required. The security requirements identify the properties that have to be enforced and the initial assumptions. The threat model formulates the hypothesis regarding the attacker's capabilities and its possible behavior. A common assumption is that the attacker is compliant with the Dolev-Yao threat model (Dolev & Yao, 1983) which is often used to formally analyze crypto-protocols in communication networks. According to this model, when two communicating parties communicate over an insecure channel, the attacker can gain control over the communication network to perform the following actions:

- Over hear the messages between the parties, intercept them, and prevent their delivery to the intended recipient.
- Introduce forged messages into the system using all the available information.

But this threat model also assumes that the end nodes are not themselves subject to attack. In order to take into account the distinguishing feature of WSNs that the sensors may be unattended and end nodes cannot, in general, be trusted, the following more powerful action is required to be included in the model:

- An attacker can capture a sensor node and acquire all the information stored within it.

Considering the above modified model, the attacks can be categorized as passive and active attacks. In passive attacks, eavesdropper can continuously monitor the whole sensor network and can launch two types of passive attacks: (i) cipher text attack wherein given the cipher text, the adversary tries to recover the encryption key, and (ii) chosen plain text attack wherein the attacker can feed the sensor with known data and then observe the encrypted message sent by the sensor. In active attacks, the attacker can capture a sensor, stealing all the information and keys stored in the sensor. Hence, providing, maintaining, and ensuring proper confidentiality and authenticity of data is a paramount importance within the limited inherent constraints of the underlying wireless sensor networks.

Sensor network attackers can be classified into two categories depending upon their capabilities (Karlof & Wagner, 2003). They are mote-class attacker and laptop-class attacker.

Mote-class attacker has access to a few ordinary sensor nodes with lesser capability and might only be able to jam the radio link in its immediate vicinity. They have limited range and cannot eavesdrop on entire network, moreover, cannot coordinate their efforts to bring down the network.

Laptop-class attacker has access to more powerful devices like laptops with greater battery power, more capable processor, a high-power transmitter, and a sensitive antenna. These attackers might be able to jam the entire network using a stronger transmitter and might be able to eavesdrop on an entire network. Laptop-class attackers might possess a high bandwidth, low-latency communication channel invisible to legitimate sensor nodes thereby setting up separate channels to allow such attackers to communicate and coordinate their efforts.

Further, sensor network attacks can be classified as outsider (external) attacks and insider (internal) attacks. Outsider attacks are launched by outsiders who have no special or legitimate access to the sensor network, that is, they do not have authentic keying material to participate in network operations as legitimate nodes. Insider attacks occur when an authorized participant in the sensor network has gone bad or compromised. The insider attack may be mounted from either compromised sensor nodes running a malicious code or attackers using laptop-class devices to attack the network after stealing the key material, code and data from legitimate nodes. Outsider attackers, once in full control of certain nodes, can become insider ones able to launch more subtle attacks. Insider attacks are generally more difficult to defend against than the outsider ones because of their possession of keying material.

In most of the threat models proposed in the literature, it is assumed that the environments in which the sensors deployed are risky and untrusted. Each sensor trusts itself, but sensors do not trust each other. Further, it is assumed that all the compromised sensors in the sensor network are compromised by the same attacker and thus collude to compromise the network. The attacker may compromise multiple sensor nodes in the network, and there is no upper bound on the number of compromised nodes. However, the attacker cannot compromise the base station, also termed as sink, which is typically resourceful and well protected. Once a sensor node is compromised, all the secret keys, data, and code stored on it are exposed to the attacker. The attacker can load a compromised node with secret keys obtained from other nodes, termed as collision, among compromised nodes. In other words, the goal of the attacker is to uncover the keys used in the system in order to disrupt the network operation. In order to achieve this, the attacker compromises individual nodes and fosters collusion among nodes. The main objective of node collusion is to incrementally aggregate the uncovered keys of individual nodes to a level that all encrypted traffic in the network is completely revealed. It is also assumed that the attacker cannot successfully compromise a node during the sensor deployment phase which is short, that is, the interval of tens of seconds when each sensor bootstraps itself, during which the sensor nodes obtain their location information and derive few keys. Indeed, such attacks can be prevented in many of the real-life scenarios when appropriate network planning and deployment are carried out to keep away attackers during the bootstrapping process. However, it should be noted that stronger threat (attacker) models need to be applied when we consider tactical military network deployment for war-field surveillance, whereas for noncritical commodity, sensor networks a less strong threat model may suffice.

A new threat model to communication confidentiality in WSNs termed as "smart attacker model" is introduced by Di Pietro, Mancini, and Mei (2006). All the random-key predeployment schemes (see section 5 for detailed discussion) proposed in the literature use an oblivious attacker model that at each step the attack sequence randomly chooses a sensor node to tamper without taking advantage of the information regarding the keys acquired during the previous attacks. Contrary to this, the smart attacker model greedily uses the previous attacks keys acquired information to choose the best sensor to tamper with in order to compromise the communication confidentiality. This reduces greatly the level of communication confidentiality provided by all the random key predeployment schemes

Routing Attacks and Examples

Any event that decreases or eliminates a network's capacity to perform its expected function is termed as a denial-of-service attack or commonly known as a DoS attack (Wood & Stankovic, 2002). Some of the major causes for DoS attacks are hardware failures, software bugs, resource exhaustion, malicious attacks, and environmental conditions. A significant challenge in securing large sensor networks is their inherent self-organizing, decentralized nature. Many of the network deployments are vulnerable to immensely more powerful attackers. Considering the layered network architecture of sensor networks depicted in Figure 2(b), the DoS vulnerabilities to the first four layers of the stack can be identified (Wood & Stankovic, 2002) as:

- **Physical layer attacks:** The most common attacks to the physical layer of a WSN are jamming and node physical tampering.
- **Data link layer attacks:** Collisions, unfairness, or exhaustion of resources are the attacks that can be launched against the data link layer of a sensor network.

- **Network layer attacks:** The possible routing layer attacks are routing information spoofing, alteration or replay, blackhole and selective forwarding attacks, sinkhole attacks, Sybil attacks, wormhole attacks, HELLO flood attacks, and acknowledgement spoofing.
- **Transport layer attacks:** The most common attacks to transport layer are flooding attacks and desynchronization attacks.

Since our main focus in this chapter is towards routing security, a detailed discussion on network layer or routing attacks will be presented next.

Sensor network routing attacks can be classified into the following categories (Karlof & Wagner, 2003):

- Routing information spoofing, alteration or replay
- Blackhole and selective forwarding attacks
- Sinkhole attacks
- Sybil attacks
- Wormhole attacks
- HELLO flood attacks
- Acknowledgement spoofing

Routing information spoofing, alteration, or replay: Targeting the routing information exchanged between the nodes is the most direct attack against a routing protocol. By spoofing, altering, or replaying routing information, an attacker can disrupt the network by creating routing loops, attracting or repelling network traffic, extending or shortening source routes, generating false error messages, partitioning the network, or increasing the end-to-end latency.

Most of the sensor network routing protocols such as TinyOS beaconing, directed diffusion and its multipath variant, geographic routing (e.g., GPSR, geographic and energy aware routing [GEAR]), minimum cost forwarding, rumor routing, energy conserving, and topology maintenance protocols (e.g., SPAN, GAF, CEC, AFECA) are prone to bogus routing information attacks.

For example, since routing updates are not authenticated in a TinyOS beaconing protocol, it is possible for any malicious node to claim itself to be a base station and become the destination of all traffic in the network. Mote class attackers can create very easily routing loops by spoofing routing updates. In GPSR, an adversary can forge location advertisements to create routing loops in data flows without having to actively participate in packet forwarding.

Black hole and selective forwarding attack: Multihop networks basically work on the assumption that nodes will participate faithfully in the forwarding of the received messages. In a blackhole attack, a malicious node refuses to forward every packet it receives thereby behaving like a block hole. In a selective forwarding attack, a malicious node selectively forwards the packets, that is, a malicious node may refuse to forward certain messages and simply drop them thereby ensuring that these packets are not propagated any further. The malicious node interested in suppressing or modifying the packets originating from a few selected nodes can reliably forward the remaining traffic thereby limiting the suspicion of its misbehavior. In order to launch a selective forwarding attack effectively, the attacker must follow the path of least resistance and attempt to include explicitly the attacker's self on the actual path of the data flow.

Most of the sensor network routing protocols such as TinyOS beaconing, directed diffusion and its multipath variant, geographic routing (e.g., GPSR, GEAR), minimum cost forwarding, clustering-based protocols (e.g., LEACH, TEEN, PEGASIS), and rumor routing, are highly prone to selective forwarding attacks.

For example, In LEACH protocol, nodes choose a cluster-head based on received signal strength. A laptop-class attacker can take advantage of this to send a powerful advertisement to all nodes in the network in order to mount a selective forwarding attack on the entire network using a small number of nodes if the target number of cluster-heads or the size of the network is sufficiently small.

Sinkhole attacks: In a sinkhole attack, the goal of the attacker is to attract nearly all the traffic from a particular region through a compromised node, thereby creating a metaphorical sinkhole with the attacker at the center. Sinkhole attacks typically work by making a compromised node appear especially attractive to surrounding nodes with respect to the routing algorithm.

Most of the sensor network routing protocols such as TinyOS beaconing, directed diffusion and its multipath variant, minimum cost forwarding, and rumor routing are prone to sinkhole attacks

For example, a laptop-class attacker with a powerful transmitter can actually provide a high quality route by transmitting with enough power to reach the base station in a single hop. Because of this high quality route through the compromised node, each neighboring node of the adversary will forward packets destined for a base station through the adversary and also propagate the attractiveness of the route to its neighbors. Due to the specialized communication pattern used, hierarchical sensor networks are highly susceptible to sinkhole attacks. In hierarchical sensor networks, all packets share the same ultimate destination, that is, a base station; a compromised node is required only to provide a single high quality route to the base station in order to attract a potentially large number of nodes.

Wormhole attacks: In the wormhole attack, an attacker tunnels messages received in one region of the network over a low-latency link and replays them in a different region. Wormhole attacks more commonly involve two distant malicious nodes colluding to understate their distance from each other by relaying packets along an out-of-bound channel available only to the attacker. An adversary situated close to a base station may be able to completely disrupt routing by creating a well-placed wormhole. An adversary could convince nodes who would normally be multiple hops from a base station that they are only one or two hops away via the wormhole. This can create a sinkhole since the adversary on the other side of the wormhole can artificially provide a high quality route to the base station thereby drawing through it potentially all traffic in the surrounding area if alternate routes are significantly less attractive. This will be the case always when the endpoint of the wormhole is relatively far from a base station. Wormholes are normally used in combination with selective forwarding. Detection becomes potentially difficult when used in conjunction with the Sybil attack.

Most of the sensor network routing protocols such as TinyOS beaconing, directed diffusion and its multipath variant, minimum cost forwarding, and rumor routing are prone to wormhole attacks

For example, protocols that construct a topology initiated by a base station are most susceptible to wormhole and sinkhole attacks. A wormhole is most effective when used to create sinkholes or artificial links that attract traffic. In sensor network employing TinyOS beaconing protocol, a powerful laptop-class attacker can launch a combined wormhole/sinkhole attack.

Sybil attacks: In a sybil attack, a single node presents multiple identities to other nodes in the network. An attacker can advertise multiple bogus nodes surrounding each target in a circle or sphere region, each claiming to have maximum energy.

Most of the sensor network routing protocols such as TinyOS beaconing, directed diffusion and its multipath variant, geographic routing (e.g., GPSR, GEAR), rumor routing, and energy conserving and topology maintenance protocols (e.g., SPAN, GAF, CEC, AFECA) are prone to Sybil attacks

For example, Sybil attacks pose a significant threat to geographic routing protocols. Location aware routing often requires nodes to exchange coordinate information with their neighbors to efficiently route geographically addressed packets. It is only reasonable to expect a node to accept but a single set of coordinates from each of its neighbors, but by using the Sybil attack an adversary can be in more than one location at once. GEAR tries to distribute the responsibility of routing based on remaining energy, so an appropriate attack would be to always advertise maximum energy as well.

HELLO flood attacks: In many routing protocols, nodes broadcast HELLO packets to announce their presence to their neighbors, and a node receiving such a packet may assume that it is within the normal radio range of the sender. This assumption can be exploited by a laptop-class attacker by broadcasting routing and other information with large enough transmission power in order to convince every node in the network that the adversary is its neighbor. The HELLO flood attack uses a single hop broadcast to transmit a message to a large number of receivers.

For example, an attacker advertising a very high-quality route to the base station to every node in the network can cause a large number of nodes to attempt to use this route, but those nodes sufficiently far away from the attacker will be sending their messages into oblivion thereby putting the network in a state of confusion. An attacker does not necessarily need to be able to construct legitimate traffic in order to use the HELLO flood attack. The attacker can simply rebroadcast overhead packets with enough power to be received by every node in the network. HELLLO floods can also be thought of as one-way, broadcast wormholes.

Most of the sensor network routing protocols such as TinyOS beaconing, directed diffusion and its multipath variant, minimum cost forwarding, clustering-based protocols (e.g., LEACH, TEEN, PEGASIS), and energy conserving and topology maintenance protocols (e.g., SPAN, GAF, CEC, AFECA) are prone to HELLO Flood attacks.

For example, in sensor network employing TinyOS beaconing protocol, a laptop-class attacker with a powerful transmitter can launch a HELLO flood attack to broadcast a routing update powerful enough to reach the entire network thereby causing every node to mark the attacker as its parent. Most nodes will be likely out of normal radio range of both a true base station and the attacker. The network is crippled as the majority of the nodes are stranded sending packets into oblivion.

In a sensor network employing minimum cost forwarding protocol, a laptop-class attacker can use the HELLO flood attack to disable the entire network by transmitting an advertisement with cost zero powerful enough to be received by every node in the network. In LEACH protocol, nodes choose a cluster-head based on received signal strength. A laptop-class attacker can take advantage of this to make every node choose the attacker as its cluster-head by using the HELLO flood attack to send a powerful advertisement to all nodes in order to disable the entire network.

Acknowledgement spoofing: Several sensor network routing algorithms that rely on implicit or explicit link layer acknowledgements are susceptible to acknowledgement spoofing. Due to the inherent broadcast medium, an attacker can spoof link layer acknowledgements for "overheard" packets addressed to the neighboring nodes with the main objective to convince the sender that a weak link is strong or that a dead or disabled node is alive.

For example, a routing protocol may select the next hop in a path using link reliability. A subtle way of manipulating such a scheme is to artificially reinforce a weak or dead link. Since packets sent along weak or dead links are lost, an adversary can effectively mount a selective forwarding attack using acknowledgement spoofing by encouraging the target node to transmit packets on those links.

Countermeasures Against Routing Attacks

Countermeasures against outsider attacks: The majority of outsider attacks against sensor network routing protocols can be prevented by simple link layer encryption and authentication using a globally shared key. Link layer acknowledgements can also be authenticated. In this case, Sybil attack is no longer relevant since nodes are unwilling to accept a single identity of the attacker. The majority of selective forwarding and sinkhole attacks are not possible because the attacker is prevented from joining the network. Although an attacker is prevented from joining the network, nothing prevents the attacker from using a wormhole to tunnel packets sent by legitimate nodes in one part of the network to legitimate nodes in another part to convince them that they are neighbors or amplifying an overheard broadcast packet with sufficient power to be received by every node in the network.

Countermeasures against insider attacks: In the presence of insider attacks or compromised nodes, link layer security techniques using a globally shared key proves completely useless. Insiders can attack the network by spoofing or injecting bogus routing information, creating sinkholes, selectively forwarding packets, using Sybil attack, and broadcasting HELLO floods. More sophisticated defense mechanisms are needed to provide reasonable protection against wormholes and insider attacks.

We present some of the countermeasures suggested in the literature (Karlof & Wagner, 2003; Mun & Shin, 2005) for routing attacks discussed in section 4.2. Note that the full effectiveness of these suggested countermeasures are yet to be proven.

Selective forwarding: The countermeasure used against selective forwarding attacks is to introduce redundancy to the network in the form of multipath routing. In this case, "n" disjoint paths are required to protect completely against selective forwarding attacks involving at most "n" compromised nodes, which is very difficult to create. The use of multiple braided paths (i.e., paths having common nodes but no common links) may provide probabilistic protection against selective forwarding. Allowing nodes to dynamically choose a packet's next hop probabilistically from a set of possible candidates can further reduce the chances of an attacker gaining complete control of a data flow.

Sybil attack: An insider attacker can participate in the network by using the identities of the nodes the attacker has compromised. The countermeasure here is to verify the identities of the participating nodes. One solution is to have every node share a unique symmetric key with a trusted base station. Two nodes can then communicate with each other by establishing a shared key after verifying each other's identity to implement an authenticated and encrypted link between them.

Wormhole and sinkhole attacks: Wormhole and sinkhole attacks are the hardest attacks to defend against, especially when the two are used in combination. Wormholes are hard to detect because they use a private, out-of-band channel invisible to the underlying sensor network. Sinkholes are difficult to defend against in protocols that use advertised information such as remaining energy or an estimate of end-to-end reliability to construct a routing topology because this information is hard to verify. Geographic routing protocols, however, are resistant to these attacks because messages are routed towards the physical location of the base station. False links will be detected by the neighboring nodes that figure out that the physical distance of an advertised route exceeds the radio signal range of motes. Though geographic routing can be relatively secure against wormhole, sinkhole, and Sybil attacks, a major problem lies with the trustworthiness of the location information advertised from neighboring nodes. Probabilistic selection of a next hop from several acceptable destinations or multipath routing to multiple base stations can tackle this problem to some extent.

A technique for detecting wormhole attacks known as "packet leashes" is presented by Hu, Perrig, and Johnson (2003) for MANETs. A leash is any information that is added to a packet designed to restrict the packet's maximum allowed transmission distance. There are two types of leashes: geographical leash and temporal leash. A geographical leash ensures that the recipient of the packet is within a certain distance from the sender. A temporal leash ensures that the packet has an upper bound on its lifetime, which restricts the maximum travel distance. Either type of leash can prevent wormhole attack because it allows the receiver of a packet to detect if the packet has traveled further than the leash allows. But it requires extremely tight time synchronization and is thus infeasible for most sensor networks. The best solution is to carefully design routing protocols which avoid routing race conditions and make these attacks less meaningful.

HELLO flood attacks: The simplest countermeasure against HELLO flood attacks is to verify the bidirectionality of a link before taking meaningful action based on a message received over that link. However, this countermeasure is less effective when an attacker has a highly sensitive receiver as well as a powerful transmitter. Such an attacker can effectively create a wormhole to every node within the range of its transmitter/receiver. One possible solution to this problem is for every node to authenticate each of its neighbors with an identity verification protocol using a trusted base station. In such a case, adversary is required to authenticate itself to every victim before it can mount an attack; an attacker claiming to be a neighbor of an unusually large number of the nodes will raise an alarm.

Acknowledgement spoofing: The most obvious solution to this problem would be authentication via encryption of all sent packets and also packet headers. Since base stations are trustworthy, attackers must not be able to spoof broadcast or flooded messages from any base station. This requires some level of asymmetry wherein no node should be able to spoof messages from a base station; at the same time every node should be able to verify them. Authenticated broadcast is also useful for localized node interactions.

The clustering algorithms used for WSNs rely on the honesty of all participating nodes, allowing a malicious node to generate false information to ensure its selection as cluster head thereby allowing an adversary to launch a sleep deprivation attack. To counteract this, Pirretti, Zhu, Narayanan, McDaniel, Kandemir, and Brooks (2005) proposes three separate defense mechanisms, the random vote scheme (randomizing the cluster head selection), the round robbin scheme (cluster head selection in a round robbin fashion), and the hash-based cluster head selection scheme (dynamic clustering) to form clusters and the selection of cluster heads.

When the network size is limited or topology is well-structured or controlled, global topology knowledge can be leveraged in security mechanisms. Drastic or suspicious changes to the topology might indicate a node compromise, and appropriate action can be taken.

Countermeasures such as link-layer encryption and authentication, multipath routing, identity verification, bidirectional link verification, and authenticated broadcast can protect sensor network routing protocols against outsiders, bogus routing information, Sybil attacks, HELLO floods, and acknowledgement spoofing and it is possible to augment existing protocols with these mechanisms. Sinkhole attacks and wormhole attacks pose significant challenges to secure routing protocol design, and it is unlikely that there exists effective countermeasures against these attacks that can be applied after the design of the protocol has completed. Hence, it is very crucial to design routing protocols by considering these attacks initially in the design so that with proper implementation of various security mechanisms, that is, robust countermeasures, these attacks can become ineffective.

ROUTING SECURITY SOLUTIONS AND TECHNIQUES

Security Goals

Security problems in WSN at the network layer can be related to router identity and router behavior. Theses two issues highlight two main tasks when we consider designing a secure routing solution.

- **Securing packet content:** This task is concerned with identity related security problems. The goal of this task is to assure that the packet is not accessed by unauthorized nodes as it travels from the source to the destination. This task can be achieved if we can provide the following services:
  - **Data confidentiality:** In this service, only the destination node should be able to access the packet content initiated

from the source node. Any intermediate router must not have any access to such information. As we can see here, the access of the packet is restricted to the destination node. Thus, if a node other than the destination accesses the packet, it means that the destination identity has been compromised.
    - **Data integrity:** When a destination node receives a message from a source, the destination should be able to detect any change that could occur in the massage.
- **Securing packet delivery:** This task deals mainly with behavior related security problems. Its objective is to guarantee that any packet transmitted will be ultimately received at the target destination. Thus, a misbehaving router node should not be able to drop a packet, misroute the packet, or deny the ability of routing of other nodes by denial-of-service attacks. This task can be interpreted in terms of a security service called data availability.
    - **Data availability:** If a node A is authorized to get information from another node B, node A should acquire this information at any time and without unreasonable delay.

There are many solutions and different approaches to achieve these tasks. However, the designer should be aware of the suitability of the solution with WSN tight constraints.

Some important guidelines that a designer should consider in the solution are:

- The solution should conserve energy as it is the rarest resource in WSN nodes. Energy conservation can be achieved by modifying an existing solution to reach the least possible energy consuming solution that can guarantee a certain level of security. Another possible approach is to make the design energy-aware in the sense that it adapts itself to the energy demands.
- The solution should not consume much memory or processing cycles. This is because security is considered as an added feature that must not compete with the application tasks in sharing memory or CPU usage. Thus, security overhead must be reasonable to keep WSN service throughput and general performance almost not affected.
- The solution proposed does not need to solve all security problems. However, a problem tackled by the proposed solution should be effective and robust. Such focused solutions should have the ability to be integrated with other solutions. This means that the design has to be flexible and has reasonable assumptions regarding the impact of nonsolved problems.
- The more modular the design is, the more robust the solution will be against future attacks. Moreover, modular design approach introduces some dynamicity that meets the very active and dynamic nature of WSN.
- Any proposed solution should be considered from an implementation point of view. If a solution cannot be implemented, it will be useless. Implementability considers issues like affordability, technology availability, application needs for such a solution, and so forth.

Intrusion Prevention and Detection Approaches

An intrusion can be defined as a set of actions that can lead to an unauthorized access or alteration of a certain system. Security is one of the key challenges to creating a robust and reliable network. Network security solutions can be generally grouped into two main categories: intrusion prevention-based techniques and intrusion detection-based techniques. Intrusion prevention-based techniques such as encryption and authentication are often the first line of defense against attacks. These intrusion prevention techniques can deter attackers from malicious behavior and reduce intrusions effectively, but cannot totally eliminate intrusions. The task of intrusion detection systems (IDS) is to monitor computer networks and systems with the main objective of detecting possible intrusions in the network, alerting users after intrusion detection, and excluding the attacker, that is, reconfiguring the network. Thus, intrusion detection systems are considered to be serving as the second line of defense by detecting existing intruders and are important in constructing highly survivable networks and have been accepted as an indispensable part of today's computer security systems.

Intrusion prevention-based techniques: Authentication and encryption-based security schemes for sensor networks are adaptations of security algorithms developed for MANETs. These adaptations aim to decrease the computation and communication overhead of these methods which were originally designed for more capable and less resource-constrained MANET nodes. Section 5 of this chapter briefly discusses a variety of approaches proposed in the literature for key agreement and key distribution for sensor networks. Prevention-based security schemes are difficult to implement especially over large scale sensor networks as these techniques are vulnerable to wireless networking challenges. Shared broadcast medium, the possibility of passive listening, and resource-limited network elements decrease the effectiveness of prevention mechanisms. The multi-hop nature of a network also necessitates additional trust requirements among the nodes and increases the vulnerabilities. It is not flexible to implement a dynamic public key cryptography scheme and to provide key exchanges with trusted central authority. On the other hand, symmetric key cryptography can be used to authenticate neighbors. In any case, powerful encryption schemes will not be available because of the computational capacity of the nodes. Thus, security provided to sensor networks with prevention only techniques is not always sufficient, or practical, because of the scalability problems, the computation, communication, and storage overhead associated with these methods. Hence, intrusion detection techniques are required to be incorporated to build a robust and reliable secure system. Numerous prevention-based mechanisms for wireless sensor networks have been proposed, but there are only a few recently proposed detection-based mechanisms for senor networks.

Intrusion detection-based techniques: Detection-based techniques are divided into two major categories: signature detection (a.k.a. misuse detection) and anomaly detection. Signature detection techniques match the known attack profiles with the current changes, whereas anomaly detection uses established normal profiles and detects unusual deviations from this normal behavior.

Misuse-based detection systems work based on a database of known attack signatures and system vulnerabilities and raise alarms when an activity matching a signature in the database is identified. Several methods have been proposed in the literature to model attack signatures and to search for a match, such as expert system, pattern matching, colored Petri-nets, and state transition analysis. Kaplantzis (2004) presents a comparative study of three classification techniques (i.e., k-means nearest neighbor classifiers, artificial neural networks, and support vector machines) in order to find the best performing classifiers in terms of speed and accuracy for an intrusion detection system using pattern matching. Classifiers are tools that partition sets of data into different classifications on the basis of specified features in that data. They showed that the support vector machines train in the shortest amount of time with an acceptable accuracy, while neural networks exhibit high accuracy at the cost of long training times. The benefits of misuse-based detection techniques are simplicity of these systems, the ability to detect attacks immediately after installation and that the signatures are based on well known intrusion activities and hence the detections are usually very accurate. In contrast, the major drawback of misuse-based detection system is the ability to manage effectively the signature database containing a huge amount of known attack patterns and updating the attack signatures as new attacks are published. These systems cannot detect unpublished attacks and are prone to circumventing false negative alarms and also the cost of generating signatures for all known attacks is very high.

Anomaly-based detection systems work based on the assumption that an intrusion can be detected by observing a deviation from normal or expected behavior of the system. These techniques buildup normal profiles from previously observed subject behavior and signal intrusions when the observed activities differ significantly from the normal behavior of the system. The major benefit of the anomaly-based detection technique is that it is very effective in defending against new attacks as it does not distinguish between known attacks and unknown attacks. However, such techniques are complex as they require a periodic online learning process in order to buildup up-to-date normal profiles and resource hungry as they are constantly generating logs and checking audit files. Moreover, anomaly-based intrusion detection systems tend to have high false alarm rates because the comprehensive knowledge of expected normal behavior of a system is hard to model. Hence, a major challenge in building anomaly-based IDS is to control effectively the false alarms. Another key challenge in identifying misbehaviors in wireless sensor networks is to develop techniques for detecting anomalies in the network, such that these techniques minimize their communication overhead and energy consumption in the network. Misbehaviors can be identified by analyzing sensor or traffic measurements to discriminate normal behavior from anomalous behavior. Anomaly detection in data with an unknown distribution is an important problem to be addressed in wireless sensor networks.

Several techniques have been proposed in the literature to identify anomalies and perform distributed data clustering in the context of MANETs. Clustering is the process of finding groups of similar data points, such that each group of data points is well separated (Han & Kamber, 2001). The Euclidean distance is used as the dissimilarity measure between pairs of data. A distributed k-means clustering algorithm has been proposed by Bandoyopadhya and Coyle (2006). An intrusion detection scheme has been proposed by Loo, Ng, Leckie, and Palaniswami (2006) to identify abnormal traffic patterns using fixed-width clustering. There has not been much work on the design of general intrusion detection system for wireless sensor networks, though the published works on intrusion detection in this area deal with specific kind of attacks or particular operations.

The work by Onat and Miri (2005) propose a predefined statistical model to identify anomalies in a distributed fashion wherein sensor nodes will have the ability to record simple statistics about the neighbors' behavior and detect anomalies in them. To make a sensor node capable of detecting an intruder, a simple dynamic statistical model of the neighboring nodes is built in conjunction with a low-complexity detection algorithm by monitoring received packet power levels and arrival rates. This work considers two types of attacks: node impersonation and resource depletion attack. These attacks reveal themselves by deviations from the normal transceiver and traffic behaviors.

A distributed, nonparametric anomaly detection scheme to identify anomalous measurements in sensor nodes has been proposed by Rajasegarar, Leckie, Palaniswami, and Bezdek (2006). In this approach, in order to minimize communication overhead, which is a major source of energy consumption, each individual sensor measurement is not sent to a central node for analysis. Instead the measurements are clustered and only cluster summaries are sent by the sensor nodes. Further, intermediate sensor nodes merge cluster summaries before communicating with other nodes. The clustering approach used here is based on the fixed-width clustering algorithm which produces a set of disjoint, fixed-width (or radius) clusters in the feature space. The anomaly detection algorithm used the average inter-cluster distance of the k-nearest neighbor (KNN) clusters to identify the anomalous clusters.

The work by Da Silva, Martins, Rocha, Loureiro, Ruiz, and Wong (2005) proposes a decentralized intrusion detection system for WSN which is based on the inference of the network behavior obtained from the analysis of events detected by a monitoring node. The only events considered are data messages listened to by the monitoring node that is not addressed to it and message collision when the monitoring node tries to send a message.

The approach proposed consists of three phases: data acquisition, rule application, and intrusion detection. In the data acquisition phase, messages are collected in a promiscuous mode and the important information is filtered before being stored for subsequent analysis. In the rule application phase the seven rules, interval rule, retransmission rule, integrity rule, delay rule, repetition rule, radio transmission rule, and Jamming rule are applied to the stored data and if the message fails the tests being applied, a failure is raised. In the intrusion detection phase, the number of raised failures is compared with the expected amount of occasional failures in the network and if the raised failures are found to be higher than the expected, then an intrusion detection is raised.

Agah, Das, and Basu (2004) introduce an intrusion detection scheme based on a noncooperative game approach. In every game theory problem, a payoff or utility function is required to be defined between players. A payoff function used in this work is based on two fundamental issues: cooperation and reputation. Payoff between two sensors is dependent on their distance and each node's transmitter signal strength. The more transmitter signal strength a node has, the more likely it cooperates with its close neighbors. In order to show how much each individual node is useful for the whole network, and gain better reputation among others, payoff between two sensor nodes should also represent how many packets each node receives and forwards at each time slot for the sake of others. Here, a nonzero sum, noncooperative game is defined between attacker and sensor nodes, where an attack is a denial-of–service attack, which is intended to be prevented. By using the game theory framework, authors show that the game achieves Nash equilibrium for both attacker and the network.

In the work proposed by Agah and Das (2007), the prevention of passive denial of service attack at routing layer in wireless sensor networks is formulated as a repeated game between an intrusion detector and nodes of a sensor network where some of these nodes act maliciously. The repeated games are associated with sequences of history-dependent game strategies. In order to prevent DoS attack, the interaction between a normal and a malicious node in forwarding incoming packets is captured as a noncooperative N player game. A framework is proposed to enforce cooperation among nodes and punishment for noncooperative behavior. The intrusion detector residing at the base station keeps track of node's collaboration by monitoring them. If performances are lower than some trigger thresholds, it means that some nodes act maliciously by deviation. The intrusion detection system rates all the nodes, which is known as subjective reputation (Michiardi & Molva, 2002), and the positive rating accumulates for each node as it gets rewarded.

An emotional ant-based approach to identify possible preattack activities in sensor networks is presented in the work IDEAS proposed by Bannerjee, Grosan, and Abraham (2005). Security monitoring in the sensor network is achieved by the foraging behavior of natural ant colonies. The work emphasizes the emotional aspects of agents where they can communicate the characteristics of a particular path among them through pheromone update. Therefore, in a sensor network if the emotional ants are placed, they could keep track of the changes in the network path, following a certain knowledge base of rules depicting the probable possibilities of attack. Once the particular path among the nodes is detected by the emotional ant, it can communicate the characteristics of the path through pheromone balancing to the other ants, and thereafter intrusion alarm can be raised.

Localization anomaly detection (LAD) proposed by Du, Fang, and Ning (2005) is a general scheme to detect localization anomalies that are caused by adversaries compromising the beacon nodes. Here localization anomaly problem is formulated as an anomaly intrusion detection problem, and a number of ways to detect localization anomalies are proposed. In the proposed work by Deng, Han, and Mishra (2004), a secure multipath routing to multiple destination base stations is designed to provide intrusion tolerance against isolation of a base station. Antitraffic analysis strategies are also proposed to help disguise the location of the base station from eavesdroppers.

Cryptography-Based Solutions

In order to achieve secure communication, robust cryptography schemes are required to ensure confidentiality (nondisclosure of secret information), integrity (prevention of data alteration), authentication (proof of identity), and nonrepudiation (unique, noncontestable message origin). To accomplish these goals, a combination of symmetric key algorithms (e.g., AES, DES, RC4), public key algorithms (e.g., RSA, ECC), and cryptographic hash functions (e.g., MD5, SHA) is commonly employed and RSA is by far the most widely used public key algorithm on the internet today. In general, they are three types of key establishment schemes: (1) trusted server based schemes, (2) self-enforcing schemes (3), and key predeployment or key predistribution schemes. Trusted-server-based schemes, for example, arbitrated keying protocol and Kerberos depend on a trusted server for key agreement among nodes. Because of the lack of trusted infrastructures, this type of scheme is not suitable for sensor networks. Self-enforcing schemes use the asymmetric encryption cryptography, such as the use of public key certificates which is limited by the current computation abilities and energy resources of sensor network technologies. Hence, key predistribution schemes are mainly considered for WSNs where the secret keys are distributed to all sensors before deployment on the ground.

An important challenge in the design of security for sensor networks is the problem of efficient key management. Efficient key management is still an important research area. Key management in WSNs involves the process of efficiently generating, storing, protecting, distributing, using, and destroying the cryptographic keys taking into account the prevailing conditions and constraints imposed by the sensor networks. The inherent characteristics of sensor nodes complicate the design of secure protocols for sensor networks and make the key management problem highly challenging. The limited computation, memory, and power resources of sensor nodes make it undesirable to use public-key cryptographic algorithms, such as Diffie-Hellman ley agreement or RSA signatures. Asymmetric cryptography may often require expensive computation which could expose power-constrained sensor nodes to denial-of-service (DoS) attacks. An attacker can perform battery-draining denial-of-service attacks by using digital signatures. In real-life challenging applications, networks of thousands of sensor nodes are deployed widely in public and hostile areas. Since these sensor nodes are low cost, and hence, not tamper-resistant, they are vulnerable to physical capture. An adversary may be able to undetectably take control of a sensor node and compromise the cryptographic keys. Since sensor nodes are usually deployed using random scattering, network protocols lack prior knowledge of which nodes will be within the communication range of each other after deployment. Due to the limited memory resources, the amount of memory required to establish unique keys with every one of the other nodes in the network is highly constrained. Due to the limited bandwidth and transmission power, the communication of large blocks of data becomes particularly too expensive.

WSN Key Management Schemes

Numerous key management schemes have been proposed for sensor networks. The objective of key management is to dynamically establish and maintain secure channels among communicating nodes. A key management framework for WSNs must deal with the following important issues:

- **Key peployment/key predistribution:** Deals with number of keys administrative keys (a.k.a. key encryption keys) required; the way the keys should be distributed before the nodes are deployed.
- **Key discovery:** Enables an arbitrary pair of sensors to discover the set of keys they share.
- **Key establishment/key setup:** Deals with establishing a secure session between a pair of nodes or a group of nodes
- **Node addition/rekeying:** Deals with how a node should be added to the network such that it is able to establish secure sessions with

existing nodes in the network in such a way that the backward secrecy is still preserved. Backward secrecy is maintained by means of periodically changing the traffic encryption keys. A newly added node can only obtain new traffic encryption key and not the previous encryption keys being used in the network and hence is not able to decipher previous traffic.

- **Node eviction/key revocation:** Deals with how a node should be evicted from the network such that it will not again be able to establish secure sessions with any one of the existing nodes in the network, and not be able to decipher future traffic in the network thereby preserving forward secrecy.

The metrics most commonly used in the published approaches to evaluate the performance of the key management schemes are local/global connectivity, resilience to sensor nodes capture, scalability, and memory efficiency. The distribution of keys to sensor nodes of large scale WSNs where physical topology is unknown prior to deployment would have to rely on key predistribution. Keys would have to be installed in sensor nodes before deployment in order to establish secure connectivity between nodes after deployment. Establishing secure pair-wise communications is a prerequisite for the implementation of secure routing and also useful for secure group communication.

Key management schemes in sensor networks can be classified broadly into dynamic or static key management schemes based on whether rekeying or updating of the administrative keys is enabled or not after network deployment. Static schemes assume a relatively static, short lived network where node replenishments are rare and keys outlive the network. Static key management schemes assume that once administrative keys are predeployed in the nodes, they will not be changed. Administrative keys are generated prior to deployment, assigned to nodes either randomly, or based on some deployment information. Another emerging class of schemes, termed dynamic key management schemes, assume long-lived networks with more frequent addition of new nodes, thus requiring network rekeying for sustained security and survivability after deployment. Dynamic key management schemes may change administrative keys periodically on demand or upon detection of node capture. The major advantage of dynamic keying is enhanced network survivability, since any captured key is replaced in a timely manner in a process known as rekeying. Another advantage of dynamic keying is providing better support for network expansion, upon adding new nodes, unlike static keying which use a fixed pool of keys and the probability of network capture does not necessarily increase.

Based on the key generation/distribution techniques used, these key management schemes can further be categorized into the following: pure probabilistic or random key predistribution schemes, polynomial-based key predistribution schemes, Blom's matrix-based key predistribution schemes, deterministic key predistribution schemes, and group key management schemes (combinatorial formulation) using exclusion-basis systems. Most of the existing schemes based upon the basic random key predistribution scheme introduced by Eschenauer and Gilgor (2002) are static key management schemes. Many of these schemes propose improvements over the basic scheme by using key polynomials, deployment knowledge (node locations, node clusters or group), as well as attack probabilities in certain portions of the network to enhance scalability and resilience to attacks.

The basic key predistribution scheme first proposed by Eschenauer and Gilgor (2002) is based on probabilistic key sharing among the nodes of a sensor network. This scheme uses a simple shared-key discovery protocol for key distribution, revocation, and node rekeying. In this scheme, before sensor nodes are deployed, an initialization phase, also known as key predistribution phase, is performed offline to generate a large random pool of P keys out of the total possible key space. For each node, k keys (also known as node's key ring) are randomly selected out of the large key pool P without replacement and loaded into each sensor node's memory. This key predistribution phase must ensure that only a small number of k keys need to be stored in each sensor node's ring such that any two nodes will share at least one key with a chosen probability.

Q-composite random key predistribution scheme proposed by Chan, Perrig, and Song (2003) is a modification to the method proposed by Eschenauer and Gilgor (2002) where q common keys are required to match between neighboring nodes instead of a single common key, thereby making it more difficult to compromise communications. The motivating factor is that as the amount of required q common keys increases, it becomes exponentially harder for an attacker with a given set of keys to break a link.

Pseudo random key deployment scheme proposed by Di Pietro, Mancini, and Mei (2003) is a further improvement to the basic random predistribution scheme proposed by Eschenauer and Gilgor (2002), which allows more efficient key discovery procedure. In the key deployment phase, keys are assigned to a node according to the output of a pseudorandom generator with a publicly known seed and the node's ID as inputs. Di Pietro et al. (2006) further improve the above work by proposing a novel efficient and secure pseudo (ESP) random key deployment scheme that requires no message exchange between nodes to establish pair-wise keys, but only k applications of the pseudo-random function where k is the number of keys in the node's key ring, thereby minimizing energy consumption in the key discovery phase.

The work proposed by Du, Deng, Han, Chen, and Varshney (2004) presents a novel random key predistribution scheme that uses deployment knowledge. With such deployment knowledge, it is shown that each node only needs to carry a fraction of the keys required by the other random key predistribution schemes (Eschenauer & Gligor, 2002; Chan et al., 2003), while achieving improved level of connectivity (in terms of secure links), higher resilience against node capture, and reduction in amount of memory required.

The random key management scheme using both attack probabilities and deployment knowledge proposed by Chan, Poovendran, and Sun (2005) uses a subgrouping approach to isolate the effect of node captures into one specific subgroup, and to provide scalability for random key predistribution in clustered distributed sensor networks. This works considers the probability of node capture for each subgroup in order to design a scalable security mechanism that improves resilience to the attacks for the sensor subgroup with larger probability of node compromise. Hence this scheme can maintain flexibility in providing different security concerns for different sensor groups. Since sensor subgroups are located in different areas, they may have different chances of being attacked by the adversaries.

The work by Yang, Zhou, Zhang, and Wong (2006) proposes a polynomial-based key management scheme called group-to-group (G2G) pairwise key establishment which enables a node to communicate securely with nodes near a certain location using their location knowledge and the communicating nodes need not know each other's ID or need not be located within each other's communication range.

Zhang, Liu, Lou, and Fang (2006) propose a suite of location-based compromise-tolerant security mechanisms to mitigate the impact of compromised nodes. They propose the novel notion of location-based keys (LBKs) based on a new cryptographic concept called pairing which binds private keys of individual nodes to both their node IDs and geographic locations. The LBKs used in this approach can act as efficient countermeasures against a Sybil attack, identity replication attack, wormhole, and sinkhole attack. This work uses an identity-based cryptography (IBC) which is receiving extensive attention as a powerful alternative to traditional certificate-based cryptography (CBC). Its main idea is to make an entity's public key directly derivable from its publicly known identity information. Eliminating the need for public-key certificates and their distribution makes IBC much more appealing for securing WSNs, where the need to transmit and check certificates has been identified as a significant limitation.

The major challenge in dynamic keying is to design a secure yet efficient rekeying mechanism. A dynamic key management scheme is proposed by Jolly, Kuscu, Kokate, and Younis (2003) that is based on the identity-based symmetric keying.

In the work proposed by Eltoweissy, Heydaru, Morales, and Sadborough (2004), a combinatorial formulation of group key management problem is developed using exclusion-basis systems (EBS). A drawback of the basic EBS-based solution is that a small number of nodes may collude and collectively reveal all the network keys. In SHELL, Younis, Ghumman, and Eltoweissy (2006) propose a modification to EBS approach to address the collusion problem which performs location-based key assignments to minimize the number of keys revealed by capturing collocated nodes. LOCK proposed by Eltoweissy, Moharrum, and Mukkamala (2006) is an EBS-based dynamic key management scheme for clustered sensor network which uses key polynomials to improve the network resilience to collusion instead of location-based key assignment as in SHELL.

An important observation that can be drawn from these proposed schemes is that the location knowledge can be used to improve the performance of the key management schemes, such as the connectivity, resilience against nodes capture, and memory efficiency. Also it is evident that most of the key management solutions for wireless sensor networks are trying to find the better tradeoffs between system security (e.g., resilience to node capture) and network connectivity. All of them have weak and strong points. The diverse usages of wireless sensor networks make it unreasonable to try to find the single perfect scheme suitable for all situations.

Having provided a brief survey of the key management schemes to highlight the current developments in this key area, in the following section we briefly describe some protocols based on key predistribution schemes designed for WSN.

Protocols Based on Key Predistribution Schemes

Security protocols for sensor networks (SPINS): This work proposed in by Perrig et al. (2001) uses a hierarchical/centralized communication architecture that assumes a forest-like network formed around one or more base stations, which interfaces the sensor network to the outside network. SPINS has two security building blocks: secure network encryption protocol (SNEP) and the micro-version of the timed efficient streaming loss-tolerant authentication protocol (μTESLA). SNEP provides data confidentiality, two-party authentication, integrity, replay protection, and message freshness. μTESLA provides authentication for data broadcast. In the key deployment phase, every node shares a unique master key with the base station. In key establishment phase, two kinds of traffic (node to node communications and broadcasts by the base station) are secured. SNEP allows two nodes to establish a session key through the base station. μTESLA allows messages broadcast by the base station to be authenticated. When a new node is added, it is loaded with a unique master key that it shares with the base station. During a node eviction, the evicted node's master key is removed from the base station.

SNEP achieves semantic security with no additional transmission overhead by sharing a counter between the sender and receiver for the block cipher in counter mode. Since the counter value is incremented after each message, the same message is encrypted differently each time thereby preventing replay of old messages. To achieve two-party authentication and data integrity, a message authentication code (MAC) is used which enforces a message ordering and weak freshness. In μTESLA, the requirement of asymmetric mechanism for authenticated broadcast is achieved through a delayed disclosure of symmetric keys. μTESLA is based on one-way key chain. During initial set-up, the base station generates a one-way key chain of n keys by choosing the last key K_n randomly and generating the remaining values by successively applying a one way hash function F (such as MD5), that is, $K_i = F(K_{i+1})$. Every node synchronizes its time with the base station. Time is divided into uniform time intervals and the base station associates each key of the key chain with one time interval. To bootstrap μTESLA, the base station distributes the root of the key chain, K_n, to the sensor nodes. During the time interval i, the base station uses the key of the current time interval K_i to compute the MAC of the packets to be broadcast in that interval and discloses the key $K_{i-\delta}$ that authenticates all the messages broadcast in and before the time interval i-δ. When a node receives the disclosed key $K_{i-\delta}$, it verifies the correctness of the key and then uses it to authenticate the message stored in its buffer received during the time interval i-δ.

Some remarks about this scheme are:

1. This scheme requires key server (base station) for the establishment of pair-wise keys compared to other schemes which only needs a bootstrap server (or base station) to initialize and deploy nodes.
2. SNEP protocol has low communication overhead, only 8 extra bytes per message.
3. SNEP is an end-to-end security protocol and cannot prevent routing misbehavior.
4. μTESLA provides a secure broadcast communication, which is a common and important communication pattern in almost all WSN applications.
5. μTESLA obtains routing security by authenticated routing that is achieved by deriving the operation on routing update packets and checking the correctness of the claiming parents by delayed key disclosure.
6. In μTESLA protocol, the base station and nodes are required to be loosely time synchronized.
7. The periodic key disclosure of μTESLA ensures compromising a single sensor does not reveal the keys of all the sensors in the network.

SeFER: secure, flexible and efficient routing protocol for distributed sensor networks: This work by Oniz, Tasci, Savas, Ercetin, and Levi (2005) presents a secure, flexible, and efficient routing protocol for sensor networks based on the basic random key predistribution scheme Eschenauer and Gilgor (2002) discussed in section 5.3.1. This protocol aims to establish secure paths in a sensor network between a controller and a set of nodes where each node has been assigned a set of randomly chosen keys out of a key pool. The primary aim of this protocol is to find routes from each sensor node to the controller with all routes from each sensor node to the controller secured. First, keys are predistributed to the sensor nodes and shared keys are discovered by the methods proposed in Eschenauer and Gilgor (2002). The protocol has six phases and it starts after each node discovers shared keys with its neighbors. The controller can communicate with the rest of the nodes indirectly via the level-one nodes, which are defined as the nodes in the wireless range of the controller. A route is formed using four phases: (1) level-one initialization phase (2), route learning phase (3), authenticate neighbor and shorter path discovery phase, and (4) key distribution phase. In the level-one initialization phase, the controller and nodes in the wireless range of the controller mutually authenticate themselves and the controller distributes the session key to be used in further phases. In the route learning phase, each node forwards messages containing route information to their downstream nodes and an initial set of routes is established. In the authentic neighbor and shorter path discovery phase, nodes broadcast messages in order to discover shorter paths to the controller. If shorter paths are found, these paths must be secured by assigning a key to that path which is performed by the controller in the key exchange phase. If the controller detects that a security breach has occurred during the execution of the routing protocol, it starts the session key expiration phase to invalidate the session key. If a legitimate sensor node is compromised, the controller starts the revocation phase in order to invalidate the key ring of the compromised node. Resilient to replay attacks is achieved in this protocol by associating each message with a nounce value proposed by Perrig et al. (2001) and a time stamp indicating expiration date.

Some remarks about this scheme are:

1. The proposed protocol is flexible such that it allows a tradeoff between route length and the route setup cost in terms of processing power and storage.
2. Different security needs for different location of nodes are not considered.

3. This scheme is less energy efficient because of the key discovery phase which requires a number of messages proportional to the number of keys assigned to each sensor.
4. The scalability of random key predistribution is a concern and not addressed.

Localized encryption and authentication protocol (LEAP): This work by Zhu, Setia, and Jajodia (2003) presents a complete key management framework for static WSNs which supports in-network processing, while at the same time restricts the security impact of a node compromise to the immediate network neighborhood of the compromised node. In order to meet the different security requirements of the messages exchanged between the sensor nodes, this protocol provides multiple keying mechanisms for securing node-to-base station traffic, base station to nodes traffic, local broadcasts, as well as node-to-node (pair-wise) communications. It also includes an efficient protocol for local broadcast authentication based on the use of one-way key chains. Each node shares four types of keys: an individual key, a group key, cluster keys, and pair-wise shared keys. In addition to these keys, a node also has to store a one-way key chain it creates, the commitments of the key chains its neighbors create, and the commitment of the base station's key chain.

Every node has a unique individual key that is only shared with the base station. This key is used for secure communication between the node and the base station. This key is generated and preloaded into each node prior to its deployment. A node uses its individual key to encrypt messages it sends to the base station. A group key is a global key shared by all the nodes in the network. This key is used by the base station for encrypting messages that are broadcast to the whole group. Messages broadcast by the base station are encrypted with the group key, but authenticated with μTESLA. A cluster key is a key shared by a node and all its neighbors and it is mainly used for securing locally broadcast messages such as routing control information or securing sensor messages which can benefit from passive participation. In passive participation, a node that overhears a neighboring sensor node transmitting the same reading as its own current reading can elect not to transmit the same, thereby saving energy consumption in sensor networks. A node's cluster key and one-way key chain allow its neighbors to authenticate its locally broadcast messages. The combination of cluster key and one-way key chain is interesting because the cluster key can be used to hide the keys in the key chain from cluster-outsiders, so that the keys do not need to be disclosed according to a schedule as in SPINS, and the keys in the key chain can be used for authentication as usual. Pair-wise shared key is a key each node shares with each of its immediate neighbors. Pair-wise keys are used for securing communications that require privacy or secure authentication. For example, a node can use its pair-wise keys to secure the distribution of cluster key to its neighbors, or to secure the transmission of its sensor readings to an aggregation node.

Some remarks about this scheme are:

1. This scheme is very effective in defending against many sophisticated attacks such as HELLO flood attack, Sybil attack and wormhole attack.
2. In this scheme, knowledge of node IDs is required to establish pair-wise keys among neighboring nodes.
3. It is assumed that all nodes are innocent within a short period after deployment.
4. This scheme is scalable and efficient in computation, communication, and storage.
5. Sensor deployment is considered to be static.
6. Different security needs for message exchanged between nodes for the hierarchical WSN are considered.
7. The issues regarding the impact of key sharing approach on in-network processing are addressed using cluster-keying mechanism.

Intrusion tolerant routing protocol for wireless sensor networks (INSENS): INSENS by Deng, Han, and Mishra (2002) constructs a tree-structured routing for wireless sensor networks. It aims to tolerate damage caused by an intruder who has compromised deployed sensor nodes and is intent on injecting, modifying, or blocking packets. INSENS incorporates distributed lightweight security mechanisms, including one-way hash chains and nested keyed message authentication codes to defend against routing attacks such as wormhole attack. Adapting to WSN characteristics, the design of INSENS also pushes complexity away from resource-poor sensor nodes towards resource-rich base stations. It constructs forwarding tables at each node to facilitate communication between sensor nodes and a base station. The INSENS secure routing system is designed to prevent flooding attacks by allowing only base station to broadcast. The authentication of the base station is achieved via one-way hashes, so that individual nodes cannot spoof the base station and thereby flood the network. For unicast packets, nodes must first communicate through the base station, allowing the base station to act as a packet filter to prevent denial-of-service via a single node. To prevent advertisement of false routing data, control routing information is authenticated. A key consequence of this approach is that the base station always receives correct partial knowledge of the topology. Symmetric key cryptography is chosen for confidentiality and authentication between the base station and each resource-constrained sensor nodes, since it is considerably less compute-intensive than public key cryptography, and the base station is chosen as the central point for computation and dissemination of the routing tables. To address the notion of compromised nodes, redundant multipath routing is built into INSENS to achieve secure routing. The paths are designed to be disjoint, so that even if an intruder brings down a single node or path, secondary paths will exist to forward the packet to the correct destination.

Some remarks about this scheme are:

1. This scheme minimizes computation, communication, storage, and bandwidth requirements at the sensor nodes at the expense of increased computation, communication, storage, and bandwidth requirements at the base station.
2. Rather than rely on traditional intrusion-detection techniques, INSENS's strategy is to design a routing mechanism that is intrusion-tolerant.
3. An important property of INSENS is that while a malicious node may be able to compromise a small number of nodes in its vicinity, it cannot cause widespread damage in the network.

Non Cryptography-Based Solutions

Noncryptographic-based solutions are mainly concerned with behavior-related security attacks, such as nonforwarding, selective forwarding, and denial-of-service attacks. The basic concept in this general approach is to enable sensor nodes to be aware of their neighbors' behavior. When misbehavior is detected, the misbehaving node is avoided in routing.

The main features of such solutions as compared with crypto-based approach are:

- Crypto-based solutions are not robust against insider attacks in which the attacker is an identifiable and authorized member of the network. Such attacks can be done by compromised nodes or selfish nodes. Others are unintentionally done by faulty nodes (Josang & Ismail, 2002). However, security systems that depend on treating nodes' behavior instead of their identities are more robust. This is especially true in networks where such misbehavior is very possible or sometimes can be the dominant type of attacks, which is the case in WSN.
- Unlike node's identity, node' behavior is not fixed and can be dynamically changed in intelligent ways. Crypto-based solutions do not have the provision of this dynamic treatment. However, behavior-based approaches provide a means for an adaptive and dynamic decision making and reaction at the individual node level behavior.
- Cryptography overhead at the node level structure such as memory consumption and

computation complexities are all relieved in this approach. However, communication overhead and behavior knowledge exchange is more complicated here.

In literature, noncrypto approach is realized by the adoption of reputation systems. A reputation system is a type of cooperative filtering algorithm which attempts to determine ratings for a collection of entities that belong to the same community. Every entity rates other entities of interest based on a given collection of opinions that those entities hold about each other (Michiardi & Molva, 2002).

Reputation systems have recently received considerable attention in different fields such as distributed artificial intelligence, economics, evolutionary biology, and so forth. Most of the concepts in reputation systems depend on social networks analogy. As expected, reputation systems are complex in the sense that they do not have a single notion, but a single system will consist of multiple parts of notions. Thus, comparing reputation systems is, in fact, a very difficult problem. All known trials on such problem were based on qualitative approach. The work proposed by Mui, Halberstadt, and Mohtashemi (2002) makes an attempt on comparing reputation systems quantitatively based on game theory. The authors, thus, identify different notions of reputation systems like, contextualization, personalization, individual and group reputation, and direct and indirect reputation.

In the context of MANET and WSN (Buchegger & Boudec, 2003; Michiardi & Molva, 2002), the reputation of a node is the amount of trust the other nodes grant to it regarding its cooperation and participation in forwarding packets. Hence, each node keeps track of each other's reputation according to the behavior it observes, and the reputation information may be exchanged between nodes to help each other to infer the accurate values. There is a trade-off between efficiency in using available information and robustness against misinformation. If ratings made by others are considered, the reputation system can be vulnerable to false accusations or false praise. However, if only one's own experience is considered, the potential for learning from the experiences of others goes unused, which decreases efficiency.

Any reputation system in the context of MANET and WSN should, generally, exhibit three main functions (Djenouri et al., 2005):

- **Monitoring:** This function is responsible for observing the activities of the nodes of its interest set.
- **Rating:** A node will rate its interest set nodes based on the node's own observation (termed as first hand information), other nodes' observations that are exchanged among themselves (termed as second hand information), the history of the observed node, and certain threshold values.
- **Response:** Once a node builds knowledge about others' reputations, it should be able to decide upon different possible reactions it can take, like, avoiding bad nodes or even punishing them.

For secure routing problem in WSN, a reputation system can be a good solution for behavior-related problems. The efficiency of a proposed solution will depend on:

- The ability to monitor misbehavior events correctly.
- Using a good rating model that closely reflects the behavior of nodes.
- Developing good routing algorithms and decision criteria that try to select the most trusted routers and follow the least risky paths.

In literature, there are reputation-based solutions proposed for MANET such as CONFIDANT and CORE. The work, however, in WSN is not heavily studied. When considering WSN, reputation systems become more challenging for the first two phases, that is, monitoring and rating. Good monitoring requires a sensor node to be always awake overhearing others' packets which is an energy consuming operation. A possible approach is to make the responsibility of monitoring for a specific set of sensor nodes. However, this yields a poor rating mechanism. Moreover, the rating model should be able to mathematically track the node behavior. Complex models may require a heavy processing task and memory usages. Theses resources are more in demand for data processing in the constrained WSN node.

In the following sections, we briefly describe some reputation-based solution designed for WSN.

Reputation-Based Solutions

SAR: Security-aware routing (SAR) proposed by Naldurg, Yi, and Kravets (2001) is a protocol derived from AODV and based on authentication and a metric called the hierarchal trust values metric. The hierarchal trust values metric governs routing protocol behavior. This metric is embedded into control packets to reflect the minimum trust value a router should have to be able to forward the received packet. This value is determined by the sender. A node that receives any packet can neither process it nor forward it unless it provides the required trust level present in the packet. Moreover, this metric is also used as a criterion to select routes when many routes satisfying the required trust value are available.

There are some problems and limitations in SAR:

- The routing operation needs to encounter a trusted route setup phase that is done using cryptographic authentication. This setup contributes some initial delay and requires some sophisticated crypto mechanisms.
- The trust metric used in SAR does not reflect exactly nodes' behavior; rather, they represent a "rank" that a node exhibits based on its identity and various security service provision. Thus, a trusted node in SAR is a node that has the appropriate rank that meets the routing requirements. To rank a node is another problem by itself that is not addressed very well.
- The routing decision rules in SAR are governed by the source, which makes the protocol less flexible.
- The routing decision is not to select the next hop but to decide to participate in the trusted route. As a result, selfish behavior is not addressed well in SAR.

TRANS: Proposed by Tanachaiwiwat, Dave, Bhindwale, and Helmy (2004), TRANS is a geographic routing protocol (GPSR-based) that provides security services using trust metric. It can be considered as a tight trust-based routing due to its specific targets and assumptions. It basically targets a misbehavior model in which an attacker selectively participates in routing signaling and control packets, but drops consistently queries and data packets. The protocol also assumes static sensor networks in which a tight mapping can be done between the nodes' identities and their locations. TRANS assumes a location-centric architecture that helps it in isolating misbehavior and establishing trust routing in sensor networks. As a result of that, the protocol assumes a certain communication model in which a single or multiple sinks initiate communication requests with various locations. During that phase, insecure locations are identified and blacklisted. The trust metric used to judge on location security is calculated based on nodes' experience among each other regarding their identities, link availability, and packet forwarding.

There are some problems and limitations in TRANS:

- In TRANS, the trust, in fact, is associated with locations rather than the nodes. The problem is that a location can be infected by a single node. The detour, then, will be around a larger area rather than a single node.
- Nodes located in proximity of an infected location might be also isolated. If not, they are also exposed to heavy routing duties that may induce selfish behavior.
- TRANS is limited by single or multiple sink communication models. This assumption is necessary for the efficient operation of the protocol.
- TRANS discusses approaches to decrease energy consumption due to the security

provision overhead. However, the protocol does not provide energy efficient techniques in the routing operation itself.

RGR: Resilient geographic routing (RGR) protocol proposed by AbuGhazaleh, Kang, and Liu (2005) is also a trust-based routing protocol that relies on a modified routing operation in GPSR. The basic idea in RGR is to assign an initial trust value for each node. Then, this value is incremented or decremented depending on the forwarding activity of the monitored node using a step function. The source node selects probabilistically a subset among its neighbors to forward its packet. This subset is selected from the node's forwarding set that exhibit trust values greater than a threshold.

There are some issues that are not considered in RGR.

- The protocol has no provision for energy efficiency.
- The protocol totally relies on trust-based forwarding. If a node is completely surrounded by misbehaving nodes, there is no other mechanism proposed to select a next hop since all nodes will be eliminated from the node's forwarding list.
- RGR is a multipath trust-based routing. Although multipath is important for reliable services, it is also believed that multipath routing is energy consuming, which is a very important issue to consider in WSN

Reputation-based framework for high integrity sensor networks: Ganeriwal and Srivastava (2004) propose a reputation-based framework for sensor networks where nodes maintain a reputation for other nodes and use it to evaluate their trustworthiness. The authors tried to focus on an abstract view that provides a scalable, diverse, and a generalized approach hoping to tackle all types of misbehaviors. They also designed a system within this framework and employed a Bayesian formulation, using a beta distribution model for reputation representation.

In this system, monitoring mechanism follows the classic watchdog methodology in which a node is assumed to be in a promiscuous mode to overhear neighbors' packets. Monitoring behavioral events can result in either cooperative event, α, in which a node is behaving well or noncooperative behavior, β, in which a node misbehaves. The count of each type is injected into the beta distribution formula as the distribution parameters to calculate the node reputation R. This formula calculates node's reputation based on first hand information. The reputation is updated as new monitoring events are obtained; second hand information is obtained and according to the age of the current reputation value. Any response action is based on selecting the most trusted node. The trust value of a node that is used for decision making is calculated as the statistical expectation of the reputation value.

This work, however; lacks some important points:

- The monitoring mechanism uses a normal watchdog mechanism that assumes a promiscuous mode operation for every node. This is not suitable for the WSN conditions in terms of energy scarcity as discussed earlier.
- The system does not show a practical solution implementation of monitoring and rating phases. From an implementation point of view, the study should provide an example of how monitoring and rating will be done under some application assumptions.
- The work does not propose a response methodology, for example, a routing algorithm. Instead, it leaves it an open issue. Therefore, the work lacks performance figures that can show the efficiency and security gain and benefits in routing operation that can be obtained in adopting this solution.

Reputation system-based solution for trust-aware routing: This work proposed by Maarouf and Naseer (2007) provides a reputation system-based solution for trust aware routing as a main security concern in WSN. In contrast to similar existing solutions for ad hoc networks like CORE (Michiardi & Molva, 2002) and CONFIDANT (Buchegger & Boudec, 2003) or those for WSN like RFSN (Josang & Ismail, 2002), this work proposes solutions to focusing on satisfying WSN resources constraints and conditions, while maintaining the security requirements. Thus, the solution proposes new mechanisms and approaches that are customized for WSN constraints.

The work adopts a modular design approach by which it treats every individual component as a separate problem and studies it in the lights of WSN conditions adaptation and customization. The integrated reputation system is termed as senor node attached reputation evaluator (SNARE) (Maarouf & Naseer, 2006) which consists of three main components: monitoring component, rating component, and response component.

For the monitoring part, the work proposes a new monitoring strategy called efficient monitoring procedure in reputation system (EMPIRE) to solve the problem of efficient monitoring in WSN. Efficient monitoring should guarantee a satisfying level of capturing neighborhood activities, while trying to minimize power consumption, memory usage, processing activities, communication overhead, and so forth. In this work, monitoring efficiency is realized by the association between the nodal monitoring activity (NMA) and various performance measures. NMA is determined by the frequency of monitoring actions that a node takes to collect direct observation information. Reducing the frequency of monitoring, that is, reducing NMA, will affect the quantity and/or the quality of the obtained information. Thus, the performance measures will be affected. However, on the other hand, this reduction implies a saving in node's resources such as power, processing, and memory, which are the constraints that are faced in WSN. EMPIRE provides a probabilistic approach to reduce nodal monitoring activities, while keeping the performance of the system, from the behavior and trust awareness perspective, at a desirable level.

The rating component proposed in this work is called cautious rating for trust enabled routing (CRATER). Basically, this technique identifies three rating factors: first hand information (FHI), second hand information (SHI), and a defined period called neutral behavior period (NBP) during which a node is not doing any activity. The new contribution in CRATER is its mathematical approach that is used to rate nodes based on what is called cautious assumptions, which are very true in most WSN. These assumptions basically introduce the cases in which WSN nodes are very sensitive to hearing SHI and are concerned with their immediate neighbors.

Moreover, the rating component is evaluated by a novel and promising mechanism proposed to evaluate different reputation systems. The evaluation scheme is called reputation systems-independent scale for trust on routing (RESISTOR). RESISTOR is based on the analogy of the resistance phenomenon in electric circuits. It defines a metric called "resistance" to represent how much a node is resisting its malicious neighbors by finding the ratio between the risk value for the malicious node, which is computed by the monitoring node using CRATER, and the number of packets flowed into that malicious node. Then, based on that figure, which is called the resistance figure, the system performance is analyzed for evaluation.

Finally, the response component of the reputation system suggests a new routing protocol that aims to provide a secure packet delivery service guarantee by incorporating the behavior trust concept into the routing decision. The proposed geographic, energy and trust aware routing (GETAR) protocol is an enhanced version of the GEAR protocol (Yu, Govindan, and Estrin 2001). GEAR is basically a geographic routing protocol in which the next hop is selected based on two metrics: the distance between the next hop and the destination and the remaining energy level the next hop owns. The new contribution of this work is to add a third metric in the next-hop selection process, that is, the risk level of a node defined as the amount of risk the sender may encounter by selecting a particular node as a next hop. The risk value a sender knows about a node reflects the "trustworthiness" that it has towards that node.

FUTURE RESEARCH DIRECTIONS

Recent research work focuses on energy-aware design and efficient communication and networking within the WSN. On the physical layer level, techniques for low-power hardware design, overcoming signal propagation, and optimized modulation schemes are of great interest. Another very important area of open research is the design of energy-aware and efficient medium access control protocol for enhanced WSN performance and prolonged network lifetime. On the network level, new integrated identity and behavior trust aware routing algorithms that are tailored for operation given the limitations of the WSN are necessary. Finally, at the application layer, protocols necessary for sensor management, task assignment and data advertisement, and sensor query and data decimation are being developed.

Node mobility is an important issue to be considered when developing secure routing protocols. Most of the current protocols assume that the senor nodes and the base stations are stationary. However, there might be situations such as battle environments where the base station and possibly the sensors need to be mobile. In such cases, frequent update of the position of the base station and sensor nodes and propagation of that information through the network and rekeying operation may excessively drain the energy of nodes. New secure routing algorithms are needed in order to handle the overhead of mobility, rekeying, and topology changes in such an energy-constrained environment. A feature that is important in every routing protocol is to adapt to topology changes very quickly and to maintain the network functions.

One aspect of sensor networks that complicates the design of a secure routing protocol is in network aggregation. In WSNs, in-network processing makes end-to-end security mechanisms harder to deploy because intermediate nodes need direct access to the contents of the messages. Finding efficiently and optimally the processing points in WSNs is still an open research issue.

There are not many published work on the general intrusion detection techniques for wireless sensor networks. There are some works on intrusion detection targeted for specific kind of attacks. Wireless sensor networks require a solution that is fully distributed and inexpensive in terms of communication, energy, and memory requirements. In order to look for anomalies, applications and typical threat models must be understood. It is particularly important for researchers and practitioners to understand how cooperating adversaries might attack the system. The promising approach for decentralized intrusion detection is the use of secure groups. More research is needed to determine better node features addressing specific vulnerabilities and to develop improved detection algorithms taking into account sensor node capabilities.

Novel techniques of network clustering that maximize the network lifetime are also a hot area of research in WSNs (Bandyopadhya & Coyle, 2003). Since sensor nodes are prone to failure, fault tolerance techniques come into the picture to keep the network operating and performing its tasks. Routing techniques that explicitly employ fault tolerance techniques in an efficient manner are still under investigation (Dulman et al., 2003).

Another area which needs extensive research is the study of survivability issues in wireless sensor networks. Survivability of a system can be defined as the capability to fulfill its mission, in a timely manner, and in the presence of intrusions, attacks, accidents, and failures. A framework of survivability model for WSN with software rejuvenation methodology, which is applicable in security, has been proposed by (Kim, Shazzad, and Park (2006).

Most of the currently proposed key management schemes are based on the assumption that all the nodes in the sensor networks are homogeneous and with similar capabilities, such as memory and radio range. It has been found that by applying heterogeneous sensor nodes in a sensor network, the small percentage of more capable sensor nodes can provide an equal level of security, and meanwhile improve the resilience of node compromise. The unbalanced scheme proposed by Traynor, Choi, Cao, Zhu, and La Porta (2004) not only reduces the number of transmissions necessary to establish session-keys but also reduces the effect of both single and multiple node captures. Another area which needs intensive research is the development of path-key establishment phase of key management scheme. Some special protocols combined with routing information may be considered to achieve the secure and efficient path-key establishment. Furthermore, based on the current research on the coverage and connectivity in the sensor networks, some random distribution model (Bettstetter, 2002) should also be considered when modeling a secure communication model in wireless sensor networks.

An important area which needs extensive research is the development of efficient node monitoring and rating approaches in reputation system-based solutions. Another problem which needs extensive research is a bootstrapping problem in sensor networks. This the startup period which is required to build reputation and trust among nodes in a network in noncryptographic-based solutions and to discover shared keys and perform key-setup among sensor nodes in cryptographic-based solution. Minimizing this startup period to prevent node compromise during bootstrapping is an open issue.

Public-key solutions built upon the pairing-based identity-based cryptography (IBC) is emerging as an alternative (more appropriate than traditional public key cryptography for WSNs) with the efficient hardware implementation of Tate pairing (Barreto, Lynn, & Scott, 2004) on smartcard (Bertoni, Chen, Fragneto, Harrison, & Pelosi, 2005), PDAs (Scott, 2005), and FPGAs (Kerins, Marnane, Popovici, & Barreto, 2005).

Another issue which has triggered a growing debate is on the use of symmetric-key vs. public-key cryptography (PKC) in WSNs. How to modify the public key cryptography and apply it to the key management issues in resource-constrained WSNs is a major challenge. Recent studies show that it is still possible to apply public key cryptography to sensor networks by judiciously selecting right algorithms and associated parameters (Arazi, Elhanany, Arazi, & Qi, 2005; Gaubatz, Kaps, & Sunar, 2004). ECC (Malan, Welsh, & Smith, 2004) is especially attractive for constrained wireless devices because the smaller keys in ECC result in memory, bandwidth, and computational savings. With the advancements of hardware and software, public key infrastructure in WSN is not only possible but also necessary (Gura, Patel, Wander, Eberle, & Shantz, 2004).

CONCLUSION

In this chapter, we have presented a comprehensive treatment of the routing security problem in wireless sensor networks. We have provided an overview of WSN architecture, possible applications, and indicated the special characteristics of wireless sensor networks from routing perspective. We have highlighted the importance of secure routing problem considering the different network aspects and special conditions of WSN. We have provided a detailed analysis of routing threats and attacks that are more specific to routing operation in wireless sensor networks and also indicated possible countermeasure against these attacks. We have provided a comprehensive review and an in-depth discussion of different intrusion prevention and detection techniques, cryptographic-based solutions (with emphasis on key management schemes), and noncryptographic-based solutions (with emphasis on trust and reputation of sensor nodes) for the secure routing problem highlighting their pros and cons. We have also presented some open problems that are currently being researched.

REFERENCES

AbuGhazaleh, N., Kang, K.D., & Liu, K. (2005, October 10-13). *Towards resilient geographic routing in WSNs*. Paper presented at MSWiM'05.

Agah, A., & Das, S.K. (2007, September). Preventing DoS attacks in wireless sensor networks: A repeated game theory approach. *International Journal of Network Security, 5*(2), 145-153.

Agah, A., Das, S.K., & Basu, K. (2004). Intrusion detection in sensor networks: A non-cooperative game approach. In *Proceedings of the Third IEEE International Symposium on Network Computing and Applications (NCA'04).*

Al-Karaki, J. N., et al. (2004, April 18-21). Data aggregation in wireless sensor networks: Exact and approximate algorithms. In *Proceedings of IEEE Workshop on High Performance Switching and Routing*, Phoenix.

Arazi, B., Elhanany, I., Arazi, O., & Qi, H. (2005). Revisiting public-key cryptography for wireless sensor networks. *Computer, 38*(11), 103-105.

Bandoyopadhya, S., et al., (2006).Clustering distributed data streams in peer-topeer environments. *Information Sciences, 176*(14), 1952-1955. Elsevier.

Bandyopadhya, S., & Coyle, E. (2003). An energy efficient hierarchical clustering algorithm for wireless sensor networks. In *Proceedings of INFOCOM 2003* (Vol. 3, 1713-1723).

Bannerjee, S., Grosan, C., & Abraham, A. (2005). *IDEAS intrusion detection based on emotional ants.* Paper presented at the 5th International Conference on Intelligent Systems Design and Applications (ISDA '05) (pp. 344-349).

Barreto, P., Lynn, B., & Scott, M. (2004). On the selection of pairing-friendly groups. In *Proceeding of Selected Areas Cryptography* (LNCS 3006, pp. 17-25). New York: Springer Verlag.

Bertoni, G., Chen, L., Fragneto, P., Harrison, K., & Pelosi, G. (2005). *Computing Tate pairing on smartcards* (White paper STMicroelectronics). Retrieved October 27, 2007, from http://www.st.com/stonline/products/families/smartcard/ast_ibe.htm

Bettstetter, C. (2002). On the minimum node degree and connectivity of a wireless multi-hop network. In *Proceedings of the 3rd ACM International Symposium on Mobile Adhoc Networking and Computing'02*, EPF Lausanne, Switzerland, (pp.80-91). ACM Press.

Buchegger, S., & Boudec, J.Y.L. (2003, July). *A robust reputation system for mobile ad-hoc networks* (Tech. Rep. IC/2003/50). EPFL IC.

Chan, H., Perrig, A., & Song, D. (2003, May 11-14). Random key predistribution schemes for sensor networks. In *Proceedings of the IEEE Symposium on Security and Privacy,* Oakland, CA, (pp.197-213).

Chan, S., Poovendran, R., & Sun, M. (2005, November 28-December 2). *A key management scheme in distributed sensor networks using attack probabilities.* Paper presented at the Global Telecommunications Conference, GLOBECOM '05 (Vol. 2, pp. 5-). IEEE.

Da Silva, A.P.R., Martins, M.H.T., Rocha, B.P.S., Loureiro, A.A.F., Ruiz, L.B., & Wong, H.C. (2005, October 13). *Decentralized intrusion detection in wireless sensor networks.* Paper presented at Q2SWinet'05, Montreal, Quebec, Canada.

Deng, J., Han, R., & Mishra, S. (2002, November). *INSENS: Intrusion-tolerant routing in wireless sensor networks* (Tech. Rep. CU-CS-939-02). University of Colorado, Department of Computer Science.

Deng, J., Han, R., & Mishra, S. (2004). *Intrusion tolerance and anti-traffic analysis strategies for wireless sensor networks.* Paper presented at the IEEE International Conference on Dependable Systems & Networks (DSN) (pp. 594-603).

Di Pietro, R., Mancini, L.V., & Mei, A. (2003). Random key-assignment for secure wireless sensor networks. In *Proceedings of the 1st ACM Workshop on Security of Ad Hoc and Sensor Networks*, Fairfax, VA, (pp. 62-71).

Di Pietro, R., Mancini, L.V., & Mei, A. (2006, December). Energy efficient node-to-node authentication and communication confidentiality in wireless sensor networks. *Springer Journal on Wireless Networking, 12*(6), 709-721.

Dolev, D., & Yao, A.C. (1983). On the security of public-key protocols. *IEEE Transactions on Information Theory, 29*(2), 198-208.

Du, W., Deng, J., Han Y.S., Chen, S., & Varshney, P.K. (2004, March). *A key management scheme for wireless sensor networks using deployment knowledge.* Paper presented at the IEEE INFOCOM.

Du, W., Fang, L., & Ning, P. (2005). *LAD: Localization anomaly detection for wireless sensor networks.* Paper presented at the IPDPS.

Dulman, S., et al. (2003, March). *Trade-off between traffic overhead and reliability in multipath routing for wireless sensor networks*. Paper presented at the WCNC Workshop, New Orleans.

Eltoweissy, M., Heydaru, H., Morales, L., & Sadborough, H. (2004, March). Combinatorial optimization of key management in group communications. *Journal of Network and Systems Management: Special Issue on Network Security*, 332.

Eltoweissy, M., Moharrum, M., & Mukkamala, R. (2006, April). Dynamic key managements in sensor networks. *IEEE Communications Magazine*, 122-130.

Eschenauer, L., & Gligor, V.D. (2002). A key-management scheme for distributed sensor networks. In *Proceedings of the 9th ACM Conference on Computer and Communications Security* (pp. 41-47). Washington D.C.: ACM Press.

Eskin, E., Arnold, A., Pereau, M., Portnoy, L., & Stolfo, S. (2002). A geometric framework for unsupervised anomaly detection: Detecting intrusion in unlabeled data. *Data Mining for Security Applications*. Kluwer.

Ganeriwal, S., & Srivastava, M. (2004). Reputation-based framework for high integrity sensor networks. In *Proceedings of the 2nd ACM Workshop on Security of Ad Hoc and Sensor Networks*, Washington, D.C.

Gaubatz, G., Kaps, J., & Sunar, B. (2004). Public key cryptography in sensor networks: Revised. In *Proceedings of 1st European Workshop on Security in Ad-hoc and Sensor Networks (ESAS 2004)*, Heidelberg, Germany, (pp. 2-18). Springer.

Gura, N., Patel, A., Wander, A., Eberle, H., & Shantz, S. C. (2004, April). Comparing elliptic curve cryptography and RSA on 8-bit CPUs. In *Proceedings of CHES*, Boston, (pp. 119-132).

Han, J., & Kamber, M. (2001). *Data mining: Concepts and techniques*. Morgan Kauffmann Publishers.

Hu, Y.C., Perrig, A., & Johnson, D. B. (2003, April). Packet leashes: A defense against wormhole attacks in wireless networks. In *Proceedings of IEEE INFOCOMM 2003.*

Jolly, G., Kuscu, M., Kokate, P., & Younis, M. (2003, June). A low-energy key management protocol for wireless sensor networks. In *Proceedings of the IEEE Symposium on Computers and Communications, ISCC'2003* (p. 335).

Josang, A., & Ismail, R. (2002, June). *The beta reputation system*. Paper presented at the 15th Bled Electronic Commerce Conference, e-Reality: Constructing the e-Economy, Bled, Slovenia.

Kaplantzis, S. (2004, October). *Classification techniques for network intrusion detection* (Tech. Rep.). Monash University, ECSE.

Karlof, C., & Wagner, D. (2003). Secure routing in wireless sensor networks: Attacks and countermeasures. *Ad Hoc Networks, 1*(2-3), 293-315.

Kerins, T., Marnane, W., Popovici, E., & Barreto, P. (2005, August-September). Efficient hardware for the for Tate pairing calculation in characteristic three. In *Proceedings of Workshop on Cryptographic Hardware and Embedded Systems*, Edinburgh, Scotland, (pp. 412-426).

Kim, D.S., Shazzad, K.M., & Park, J. S. (2006). A framework for survivability model for wireless sensor network. In *Proceedings of First International Conference on Availability, Reliability and Security (ARES'06).*

Loo, C.E., Ng, M.Y., Leckie, C., & Palaniswami, M. (2006). Intrusion detection for sensor networks. *International Journal of Distributed Sebsor Networks.*

Maarouf, I.K., & Naseer, A.R. (2007, May). *WSNodeRater: An optimized reputation system framework for security aware energy efficient geographic routing in WSNs*. Paper presented at the ACS/IEEE International Conference on Computer Systems and Applications, Amman, Jordan.

Malan, D. J., Welsh, M., & Smith, M.D. (2004, October). A public-key infrastructure for key distribution in tinyOS based on elliptic curve cryptography. In *Proceedings of IEEE SECON*, Santa Clara, CA, (pp.71-80).

Marouf, I.K., & Naseer, A.R. (2006, December). *SNARE: Sensor node attached reputation evalua-*

tor. Paper presented at the CONEXT '06, LIsboa, Portugal.

Michiardi, P., & Molva, R. (2002, September). *Core: A collaborative reputation mechanism to enforce node cooperation in mobile ad hoc networks*. Paper presented Communication and Multimedia Security Conference, Portoroz, Slovenia, (pp. 26-27).

Mui, L., Halberstadt, A., & Mohtashemi, M. (2002, July). Notions of reputation in multi-agents systems: A review. In *Proceedings of First International Joint Conference Autonomous Agents and Multi-Agent Systems* (pp. 280-287).

Mun, Y., & Shin, C. (2005, May 9-12). *Secure routing in sensor networks: Security problem analysis and countermeasures*. Paper presented at the International Conference on Computational Science and Its Applications – ICCSA 2005, Singapore, (LNCS 3480, pp. 459-467). Heidelberg, Germany: Springer Verlag.

Naldurg, S., Yi, R., & Kravets, R. (2001). *Security-aware ad-hoc routing for wireless networks*. Paper presented at the ACM Workshop on Mobile Ad Hoc Networks, MOBIHOC.

Onat, I., & Miri, A. (2005, August). An intrusion detection system for wireless sensor networks. *Wireless and Mobile computing Networking and Communications, 3*, 253-259.

Oniz, C.C., Tasci, S.E., Savas, E., Ercetin, O., & Levi, A. (2005). *SeFER: Secure, flexible and efficient routing protocol for distributed sensor networks*. Paper presented at the IEEE 2005 (pp. 246-255).

Perrig, A., Szewezyk, R., Wen, V., Culler, D., & Tygar, J. (2001). SPINS: Security protocols for sensor networks. In *Proceedings of Mobile Networking and Computing 2001*.

Pirretti, M., Zhu, S., Narayanan, V., McDaniel, P., Kandemir, M., & Brooks, R. (2005, October). *The sleep deprivation attack in sensor networks: Analysis and methods of defense*. Paper presented at the Conference on Innovations and Commercial Applications of Distributed Sensor Networks.

Pottie, G., & Kaiser, W. (2000). Wireless integrated network sensors. *Communications of the ACM, 43*(5), 551-558.

Rajasegarar, S., Leckie, C., Palaniswami, M., & Bezdek, J.C. (2006, October 30-November 1). Distributed anomaly detection in wireless sensor networks. In *Proceedings of Tenth IEEE International Conference on Communications Systems (IEEE ICCS 2006)*, Singapore.

Savvides, A., Han, C., & Srivastava, M. (2001, July). Dynamic fine-grained localization in ad-hoc networks of sensors. In *Proceeding of 7th ACM MobiCom* (pp. 166-179).

Scott, M. (2005, February). Computing the Tate pairing. In *Proceedings of Cryptographers' Track at the RSA Conference*, San Francisco, (pp. 293-304).

Tanachaiwiwat, S., Dave, P., Bhindwale, R., & Helmy, A. (2004, April). *Location-centric isolation of misbehavior and trust routing in energy-constrained sensor networks*.

Traynor, P., Choi, H., Cao, G., Zhu, S., & La Porta, T. F. (2004). *Establishing pair-wise keys in heterogeneous sensor networks* (Networking and Security Center, Tech. Rep. NAS-TR-0001-2004). Penn State University, Dept of Computer Science & Engineering.

Wood, A., & Stankovic, J. (2002, October). Denial of service in sensor networks. *IEEE Computers*, 54-62.

Yang, C., Zhou, J., Zhang, W., & Wong, J. (2006, May 29- June 1). Pairwise key establishment for large scale sensor networks: From identifier based to location based. In *Proceedings of the first International Conference on Scalable Information Systems, INFOSCALE'06*, HongKong.

Younis, M., Ghumman, K., & Eltoweissy, M. (2006). Location-aware combinatorial key management for clustered sensor networks. *IEEE Transactions on Parallel and Distributed Systems*.

Yu, Y., Govindan, R., & Estrin, D. (2001, May). *Geographical and energy-aware routing: A re-*

cursive data dissemination protocol for wireless sensor networks (Tech. Rep. UCLA/CSD-TR-01-0023). University of Southern California.

Zhang, Y., Liu, W., Lou, W., & Fang, Y. (2006, February). Location based compromise-tolerant security mechanisms for wireless sensor networks. *IEEE Journal on Selected Areas in Communications, 24*(2).

Zhu, S., Setia, S., & Jajodia, S. (2003). LEAP: Efficient security mechanisms for large-scale distributed sensor networks. In *Proceedings of ACM CCS, 2003.*

KEY TERMS

DoS Attack: Any event that decreases or eliminates a network's capacity to perform its expected function is termed as a denial-of-service attack or commonly known as DoS attack.

Intrusion: Can be defined as a set of actions that can lead to an unauthorized access or alteration of a certain system.

Key Management: A scheme to dynamically establish and maintain secure channels among communicating nodes. In wireless sensor networks, a key management scheme must deal with the following important issues: key deployment/key predistribution, key discovery, key establishment/key setup, node addition/rekeying, and node eviction/key revocation.

Reputation System: A type of collaborative filtering algorithm which attempts to determine ratings for a collection of entities, given a collection of opinions that those entities hold about each other.

Routing Attacks: Network layer attacks such as routing information spoofing, alteration or replay, blackhole and selective forwarding attacks, sinkhole attacks, Sybil attacks, wormhole attacks, HELLO flood attacks, and acknowledgement spoofing.

Routing Security: Securing routing operation from attacks in a network by deploying appropriate defense.

Trust: A relationship of reliance. Trust is a prediction of reliance on an action, based on what a node knows about the other node, in the context of wireless sensor networks. The notion of trust is increasingly adopted to predict acceptance of behaviors by others.

Wireless Sensor Network (WSN): A wireless network consisting of spatially distributed autonomous devices using sensors to cooperatively monitor physical or environmental conditions, such as temperature, sound, vibration, pressure, motion, or pollutants at different locations.

Chapter XXXVII
Localization Security in Wireless Sensor Networks

Yawen Wei
Iowa State University, USA

Zhen Yu
Iowa State University, USA

Yong Guan
Iowa State University, USA

ABSTRACT

Localization of sensor nodes is very important for many applications proposed for wireless sensor networks (WSN), such as environment monitoring, geographical routing, and target tracking. Because sensor networks may be deployed in hostile environments, localization approaches can be compromised by many malicious attacks. The adversaries can broadcast corrupted location information; they can jam or modify the transmitting signals between sensors to mislead them to obtain incorrect distance measurements or nonexistent connectivity links. All these malicious attacks will cause sensors not able to or wrongly estimate their locations. In this chapter, we summarize the threat models and provide a comprehensive survey and taxonomy of existing secure localization and verification schemes for wireless sensor networks.

INTRODUCTION

In recent years, the availability of low-cost, low-power, multifunctional, small-size autonomous devices equipped with various sensors has expedited the development of wireless sensor networks (WSN). Wireless sensor networks have both military applications (e.g., battlefield surveillance) and civilian applications (e.g., environment and habitat monitoring, target tracking, seismic detection, smart-home automation, and traffic control). To facilitate the cooperation between sensors and achieve different application goals, network and application protocols such as routing protocol, data aggregation algorithm, and localization algorithm need to be properly designed.

Among these research issues, localization of sensor nodes is very important to some applications. For example, in environment surveillance applications, a sensor must report its location to the monitoring center when it detects some enemy force (e.g., a tank); in geographical routing protocol, a sensor should know the locations of its neighbors and forwards data packets to the neighbor who is closest to the destination.

Traditional localization approaches require the sensor nodes to equip with expensive global positioning system (GPS) devices, which are not affordable in some cases, especially in large-scale sensor networks. Hence, many localization schemes (Bahl & Padmanabhan, 2000; Bulusu, Heidemann, & Estrin, 2000; Doherty, Pister, & Ghaoui, 2001; Fang, Du, & Ning, 2005; Harter, Hopper, Steggles, Ward, & Webster, 1999; He, Huang, Blum, Stankovic, & Abdelzaher, 2003; Lazos & Poovendran, 2004; Nicolescu & Nath, 2001, 2003; Priyantha, Chakraborty, & Balakrishnan, 2000; Savvides, Han, & Srivastava, 2001; Shang, Ruml, Zhang, & Fromherz, 2003; Smith, Balakrishnan, Goraczko, & Priyantha, 2004) have been proposed. These schemes assume that some special sensor nodes (named anchors) can obtain their absolute locations through GPS device. Thus other sensors can use the measured distance or connectivity information between them and the beacon messages sent from the anchors to calculate their locations.

When sensor networks are deployed in hostile environments, localization approaches are vulnerable to many malicious attacks. For example, the adversaries can compromise a sensor node and send out false location information to disturb the localization of other nodes. Sensor nodes are constrained by limited energy resources, memory resource, computation ability, and communication bandwidth, therefore, traditional cryptography mechanisms such as a public key system cannot be applied to wireless sensor networks. Moreover, localization approaches utilize the physical features of the transmitting signals between sensors (e.g., transmitting time or signal strength), thus they are vulnerable to many localization-specific attacks (e.g., distance-modification attack) that cannot be prevented by traditional security mechanisms. All these attacks can cause the sensors to be not able to or wrongly estimate their locations.

In this chapter, we provide a comprehensive survey and taxonomy of existing countermeasures that secure the localization in wireless sensor networks. We classify the secure countermeasures into secure localization schemes, which enhance sensors' attack-resistant ability, and location verification schemes, which verify sensors' locations (accept the correct location estimations and discard the abnormal ones) after the sensors have obtained their locations. We also classify these secure localization (or verification) schemes on whether they use precise (with nanosecond precision) time-measuring hardware, sectored antenna, or not use any special hardware.

The rest of chapter is organized as following: In the following section, we take an overview and give a classification of current localization approaches. We describe the threat models, and we provide the taxonomy of existing secure localization approaches. Finally, we discuss some future trends and conclude the chapter.

BACKGROUND: LOCALIZATION IN WIRELESS SENSOR NETWORKS

In recent years, many localization approaches have been proposed for wireless sensor networks. Before we talk about the security issues, let us take an overview of the localization systems and the techniques involved in different localization approaches.

The most traditional and widely-used localization system is the global positioning system. The earth-based GPS receivers can provide users with location, speed, and time by calculating the distances from at least three satellites. However, it is not feasible to equip the relatively expensive GPS receiver on each node in large scale sensor networks. Most localization algorithms assume that only a fraction of sensor nodes in the field can obtain their locations through GPS receivers (or through manual configurations). These nodes are called anchors and serve as location references for other nodes to localize. Depending on the availability of such anchors, we classify the localization approaches as anchor-based or anchor-free ones. We also classify them into range-based or range-free ones on whether they require distance measurements between sensors, and centralized or distributed on whether the localization is performed by a computing center or by sensors themselves. The classification is given in Table 1, where "(c)" means that the approach is centralized.

Anchor-Based Range-Based Approaches

In anchor-based localization approaches, some anchors are deployed whose positions are known from GPS device or manual configuration. In range-based approaches, the sensors' locations are determined by using trilateration technique based on the distances between sensors. The trilateration method solves a set of equations and estimates sensor's location that best satisfy the distance constraints (according to some optimization criteria, e.g., least square error criteria). Active Bat (Harter et al., 1999), RADAR (Bahl & Padmanabhan, 2000), AHLoS (Savvides et al., 2001), and SDP (So & Yu, 2005) are all anchor-based range-based approaches. Besides the least mean square criteria, Kalman-filter (KF) and least mean square (LMS) (Smith et al., 2004) can also be applied to obtain the optimal solution for sensors' location.

Anchor-Based Range-Free Approaches

Since range-based approaches require special hardware to measure the distances between sensor nodes, range-free approaches attracted more research interests recently. In anchor-based range-free approaches, no distance measurements are needed and sensors determine their locations using the beacon messages from anchors. Both Active Badge (Want et al., 1992) and Cricket (Priyantha et al., 2000) belong to this category. Centroid method (Bulusu et al., 2000) calculates a sensor's location as the mean value of the locations of anchors from which this sensor hears beacon messages. APIT (He et al., 2003) determines some triangles in which a sensor may reside, and estimates the sensor's location as the overlapping region of theses triangles. SeRLoc (Lazos & Poovendran, 2004) uses sectored antennas equipped on anchors and computes sensor's location as the centroid of the overlapping region of multiple sectors. DV-hop (Nicolescu & Nath, 2001) and DV-based AoA (Nicolescu & Nath, 2003) first obtain the hop

Table 1. Classification of localization approaches

	Range-Based	Range-Free
Anchor-Based	**Active Bat(c)** (Harter et al., 1999) **RADAR(c)** (Bahl & Padmanabhan, 2000) **AHLoS** (Savvides, Han, & Srivastava, 2001) **LMS/KF** (Smith et al., 2004) **SDP(c)** (So & Yu, 2005)	**Active Badge(c)** (Want et al., 1992) **Centroid** (Bulusu et al., 2000) **Cricket** (Priyantha et al., 2000) **Convex(c)** (Doherty et al., 2001) **DV-hop** (Nicolescu & Nath, 2001) **DV-based AoA** (Nicolescu & Nath, 2003) **APIT** (He et al., 2003) **Amorphous** (Nagpal, Shrobe, & Bachrach, 2003) **SeRLoc** (Lazos & Poovendran, 2004)
Anchor-Free	**MDS-MAP(c)** (Shang et al., 2003)	**MDS-MAP(c)** (Shang et al., 2003) **Deployment Knowledge** (Fang, Du, & Ning, 2005)

counts from sensors to anchors by flooding through the sensor field, then estimates the average hop distance and translates the hop-count distances to real distances to determine sensors' locations. Amorphous (Nagpal et al., 2003) employs a similar strategy as DV-hop but estimates the average hop distance offline. Convex (Doherty et al., 2001) utilizes a linear programming (LP) method to solve the linear equations and obtain the optimal solutions for the sensors' locations.

Anchor-Free Range-Based Approaches

There are relatively fewer anchor-free range-based localization approaches. One is MDS-MAP (Shang et al., 2003), which is based on multidimensional scaling technique to derive the locations of all sensors. It can also work as a range-free approach when only using the connectivity information between sensors instead of the distance measurements, which may cause some degradations of the localization performance.

Anchor-Free Range-Free Approaches

MDS-MAP (Shang et al., 2003) is a centralized anchor-free range-free localization approach. Besides, Fang et al. (2005) proposed a decentralized approach, which assumed that sensors are deployed in groups and the sensors in the same group can land in different locations following a known probability distribution. With this prior deployment knowledge, a sensor utilizes the observation of the group memberships of its neighbors, and utilizes the maximum likelihood estimation method to determine its location.

THREATS TO LOCALIZATION APPROACHES

Since sensor networks may be deployed in hostile environments, the localization approaches are subject to many malicious attacks. In this section, we classify and discuss the possible attacks launched to the current localization approaches.

Attackers can compromise anchors or sensors and send out false location information. They can jam the communications between sensors and replay the messages, which maes sensors wrongly estimate the time-of-flight value and obtain wrong distance measurements. They can strengthen or weaken the signal strength, which also makes the sensors obtain wrong distance measurements. Finally, the attackers can use a wired link (called wormhole) to transmit messages received from one location and broadcast at the other location, thus making sensors build nonexistent neighboring connectivity, which results in wrong estimations of the sensors' locations.

We can classify the attackers into internal attackers and external attackers. An internal attacker can compromise a sensor, obtain its key materials, and authenticate itself to others. An external attacker cannot obtain any cryptographic secrets or authenticate itself, but it can corrupt the physical features of the communications between sensors, for example, they can corrupt the distance measurements or neighboring connectivity by jamming the communications between sensors. In Table 2, we list the threat models and the corresponding attackers that can launch the threat models. We then describe them in more details in the following subsections.

Fake Location

Fake locations information can be generated by the internal attackers who compromise sensors and authenticate themselves as legitimate ones. The impact of this attack is twofold. First, many location-based applications such as environment monitoring and target tracking will be fooled by the wrong location of some specific events, for example, high-temperature area and location of an enemy tank. Second, other sensors' locations will be polluted if they refer to these fake locations when localizing themselves.

Wormhole

Wormhole attack was first discussed by Hu, Perrig, and Johnson (2003). In the wormhole attack, the

Table 2. Classification of threat models

	Fake Location	Wormhole	Range Englargement	Range Reduction
Internal Attackers	X		X	X
External Attackers		X	X	X

Figure 1. A wormhole attack on sensor localization

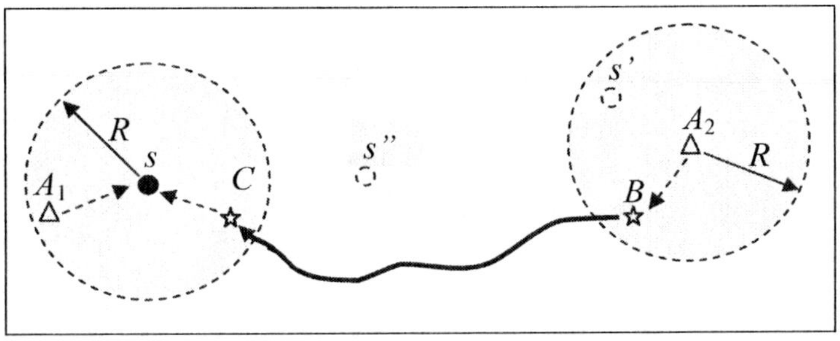

adversaries copy the messages heard at one location and replay them at another location.

Figure 1 illustrates how a wormhole attack can damage a sensor's localization. As shown in the figure, sensor *s* can directly hear the beacon message of anchor A1, but not of anchor A2. To attack the localization of s, an adversary establishes a wormhole between position B and C, which are near A2 and *s*, respectively. Then, the adversary records A2's beacon message at position B, transmits it through the wormhole tunnel, and replays it at position C. If *s* determines its location only based on A2's beacon message, it may assume it is near anchor A2 (at some location within the transmission region of A2). If it uses both messages of A1 and A2, it may either believe it is located somewhere between A1 and A2 (e.g., at location *s''*) or it may not be able to determine its location at all because it is not expected to receive the beacon messages from two anchors so far away from each other.

In such a wormhole attack, the adversaries do not need to compromise any sensor or anchor to understand the meaning of the messages, they just copy and transmit the messages through the established wormhole tunnel to corrupt the localization approaches.

Range Enlargement and Reduction

The range modification attacks are detrimental to range-based localization approaches.

(1) If a time-of-flight method is used to estimate distance, external attackers can jam and replay the signal or transmit it through multipaths to prolong the transmitting time (range enlargement attack). Or they can speed-up the signals to shorten the transmitting time (range reduction attack). For example, they transform the ultrasound signal into radio frequency signal whose transmitting speed is faster, and transform the signal back to ultrasound and broadcast the signals at the end point. Internal attackers can fully control the compromised sensors, thus they may hold on to the signal for a short period of time before transmitting to launch a range enlargement attack. (2) If a signal strength method is used to estimate distance, external attacker can jam and strengthen or weaken the signal before replaying it; internal attackers can directly broadcast signals with strengthened or weakened signals.

A TAXONOMY OF SECURE MECHANISMS

In recent years, many secure mechanisms have been proposed to defend against the attacks to localization in wireless sensor networks. We provide a taxonomy in Table 3. Secure location schemes can help the sensors to correctly localize themselves; location verification schemes can detect and discard abnormal locations of sensors after their locations have been determined. We classify these secure mechanisms on whether they use delicate hardware, directional antenna, or no special hardware. In the following subsections, we discuss each category of the secure mechanisms in more details.

Secure Localization Schemes Against Wormholes

Hu et al. (2003) propose the first work called pack leashes to defend against wormhole attacks. In their work, a temporal packet leash is established by restricting an upper bound on the lifetime of a packet. When receiving a packet, the receiver checks if it has been expired and discards the expired packets that are transmitted through wormholes and incur long processing and transmitting time. A geographical packet leash is established by calculating the distance between two sensors' geographical positions. The receiver can recognize the wormhole packets that travel a distance longer than a certain threshold. In the temporal leash, highly precise synchronization (hundreds of nanoseconds) is required, since a radio signal travels at the speed of light and the mutual distance between sensors are of only several meters. In the geographical leash, correct geographical locations are necessary, thus it cannot be used to defend against wormhole attacks that make the sensors' locations not trustworthy.

Hu and Evans (2003) utilize sector antennas equipped on sensors to detect wormholes. They assume that each antenna has N equally divided zones (numbered from 1 to N). A sensor listens to the carrier in omnimode, and receives signals through the zone in which the signal power is maximal. By using a magnetic needle, it can be ensured that the antenna zones of the same number (e.g., zone of number 1) on all sensors face the same direction. In Figure 2, we see that the signals between true neighbors are sent and received in the opposite zones (e.g., Zone 4 and Zone 1). Therefore, if a sensor receives a message in Zone i, and the mes-

Table 3. A taxonomy of secure localization and location verification schemes

	Secure localization schemes		Location verification schemes
	Against wormholes	Against all attacks	
Delicate hardware required	**Packet Leashes** (Hu et al., 2003)		**Distance-bounding** (Brands & Chaum, 1993) **Claim** (Sastry, Shankar, & Wagner, 2003) **Verifiable Multilateration(L)** (Capkun & Hubaux, 2005) **Covert Base-station** Capkun, Cagalj, & Srivastava, 2006)
Sector antenna required	**Sectored antenna** (Hu & Evans, 2003) **SeRLoc** (Lazos & Poovendran, 2004)		
No special hardware required		**MMSE-Outlier** (Liu, Ning, & Du, 2005) **LMS-Outlier** (Li, Trappe, Zhang, & Nath, 2005) **COTA** (Wei, Yu, & Guan, 2006)	**LAD(L)** (Du, Fang, & Ning, 2005) **PLV** (Ekici, McNair, & Al-Abri, 2006)

sage is sent from Zone j of the sender node, and i and j are not opposite to each other, we can detect that messages may be transmitted through some wormholes. Besides this basic detecting method, the authors propose a verified-neighbor-discovery protocol and a strict-neighbor-discovery protocol to detect the sophisticated wormholes. These protocols require some *potential verifier nodes* to help a sensor to distinguish legitimate neighbors from the wormhole ones. Thus the lack of sufficient verifier nodes will result in the lost of some legitimate connectivity links and degradation of the localization performance.

Lazos and Poovendran (2004) propose another secure localization scheme called SeRLoc that also uses sectored antennas. An anchor transmits different beacons at each antenna sector containing the anchor's location and the angles of the antenna boundary lines. Each sensor determines its location as the center of gravity of the overlapping region of all sectors it hears. During this localization process, wormholes can be detected using two properties: the sector uniqueness property and the communication range violation property. If two sectors of a single anchor are heard, or if two anchors heard by the sensor have a mutual distance greater than 2R (R is the communication range), the sensor can detect that it is under wormhole attacks. After detecting the wormhole, the sensor broadcasts a random nonce and identifies the closest anchor, Li, by the first reply, then takes the center of gravity closest to Li as its estimated location. This technique is named attach to closer locator algorithm (ACLA). One problem of ACLA is that innocent packets may sometimes arrive later than the ones

Figure 2. Detect wormholes using sector antennas

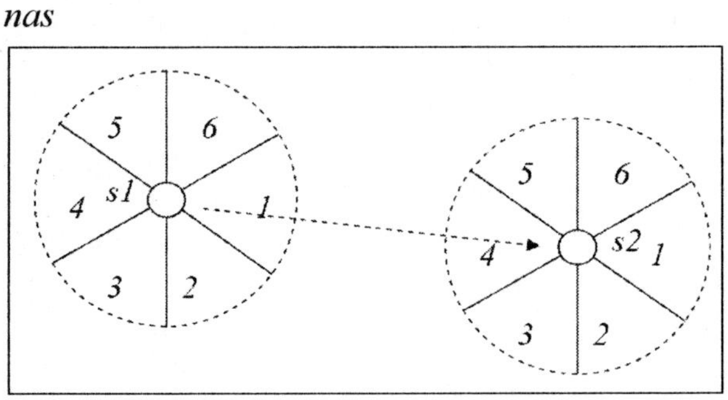

through wormholes because the communications are unreliable in reality and the messages may need to be retransmitted multiple times before the receiver can actually receive them.

Secure Localization Schemes Against All Attacks

All malicious attacks to localization including fake locations, wormholes, and range modifications have a common feature: they all provide inconsistent location references, namely, the sending sensor's location and the measured distance between the sender and the receiver are inconsistent. Therefore, some experts suggested using statistical outlier-removing methods to filter out inconsistent references.

Liu et al. (2005) take the mean square error (MSE) as an indicator of the degree of inconsistency among location references. They propose a greedy algorithm that starts with the set of all location references, and each time considers the subsets with one fewer reference and chooses one subset with the least MSE as the input to the next round, until the MSE value drops below a reasonable threshold. This scheme can effectively enhance sensors' attack-resistant ability, but it launches relatively high computation overheads on sensors. Another problem is that it requires benign references to be the majority among all location references, and may not work well when corrupted location references collude together and take a larger percentage (e.g., around 50%) among all references.

Instead of identifying and eliminating inconsistent references before localization, Li et al. (2005) propose a scheme that lives with these inconsistent references and estimates reasonable locations for sensors using least median of the squares (LMS) technique. LMS is one of the most commonly used robust fitting algorithms and can tolerate up to 50% outliers among the total references. Since the exact LMS solutions are computationally prohibitive, the authors adopted an efficient alternative technique (Rousseeuw & Leroy, 2003) to first get several candidate reference subsets, then choose the one with the least median squares to estimate a sensor's location.

Both of the above schemes try to prevent sensors from wrongly localizing themselves, however, when a sensor fails to filter out the inconsistent references, its corrupted location would "pollute" the localization of many downstream sensors and cascade through the entire sensor network. Wei, Yu, and Guan (2006) propose a scheme named COTA that uses confidence tags to identify spurious localizations of sensors. COTA consists of a tag generation process and a reference filtering process. In the tag generation phase, two methods (the statistic indicator and the geographical indicator) can be used to calculate the sensors' confidence tags based on the positions of their neighbors, distance measurements, and the confidence tags of their neighbors. In the reference filtering phase, bad references can be filtered out by comparing their confidence tags to the absolute and relative metrics. COTA can effectively prevent the proliferation of location errors in the sensor field.

Location Verification Schemes

Although many secure localization schemes have been proposed to provide robust localization performance, they require special hardware or assume some limitations on the adversaries' abilities, and cannot guarantee that all sensors can calculate correct location estimations. Moreover, a compromised sensor (internal attacker) can directly report corrupted locations to the base station; meanwhile it provides a correct location to its neighbors and cannot be detected. These corrupted locations can cause severe consequences to many location-based applications. For example, wrong locations of enemy force will make the surveillance center not able to locate or track the real target, and thus the location verification is a necessary second-line to defend against the adversaries. Note that some verification schemes can also be used as secure localization schemes if sensors' locations have not been determined, and we denote them by "(L)" in Table 3.

Verification Using Special Hardware

The location verification problem was first introduced by Sastry et al. (2003), where the authors propose the echo protocol to verify if a device is inside some specific region (e.g., a room or a football stadium) to facilitate location-based access control. Their protocol is very simple in that the verifier node sends a packet containing a nonce using RF and the device echoes the packet back using ultrasound. Then by checking the packet transmission time and the processing delay, the verifier can verify if the device locates inside the circle region centered at the verifier.

If RF time-of-flight method can be used to measure distance, distance-bounding protocol (Brands & Chaum, 1993) can upper bound the measured distance from one device to another. The important assumption of this protocol is that the device can bound its xor processing to a few nanoseconds and the verifier can measure time with nanosecond precision. Based on this distance-bounding protocol, Capkun and Hubaux (2005) propose a location verification scheme for wireless sensor networks using a verifiable multilateration (VM) technique. The rationale behind VM technique is that when a sensor claims to locate somewhere within a triangle region formed by three verifiers, then its location can be verified only when all three distances from the sensor to the verifiers are consistent with the calculated ones. The limitations of the VM technique are the requirement of delicate hardware to perform distance-bounding protocol and the requirement of dense deployment of verifiers.

Lazos, Poovendran, and Capkun (2005) propose a secure localization and verification system called ROPE, which combines the secure properties of the verifiable multilateration technique (Capkun & Hubaux, 2005) and SeRLoc (Lazos & Poovendran, 2004).

Capkun et al. (2006) propose a verification scheme using covert base stations. The covert base stations (CBS) are silent to the on-going communications and their positions are only known to the verification infrastructure. Upon receiving location messages from a sensor, several CBS cooperate (through wired links) and check if their location is consistent with the difference of time-of-arrival to each CBS. Because sensors do not know the positions of CBS, their success rate to achieve consistency through guessing is very small. A mobile base station (MBS) can also play the role of verifier, by sending a verification request from one location, moving, and waiting for the response at a different location. Therefore, at the time of performing verification, a sensor does not know the positions of the MBS.

Verification Without Special Hardware

Unlike other verification schemes that use some special hardware, Du et al. (2005) propose a scheme that verifies sensors' locations by checking the consistency of the locations with the deployment knowledge. They assume that all sensors are deployed in groups (each group has a unique group ID) following a known probability distribution. A sensor's location can be verified only when its neighborhood observation is consistent with that derived from the deployment knowledge. The difference between this scheme and the previous works is that in this scheme, the sensors are verified if their locations are within an anomaly degree from their true locations, rather than exactly at the true locations.

Recently, Ekici et al. (2006) proposed probabilistic a location verification (PLV) algorithm to verify sensors' locations in densely deployed sensor networks. PLV explores the probabilistic relation between the number of hops a packet traverses to reach a destination and the Euclidean distance between source and destination. Then the verifier can determine plausibility (between 0 and 1) and create a trust level for each sensor's location claim.

FUTURE TRENDS

Although various secure mechanisms have been proposed for localization in wireless sensor networks, there is still a large space for future improvements.

First, very few works have been done to secure range-free localization approaches which deserve more research efforts. For example, in DV-hop approach, if the adversaries compromise a single node and send out a false hop count, then all down-steaming nodes will be influenced and estimate false hop counts from them to the anchor, resulting in a biased estimation of the average hop-distance.

Another issue is that current location verification schemes either verify if a sensor exactly locates at its claimed location, or verify if it locates within the anomaly degree of its true location. However, verification regions can be arbitrary and should be related to the specific application. For example, in a military surveillance application, the monitoring center decides to project a missile at the location reported by the sensor who detects the enemy force, thus it should determine a specific verification region in which the detecting sensor should reside to guarantee that the target can be destroyed.

CONCLUSION

In this chapter, we provide a taxonomy of the research efforts devoted to secure localization in wireless sensor networks. We classify them into secure localization schemes that aim to provide correct location estimations for sensors at the front-line, and location verification schemes that aim to detect abnormal locations of sensors at the second-line, that is, after sensors' locations have been determined using any other (insecure or secure) localization approaches. We also classify the security localization mechanisms on whether they require any special hardware. Generally, localization for sensor networks becomes more robust with the availability of more advanced hardware, for example, sectored antennas, fast processing hardware, or even nanosecond-precision clocks. If there is no such special hardware, other information such as deployment knowledge is needed to detect the inconsistent information injected by adversaries.

REFERENCES

Bahl, P., & Padmanabhan, V. N. (2000). *RADAR: An in-building RF-based user location and tracking system.* Paper presented at the IEEE Conference on Computer Communications (INFOCOM).

Brands, S., & Chaum, D. (1993). Distance-bounding protocols. *Theory and application of cryptographic techniques* (pp. 344-359).

Bulusu, N., Heidemann, J., & Estrin, D. (2000). GPS-less low cost outdoor localization for very small devices. *IEEE Personal Communications, 7*(5), 284.

Capkun, S., Cagalj, M., & Srivastava, M. (2006). *Secure localization with hidden and mobile base stations.* Paper presented at the IEEE Conference on Computer Communications (INFOCOM).

Capkun, S., & Hubaux, J. (2005). *Secure positioning of wireless devices with application to sensor networks.* Paper presented at the IEEE Conference on Computer Communications (INFOCOM).

Doherty, L., Pister, K. S., & Ghaoui, L. (2001). *Convex position estimation in wireless sensor networks.* Paper presented at the IEEE Conference on Computer Communications (INFOCOM).

Du, W., Fang, L., & Ning, P. (2005). LAD: Localization anomaly detection for wireless sensor networks. In *Proceedings of IEEE International Parallel and Distributed Processing Symposium (IPDPS).*

Ekici, E., McNair, J., & Al-Abri, D. (2006). A probabilistic approach to location verification in wireless sensor networks. In *Proceedings of IEEE International Conference on Communications (ICC).*

Fang, L., Du, W., & Ning, P. (2005). *A beacon-less location discovery scheme for wireless sensor networks.* Paper presented at the IEEE Conference on Computer Communications (INFOCOM).

Harter, A., Hopper, A., Steggles, P., Ward, A., & Webster, P. (1999). *The anatomy of a context-aware application.* Paper presented at the Annual International Conference on Mobile Computing and Networking (ACM Mobicom).

He, T., Huang, C., Blum, B., Stankovic, J., & Abdelzaher, T. (2003). *Range-free localization schemes in large scale sensor network.* Paper presented at the Annual International Conference on Mobile Computing and Networking (ACM Mobicom).

Hu, L., & Evans, D. (2003). Using directional antennas to prevent wormhole attacks. In *Proceedings of the 11th Network and Distributed System Security Symposium* (pp. 131-141).

Hu, Y., Perrig, A., & Johnson, D. (2003). *Packet leashes: A defense against wormhole attacks in wireless ad hoc networks.* Paper presented at the IEEE Conference on Computer Communications (INFOCOM).

Lazos, L., & Poovendran, R. (2004). *SeRLoc: Secure range-independent localization for wireless sensor networks.* Paper presented at the ACM Workshop on Wireless Security.

Lazos, L., Poovendran, R., & Capkun, S. (2005). *Rope: Robust position estimation in wireless sensor networks.* Paper presented at the ACM/IEEE *Information Processing in Sensor Networks (*IPSN).

Li, Z., Trappe, W., Zhang, Y., & Nath, B. (2005). *Robust statistical methods for securing wireless localization in sensor networks.* Paper presented at the ACM/IEEE *Information Processing in Sensor Networks (*IPSN).

Liu, D., Ning, P., & Du, W. (2005). *Attack-resistant location estimation in sensor networks.* Paper presented at the ACM/IEEE Information Processing in Sensor Networks (IPSN).

Nagpal, R., Shrobe, H., & Bachrach, J. (2003). *Organizing a global coordinate system from local information on an ad hoc sensor network.* Paper presented at the ACM/IEEE Information Processing in Sensor Networks (IPSN).

Nicolescu, D., & Nath, B. (2001). *Ad-hoc positioning systems* (APS). Paper presented at the IEEE Global Telecommunications Conference (GLOBECOM).

Nicolescu, D., & Nath, B. (2003). *Ad hoc positioning system (APS) using AoA.* Paper presented at the IEEE Conference on Computer Communications (INFOCOM).

Priyantha, N., Chakraborty, A., & Balakrishnan, H. (2000). *The cricket location-support system.* Paper presented at the Annual International Conference on Mobile Computing and Networking (ACM Mobicom).

Rousseeuw, P., & Leroy, A. (2003). *Robust regression and outlier detection.* John Wiley & Sons, Inc.

Sastry, N., Shankar, U., & Wagner, D. (2003). *Secure verification of location claims.* Paper presented at the ACM Workshop on Wireless Security (WiSe).

Savvides, A., Han, C.-C., & Srivastava, M. (2001). *Dynamic fine-grained localization in ad-hoc networks of sensors.* Paper presented at the Annual International Conference on Mobile Computing and Networking (ACM Mobicom).

Shang, Y., Ruml, W., Zhang, Y., & Fromherz, M. (2003). *Localization from mere connectivity.* Paper presented at The ACM International Symposium on Mobile Ad Hoc Networking and Computing (MobiHoc).

Smith, A., Balakrishnan, H., Goraczko, M., & Priyantha, N. (2004). *Tracking moving devices with the Cricket location system.* Paper presented at the International Conference on Mobile Systems, Applications, and Services (MobiSys).

So, A., & Yu, Y. (2005). *Theory of semidefite programming for sensor network localization.* Paper presented at the ACM-SIAM Symposium on Discrete Algorithms (SODA).

Want, R., Hopper, A., Falcao, V., & Gibbons, J. (1992). The active badge location system. *ACM Transactions on Information Systems, 10*(1), 91-102.

Wei, Y., Yu, Z., & Guan, Y. (2006). COTA: A robust multi-hop localization scheme in wireless sensor networks. In *Proceedings of IEEE/ACM International Conference on Distributed Computing in Sensor Systems (DCOSS).*

KEY TERMS

Anchors: Anchors are special sensors that know their locations before localization through a GPS device equipped on them or through manual configurations.

Localization: Localization in wireless sensor networks is the process that all sensors obtain their relative or absolute locations, by themselves or by network computing center.

Location Verification: Location verification in wireless sensor networks is the process that correctly estimated locations of sensors can be verified and corrupted locations can be detected.

Range-Based/Range-Free: A localization approach is range-based (or range-free) if it does (or does not) use the measured distance between sensors to estimation their locations.

Secure Localization: Secure localization in wireless sensor networks is the process that sensors can obtain their locations in the presence of malicious attacks.

Wireless Sensor Network (WSN): A wireless sensor network (WSN) is a wireless network consisting of autonomous devices that cooperatively monitor environmental conditions, such as temperature, sound, pollutants, and so forth.

Wormholes: Wormholes in wireless sensor networks are nonexisting communication tunnels (usually wired links) created by adversaries. The messages received at one end of a wormhole can be transmitted through the tunnel, and broadcasted at the other end.

Chapter XXXVIII
Resilience Against False Data Injection Attack in Wireless Sensor Networks

Miao Ma
The Hong Kong University of Science and Technology, Hong Kong

ABSTRACT

One of the severe security threats in wireless sensor network is false data injection attack, that is, the compromised sensors forge the events that do not occur. To defend against false data injection attack, six en-route filtering schemes in a homogeneous sensor network are described. Furthermore, one sink filtering scheme in a heterogeneous sensor network is also presented. We find that deploying heterogeneous nodes in a sensor network is an attractive approach because of its potential to increase network lifetime, reliability, and resiliency.

INTRODUCTION

Wireless sensor networks (WSN) usually consist of a large number of inexpensive and small nodes with sensing, data processing, and communication capabilities. These nodes are densely deployed in a region of interest and collaborate to accomplish a common task, such as environmental monitoring, military surveillance, and industry process control. Distinguished from traditional wireless networks and ad hoc networks, WSN are featured in dense node deployment, unreliable sensor node, frequent topology change, limited power resource, and limited computation capacity, restricted memory space. These unique characteristics and constraints present many new challenges to the design and implementation of WSN.

For many mission-critical applications, the sensor nodes are deployed in an unattended or often hostile environment and WSN face many security and privacy challenges. One challenge is that when deployed in hostile environments, sensor nodes may be captured or compromised by the adversaries. Then the adversaries can obtain the secret keys stored in the compromised nodes, and misuse them to launch *insider attacks*. Therefore, a nonresilient security protection scheme will exhibit a *threshold breakdown* problem. That is, the design is secure against t or less compromised nodes, but once more than t nodes are compromised the security design completely breaks down, where t is a fixed threshold. Since in reality nobody can prevent an attacker from compromising more than t nodes, such a security protection solution cannot meet the resilience requirement. Our expectation in terms of resilience is that, compromising t nodes in a certain area can only enable an adversary to forge nonexisting events in that specific area, rather than any other location at all. Put in other words, for an attacker, the only way to generate a valid report on a nonexisting event happening in a certain area is to compromise t nodes in that area.

In this chapter, we overview several schemes that have been proposed to defend against *compromised nodes*. We will show that several schemes are only resilient against a small, fixed number of compromised nodes with threshold breakdown problems, while subsequent schemes partially and completely solve the threshold breakdown problems.

The rest of this chapter is organized as follows. In the next section, we introduce the background. Several en-route filtering schemes in a homogeneous sensor network are presented. Furthermore, a sink filtering scheme in a heterogeneous sensor network is shown. Finally, the last section concludes the chapter.

BACKGROUND

False Data Injection Attacks

We consider a sensor network, which consist of hundreds or thousands of low-cost sensors. Each sensor senses and collects data from the environment. There is at least one base station (or *sink*), which is typically a resource-abundant computer equipped with sufficient computation and storage capabilities. We assume that the sensor nodes are deployed in a high density, so that once an event happens it can be detected by multiple sensors. However, it is inefficient and also unnecessary for every sensor node to report their raw data to the sink node, because: (1) every data packet usually needs to travel many hops (e.g., tens or event longer) to reach the sink; (2) each sensor node is often constrained by scarce resources in memory, computation, communication, and battery; and (3) in many cases there is high redundancy in the raw data. Hence, raw data are often fused and aggregated locally, and only the aggregated information is returned to the sink. In such a setting, certain nodes in the sensor network will function as *cluster heads* (*CHs*), to collect the raw sensing data from the sensors, process it locally, and return the aggregation report to the sink. Once the sink receives an event report, it may take action accordingly.

Unfortunately, the above event detection and reporting process can be seriously threatened by *false data injection attacks*. As we stated above, sensors are usually deployed in unattended or even hostile environments, and an adversary may capture or compromise sensor nodes. Once this happens, the compromised nodes can easily inject false data reports of nonexisting events. Even worse, when an adversary compromises more nodes and combines all the obtained secret keys, the adversary can freely forge the event reports which not only "happen" at the locations where the nodes are compromised, but also at *arbitrary* locations in the field. These fabricated reports not only produce false alarms (and lead to *false positives*), but also waste valuable network resources, such as energy and bandwidth, when delivering the forged reports to the base station. Therefore, it is important to design an effective filtering scheme to defend against such attacks and minimize their impacts.

In this chapter, we consider the following *threat* model. The attacker may compromise multiple sensor nodes in the network, but cannot compromise the sink. Once a sensor node is compromised, the attacker can obtain all secret keys, data, and codes stored in the sensor. Whenever more nodes are compromised, the attacker can combine all the secret keys that have obtained, and can also load a compromised node with the secret keys obtained from other compromised nodes. We also assume that the attacker cannot successfully compromise a node during the short deployment phase.

Besides report fabrication attack, there are various other attacks in wireless sensor networks. For example, a compromised node may simply not report an event that occurs (which leads to *false negative*), or a compromised node replays a legitimate report, and so forth. However, these threats are addressed in other related work and not the focus of this chapter. Instead, in this chapter we overview several schemes that have been proposed to reduce *false positive*, that is, prevent an attacker from fabricating reports about events that do not occur. Two main design goals of these schemes are summarized as follows:

1. **Resilience against a large number of compromised nodes:** A good protection scheme is expected to degrade gracefully as the number of compromised sensor increases, without the threshold breakdown problem.
2. **Adaptive to dynamic topology:** The scheme can deal with dynamic topology of sensor networks and is scalable for large-scale sensor networks.

En-Route Filtering Framework

Statistic en-route filtering mechanism (SEF) (Ye, Luo, Lu, & Zhang, 2004) is the first effort that addresses false data injection attacks in the presence of compromised sensors, where an en-route filtering framework was originally proposed. The en-route filtering framework has three components: report generation using message authentication codes (MACs), en-route filtering, and sink verification.

Report Generation Using MACs

To generate a valid report, multiple (say m, where $m > 1$) nodes detect the event simultaneously and agree on the content of the event report. To be forwarded by intermediate nodes and accepted by the sink, each valid report must carry m MACs; each MAC is generated by the sensing node that detects the event. Each sensor stores a few symmetric secret keys, and the MAC is generated by using one of the secret keys.

En-Route Filtering

By using a suitable key assignment scheme, any intermediate node is able to verify the report with certain probability or deterministically. Whenever an intermediate node receives a report, it first checks whether the report carries m distinct MACs; it then check if itself stores a same key with the sensing node. If yes, it checks whether the carried MAC is the same as the MAC it computes via its locally stored key. It drops the report when any of these checks fails. Otherwise (i.e., it does not have any of the keys or the MACs are correct), it forwards the reports as usual. Notice that though the filtering power of any single node is limited, the collective filtering power along the forwarding path is significant. The more hops a forged report travels, the higher chance it is dropped en-route.

Sink Verification

The en-route filtering performed by the intermediate nodes may be probabilistic in nature, thus cannot guarantee to detect and drop all forged reports. The sink serves as the final guard in rejecting any escaping ones. Because the sink knows all the keys, it can verify each MAC carried in a report. On the basis of the number of correct MACs each report carries, the sink decides whether to accept the event or not.

Besides a SEF scheme, five more designs including interleaved hop-by-hop authentication (IHA) (Zhu, Setia, Jajodia, & Ning, 2004), commutative cipher-based en-route filtering (CCEF) (Yang & Lu, 2004), location-based resilient security (LBRS) (Yang, Ye, Yuan, Lu, & Arbaugh, 2005), location-aware end-to-end data security (LEDS) (Ren, Lou, & Zhang, 2006), and dynamic en-route filtering (DEF) (Yu & Guan, 2006) are all specific instances within the above framework. Based on the above framework, these five proposals have adopted different key management schemes, which immediately lead to different resilience behavior of their designs. We will describe their methodologies in details in the subsequent sections.

SCHEMES IN EN-ROUTE FILTERING FRAMEWORK

Statistic En-Route Filtering

Methodology

In statistic en-route filtering (Ye et al., 2004), there is a global key pool which is divided into multiple T nonoverlapping partitions. Before deployment, each node randomly selects a few keys from a single partition, and is then loaded with these keys and associated key indices. Once an event occurs, each sensing node generates a MAC by using a key in a different partition. The cluster head (CH) node collects the MACs and attaches them to the report. Any intermediate node has a same predetermined probability to detect and filter false reports, and hence SEF filters the forged reports en-route in a probabilistic manner. The sink can always verify every report because it knows the entire key pool. As a result, most of the forged reports are quickly dropped by the forwarding nodes, and the few escaping ones are further rejected at the sink.

Features

First, SEF suffers from the threshold breakdown problem. Second, SEF is independent of dynamic topology changes of sensor networks, and hence is robust against node failures and routing path changes.

Interleaved Hop-by-Hop Authentication

Methodology

Distinguished from SEF, interleaved hop-by-hop authentication (Zhu et al., 2004) verifies the reports in a deterministic and *an interleaved, hop-by-hop* fashion. In the deployment phase, each node is preloaded with a unique ID and keying materials that can allow it to establish a pair-wise key with another node. The nodes form multiple clusters and each cluster has at least $(t + 1)$ nodes, where t is a design parameter. Each cluster head discovers a path to the sink. Along the path, two nodes that are $(t + 1)$ hops away are *associated* by establishing a pair-wise key. Upon an event, each detecting node computes two MACs, one using its key shared with the sink and the other using its pair-wise key shared with its downstream associated node. The cluster head sends out a final report that carries the MACs from $(t+1)$ detecting nodes. In the en-route filtering phase, each forwarding node verifies the MAC from its upstream associated node. Upon successful verification, it replaces the old MAC with a new one using its pair-wise key shared with its downstream associated node. The sink performs a final verification on the report. IHA guarantees that if no more than t nodes are compromised, the base station will detect any false data packets injected by the compromised sensors.

Features

First, IHA suffers from threshold breakdown problem, similarly to SEF. Second, since IHA requires that the messages transmitted from the base station to a cluster head and from the cluster head to the base station follow the same fixed path, IHA scheme is not suitable for the sensor networks with dynamic topology.

Commutative Cipher-Based En-Route Filtering

As we discussed above, both SEF and IHA schemes suffer from a threshold breakdown problem. To solve this problem, a commutative cipher-based en-route filtering scheme (Yang & Lu, 2004) was presented on the basis of public-key algorithms.

Methodology

CCEF exploits the typical operational mode of query-response in sensor networks, and installs security states in the nodes in an on-demand manner. Specifically, in CCEF, each node has a unique ID and is preloaded with a unique node key before deployment. When reports are needed, the base station sends an encrypted session key to the desired cluster head and a witness key in plain-text to all forwarding nodes along the path, through a query message. A legitimate report is endorsed by *a node MAC* jointly generated by the detecting nodes using their node keys, and *a session MAC* generated by the source node using the session key. Through the usage of a commutative cipher, a forwarding node can use the witness key to verify *the session MAC*, without knowing the session key, and drop the fabricated reports. The base station further verifies *the node MAC* in the report that it receives, and refreshes the session key upon detection of compromised nodes.

Features

First, CCEF solves the threshold break down problem. Second, CCEF suffers the dynamic topology problem, similarly to IHA scheme, since it requires the same fixed path for messages in both directions between the base station and the cluster head. Third, CCEF uses the commutative ciphers that are based on public-key algorithms, which have been reported not suitable for sensor networks (Eschenauer & Gligor, 2002).

Location-Based Resilient Security

To mitigate the threshold breakdown problem identified in IHA and SEF schemes, a location-based resilient security scheme (Yang et al., 2005) was proposed which exploited a location-based approach as the fundamental mechanism.

Methodology

In LBRS, the terrain is divided into a geographic grid and binds multiple keys to each cell on it. Such keys are termed as *location-binding keys*. Each node stores two types of keys. The first type is for the local cells within its sensing range, called *sensing cells*. Each node stores *one* key for each of its sensing cells. Such keys are used to endorse events detected in those cells. The second type is for a few randomly chosen remote cells, called *verifiable cells*. Each node also stores *one* key for each of its verifiable cells. Such keys are used to verify events claimed to happen in those cells. Each legitimate report carries m distinct MACs, jointly generated by the detecting nodes using the keys bound to the event's cell. When an intermediate node receives a report, it retrieves the event's location from the report and checks whether the location is in one of its verifiable cells. If so, it checks whether it has one of the keys hose indices are carried in the report. If it has such a key, it recomputes the MAC and compares to the carried one. If the two MACs do not match, the report is dropped. Otherwise, it forwards the report. The sink performs final verification on the received reports. It knows all location-binding keys, thus able to verify every MAC in the report.

Features

First, compared with SEF and IHA schemes, LBRS partly solves the threshold breakdown problem, since compromising a certain number of nodes only enables the attacker to fabricate events "appearing" at certain areas without being detected. However, it is still far from achieving the *expected* data authenticity requirement: to generate a valid report on a nonexisting event happening in a certain area, the only way is to compromise T nodes in that area, and otherwise impossible. Second, LBRS is suitable for the sensor networks with dynamic topology.

Location-Aware End-to-End Data Security

Later on, Ren et al. (2006) came up with a location-aware end-to-end data security to address the vulnerabilities in existing security designs, by exploiting the static and location-aware nature of WSNs.

Methodology

In LEDS, each node computes three different types of location-aware keys: (a) two *unique secret keys* shared between the node and the sink and used to provide node-to-sink authentication; (b) one *cell key* shared with other nodes in the same cell that is used to provide data confidentiality; and (c) a set of *authentication keys* shared with the nodes in its *report-auth cells* and used to provide cell-to-cell authentication and en-route bogus data filtering. All these keys are computed by each node locally and independently. In addition, LEDS adopts a (t, T) threshold linear secret sharing scheme (LSSS) so that the sink can recover the original report from any t out of T legitimate report shares. Moreover, LEDS adopts a one-to-many data forwarding approach, that is, all reports in LEDS can be authenticated by multiple next-hop nodes independently so that no reports could be dropped by a single node(s).

Features

First, LEDS meets the expected requirement in terms of resilience, with totally solving the threshold breakdown problem. Second, LEDS is suitable for the sensor networks with dynamic topology.

Dynamic En-Route Filtering

At the meantime, Yu et al. (2006) presente a dynamic en-route filtering.

Methodology

In DEF (Yu et al., 2006), a legitimate report is endorsed by multiple sensing nodes using their own authentication keys generated from one-way hash chains. A cluster head (CH) uses a *hill climbing* approach to disseminate the authentication keys of sensing nodes to the forwarding nodes along multiple paths towards the base station. In filtering phase, each forwarding node validates the authenticity of the reports and drops those false ones.

Features

First, compared with SEF and IHA schemes, DEF can tolerate a larger number of compromised nodes. However, DEF scheme still cannot meet the expected requirement in terms of resilience, as LEDS does. Second, LEDS can deal with dynamic topology of sensor networks; but the overhead incurred (on disseminating the authentication keys to forwarding nodes) is high. Third, as the authors discussed in the chapter, LEDS raises some new types of attacks specific to its scheme.

SCHEMES IN HETEROGENEOUS SENSOR NETWORKS

In the previous section, we looked at some of the security protection schemes in homogeneous sensor networks. However, there is another class of sensor networks, heterogeneous sensor networks, which use two or more type of nodes. It is known that the presence of heterogeneous nodes in a sensor network helps to increase network lifetime and reliability. In this section, we present the design of a sink filtering scheme (SFS) (Ma, 2006a, 2006b) in a heterogeneous network, showing that the presence of heterogeneous nodes in a sensor network also helps to improve the resiliency.

Model of a Heterogeneous Sensor Network

We consider a heterogeneous sensor network where two types of sensors are deployed: basic sensor and cluster head. A basic sensor is simple, inexpensive and power-limited, while a CH has more capabilities on processing and communication, richer power supply, and is more compromise-resilient.

We regard a target deployment area as a two-dimensional square region with size A^2. The sink is located at the center (0, 0). The deployment area is divided into C equal size grids (i.e., clusters), with each grid's size as a^2. The basic sensors are uniformly distributed across the entire deployment area. Without loss of generality, we assume that each CH is deployed at the center of each grid. Each basic sensor is assigned to the nearest CH. Every basic sensor and CH has a unique identification (ID).

Sink Filtering Scheme (SFS)

The two types of sensors and sink node implement different tasks in SFS. A basic sensor senses events and provides its CH a proof for any aggregation report it has agreed. A CH collects raw sensing data from basic sensors, generates an aggregation report, and relays the report to the sink node. A sink node checks the validity of the carried MACs in an aggregation report and filters out the forged report.

Methodology

We assume that the basic sensors are deployed in a high density. Once a real event occurs, *n* basic sensors within the sensing range can sense it. Instead of communicating directly with the sink, each basic sensor only communicates with its CH. Each CH collects raw sensing data from the basic sensors within the cluster, generates an aggregation report, and then relays the report to the sink node. Since the average number of hops each report has to travel from a CH to the sink in *heterogeneous* sensor networks is definitely much smaller than that in *homogeneous* sensor networks, the design is more energy-saving, efficient, and scalable.

The key management for the clusters of heterogeneous sensor network is as follows. (a) Before deployment, each sensor (either a basic sensor or a CH) shares a secret key with the sink. (b) We assume the neighborhood relationship among CHs is known in advance. Before deployment, each CH simply preloads eight pair-wise keys with its eight immediate neighboring CHs, respectively. The CHs, therefore, organize themselves into a static ad hoc network. (c) Upon deployment, each basic sensor establishes a pair-wise key with its one-hop neighboring basic sensors; the one-hop pair-wise key establishment scheme in LEAP (Zhu, Setia, & Jajodia, 2003) is adopted to achieve this goal. (d) Upon deployment, each CH establishes a pair-wise key with every basic sensor within its cluster; the one-hop (or multi-hop whenever a basic sensor cannot reach its CH in a single hop) pair-wise key establishment scheme in LEAP (Zhu et al., 2003) is used for this purpose.

Features

First, SFS meets the expected requirement in terms of resilience, with totally solving the threshold breakdown problem. Second, SFS is adaptive to the dynamic topology. Third, compared with all the schemes in homogeneous sensor networks, SFS in heterogeneous sensor networks is more efficient and scalable. Interested readers may refer works by Ma (2006a, 2006b) for more details on resiliency study and overhead evaluation.

CONCLUSION

In this chapter, we presented six en-route filtering schemes in a homogeneous sensor network, including statistic en-route filtering (SEF), interleaved hop-by-hop authentication (IHA), commutative cipher-based en-route filtering (CCEF), location-based resilient security (LBRS), location-aware end-to-end data security (LEDS), and dynamic en-route filtering (DEF). Furthermore, a sink filtering scheme in a heterogeneous sensor network is also introduced. Our study demonstrates that exploiting heterogeneity in sensor networks also helps to improve the resiliency.

REFERENCES

Eschenauer, L., & Gligor, V. D. (2002). *A key-management scheme for distributed sensor networks.* Paper presented at the ACM CCS.

Ma, M. (2006a). *Resilient against report fabrication attack in clusters of heterogeneous sensor networks.* Paper presented at the IEEE WCNC.

Ma, M. (2006b, December). Resilience of sink filtering scheme in wireless sensor networks. *Computer Communications, 30*(1), pp. 55-65. Elsevier

Ren, K., Lou, W., & Zhang, Y. (2006). *LEDS: Providing location-aware end-to-end data security in wireless sensor networks.* Paper presented at the IEEE INFOCOM.

Yang, H., & Lu, S. (2004). *Commutative cipher based en-route filtering in wireless sensor networks.* Paper presented at the IEEE VTC.

Yang, H., Ye, F., Yuan, Y., Lu, S., & Arbaugh, W. (2005). *Toward resilient security in wireless sensor networks.* Paper presented at the ACM MobiHoc (pp. 34-45).

Ye, F., Luo, H., Lu, S., & Zhang, L. (2004). *Statistical en-route filtering of injected false data in sensor networks.* Paper presented at the IEEE INFOCOM.

Yu, Z., & Guan, Y. (2006). *A dynamic en-route scheme for filtering false data injection in wireless sensor networks.* Paper presented at the IEEE INFOCOM.

Zhu, S., Setia, S., & Jajodia, S, (2003). *LEAP: Efficient security mechanisms for large-scale distributed sensor networks.* Paper presented at the ACM CCS.

Zhu, S., Setia, S., Jajodia, S., & Ning, P. (2004). *An interleaved hop-by-hop authentication scheme for filtering of injected false data in sensor networks.* Paper presented at the IEEE Symposium on Security and Privacy (S&P).

KEY TERMS

Aggregation Report: A data structure that synthesizes the state of the phenomena that the wireless sensor network is monitoring.

Compromised Nodes: Nodes on which an attacker has gained control after network deployment.

False Data Injection Attack: The type of attack when the compromised sensors forge the events that do not occur.

Key Management: The process of managing key materials (e.g., key generation, key distribution, etc.) in a cryptosystem.

Message Authentication Code (MAC): It is a short piece of information used to authenticate a message.

Threshold Breakdown Problem: We say a security design has threshold breakdown problem if the design is secure against t or less compromised nodes, but once more than t nodes are compromised the security design completely breaks down, where t is a fixed threshold.

Wireless Sensor Network (WSN): The wireless networks consisting of small sensors that cooperatively monitor environmental conditions, such as temperature, humidity, and so forth.

Chapter XXXIX
Survivability of Sensors with Key and Trust Management

Jean-Marc Seigneur
University of Genev, Switzerland

Luminita Moraru
University of Genev, Switzerland

Olivier Powell
University of Patras, Greece

ABSTRACT

Weiser (1991) envisioned ubiquitous computing with computing and communicating entities woven into the fabrics of every day life. This chapter deals with the survivability of ambient resource-constrained wireless computing nodes, from fixed sensor network nodes to small devices carried out by roaming entities, for example, as part of a personal area network of a moving person. First, we review the assets that need to be protected, especially the energy of these unplugged devices. There are also a number of specific attacks that are described, for example, direct physical attacks are facilitated by the disappearing security perimeter. Finally, we survey the protection mechanisms that have been proposed with an emphasis on cryptographic keying material and trust management.

INTRODUCTION

Weiser (1991) envisioned a ubiquitous computing world where intelligent computing and communicating devices are pervasive and woven into the fabrics of every day artifacts. His vision is being materialised: the market of large scale sensors and hand-held devices networks has been gaining momentum. However, one may question whether or not these computing and communicating entities will be able to survive in an open environment. These computing entities are no more protected by a physical security perimeter; foreign, potentially malicious, entities can tamper with them. Another challenge for the real deployment of these networks of sensors and portable devices is to provide them with enough energy for long term functioning because it is assumed that they are unplugged from the main electrical power supply and can rarely recharge themselves by this means. Any action carried out by these entities depletes their energy. In addition to being resource-constrained in terms of energy, these entities are resource-constrained in terms of memory and processing, which limit what they can do, especially when these entities are small, such as the sensors deployed in sensors networks.

Usually, sensors are performing two important types of actions or tasks: they have to sense the environment and to send information to a specific target entity, sometimes called sink. For example, the sink may be an Internet gateway that will propagate the information for persistent storage and analysis. Security problems exist both when messages are generated and when they are relayed. Working most of the time in an unattended environment without tamper-proof hardware makes the sensors very vulnerable to attacks.

Generally, mobile ad hoc networks (MANETs) are thought to be composed of nodes bigger than the sensors of sensors networks. Also, whereas sensors are considered (after their deployment) rather fixed concerning their location, MANETs imply that the nodes move. If we assume that the MANET nodes are also unplugged from the main power supply, the nodes have also limited energy. Another difference between sensors and MANET nodes is that instead of just having to sense and forward simple information, MANET nodes are expected to run much more complicated operations that surely require more energy than simple tasks. In this chapter, we consider all ad hoc networks where the wireless nodes are resource-constrained, especially in terms of energy. Thus, as introduced above, the nodes may go from the tiny fixed deployed sensor to the mobile unplugged mobile device.

In this chapter, we first survey the different assets of these entities and then delve into specific attacks on these assets. We present further two main protection mechanisms: cryptographic keying material and evidence-based trust management. Finally, we discuss future trends and draw our conclusion.

BACKGROUND ASPECTS OF NODES SURVIVABILITY

In this section, we first discuss what we mean by nodes survivability, their assets, and especially their energy. Then, we focus on the routing asset, which is an important asset that enables the nodes to communicate beyond their own wireless communication range. It shows that the routing has been initially engineered without attackers in mind, which is also the case for most of the other enabling mechanisms and assets. However, there are a number of attacks that can be carried out on these assets. We survey them at the end of the section.

Node(s) Survivability

First, it is important to note we use the plural in the heading of this section, *nodes survivability,* because it emphasises that the scope of the node's mission may span more than one node. On one hand, it may be a scenario where the survivability of the node itself is more important than the survivability of the other nodes. For example, a user who carries a mobile phone in the mountains may be selfish and would not bother forwarding the messages of other users as they are met on the way to the top of the mountain. The forwarding of a message from another user would deplete the energy of the mobile phone and endanger the survivability of the device and its mission lifetime. On the other hand, the mission may be that the majority of the nodes survive at the expense of the survival of one specific node. It is usually the case in sensors networks where the goal is to sense and monitor a region thanks to the collaboration of many nodes. If a part of the monitored region is quite active, it is possible that the nodes in this active region take over the work of another node, for example, to forward the sensed information in order to maximise the lifetime of the monitoring of the whole region. That type of scenario requires that there is some sort of control on the nodes; an authority is needed to guarantee that the nodes will collaborate and follow the rules. For example, in a military environment, the nodes that are deployed are configured before the deployment and we can assume that they will all follow the rules (until they are captured or they fail). A military environment is said to be a controlled environment, its extreme being an open environment where every node is free to behave as it wishes. In the middle of these extremes, there are scenarios where a selfish node may increase its lifetime by sporadically helping other nodes: the nodes being interdependent.

The Assets Underlying the Survivability

As explained in the previous subsection, the first type of asset underlying the survivability is the survivability of the node or the survivability of the network of nodes. In interdependent settings, the neighbouring nodes are an asset that enables the node to achieve more than what it could achieve alone. An underlying asset is the trustworthiness of these neighbouring nodes. The node may be able to choose or influence whose nodes are its neighbours, for example, by moving to another location or by mere selection (the third section explains how evidence-based trust management can be used to optimise the selection process). In sensor networks, it is less often assumed that new nodes can be added after the initial deployment of the nodes. Sensors are mainly used to monitor the characteristics of a specific fixed area or targets. However, in MANET scenarios, when the nodes correspond to devices carried by people, it is clear that the nodes can come and go as the people roam from place to place. If the adjunct nodes and their location can be chosen, the survivability of the network of nodes may be prolonged. All the nodes may participate for the survivability of the network of nodes. Therefore, the nodes' participation can be considered as an asset.

Another asset for a node is to be highly tamper-proof. However, as said above, nodes in open environments can be captured and highly tamper-proof hardware may be too costly. An asset may be the possibility to communicate the evidence of tampering, for example, an alert being sent at time of capture or creating some form of tamper-evidence.

Besides being captured by an adversary and removed from the network, the lifetime of the node and its mission mainly depend on its consumption of energy. The design of the node is an inherent asset of the node because both hardware and software designs may end up in consuming more or less energy. Current sensor nodes are battery-enabled devices. The energy consumption of a sensor is due to the computation and the communication modules. Energy is depleted mainly by the communication module. Radio transmission consumes most of the energy spent for security mechanisms and encryption only consumes 3% (Hwang, Lai, & Verbauwhede, 2004). Thus, minimising security transmission is important with regard to energy saving. However, sensors are constrained devices and the security mechanisms available for conventional networks are not suitable due to their limited computational, memory, and energy resources. They are often deployed in hostile environments with no physical access to nodes after deployment. Thus, the sensor lifetime is limited by the initial energy of the battery. Different energy harvesting mechanisms are being researched for sensors, such as the use of solar cells, but they are still not very common. The user's mobile computing devices like personal digital assistants and mobile phones are also battery-powered but they can be more easily recharged.

While energy recharging and harvesting mechanisms are put in place, the nodes need energy saving mechanisms. Usually, energy saving mechanisms are based on the difference of energy consumption between the active and the idle states. Energy is saved by minimising the fraction of time while the device is active. These energy saving mechanisms are normally targeting the subsystem of the device that has the greatest difference between the energy consumption of the two states. Once no application is running on the device, it is put into a low power state, extending the battery life. A sensor node can usually be in four different states: transmit, receive, idle, and sleep. The sleep state is significantly less energy consuming than the other states. Preserving network longevity is one of the main issues in sensor networks. The topology of the network is not known a priori because the nodes are randomly scattered to a target area. Sensor networks are often dense networks. Not all the nodes are necessary to accomplish a specific request. One method to save energy is to put nodes to sleep in a manner that does not interfere with the functionality of the network. In a sensor network the lifetime of the network is more important than each sensor. Thus, the protocols developed for sensor networks consider the optimisation of network lifetime.

The topology of the network may be dynamic. Nodes may become temporary inactive to save their energy or they drain out of battery. At the same time, new nodes may be deployed in the same area. Energy is limited in the network. However, the nodes may have to repeatedly communicate with a base station on a hop-by-hop basis. To minimise the energy spent in the network, energy-preserving secure routing protocols (surveyed in the following subsection) have been developed. The communication patterns are concerned with balancing energy consumption and preserving network lifetime and purpose. Usually, the whole region needs to be covered by the nodes. The purpose of the network requires that the sensing coverage works for all localisations. At any location, the nodes should be able to send the collected data to the base station. Energy saving should not deteriorate the connectivity and the coverage of the network. An energy optimisation scheme should also maintain the initial coverage. Energy efficient schemes group sensors in different sets that are alternatively active (Cardei, 2005; Ramchurn, Jennings, Sierra, & Godo, 2004).

Another solution is to enforce clustering algorithms (Handy, 1995). An example of energy at the nodes level occurs with cluster-based sensors network topology. In this case, energy efficient routing protocols use hierarchical structures like clusters among the nodes forming the network. The nodes in the cluster only communicate with the cluster head. The cluster head is the only one to communicate with the other cluster heads and provides aggregation of data for the nodes forming the cluster.

The nodes that are not cluster head may receive the information later. The responsiveness of the node, concerning computation and communication, is also important. In addition to consuming more energy, the use of security mechanisms may also require more storage space, for example, for the keying material, and may slow down the processes due to the additional security steps, such as, encryption, decryption, and signatures.

Besides the above special assets, there are also more basic security assets, namely, the confidentiality/privacy, integrity, and availability properties of each node and their messages. When these basic assets are compromised, the other assets may be more easily compromised.

The list below summarises the different assets that we have discussed in this section:

- **Node-level assets:**
    - **Node mission lifetime:**
        - Node energy
            - Harvesting source
            - States and actions management
        - Node tamper-proof and tamper-evidence
    - Node localisation
    - Node mobility (in case of non-fixed settings)
    - Node computing performance
    - Node neighbours presence in interdependent settings
    - **Node communication:**
        - Ability
            - Reception
            - Transmission
            - Coverage range
        - Confidentiality
        - Integrity
        - Speed
- **Network of nodes-level assets:**
    - Network mission lifetime
    - Deployment of new nodes
    - Nodes participation and trustworthiness
    - Network connectivity, performance and coverage

The Routing Asset Case-Study

The nodes can use their wireless link to directly communicate with the other nodes in range. Sometimes the nodes can increase their transmission energy to reach farther nodes. However, as said above, communication tasks use a lot of energy; for example, if we assumes Friss' (1946) free-space attenuation, the energy needed for wireless transmission over a distance *d* is proportional to *d square*. Thus, the nodes may save energy by using other closer nodes to forward their message to farther nodes. In addition, if the nodes cannot increase their transmission energy to reach a specific far-away node, the only remaining solution is to use intermediate nodes to forward the message. It is why routing algorithms have been researched. In this subsection, we survey the most well-known protocols that allow the nodes to exchange messages. We start by the MANET protocols and then the protocols said to be specific to sensor networks, which are explicitly energy-aware.

MANET Routing Protocols

Maltz (2001) depicts the history of MANETs. The first significant project towards MANETs is called the DARPA-sponsored military packet radio network (PRNET) in 1972. Now, MANETs seem to be used on battlefields. The goal of researchers, like Maltz, was to outperform the performance of the military protocols. They reached efficient and good performance for routing in MANETs with ad hoc on-demand distance vector routing (AODV) (Perkins & Royer, 1999) or dynamic source routing (DSR) (Maltz, 2001).

Both AODV and DSR are *reactive routing protocols* because they compute the route between two nodes only when the route is needed, that is, 'on demand.' In doing so, there are far fewer tasks to be carried out because all the routes do not have to be maintained all the time. It is very important from an energy point-of-view in mobile settings where the nodes come and go very quickly and where the routing information would need to be updated very often. However, neither AODV nor DSR integrate further specific mechanisms to minimise the energy consumption along the route.

Another limitation comes from the fact that Maltz and colleagues designed their protocols with the same assumption as for military MANETs. In military MANETs, it is often assumed that the deployed nodes are controlled; it is a controlled environment where it is understandably supposed that nodes are not free to do whatever they can do. In open MANETs, where any user's node can come and go depending on the user's will, the nodes might not follow the rules and they challenge the correct functioning of these routing protocols. Thus, the researchers had to revise their protocol approach (not to say restart from scratch) because all was working well under the assumption that the nodes do collaborate, but in open MANETs, where nodes are owned by free people, assuming that everyone collaborates is simply not realistic. In 2001, the conclusion was that security in MANETs is particularly difficult due to their specificities (Hubaux, Buttyán, & Capkun, 2001): vulnerability of channels and nodes (i.e., less physical security); resource-constrained nodes; high probability of absence of infrastructure; and dynamically changing topologies and high uncertainty. An interesting issue is the question of collaboration, which is vital for some MANETs to stay up: the nodes are neither dependent nor independent but interdependent. If too many nodes are too selfish, the overall availability is endangered (Miranda & Rodrigues, 2003).

Sensors Network Protocols

As mentioned above, in sensors networks, the deployed nodes are usually supposed to collaborate. However, due to their small size and the assumption that they can never be recharged, the MANET protocols are not sufficient to optimise the use of the energy of the nodes. This is why other researchers have researched new routing protocols with an emphasis on energy consumption optimisation. *Energy-aware routing protocols* explicitly take into account the energy consumption as a parameter. This subsection surveys seven of these new protocols that use one or several of these following energy saving basic techniques:

- Keeping short range transmissions
- Aggregating data
- Building efficient paths

- Switching between sleep/awake states
- Efficiently controlling multi-paths

During the set-up phase of the minimum cost forwarding algorithm (MCFA) (Ye, Chen, Liu, & Zhang, 2001), each node initiates its least cost to the sink estimated to be at an infinite distance. The sink broadcasts to its neighbours a setup message. Each of the neighbours computes and updates its least cost estimate to the sink and eventually broadcasts further to its own neighbours. When receiving a broadcast message, a node computes its new least-cost estimate. If it is lower than the current least cost estimate, the node updates it and broadcasts its new estimated least cost to its neighbours, and so on and so forth. In order to avoid collision as well as duplication of unnecessary message, that is, in order to optimise the flooding involved, MCFA introduces a back-off mechanism which is basically a timeout before propagating the updated values of the estimated least cost. During the propagation phase, when a node needs to send a message to the sink, it broadcasts to its neighbours. When a node receives a message, it checks if it is on the least cost path, and if so, propagates the message further. Otherwise it just drops the message.

Gradient-based routing (GBR) (Schurgers & Srivastava, 2001) is somehow similar to MCFA. It proposes to slide messages along a gradient towards the sink. GBR is a general scheme; it proposes a few gradients but it is open to other possible gradients. The gradient can be computed similarly as in MCFA, that is, using the back-off mechanism. If one wants to introduce the hop-count in the gradient, the hop-count is included in the gradient formula. MIX (Powell, Jarry, Leone, & Rolim, 2006) is a variant of GBR that allows the node to eject a message directly to the sink in case of high remaining energy on the current node compared to the energy remaining on its neighbours. In MIX, a sensor can choose to eject a message when all its short-range neighbours have lower energy than itself. To eject means that the sensor increases the power of transmission to be able to reach the base station in one transmission. As said above, the energy spent increases a lot, nonlinearly, with the distance. The ejection feature of MIX can be seen as a dynamic clustering technique: the ejecting nodes are cluster heads and the cluster members are nodes which propagate towards the ejectors. The cluster heads (and thus the clusters) are automatically updated by the distributed algorithm.

The low-energy adaptive clustering hierarchy (LEACH) (Handy, 1995) is a more well-known distributed randomised cluster formation algorithm. Many more complicated and optimised algorithms that exist in the literature have been inspired by LEACH. LEACH is based on partitioning the network into clusters. It features two distinct phases:

1. **Cluster formation:** Cluster heads are self-elected according to a very simple random rule: each node decides to become a cluster head with probability *p*, where *p* depends on a threshold value. This threshold function is dependent of parameters such as the remaining energy, the time elapsed since the network started, and the number of times it has been a cluster head before. Thus, energy balancing is possible through the fine-tuning of the parameters. Once self-elected, the cluster heads advertise themselves to noncluster heads by broadcasting an announcement message. Noncluster head nodes then decide to which cluster they will attach themselves. Basically, they attach themselves to the closest cluster head, although *closest* could have different meanings. Once the cluster head is aware of all of its cluster members, it computes a time division multiple access (TDMA) scheme and assigns a time slot to each of the members of the cluster. The cluster members are only allowed to transmit data to the cluster head during the time slot that they have been assigned to. Hence, no message collision occurs.
2. **Data propagation phase (once the clusters have been formed):** Data are sent by the cluster members directly to their cluster head. The cluster head then aggregates the data before sending them directly to the sink. Other protocols inspired by LEACH propose to run a more sophisticated algorithm than

direct transmission to the sink by using another routing protocol on a subgraph of the communication graph, sometimes called a 'backbone network,' where nodes are typically the cluster heads. The distributed cluster formation phase has to be rerun from time to time in order to avoid early energy depletion of cluster heads.

The directed diffusion (DD) (Intanagonwiwat, Govindan, Estrin, Heidemann, & Silva, 2003) protocol is also very famous but a bit more complicated than LEACH. It works roughly in the following way. The sink(s) sends *interests* (consisting of attribute value pairs) which are flooded across the network. The sources are being discovered as the nodes are capable of satisfying the interests. In a second stage, the gradients are being set (the exact way in which gradients are being set is application specific). Different gradients permit to build routes between the sources and the sink with different properties, for example, high robustness, low latency, low energy cost, and so forth. Information is propagated along the paths following the gradients. Energy can be further spared by carrying-out network data-aggregation as well as using caching techniques. DD seems to be very suitable for the so called *continuous monitoring* application domain (as opposed to *event driven* monitoring), because there is a cost in establishing the routes (or interests gradients, more precisely), and once they have been established they should be used for some time, that is, *continuously*.

In the sensor protocols for information via negotiation (SPIN) (Kulik, Heinzelman, & Balakrishnan, 2002), data are disseminated throughout the network, assuming that each sensor node is a potential sink. This is particularly efficient in the case of mobile networks. The protocol works in the following way. When a node detects an event, it sends an *ADV* message advertising the detected event. The nodes receiving the ADV decide if they are interested in the information, and if so, send a request (*REQ*) message, following which the actual data will be sent. The idea is to avoid unnecessary flooding of long data messages in the network when the information is redundant. The nodes decide to send REQ messages by analysing the ADV message which is assumed to contain the necessary information to take this decision. The actual encoding of the ADV message is application specific and not specified by the SPIN protocol. Variants of SPIN offer different optimisation for different contexts. For example, nodes can decide to stop participating when they believe that they do not have enough energy to complete all stages of the protocol.

The final protocol that we survey is probabilistic forwarding (PFR) (Chatzigiannakis, Dimitriou, Nikoletseas, & Spirakis, 2006). It uses the following approach which is common in routing protocols. In order to ensure robustness (due to link failure, message collision, etc.), data are propagated in a multipath way. However, the total number of paths has to be controlled. In PFR, it is assumed (although this assumption can be relaxed) that the angle of reception as well as the direction to the destination of a message can be computed. For example, this assumption is possible in a localised network. Using this information and by piggy-backing/adding $O(1)$ bits of information to the message, the nodes can probabilistically decide to forward or drop the message. The messages are propagated along a multipath beam whose width can be controlled by a parameter of the algorithm. The fact of being able to control the width of the beam enables the control of the energy cost overhead implied by multipath routing, and the fact of having a multipath beam ensures the robustness to link failure.

In the next subsection, we cover a specific form of failure, namely, failure due to malicious activities, also known as attacks.

Survey of the Attacks on the Nodes Assets

Above, we have surveyed the protocols that have been proposed for routing. These protocols are more or less energy-aware. However, most of the times these protocols assume that there is no attacker. As we are moving to a ubiquitous computing world, assuming that there is no attacker is unrealistic because the nodes are deployed in open environments where anybody can deploy nodes or try to tamper with any nodes. We consider the cost of a physical attack as low because the nodes are assumed to not have significant tamper resistance due to the cost of such protection for devices that are supposed to be affordable for large scale deployment (Pirretti, Zhu, Narayanan, McDaniel, Kandemir, & Brooks, 2005). In a *node capture/tampering attack*, an adversary has physical access to the node. Current security solutions are evaluated by taking into account the resistance of the network to nodes capture, that is, the number of nodes needed to be captured in order to corrupt the entire network. Time is the factor used to evaluate the attacks that are in progress.

Another type of attack may especially target the energy asset. That form of attack is usually called the *energy starvation attack*. For example, Martin, Hsiao, Ha, and Krishnaswami, (2004) depict a denial-of-service attack targeting battery powered devices. Its purpose is to drain out the battery of the device, for example, by obliging the nodes to consume more energy than necessary. In the case of mobile computing devices, the attack leads to an inoperable device. It may only be temporary for a mobile device but it is usually not the case in sensors networks where the nodes cannot be recharged. In addition, the inoperability of several sensors can disrupt the functionality of an entire network region. An energy starvation attack may prevent the device from entering into its low power state, thus increasing the time while the device is active. This attack can be carried out in the case of the use of energy saving schemes. As said above, an energy saving scheme schedules for each node an awake/sleeping cycle. In a *sleep deprivation attack* a node is forced to remain in the awake state. We start by two types of energy starvation attacks. The first type is the sleep deprivation attack that targets the communication subsystem and prevents the sleep state. The second type called the barrage attack is enforced by demanding energy intensive operations. A node receives successive task requests.

Another possibility of increasing the energy consumption is to increase the energy needed for executing a task. The measure of the success of the attack may be the increase in overall energy consumption. The attack may be detected by the owner because the battery is expected to have a certain lifetime. Ultimately, the measure of this attack may be the report between the real and the expected lifetime of the battery. It has been reported that for mobile devices the report may be from one to two orders of magnitude (Pirretti et al., 2005). Martin et al. (2004) identify three types of sleep deprivation attacks on mobile devices. In a service request power attack, a device must repeatedly execute a network service on a remote entity. Even if the service is not available, the process of authentication consumes time and energy. Another type of power attack may be to request the devices to repeatedly execute an energy-hungry task. On mobile devices, power attacks may be detected by scanning software that compares the current energy consumption to normal energy consumption. On small sensors, it may be infeasible to run such scanning software. Other solutions analyse the energy consumption pattern because power attacks modify the energy consumption signature of the applications. Another solution may be to define and impose an energy limit for an application or a task.

The nodes executing important tasks, like cluster heads, are perfect targets to initiate stronger attacks over the other nodes in the cluster. These attacks are prevented by preventing the misbehaving nodes from becoming a cluster head. The solution evaluated by Pirretti et al. (2005) as the best to prevent this type of attack is a hash-based cluster head selection. The cluster head does not decide itself to be the next cluster head, but it is selected by random vote by the neighbours. This attack can be categorised as *topologically-inspired attack* (Seigneur, 2005), where the knowledge of the topology of the network of nodes is used to carry out more harmful attacks. This knowledge can be extracted by standard attacks that are also possible in our settings.

The messages sent by the nodes can be captured and read by attackers, which constitutes a confidentiality attack. A confidentiality attack may also be carried out to infer message provenance, route analysis, and activity monitoring. In some sensors network scenarios, it is crucial that the location of the nodes that sensed the information is not known. For example, sensors network have been deployed to monitor the location of pandas in their natural habitat (Ozturk, Zhang, & Trappe, 2004). Due to the presence of hunters, source-location privacy is crucial. If we extend the scenario to the location of people, we can really talk of *source-location privacy attacks*. The network topology can be inferred from this information. More knowledge can help the attacker to carry out more harmful attacks targeting specific active/low-energy zones or traffic control. Among the other standard attacks, there are also the attacks that target the integrity of the messages as well as of the traffic or specific zones. The messages may be change replayed, delayed, or even destroyed.

As said above, the routing protocols work well when all nodes cooperate. However, in real settings, the cooperation assumption may not be valid. If the nodes are small, low-power devices, they are limited in energy and may be motivated to have a selfish, noncooperative behaviour when it comes to relaying the messages from other nodes. They can save power by not forwarding the messages received from the neighbours. Furthermore, selfishness is not the only misbehaviour that has to be addressed. An attacker can compromise nodes and then prevent packets to reach their destination. For example, in MIX, a few neighbour nodes may lie about their current energy level to avoid having to forward messages, or worse, they may not forward messages when asked to do so. In the latter case, these misbehaving nodes carry out an attack commonly called *sink hole attack* (Pirzada & McDonald, 2005). A sensor behaving like a sinkhole will drop any packet it receives. In a *worm hole attack* (Hu, Perrig, & Johnson, 2002), two colluding sensors create a tunnel between them. The first node may be situated in the proximity of the base station and replays the messages received by the second one. The tunnel is a fast path and will encourage the nodes to use it for routing. This attack is hard to detect because the authenticity and confidentiality security requirements are maintained. Once the packets are routed through the wormhole, *denial-of-service attacks* can be enforced. Packets will be forwarded selectively or they will be not forwarded at all. In a *Sybil attack* (Douceur, 2002), a node uses multiple identities without revealing that it owns these different identities. If some mechanisms in the network use the majority of votes in their decision making, a node with many identities can cheat during the voting process by using more than one vote. For a routing protocol that use several paths to the destination, a Sybil attack can advertise one path as several ones. Additionally, a Sybil attack can be correlated with sink hole or worm hole attacks.

PROTECTION MECHANISMS

Different protection mechanisms have been proposed to increase the survivability of the nodes and protect their assets. For example, a few of the surveyed above routing protocols have recently been patched with security mechanisms: secure-SPIN (Xiao, Wei, & Zhou, 2006) adds cryptographic functions to SPIN that do not require too much memory and processing power; and secure directed diffusion (SDD) (Wang, Yang, & Chen, 2005) uses an efficient one-way chain rather than asymmetric cryptography, which is too complex for the resource-constrained nodes, to increase the security of the protocol. Indeed, the cost of the protection mechanisms has to really be taken into account due to the resource-constraints of the nodes. Cryptographic solutions may be used for confidentiality and integrity of data but they may be too heavy in some settings. Any protection mechanism needs to be analysed with regard to its computation cost, its memory cost, its communication cost, and its energy cost (Hwang et al., 2004). In the next subsections, we detail two fast-evolving protection mechanisms: cryptographic key deployment and management among the nodes, and computational trust management.

Key Deployment and Management

A first line of defence is the use of cryptography to encrypt the communication between the nodes. However, this requires the distribution/deployment of secrets in the nodes to allow them to encrypt the communication with this secret. The distribution of keys is usually followed by a shared key discovery phase and a path key establishment phase. Other elements that need to be considered are key revocation, rekeying, and addition of nodes. Two neighbour nodes can communicate only if they share the same key. Network resilience is defined as the number of captured nodes before an attacker is able to control the network. Network connectivity is defined as the probability that two nodes can communicate. Rekeying overhead is defined as the network traffic needed to establish a new key. Both network resilience to node capture and pair-wise connectivity depends on the size of keying material stored on the nodes. While public key cryptography is not feasible due to limited computational resources, the distribution of secret keys to each sensor is assumed to be feasible in the literature. As we underlined above, the nodes are low cost devices without strong tamper proof hardware. Thus, a captured node will, at some stage, permit access to its cryptographic material. Key management schemes (Chan, Perrig, & Song, 2003; Mohammed & Mohamed, 2005) try to increase network resilience to node capture while maintaining the performance goals and minimising the resulting cost of lower network connectivity due to sensors who do not share similar secret keys. There is a trade-off between the energy spent, the cost of used memory for protection, and the security level reached (Hwang et al., 2004).

Static keying means that the nodes have been allocated keys off-line before deployment, that is, predeployment. The existing solutions assign keys either randomly or based on deployment information, for example, the predicted neighbourhood of the nodes. A basic scheme is to generate p keys off-line and the nodes are allocated k keys randomly among these p keys. After deployment, a node broadcasts a set of identifiers of its known keys and can communicate with the nodes that have at least one common key. The advantage of static keying is no communication overhead after the deployment. The easiest way to secure a network is to give a unique key at predeployment time. However, in this case, if only one node is compromised the whole network is compromised and it seems viable to extract the key from one node as they are cheap and not so well protected (at least in nonmilitary application scenarios). The second approach is to have pairwise keys for all sensors on each sensor, which is impractical due to the memory constraints of the sensors. Saurabh and Mani (2004) argue that previous approaches relying on keying management and cryptographic means are not suitable for small nodes, such as sensors, due to their resource constraints or the fact that it is easy to recover their cryptographic material because they are cheap and not fully tamper-proof. For n nodes deployed in the network, each node would have to store n-1 keys. Even if the keys are small (e.g, 64 bits), for a network of tens of thousands of nodes the storage space required for the keys is impractical. It is worth noting that only a small fraction of the keys may be used in fixed standard sensors networks because the density of the network may be low and a sensor may only be able to communicate with few neighbouring nodes with direct communication. Eschenauer and Gligor (2002) mitigate the memory constraints problem whilst keeping the key resilience level at a target threshold level. If we consider N the number of nodes in the network and p the probability that two nodes share a common key, then each node will store a set of Np keys, called a *key ring*. The keys are selected from a larger *key pool*. Each node stores a set of keys and an identifier for each key. A *shared key discovery phase* between the neighbours is necessary after the deployment. Each node broadcasts the identifiers of the keys in its key ring. If the nodes share a common key, there is a link between them. If a common key between two nodes does not exist, then a *path key establishment* procedure takes place. An alternative is to use location information to improve connectivity. Polynomial-based key predistribution schemes (Chan et al., 2003) use a random symmetric t-degree polynomial P. A *polynomial share* is defined as a partially evaluated polynomial: $P(i,y)$ or $P(y,i)$. Based on the polynomial share, each node can compute a common key: $f(i,j)$. The scheme is resistant to t collusions.

Dynamic keying means that the keys can be (re)generated after deployment. New keys are created in order to prevent a potential attacker from using the keying information obtained by node capture. It creates more communication overhead but stronger resilience to node capture. Dynamic key management schemes are based on the exclusion basis systems (EBS) (Mohammed & Mohamed, 2005). There is an initial pool of $k+m$ keys. Each node stores k keys. Rekeying is carried out from time to time because there is the assumption that some nodes are captured from time to time. The m keys that are unknown to the captured nodes are used to encrypt replacement keys that are distributed to the safe nodes. However, since m is usually chosen quite small to limit the number of messages needed for rekeying, a few nodes may be enough to collude and unveil the keys of the whole network. To mitigate this issue, Mohammed and Mohamed (2005) propose a variant based on key polynomials instead of basic keys. Each node stores k polynomials of t-degree out of a $k+m$ pool of polynomials. In order to obtain a key, $t+1$ shares of each polynomial are needed. Another approach to mitigate the attacks of colluding nodes may be to evaluate the trust that can be placed in the involved nodes.

Computational Trust Management

In the human world, trust exists between two interacting entities and is very useful when there is uncertainty in result of the interaction. The requested entity uses the level of trust in the requesting entity as a mean to cope with uncertainty and to engage in an action with potential benefits in spite of the risk of a harmful outcome. In our settings, the nodes may be interdependent, for example, to reach far away nodes via routing and forwarding between intermediate nodes. We have seen above that it is an asset for the nodes to have trustworthy neighbours and *computational trust* is a means to compute trust in them. There are many definitions of the human notion trust in a wide range of domains with different approaches and methodologies (McKnight & Chervany, 2000), such as sociology, psychology, economics, pedagogy, and so forth. Romano's (2003) recent definition tries to encompass the previous work in all these domains.

Trust is a subjective assessment of another's influence in terms of the extent of one's perceptions about the quality and significance of another's impact over one's outcomes in a given situation, such that one's expectation of, openness to, and inclination toward such influence provide a sense of control over the potential outcomes of the situation.

A computed trust value in an entity may be seen as the digital representation of the trustworthiness or level of trust in the entity under consideration; a nonenforceable estimate of the entity's future behaviour in a given context based on evidence (Trustcomp, n.d.). A computational model of trust based on social research was first proposed by Marsh (1994). Trust in a given situation is called the *trust context*. Each trust context is assigned an importance value in the range [0,1] and utility value in the range [-1,1]. Any trust value is in the range [-1,1). Risk is used in a threshold for trusting decision making. Evidence encompasses outcome observations, recommendations, and reputation. The propagation of trust in peer-to-peer network has been studied by Despotovic and Aberer (2004) who introduce a more efficient algorithm to propagate trust and recommendations in terms of computational and communication overhead. Such overhead is especially important in networks of nodes as any communication overhead requires more energy spending.

A high level view of a trust engine is depicted in Figure 1. The decision-making component can be called whenever a trusting decision has to be made. Most related work has focused on trust decision making when a requested entity has to decide what action should be taken due to a request made by another entity, the requesting entity. It is the reason that a specific module called entity recognition (ER) (Seigneur, 2005) is represented to recognise any entities and to deal with the requests from virtual identities. In our network of nodes settings, when keying material is used, the nodes may be recognised via the secret keys that they own and use.

The decision making of the trust engine uses two subcomponents:

1. A trust module that can dynamically assess the trustworthiness of the requesting entity based on the trust evidence of any type stored in the evidence store.
2. A risk engine that can dynamically evaluate the risk involved in the interaction, again based on the available evidence in the evidence store.

A common decision-making policy is to choose (or suggest to the user) the action that would maintain the appropriate cost/benefit. For example, in the sensor network application domain, we have to balance ejecting a message or forwarding it based on how much energy has to be spent and risk of failure in each case to successfully reach the base station or sink. In the background, the evidence manager component is in charge of gathering evidence such as recommendations, comparisons between expected outcomes of the chosen actions and real outcomes, and so forth. These pieces of evidence are used to update risk and trust levels. Thus, trust and risk follow a managed lifecycle.

Although 'subjective logic' (Jøsang, 2001) does not use the notion of risk, it can be considered as a trust engine that integrates the element of ignorance and uncertainty, which cannot be reflected by mere probabilities but is part of the human aspect of trust. In order to represent imperfect knowledge, an opinion is considered to be a triplet whose elements are belief (b), disbelief (d), and uncertainty (u), such that:

$$b + d + u = 1 \qquad \{b, d, u\} \in [0,1]^3$$

The relation with trust evidence comes from the fact that an opinion about a binary event can be based on statistical evidence. Information on posterior probabilities of binary events are converted in the b, d, and u elements in a value in the range [0,1]. The trust value (w) in the virtual identity (S) of the virtual identity (T) concerning the trust context p is:

$$w^{T}_{p(S)} = \{b, d, u\}$$

The subjective logic provides more than 10 operators to combine opinions. For example, the recommendation (⊗) operator corresponds to use the recommending trustworthiness (RT) to adjust a recommended opinion. Jøsang's approach can be used in many applications since the trust context is open. In the case of our networks of nodes, we can apply this kind of triple and statistical evidence count to compute the node trust value. For example, in case of a sink base station and a network of nodes, the messages sent by a node may be acknowledged by the base station by sending an acknowledgement message with strong energy transmission. Depending on which neighbour node was used to forward the message, the sending node can count how many times the sent messages were acknowledged via this neighbour node. Each neighbour node is given a triple (b, d, u) as its trust value. If a message is acknowledged, b is increased by one. If after a timeout, the message has still not been acknowledged, d is updated by one. From the sending time of the message to the acknowledgement or the timeout, u is increased by 1 (and then decreased by 1). Concerning the memory/protection cost trade-off (Hwang et al., 2004), it seems to be a reasonable assumption be-

Figure 1. High-level view of a trust engine

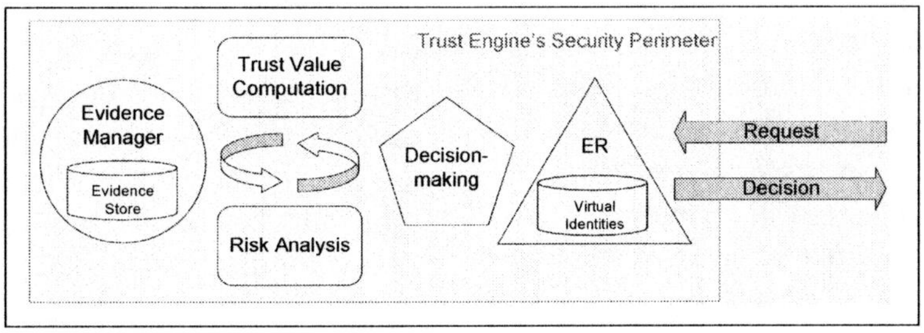

cause there are usually few neighbours and there is enough memory space for a few triples. However, each node has to maintain active the radio link, in the so-called promiscuous mode, in order to listen to the activity of the neighbours. Since sensors are low energy devices, the energy consumption due to the listening state represents a major drawback of any trust system.

Several other mechanisms have been proposed to make decisions about whether to cooperate or not with their peer nodes based on their previous behaviour. The information used to build the reputation value of the neighbours is collected mainly by direct interactions and following observations. Although it is accurate, it requires some time before enough evidence has been collected. In a scenario consisting of static nodes such as deployed sensors, there is more time to build trust with the neighbour sensors because they do not move. In this case, one may consider using a temporary ramp-up counter of 10 messages in the trust metrics to be sure of the behaviour of the node. If recommendations are used, the reputation of the nodes that provide the recommendations has to be taken into account. In this latter case, it may create a vulnerability to false report attacks. We survey below the other trust models that have been applied to the application domain of resource-constraint nodes and sensors networks.

Michiardi and Molva's (2002) core trust model builds the reputation of a sensor as a value that is increased on positive interactions and decreased otherwise. It takes also into account positive ratings from the neighbours. If the aggregated value of the reputation is positive, the sensor cooperates, otherwise it refuses cooperation. Buchegger and Le Boudec's (2004) confidant trust model considers only negative ratings from the neighbours and monitors the communication to detect the nodes that do not forward the messages. In order to compute a reputation value, different weights are assigned to personal observations and reported reputations. Concerning the memory/protection cost trade-off (Hwang et al., 2004), it seems feasible because they only forward second-hand information/recommendations, called alarms, to a limited number of nodes, called friends, maintained in a simple table. However, the monitor module of confidant still requires the consumption of a non-negligible amount of energy for resource-constrained nodes. Saurabh and Mani's (2004) trust model uses only positive ratings and models the reputation value as a probabilistic distribution by the means of a beta distribution model. A sensor will cooperate with the neighbours that have a reputation value higher than a threshold. Twigg's (2003) trust model focuses on the MANET DSR protocol and on the relation between trust value communication and energy cost. He assumes Friss' free-space attenuation to compute the risk and cost of ejecting to more or less far nodes. He proposes to consider aggregate properties including retransmissions and calculate the probability of successful transmission before a certain time. Pirzada and McDonald (2005) introduce the use of computational trust based on direct observations to mitigate both sinkhole and wormhole attacks. However, their work is also limited to the MANET DSR protocol. They cover two trust contexts: trust packet precision (TPP) for wormhole and trust packet acknowledgment (TPA) for sinkhole. They combine the two trust contexts. If the sensor is suspected to be a wormhole, the combined trust value T is 0. Otherwise TPP is equal to 1. TPA is a counter that is incremented each time a node is used to forward a packet and an acknowledgement has been received before a timeout; it is decreased otherwise. The inverse of the combined trust value simply replaces the default cost of 1 in the LINK CACHE of the standard DSR protocol. If it is a wormhole the cost is set to infinity.

FUTURE TRENDS

In the future, the importance of the energy asset of the nodes may decrease. For example, a new energy solution may come from new contact-less, distant energy transfer means to recharge nodes, such as via inductive coupling. In addition, although current energy harvesting mechanisms have performance and size limitations, the advances in nanotechnologies may allow even small nodes to effectively harvest energy after deployment. For example, solar cells in new nanomaterial are much more flexible than before. In this case, the attacks may be turned towards the external harvested energy sources.

The advances in nanotechnologies may also mean that even smaller nodes are possible. In this case, it is likely that current cryptographic mechanisms will have to be scaled down. Routing and communication between these nanoscale nodes may also change dramatically. Quantum computing may introduce even further probabilistic mechanisms with less determinism at the node level than at the nodes as a whole level. In this case, decision-based under uncertainty may still benefit from the use of computational trust.

CONCLUSION

Due to the resource-constraints of the nodes involved in mobile ad hoc or sensors networks settings, new security mechanisms are needed to guarantee the survivability of these networks of nodes. However, these new security mechanisms have a strong constraint with regard to their resource consumption. Computational trust management is one of these new schemes that are proposed because the nodes are interdependent and need to collaborate to achieve more that what they can achieve alone. There are still limitations though: both the listening mode and the communication overhead are costly in terms of energy. The cryptographic tasks involved in key management consume less energy but rekeying still necessitates extra communication. There is still some work ahead to fine-tune and combine these new security mechanisms for optimal survivability, being survivability at the node level or at the network of nodes level.

REFERENCES

Buchegger, S., & Le Boudec, J.-Y. (2004). *A robust reputation system for P2P and mobile ad-hoc networks*. Paper presented at the Second Workshop on the Economics of Peer-to-Peer Systems.

Cardei, M. (2005). *Energy-efficient target coverage in wireless sensor networks*. Paper presented at the INFOCOM.

Carle, J., & Simplot-Ryl, D. (2004). Energy-efficient area monitoring for sensor networks. *IEEE Computer, 37*(2).

Chan, H., Perrig, A., & Song, D. (2003). *Random key predistribution schemes for sensor networks*. Paper presented at the IEEE Security and Privacy Symposium.

Chatzigiannakis, I., Dimitriou, T., Nikoletseas, S., & Spirakis, P. (2006). A probabilistic algorithm for efficient and robust data propagation in smart dust networks. *Elsevier Journal of Ad-hoc Networks, 4*(5).

Despotovic, Z., & Aberer, K. (2004). *Trust and reputation management in P2P networks*. Paper presented at the International Conference on E-Commerce Technology.

Douceur, J. R. (2002). *The Sybil attack*. Paper presented at the 1st International Workshop on Peer-to-Peer Systems.

Eschenauer, L., & Gligor, V. (2002). *A key management scheme for distributed sensor networks*. Paper presented at the ACM Conference on Computer and Communications Security.

Friss, H. T. (1946). *A note on a simple transmission formula*. Paper presented at the Proceedings of IRE.

Handy, M. J., Haase, M., & Timmermann, D. (2002). *Low energy adaptive clustering hierarchy with deterministic cluster-head selection*. Paper presented at the International Conference on Mobile and Wireless Communications Networks.

Hu, Y., Perrig, A., & Johnson, D. (2002). *Wormhole detection in wireless ad hoc networks* (Tech. Rep.). Rice University.

Hubaux, J.-P., Buttyán, L., & Capkun, S. (2001). *The quest for security in mobile ad hoc networks*. Paper presented at the ACM Symposium on Mobile Ad Hoc Networking and Computing.

Hwang, D. D., Lai, B.-C. C., & Verbauwhede, I. (2004). *Energy-memory-security tradeoffs in distributed sensor networks.* Paper presented at the Ad-hoc Now Conference.

Intanagonwiwat, C., Govindan, R., Estrin, D., Heidemann, J., & Silva, F. (2003). Directed diffusion for wireless sensor networking. *IEEE/ACM Transactions on Networking, 11.*

Jøsang, A. (2001). A logic for uncertain probabilities. *Fuzziness and Knowledge-Based Systems, 9*(3).

Kulik, J., Heinzelman, W. R., & Balakrishnan, H. (2002). Negotiation-based protocols for disseminating information in wireless sensor networks. *Wireless Networks, 8.*

Maltz, D. A. (2001). *On-demand routing in multi-hop wireless ad hoc networks.* Unpublished doctoral thesis, Carnegie Mellon University.

Marsh, S. (1994). *Formalising trust as a computational concept.* Unpublished doctoral thesis, University of Stirling, Department of Mathematics and Computer Science.

Martin, T., Hsiao, M., Ha, D., & Krishnaswami, J. (2004). *Denial-of-service attacks on battery-powered mobile computers.* Paper presented at the 2nd IEEE Pervasive Computing Conference.

McKnight, D. H., & Chervany, N. L. (2000). *What is trust? A conceptual analysis and an interdisciplinary model.* Paper presented at the Americas Conference on Information Systems.

Michiardi, P., & Molva, R. (2002). *Core: A collaborative reputation mechanism to enforce node cooperation in mobile ad hoc networks.* Paper presented at the IFIP TC6/TC11 Sixth Joint Working Conference on Communications and Multimedia Security.

Miranda, H., & Rodrigues, L. (2003). *Friends and foes: Preventing selfishness in open mobile ad hoc networks.* Paper presented at the 23rd International Conference on Distributed Computing Systems.

Mohammed, A. M., & Mohamed, E. (2005). *A study of static versus dynamic keying schemes in sensor networks.* Paper presented at the 2nd ACM International Workshop on Performance Evaluation of Wireless Ad hoc, Sensor, and Ubiquitous Networks, Montreal, Quebec, Canada.

Ozturk, C., Zhang, Y., & Trappe, W. (2004). *Source-location privacy in energy-constrained sensor network routing.* Paper presented at the 2nd ACM Workshop on Security of Ad hoc and Sensor Networks.

Perkins, C. E., & Royer, E. M. (1999). *Ad hoc on-demand distance vector routing.* Paper presented at the 2nd IEEE Workshop on Mobile Computing Systems and Applications.

Pirretti, M., Zhu, S., Narayanan, V., McDaniel, P., Kandemir, M., & Brooks, R. R. (2005). *The sleep deprivation attack in sensor networks: Analysis and methods of defense.* Paper presented at the Innovations and Commercial Applications of Distributed Sensor Networks Symposia.

Pirzada, A. A., & McDonald, C. (2005). *Circumventing sinkholes and wormholes in wireless sensor networks.* Paper presented at the International Workshop on Wireless Ad-hoc Networks.

Powell, O., Jarry, A., Leone, P., & Rolim, J. (2006). *Gradient based routing in wireless sensor networks: A mixed strategy* (Tech. Rep.). University of Geneva.

Romano, D. M. (2003). *The nature of trust: Conceptual and operational clarification.* Unpublished doctoral thesis, Louisiana State University.

Saurabh, G., & Mani, B. S. (2004). *Reputation-based framework for high integrity sensor networks.* Paper presented at the 2nd ACM Workshop on Security of Ad hoc and Sensor Networks, Washington D.C.

Schurgers, C., & Srivastava, M. B. (2001). *Energy efficient routing in wireless sensor networks.* Paper presented at the MILCOM Communications for Network-Centric Operations: Creating the Information Force.

Seigneur, J.-M. (2005). *Trust, security and privacy in global computing.* Unpublished doctoral thesis, Trinity College Dublin.

Trustcomp. (n.d.). Retrieved August 4, 2006, from http://www.trustcomp.org/

Twigg, A. (2003). *A subjective approach to routing in P2P and ad hoc networks.* Paper presented at the First International Conference on Trust Management.

Wang, X., Yang, L., & Chen, K. (2005). SDD: Secure directed diffusion protocol for sensor. *Security in ad-hoc and sensor networks* (Vol. 3313). Springer.

Weiser, M. (1991). *The computer for the 21st century.* Scientific American.

Xiao, D., Wei, M., & Zhou, Y. (2006). *Secure-SPIN: Secure sensor protocol for information via negotiation for wireless sensor networks.* Paper presented at the Conference on Industrial Electronics and Applications.

Ye, F., Chen, A., Liu, S., & Zhang, L. (2001). *A scalable solution to minimum cost forwarding in large sensor networks.* Paper presented at the Tenth International Conference on Computer Communications and Networks.

KEY TERMS

Node: A node may go from the tiny fixed deployed sensor to the mobile unplugged mobile device.

Node(s) Survivability: Emphasises that the scope of the nodes mission may span more than one node. The survivability of the node itself may be more important than the survivability of the other nodes or the mission may be that the majority of the nodes survive at the expense of the survival of one specific node.

Reactive Routing Protocols: Compute the route between two nodes only when the route is needed, that is, 'on demand.'

Energy-aware Routing Protocols: Explicitly take into account the energy consumption as a parameter.

To Eject: Means that the sensor increases the power of transmission to be able to reach the base station in one transmission.

Static Keying: Means that the nodes have been allocated keys off-line before deployment, that is, predeployment.

Dynamic Keying: Means that the keys can be (re)generated after-deployment.

Network Resilience: The number of captured nodes before an attacker is able to control the network.

Network Connectivity: The probability that two nodes can communicate.

Rekeying Overhead: The network traffic needed to establish a new key.

Trust: Trust 'is a subjective assessment of another's influence in terms of the extent of one's perceptions about the quality and significance of another's impact over one's outcomes in a given situation, such that one's expectation of, openness to, and inclination toward such influence provide a sense of control over the potential outcomes of the situation' (Romano, 2003).

Computed Trust Value: A nonenforceable estimate of the entity's future behaviour in a given context based on evidence ("Trustcomp," n.d.).

Chapter XL
Fault Tolerant Topology Design for Ad Hoc and Sensor Networks

Yu Wang
University of North Carolina at Charlotte, USA

ABSTRACT

Fault tolerance is one of the premier system design desiderata in wireless ad hoc and sensor networks. It is crucial to have a certain level of fault tolerance in most of ad hoc and sensor applications, especially for those used in surveillance, security, and disaster relief. In addition, several network security schemes require the underlying topology provide fault tolerance. In this chapter, we will review various fault tolerant techniques used in topology design for ad hoc and sensor networks, including those for power control, topology control, and sensor coverage.

INTRODUCTION

With great potentials in a large number of application fields, ad hoc and sensor networks have been undergoing a revolution that promises a significant impact on society. Unlike traditional fixed infrastructure networks, there are no centralized controls over wireless *ad hoc* networks, which consist of a collection of devices equipped with wireless communication and networking capability. Any communication and network service in ad hoc networks is done in a self-organized and decentralized manner. Usually connections are multihop routed via intermediate nodes to enable communication between nodes without a direct link. A wireless sensor network is a network of small, wirelessly communicating nodes where each node is equipped with computation, communication, and sensing devices. These nodes usually form a self-organized ad hoc network, observe the physical space around them, and measure some physical signals or detect various phenomena of interest. Ad hoc and sensor networks are widely deployed for environment monitoring, biomedical observation, surveillance, security, disaster relief, and so on.

Ad hoc and sensor networks trigger many challenging research problems, as they intrinsically have many special characteristics and unavoidable limitations, compared with other wired or wireless networks. An important requirement of ad hoc and sensor networks is that they should be self-organizing, that is, transmission ranges and data paths are dynamically restructured with changing topology. Energy conservation and network performance are probably the most critical issues in ad hoc and sensor networks, since wireless devices (such as tiny sensor nodes in sensor networks) are usually powered by batteries only and have limited computing capability and memory. Topology control and power control are two primary techniques with respect to energy-efficiency in ad hoc and sensor networks.

The topology control technique is to let each wireless device locally select certain neighbors for communication, while maintaining a topology that can support energy efficient routing and improve the overall network performance. Unlike traditional wired networks and cellular wireless networks, mobile devices are often moving during the communication, which could change the network topology in some extent. Hence it is more challenging to design a topology control algorithm for ad hoc and sensor networks. The power control technique is to control the network topology by adjusting the wireless device's transmission range. Reducing the transmission range can save the power consumption at each node and reduce the signal interference among neighbors, but it may hurt the connectivity of the induced topology. Power control tries to minimize the power consumption used by all nodes while maintaining a topology that is connected and has certain desired properties such as fault tolerance.

Although fault tolerance has been studied for several decades in computer and VLSI systems, limited resources on small devices, lack of centralized control, and high mobility make fault-tolerance much harder to achieve in ad hoc and sensor networks. One key characteristic of such networks is that node and link failure is an event of non-negligibility, in some cases even as a regular or common event. This is particularly the case in sensor networks where the equipment is restricted to a minimum due to limitations in cost and weight. First of all, battery driven sensor nodes may stop working because they run out of energy supply. Second, the shared wireless medium is inherently less stable than wired media. This situation results in more packet losses and lower throughput. Third, sensor networks often operate in potentially hostile or at least harsh and unconditioned environments. Tiny sensor devices with limited security techniques are usually vulnerable from various attacks. Another aspect that has an influence on the required degree of redundancy and fault-tolerance is mobility, which is a key issue in ad hoc networks. Therefore, reliability and fault-tolerance are emerging as premier and crucial system design desiderata in ad hoc and sensor networks. In addition, fault-tolerance design is also one of basic components in ad hoc and sensor network security.

Fault tolerance strongly depends on the network connectivity. To make fault tolerance possible, first of all, the underlying network topology must be k-connected for some $k > 1$, that is, given any pair of wireless devices, at least k disjoint paths are needed to connect them. With k-connectivity, the network can survive k-1 node/link failures. Traditional topology control or power control solutions cannot cope with those fault-tolerance requirements, since fault-tolerance is usually sacrificed for power efficiency. In order to be power efficient, topology control and power control algorithms try to reduce the number of links and thereby reduce the redundancy available for tolerating node and link failures. On the other hand, to achieve fault-tolerance, existing algorithms usually sacrifice power efficiency concern. Thus, topology design for ad hoc and sensor networks needs to consider both power efficiency and fault-tolerance.

This chapter is focused on fault tolerant topology design for ad hoc and sensor networks. In the second section, fault tolerant techniques used in power control protocols (such as power assignment and critical transmission range) are reviewed. In the third section, we survey fault tolerant design in topology control, that is, how to design fault tolerant geometric or hierarchical structures. In the fourth section, fault tolerant coverage and protection in sensor networks are discussed. There is a conclusion in the fifth section, while the chapter ends with references and key definitions.

FAULT TOLERANT DESIGN IN POWER CONTROL

Fault tolerant design in power control studies how to set the transmission range for each node in a network such that the induced topology is k-connected, that is, the network can survive under k-1 failures. Obviously, by setting the transmission range sufficiently larger, the induced network topology will be k-connected without doubt. However, as power is a scarce resource in ad hoc and networks, it is important to save the power consumption without losing the network connectivity. Thus, the question is how to find the minimum transmission range such that the induced topology is multiply connected. There are two sets of research in this direction: *critical transmission range* for random networks and *minimum power assignment optimization* for static networks.

Given n static wireless nodes V, each with transmission range r_n, the wireless network can be modeled by graph $G(V,r_n)$ in which two nodes are connected if their Euclidean distance is no more than r_n. The minimum range r_n used by all wireless nodes such that the induced network topology has certain property (such as connectivity) is called the *critical transmission range* (CTR). The CTR for connectivity has been studied in the literature (Gupta & Kumar, 1998; Penrose, 1997; Ramanathan & Rosales-Hain, 2000; Sanchez, Manzoni, & Haas, 1999). Characterizing the CTR for connectivity (or k-connectivity) helps the system designer to answer fundamental questions, such as: (1) given a number of nodes n to be deployed in a region, what is the minimum value of transmission range that ensures network connectivity (or k-connectivity)?; or (2) given transmission range of certain technology, how many nodes need to be distributed over a given region to ensure network connectivity (or k-connectivity)?

Recently, applying stochastic geometry, Penrose (1999), Bettstetter (2002), Li, Wang, Wan, and Yi (2003), and Wan and Yi (2004) studied CTR to achieve the k-connectivity with certain probability for a network when wireless nodes are uniformly and randomly distributed over a two-dimensional region. Penrose (1999) shows that with high probability the network becomes k-connected when the minimum node degree in the communication graph becomes k. In other words, the characterization of the CTR for k-connectivity can be derived by analyzing the probability of the relatively simpler event that every node in the network has a degree at least k. Based on results from Penrose, Li et al. (2003) first derives the upper bound and the lower bound of the CTR for k-connectivity in a two-dimensional network. They proved that, given n wireless nodes which are randomly distributed in a unit square, if the transmission range r_n of wireless devices satisfies, $n\pi \cdot r_n^2 \geq \ln n + (2k-3)\ln\ln n - 2\ln(k-1)$!$+\ \alpha + 2\ln(8(k-1)/(2^{k-1}\sqrt{\pi}))$then $G(V,r_n)$ is k-connected with probability at least $e^{-e^{-\alpha}}$ as n goes to infinity. Here α is any real number. Wan and Yi (2004) close the gap between the upper bound and the lower bound by giving an exact formula for the probability of k-connectivity when n goes to infinity. They show the CTR for k-connectivity: $r_n = \sqrt{(\log n + (2k-3)\log\log n + f(n))/\pi n}$ where $f(n)$ is an arbitrary function such that $\lim_{n\to\infty} f(n) = +\infty$. Bettstetter (2002) also investigated the minimum node degree and k-connectivity and constructed various simulations to verify his analytical expressions. However his theoretical result does not consider the boundary effects (assume the network is distributed in a very large area), which is impossible in real networks. Even though the theoretical results of the CTR for k-connectivity has been derived, the theoretical bounds only hold when n goes to infinity. How to set the transmission range in a real network where n is a small pratical integer is studied by Li et al. (2003) by conducting simulations. Another related work is about the CTR for connectivity with Bernoulli nodes. So far we assume that all nodes will always function properly, however, in certain scenarios, nodes may be fault (or put into sleep) with a certain probability $p > 0$. Wan and Yi (2005) model this scenario using Bernoulli nodes and studied the CTR for connectivity with Bernoulli nodes.

All analytical results on CTR assume wireless nodes are randomly distributed and the transmission range of every node is equal. These assumptions are not always true for ad hoc and sensor networks in practice. Another power control technique is to allow each wireless device to adjust its transmission power according to its neighbors' positions. A natural question is then, given a static network, how to assign the transmission power for each node such that the network is k-connected with optimization criteria minimizing the total (or maximum) transmission power assigned. This kind of optimization questions is called *minimum power assignment optimization*. See Figure 1 for illustrations of minimum total power assignment for k-connectivity ($k = 1$ or 2).

The *minimum maximum power assignment problem* can be solved in polynomial time by using a simple binary-search-based approach (Lloyd, Liu, Marathe, Ramanathan, & Ravi, 2002). The *minimum total power assignment for connectivity problem* was first studied and proved to be NP-hard by Chen and Huang (1989), in which the induced communication graph is strongly connected while the total power assignment is minimized. Recently, this problem has been heavily studied and many approximation algorithms have been proposed when the network is modeled using symmetric or asymmetric links (Althaus, Calinescu, Mandoiu, Prasad, Tchervenski, & Zelikovsly, 2003; Calinescu, Kapoor, Olshevsky, & Zelikovsky, 2003; Clementi, Penna, & Silvestri, 2000; Clementi, Huiban, Penna, Rossi, & Verhoeven, 2002; Kirousis, Kranakis, Krizanc, & Pelc, 2000; Ramanathan & Rosales-Hain, 2000). Along this line, Calinescu and Wan (2006), Cheriyan, Vempala, and Vetta (2002), and Hajiaghayi, Immorlica, and Mirrokni (2003) consider the minimum total power assignment while the resulting network is k-connected (or (k-1) fault tolerant). This problem has been shown to be NP-hard too. Many of the best-known approximation algorithms (e.g., Cheriyan et al., 2002) are based on linear programming (LP) approaches. However, Haijaghayi et al. (2003) show that for the minimum total power assignment for k-connectivity problem, the natural integer LP formulation has an integrality gap of $\Omega(n/k)$, implying that there is no approximation algorithm based on LP with an approximation factor better than $\Omega(n/k)$.

Some heursitics (Bahramgiri, Hajiaghayi, & Mirrokni, 2002; Ramanathan & Rosales-Hain, 2000) are proposed as well. Bahramgiri et al. (2002) show that the cone-based topology control (CBTC) algorithm by Wattenhofer, Li, Bahl, and Wang (2001) and Li, Halpern, Bahl, Wang, and Wattenhofer (2001) can be extended to slove the k-fault tolerance. Haijaghayi et al. (2003) also constructed examples which demonstrate that the approximation factor for CBTC algorithm is at least $\Omega(n/k)$. Recently, Lloyd et al. (2002) presented a centralized $8(1\text{-}1/n)$-approximation for the minimum total power assignment for 2-connectivity problem. Calinescu and Wan (2006) further show that their algorithm could achieve $2k$-approximation ratio for the minimum total power

Figure 1. Illustrations of power control: minimum total power assignment for connectivity

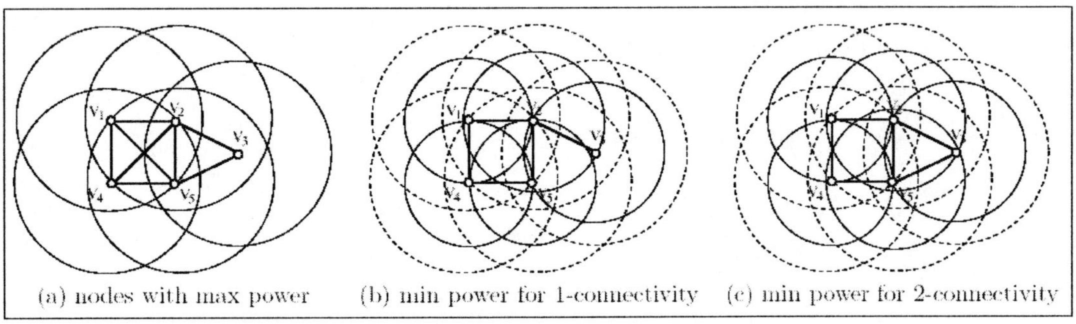

assignment for k-connectivity problem. Haijaghayi et al. (2003) present algorithms minimizing power while maintaining k-connectivity with guarantee. Their first algorithm gives an $O(k\alpha)$-approximation where α is the best approximation factor for the related problem in wired networks (the best α so far is in $O(\log k)$ by Cheriyan et al., 2002)). The second algorithm is based on an approximation algorithm introduced by Kortsarz and Nutov (1994). It is more complicated and can achieve $O(k)$ approximation for general graphs. Their first two algorithms are centralized algorithms. Then they present two distributed approximation algorithms for the cases 2- and 3-connectivity in geometric graphs with constant approximation ratios. Both these algorithms use the distributed minimum spanning tree algorithm.

FAULT TOLERANT DESIGN IN TOPOLOGY CONTROL

Topolgoy control algorithms have been proposed to maintain network connectivity while improving energy efficiency and increasing network capacity by solely keeping selected links. However, by reducing the number of links in the network, topology control actually decreases the degree of routing redundancy. As a result, the induced topology is more susceptible to node failures or departures. Thus, in this section we review the fault tolerant design which enforces k-connectivity in the topology control process. Usually, there are two sets of solutions for topology control: geometric topology (flat structure) and virtual backbone (hierarchical structure).

Geometric topology control algorithms assume each node knows the position information of itself and its neighbors and all nodes have the same transmission range. Using this geometric information, each node makes a local decision to keep some links and remove other links. Well-known geometric topologies used in ad hoc networks include local minimum spanning tree (LMST) (Li, Hou, & Sha, 2003), relative neighborhood graph (RNG) (Bose, Morin, Stojmenovic, & Urrutia, 2001; Seddigh, Gonzalez, & Stojmenovic, 2002), Gabriel graph (GG) (Bose et al., 2001; Karp & Kung, 2000), Yao graph (YG) (Li, Wan, & Wang, 2001; Li, Wan, Wang, & Frieder, 2002) and CBTC (L. Li et al., 2001; Wattenhofer et al., 2001). See Figure 2 for illustrations of their definitions. All of these topologies do guarantee the connectivity but not fault tolerance. Therefore, variations of these topologies have been proposed to improve the fault tolerance, that is, preserving k-connectivity.

Li and Hou (2004) present a variation of LMST algorithm to construct a k-connected topology, called fault-tolerant local spanning subgraph ($FLSS_k$). Similarly to LMST, algorithm to build $FLSS_k$ is composed of three phases: information exchange, topology construction, and determination of transmit power. The main difference between LMST and $FLSS_k$ is in the topology construction phase: instead of building a local MST on its neighbor (such as the two local trees for u and v in Figure 2[a]), a node builds a spanning subgraph to preserve k-connectivity using a simple greedy algorithm. Li and Hou prove that $FLSS_k$ guarantees the k-connectivity and maintains bidirectionality for all the links in the topology while reducing the power consumption.

Figure 2. Illustrations of the definitions of different topologies

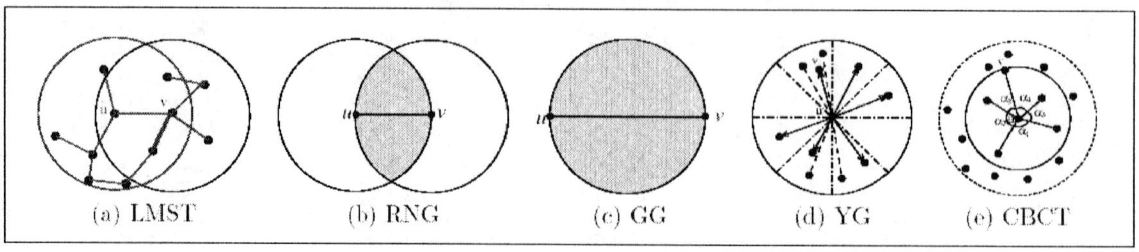

Zhou, Das, and Gupta (2005) generalize the RNG structure to *k*-RNG structure to preserve the *k*-connectivity for sensor networks. In RNG, a link *uv* exists if and only if there is no other node *w* with edges *uw* and *wv* satisfying $||uw||<||uv||$ and $||wv||<||uv||$ simultaneously. Here $||.||$ is the Euclidean distance. See Figure 2(b). In *k*-RNG, an edge exists between *u* and *v* if and only if there are at most (*k*-1) nodes *w* that satisfy $||uw||<||uv||$ and $||wv||<||uv||$. Obviously, similar to RNG, *k*-RNG can be constructed locally. Zhou et al. proved that *k*-RNG is *k*-connected if the original communication graph is *k*-connected. Notice that it is also easy to show we can use the same idea to generalize GG structure to *k*-GG while preserving the *k*-connectivity. There is an edge *uv* in *k*-GG if and only if there are at most (*k*-1) nodes inside the disk with *uv* as the diameter. See Figure 2(c). The nice property of GG and *k*-GG is that their power spanning ratios are equal to one (X.-Y. Li et al., 2001, 2002). In other words, GG/*k*-GG can keep all links on least power consumption paths in the original communication graph. Notice that LMST/$FLSS_k$ and RNG/*k*-RNG do not have this property.

X.-Y. Li et al. (2003) modifiy the Yao structure as follows such that the structure is *k*-connected. Each node *u* defines any *p* equally-separated rays originated at *u*, where $p > 6$. These rays define *p* cones inside the transmission range. Figure 2(d) shows an example with $p = 8$ cones. In each cone, *u* chooses the *k* closest nodes in that cone, if there is any, and adds directed links from *u* to these nodes. Ties are broken arbitrarily. X.-Y. Li et al. (2003) proved that the modified Yao structure ($YG_{p,k}$) can preserve the *k*-connectivity. In addition, $YG_{p,k}$ is a length/power spanner with bounded node degree even when (*k*-1) nodes fault. Here a length/power spanner has constant length spanning ratio and power spanning ratio, which indicates the topology is power efficient for unicast routing.

Bahramgiri et al. (2002) also discuss how to generalize the CBTC algorithm to ensure *k*-connectivity. Basically, for each node, it enlarges the transmission range until it reaches its maximum power or the maximum angle between two consecutive neighbors of the induced topology is at most $2\pi/(3k)$. See Figure 2(e). Finally, it eliminates one-directional edges and keeps bidirectional edges. Bahramgiri et al. (2002) proved the resulted topology is *k*-connected if the original graph is *k*-connected. We can also prove the topology is a length spanner even with (*k*-1) nodes faults. However, unlike $YG_{p,k}$, the topology does not bound the node degree. A counter example is given by X.-Y. Li et al. (2003), so is an enhancement method to bound the node degree.

While all geometric structures above are flat structures, there is another set of structures, called hierarchical structures, widely used in ad hoc and sensor networks. Instead of involving all nodes to relay packets for other nodes, the hierarchical topology control protocols pick a subset of nodes to serve as the cluster-heads. These cluster-heads form a virtual backbone and forward packets for other nodes. The structure used to build this virtual

Figure 3. Examples of dominating set and k-dominating set

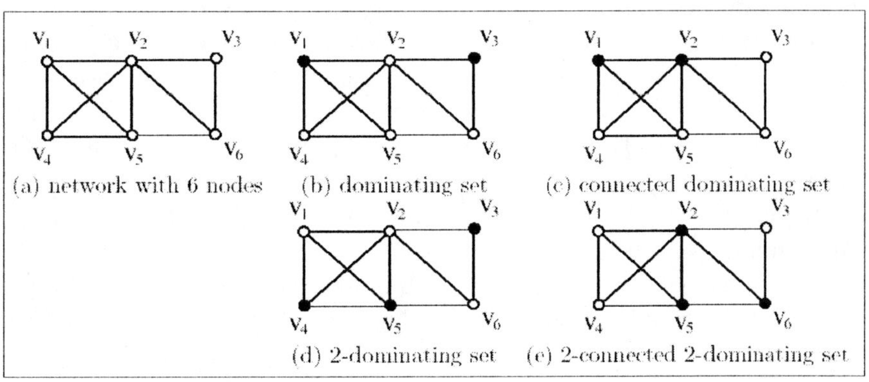

backbone is usually a (connected) dominating set. Many distributed clustering (or dominating set) algorithms have been proposed in the literature (e.g., Alzoubi, Wan, & Frieder, 2002; Das & Bharghavan, 1997; Wan, Alzoubi, & Frieder, 2002; Wu & Li, 1999, 2000). All these algorithms first form several clusters where all cluster-heads form a dominating set. Each node either is a cluster-head (or called dominator) or belongs to one cluster (i.e., it is dominated by a dominator). All the cluster-heads can then be connected via several additional gateways to form the virtual backbone. However, a single node failure may cause the backbone to be broken in these algorithms. Thus, a fault-tolerant design is needed for these backbones too.

Kuhn, Moscibroda, and Wattenhofer (2006) studied the *k*-dominating set (*k*-DS) problem: find a set of nodes such that each of the (other) nodes is dominated by at least *k* nodes from this set. The set of such nodes is called a *k*-dominating set. Thus, the backbone can survive (*k*-1) node failures in the *k*-dominating set. For example, black nodes v_1 and v_3 in Figure 3(b) form a DS for the network in Figure 3(a), while black nodes v_3, v_4, and v_5 in Figure 3(d) form a *2*-DS. Kuhn et al. (2006) give two distributed approximation algorithms for the *k*-minimum dominating set problem in two different models: general graphs and unit disk graphs (UDG). The first one is for general graphs and based on LP approximation. For an arbitrary parameter *t*, it runs in time $O(t^2)$ and achieves an approximation ratio of $O(t\Delta^{2/t}\log\Delta)$, where Δ denotes the maximal degree. The second one is a probabilistic algorithm for unit disk graphs. It runs in time $O(\log\log n)$ and achieves a constant approximation in expectation.

Dai and Wu (2005) studied how to construct a *k*-connected *k*-dominating set (*k*-CDS) as a backbone to balance efficiency and fault tolerance. Here, a *k*-DS is a *k*-CDS if its induced topology is *k*-connected. Figure 3(c) shows a CDS, and Figure 3(e) shows a *2*-CDS. Three localized *k*-CDS construction algorithms are proposed. The first one (called *k*-Gossip) randomly selects virtual backbone nodes with a given probability p_k, where p_k depends on network condition and the value of *k*. The second one is a deterministic approach based on the authors' previous method for *1*-CDS. The last algorithm (color-based *k*-CDS constriction, CBKC) is a hybrid paradigm that enables *1*-CDS algorithms to construct a *k*-CDS with high probability in relatively dense networks. It is a hybrid of probabilistic and deterministic approaches.

Besides *k*-DS and *k*-CDS, there are other techniques to enhance the fault tolerance of virtual backbones. Chen and Son (2005) present methods to add necessary redundant nodes to the simple CDS backbone, which results in a higher vertex connectivity degree. They also identify several factors and synchronization methods that may affect the redundant node selection. For example, the nodes in CDS would like to select nodes with more power or higher degree or some combination of factors. Wang, Wang, and Li (2006) propose an efficient distributed method to construct a weighted backbone with low cost. By assuming each node has a cost, they can construct a weighted CDS while the total cost of the CDS is bounded by a constant from the optimal. If each node can estimate its probability of being faulty and we treat it as the weight, we can use the algorithm by Y. Wang et al. (2006) to build a fault-tolerant backbone. Notice that building the most fault-tolerant backbone is equivalent to finding a CDS with the minimum total cost.

Most of the fault tolerant topology designs discussed so far assume the underlying communication graph is *k*-connected. This is true when the network density is large, but for sparse network it may not hold. Bredin, Demaine, Hajiaghayi, and Rus (2005) studied an interesting problem of repairing a sensor network to guarantee a specified level of connectivity. They present a generic algorithm that determines how to establish *k*-connectivity by placing minimum additional sensors geographically between existing pairs of sensors. This problem is NP-hard, and thus their algorithm is an approximation algorithm. They proved that the number of additional sensors is within a constant factor of the absolute minimum, for any fixed *k*.

A related fault-tolerant problem in two-tiered sensor network deployment is studied by Hao, Tang, and Xue (2004) and Liu, Wan, and Jia (2005). A two-tired sensor network is a cluster-based network.

Relay nodes are placed in the playing field to act as cluster-heads and to form a connected topology for data transmission in the higher tier. They are able to fuse data from sensor nodes (lower tier) in their clusters and send them to sinks through higher tier topology. Hao et al. (2004) studied a fault-tolerant relay node placement problem, where a minimum number of relay nodes are placed such that (1) each sensor node can communicate with at least two relay nodes and (2) the network of relay nodes is 2-connected. They proved the problem is NP-hard and gave a $O(D\log n)$-approximation, where D is the diameter of the network. Notice that the ratio is not a constant but a function of the size of input. Liu et al. (2005) studied a more general relay-node placement problem where a minimum number of relay-nodes are placed in a 2-tiered sensor network such that the whole network is (1) connected or (2) 2-connected. They assumed that sensor nodes do not participate in forwarding data for others. They first gave a $(6+\xi)$-approximation algorithm for a 1-connectivity case. Then they further proposed a $(24+\xi)$-approximation algorithm and a $(6/T+12+\xi)$-approximation algorithm for a 2-connectivity case, respectively, for any $\xi > 0$, where T is the ratio of the number of relay nodes placed to the number of sensors in the first case.

Thallner and Moser (2005) studied fault-tolerant overlay topology for a fully connected network. They modeled the network as a weighted complete graph, where the weight of an edge is the cost of that connection. Their proposed algorithm can build and maintain a k-regular subgraph that is k-connected and has low total weight. However, since it assumes a fully connected communication graph, the algorithm is more suitable for an overlay network (such as peer-to-peer network) than an ad hoc network.

Another fault tolerant issue in topology control is how to detect and recover from topology failures for classical topology control protocols (not the fault tolerant ones we discussed above). It focuses on the design of detection and recovering schemes instead of redundancy topology design with certain redundancy (k-connectivity). For example, Stratil (2005) presents an analysis of the requirements to tolerate crash failures in the topology with the help of failure detectors. Gupta and Younis (2003) also studied the efficient recovering mechanism for cluster-head failures. However, since fault detection and recovering are not the focus of this chapter, we do not review them in detail.

FAULT TOLERANT DESIGN IN COVERAGE AND PROTECTION

In sensor networks, coverage problem (Cardei & Wu, 2006) is also a critical issue during topology design and sensor deployment. Usually each sensor has a sensing range covering a small sensing region, and it can sense certain kinds of events happening inside its sensing region. Thus, we say the sensor covers its sensing region. The main objective of the sensor network is to cover (monitor) an area A, that is, every point in the area should be covered. Some applications may require different degrees of coverage. A network has a coverage degree k (k-coverage) if every location is within the sensing range of at least k sensors. Networks with a higher coverage degree can obtain higher sensing accuracy and be more robust to sensor failures. Given a sensor field with n sensor nodes of sensing range r deployed, and a desired coverage degree $k \geq 1$, *minimum k-coverage problem* studies how to select a minimal subset of nodes to entirely cover all locations in A such that every location is within the sensing range of at least k different nodes. The minimum k-coverage problem is also a well-known NP-hard problem. Figure 4 illustrates a set of examples of coverage set. Figure 4(a) shows the sensors and their sensing ranges. Assume that the target area A is the big square area $v_1v_3v_9v_7$. Figures 4(b) and 4(c) give two 1-coverage sets (black nodes), while Figure 4(d) gives a 2-coverage set.

Zhou, Das, and Gupta (2004) studied the minimum connected k-coverage problem and give a centralized approximation algorithm that achieves $O(\log n)$ approximation ratio. Their method is a greedy algorithm: iteratively adding a set of nodes which maximizes a measure called k-benefit to an initially empty set of nodes. The authors also present a distributed version of their algorithm.

Kumar, Lai, and Balogh (2004) studied k-coverage problem in sensor networks where many sensors are put to sleep for most of their lifetimes. They

Figure 4. Examples of k-coverage set in sensor networks

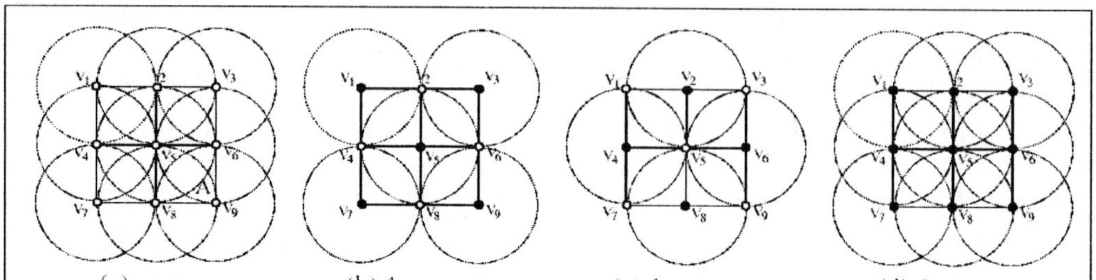

first propose a sleep/active schedule, to minimize energy consumption, in which each sensor is active with probability p, independently from the others. Then they derive the critical sensing range for their sleep scheme such that the sensor network achieves k-coverage with high probability.

Yang, Dai, Cardei, and Wu (2006) also studied the minimum connected k-coverage problem with different coverage assumption. They assumed that the network is sufficiently dense so that point coverage can approximate area coverage. Thus instead of covering the whole area A, they only required covering every sensor in area A. This k-coverage problem is also NP-hard since it is an extension of the k-dominating set problem. They propose a centralized approximation solution based on integer linear programming. The algorithm works by relaxing the problem to ordinary linear programming, where the variables may take real values. They also designed two distributed algorithms. One uses a cluster-based approach to select backbone nodes to form the active set; the other uses the pruning algorithm based on only 2-hop neighborhood information to reduce the number of active sensors.

Notice that the coverage problem studied by Yang et al. (2006) is the same problem studied by Wang, Zhang, and Liu (2006) and Wang, Li, and Zhang (2007) as *self-protection problem*. A self-protection problem focuses on using sensor nodes to provide protection to themselves instead of the objects or the area, so that they can resist the attacks targeting on them directly. A wireless sensor network is p-self-protected, if at any moment, for any wireless sensor (active or non-active), there are at least p active sensors that can monitor it. D. Wang et al. (2006) studied the minimum 1-self protection problem and give a centralized method with $2(1+log n)$ approximation ratio, using approximation algorithm for the minimum dominating set, and two randomized distributed algorithms. Wang et al. (2007) provide several efficient centralized and distributed algorithms with constant approximation ratios for the minimum p-self-protection problem in sensor networks with either homogeneous or heterogeneous sensing radius.

Not until recently have coverage and connectivity problems been studied together in sensor networks. Xing, Wang, Zhang, Lu, Pless, and Gill (2005) designed an integrated coverage configuration protocol to provide both certain degrees of coverage and connectivity guarantee. Zhang and Hou (2005) propose a decentralized density control algorithm to maintain sensing coverage and connectivity in high-density sensor networks. Both Xing et al. (2005) and Zhang and Hou (2005) prove that if the radio range is at least twice of the sensing range, complete k-coverage of a convex area implies k-connectivity among the working set of nodes. Recently, Bai, Kuma, Xua, and Lai (2006) studied the optimal deployment pattern to achieve both 1-coverage of an area and 2-connectivity of the sensors. Zhou et al. (2005) propose a set of distributed algorithms to achieve both k-connected and k-covered network by using localized Voronoi and extended relative neighborhood graphs.

CONCLUSION

Fault tolerance is one of the premier system design desiderata in wireless ad hoc and sensor networks.

It is crucial to have a certain level of fault tolerance in most of ad hoc and sensor applications, especially for those used in surveillance, security, and disaster relief. In addition, several network security schemes (such as localized intrusion detection) require that the underlying topology provide fault tolerance. In this chapter we discussed various fault tolerant techniques used in topology design, including those for power control, topology control, and sensor coverage. Due to space limit, we did not give all of the detailed algorithms, proofs, and simulation results for most techniques reviewed here. For more details, please refer to the references. Though fault tolerant topology design has attracted considerable attention and has been heavily studied recently, there are still many open problems, such as how to efficiently maintain these proposed fault tolerant topologies. We strongly believe that fault tolerant topology design remains one primary challenge and plays an important role in research of ad hoc and sensor networks.

REFERENCES

Althaus, E., Calinescu, G., Mandoiu, I., Prasad, S., Tchervenski, N., & Zelikovsly, A. (2003). Power efficient range assignment in ad-hoc wireless networks. In *Proceedings of the IEEE Wireless Communications and Networking Conference (WCNC '03).*

Alzoubi, K. M., Wan, P.-J., & Frieder, O. (2002). Message-optimal connected dominating sets in mobile ad hoc networks. In *Proceedings of the 3rd ACM International Symposium on Mobile Ad hoc Networking & Computing.*

Bahramgiri, M., Hajiaghayi, M. T., & Mirrokni, V. S. (2002). Fault-tolerant and 3-dimensional distributed topology control algorithms in wireless multi-hop networks. In *Proceedings of the 11th Annual IEEE Internation Conference on Computer Communications and Networks (ICCCN).*

Bai, X., Kuma, S., Xua, D., & Lai, T.H. (2006). Deploying wireless sensors to achieve both coverage and connectivity. In *Proceedings of the ACM MobiHoc 2006.*

Bettstetter, C. (2002). On the minimum node degree and connectivity of a wireless multihop network. In *Proceedings of the 3rd ACM International Symposium on Mobile Ad Hoc Networking and Computing (MobiHoc '02).*

Bose, P., Morin, P., Stojmenovic, I., & Urrutia, J. (2001). Routing with guaranteed delivery in ad hoc wireless networks. *ACM/Kluwer Wireless Networks*, *7*(6), 609-616.

Bredin, J. L., Demaine, E. D., Hajiaghayi, M., & Rus, D. (2005). Deploying sensor networks with guaranteed capacity and fault tolerance. In *Proceedings of the ACM Mobihoc 2005.*

Calinescu, G., Kapoor, S., Olshevsky, A., & Zelikovsky, A. (2003). Network lifetime and power assignment in ad-hoc wireless networks. In *Proceedings of the 11th Annual European Symposium on Algorithms (ESA 2003).*

Calinescu, G., & Wan, P.-J. (2006). Range assignment for biconnectivity and *k*-edge connectivity in wireless ad hoc networks. *ACM/Springer Mobile Networks and Applications (MONET)*, *11*(2), 121-128.

Cardei, M., & Wu, J. (2006). Energy-efficient coverage problems in wireless ad hoc sensor networks. *Computer Communications Journal*, *29*(4), 413-420. Elsevier.

Chen, W., & Huang, N. (1989). The strongly connecting problem on multi-hop packet radio networks. *IEEE Transactions on Communications*, *37*(3), 293-295.

Chen, Y., & Son, S. H. (2005). A fault tolerant topology control in wireless sensor networks. In *Proceedings of the 3rd ACS/IEEE International Conference on Computer Systems and Applications.*

Cheriyan, J., Vempala, S., & Vetta, A. (2002). Approximation algorithms for minimum-cost k-vertex connected subgraphs. In *Proceedings of the 34th Annual ACM Symposium on Theory of Computing.*

Clementi, A., Huiban, G., Penna, P., Rossi, G., & Verhoeven, Y.C. (2002). Some recent theoretical advances and open questions on energy consumption in ad-hoc wireless networks. In *Proceedings of the 3rd Workshop on Approximation and Randomization Algorithms in Communication Networks.*

Clementi, A., Penna, P., & Silvestri, R. (2000). The power range assignment problem in radio networks on the plane. In *Proceedings of the XVII Symposium on Theoretical Aspects of Computer Science (STACS'00)* (LNCS 1770, pp. 651-660).

Dai, F., & Wu, J. (2005). On constructing k-connected k-dominating set in wireless networks. In *Proceedings of the International Parallel and Distributed Processing Symposium.*

Das, B., & Bharghavan, V. (1997). Routing in ad-hoc networks using minimum connected dominating sets. In *Proceedings of the IEEE International Conference on Communications (ICC'97).*

Gupta, P., & Kumar, P. R. (1998). Critical power for asymptotic connectivity in wireless networks. In W. M. McEneaney, G. Yin, & Q. Zhang (Eds.), *Stochastic analysis, control, optimization and applications: A volume in honor of W.H. Fleming.* Boston: Birkhäuser.

Gupta, G., & Younis, M. (2003). Fault-tolerant clustering of wireless sensor networks. In *Proceedings of the IEEE Wireless Communications and Networking 2003.*

Hajiaghayi, M., Immorlica, N., & Mirrokni, V. S. (2003). Power optimization in fault-tolerant topology control algorithms for wireless multi-hop networks. In *Proceedings of the 9th Annual International Conference on Mobile Computing and Networking.*

Hao, B., Tang, J., & Xue, G. (2004). Fault-tolerant relay node placement in wireless sensor networks: formulation and approximation. In *Proceedings of the IEEE HPRS 2004.*

Karp, B., & Kung, H. (2000). GPSR: Greedy perimeter stateless routing for wireless networks. In *Proceedings of the ACM International Conference on Mobile Computing and Networking (MobiCom).*

Khuller, S., & Vishkin, U. (1994). Biconnectivity approximations and graph carvings. *Journal of ACM, 41*, 214-235.

Kirousis, L. M., Kranakis, E., Krizanc, D., & Pelc, A. (2000). Power consumption in packet radio networks. *Theoretical Computer Science, 243*(1-2), 289-305.

Kuhn, F., Moscibroda, T., & Wattenhofer, R. (2006). Fault-tolerant clustering in ad hoc and sensor networks. In *Proceedings of the IEEE ICDCS 2006.*

Kumar, S., Lai, T. H., & Balogh, J. (2004). On k-coverage in a mostly sleeping sensor network. In *Proceedings of the ACM MobiCom 2004.*

Li, L., Halpern, J. Y., Bahl, P., Wang, Y.-M., & Wattenhofer, R. (2001). Analysis of a cone-based distributed topology control algorithms for wireless multi-hop networks. In *Proceedings of the ACM Symposium on Principle of Distributed Computing (PODC).*

Li, N., & Hou, J. C. (2004). FLSS: A fault-tolerant topology control algorithm for wireless networks. In *Proceedings of the ACM MOBICOM 2004.*

Li, N., Hou, J. C., & Sha, L. (2003). Design and analysis of a mst-based topology control algorithm. In *Proceedings of the IEEE INFOCOM 2003.*

Li, X.-Y., Wan, P.-J., & Wang, Y. (2001). Power efficient and sparse spanner for wireless ad hoc networks. In *Proceedings of the IEEE International Conference on Computer Communications and Networks (ICCCN '01).*

Li, X.-Y., Wan, P.-J., Wang, Y., & Frieder, O. (2002). Sparse power efficient topology for wireless networks. In *Proceedings of the 35th IEEE Hawaii International Conference on System Sciences (HICSS-35).*

Li, X.-Y., Wang, Y., Wan, P.-J., & Yi, C.-W. (2003). Fault tolerant deployment and topology control for wireless ad hoc networks. In *Proceedings of the 4th*

ACM International Symposium on Mobile Ad Hoc Networking and Computing (MobiHoc '03).

Liu, H., Wan, P.-J., & Jia, X. (2005). Fault-tolerant relay node placement in wireless sensor networks. In *Proceedings of the COCOON 2005* (LNCS 3595, pp. 230-239).

Lloyd, L., Liu, R., Marathe, M. V., Ramanathan, R., & Ravi, S. S. (2002). Algorithmic aspects of topology control problems for ad hoc networks. In *Proceedings of the 3rd ACM International Symposium on Mobile Ad Hoc Networking and Computing (MobiHoc '02).*

Penrose, M. (1997). The longest edge of the random minimal spanning tree. *Annals of Applied Probability, 7,* 340-361.

Penrose, M. (1999). On k-connectivity for a geometric random graph. *Random Structures and Algorithms, 15,* 145-164.

Ramanathan, R., & Rosales-Hain, R. (2000). Topology control of multi-hop wireless networks using transmit power adjustment. In *Proceedings of the IEEE INFOCOM.*

Sanchez, M., Manzoni, P., & Haas, Z. (1999). Determination of critical transmission range in ad-hoc networks. In *Proceedings of the Multiaccess, Mobility and Teletraffic for Wireless Communications (MMT '99).*

Seddigh, M., Gonzalez, J. S., & Stojmenovic, I. (2002). RNG and internal node based broadcasting algorithms for wireless one-to-one networks. *ACM Mobile Computing and Communications Review, 5*(2), 37-44.

Stratil, H. (2005). Fault tolerant topology control with unreliable failure detectors. In *Proceedings of the 17th International Conference on Parallel and Distributed Computing and Systems.*

Thallner, B., & Moser, H. (2005). Topology control for fault-tolerant communication in highly dynamic wireless networks. In *Proceedings of the Third International Workshop on Intelligent Solutions in Embedded Systems.*

Wan, P.-J., Alzoubi, K. M., & Frieder, O. (2002). Distributed construction of connected dominating set in wireless ad hoc networks. In *Proceedings of IEEE INFOCOM 2002.*

Wan, P.-J., & Yi, C.-W. (2004). Asymptotic critical transmission radius and critical neighbor number for k-connectivity in wireless ad hoc networks. In *Proceedings of the 5th ACM International Symposium on Mobile Ad hoc Networking and Computing (MobiHoc '04).*

Wan, P.-J., & Yi, C.-W. (2005). Asymptotic critical transmission ranges for connectivity in wireless ad hoc networks with Bernoulli nodes. In *Proceedings of 2005 IEEE Wireless Communications and Networking Conference (WCNC 2005),* New Orleans.

Wang, Y., Li, X.-Y., & Zhang, Q. (2007). Efficient self protection algorithms for Static wireless sensor networks. In *Proceedings of the 50th IEEE Global Telecommunications Conference (Globecom 2007).* Extended version to appear in IEEE Transaction on Parallel and Distributed Systems (TPDS), 2008.

Wang, Y., Wang, W., & Li, X.-Y. (2006). Efficient distributed low-cost weighted backbone formation for wireless ad hoc networks. *IEEE Transaction on Parallel and Distributed Systems (TPDS), 17*(7), 681-693.

Wang, D., Zhang, Q., & Liu, J. (2006). Self-protection for wireless sensor networks. In *Proceedings of the IEEE ICDCS 2006.*

Wattenhofer, R., Li, L., Bahl, P., & Wang, Y.-M. (2001). Distributed topology control for wireless multihop ad-hoc networks. In *Proceedings of the IEEE INFOCOM 2001.*

Wu, J., & Li, H. (1999). On calculating connected dominating set for efficient routing in ad hoc wireless networks. In *Proceedings of the Third International Workshop on Discrete Algorithms and Methods for Mobile Computing and Communications.*

Wu, J., & Li, H. (2000). Domination and its applications in ad hoc wireless networks with unidirectional links. In *Proceedings of the International Conference on Parallel Processing.*

Xing, G., Wang, X., Zhang, Y., Lu, C., Pless, R., & Gill, C. (2005). Integrated coverage and connectivity configuration for energy conservation in sensor networks. *ACM Transactions on Sensor Networks, 1*(1), 36-72.

Yang, S., Dai, F., Cardei, M., & Wu J. (2006). On connected multiple point coverage in wireless sensor networks. *Journal of Wireless Information Networks, 13*(4), 289-301.

Zhang, H., & Hou, J. (2005). Maintaining sensing coverage and connectivity in large sensor networks. *Ad Hoc and Sensor Wireless Networks: An International Journal, 1*(1-2), 89-123.

Zhou, Z., Das, S., & Gupta, H. (2004). Connected k-coverage problem in sensor networks. In *Proceedings of the International Conference on Computer Communications and Networks.*

Zhou, Z., Das, S.R., & Gupta, H. (2005). Fault tolerant connected sensor cover with variable sensing and transmission ranges. In *Proceedings of the IEEE MASS 2005.*

KEY TERMS

Fault Tolerance: If a network is fault tolerant or *k*-fault tolerant it means the network can survive under single or *k* node/link failures simultaneously.

K-Connectivity: If a network (graph) has *k-connectivity,* it means the it is *k*-connected, that is, given any pair of wireless devices (nodes), there are at least *k* disjoint paths to connect them.

K-Coverage: A sensor network achieves *k*-coverage if every location is covered by at least *k* different sensor nodes, *that is,* every location is within the sensing range of at least *k* different sensor nodes.

Power Control: Controls the network topology by adjusting the wireless device's transmission range to minimum energy consumption while maintaining a topology that is connected or has certain desired properties.

Self-Protection: A sensor network is *p*-self-protected, if at any moment, for any wireless sensor (active or nonactive), there are at least *p* active sensors that can monitor it.

Topology Control: Let each wireless device locally select certain neighbors for communication, while maintaining a topology that can support energy efficient routing and improve the overall network performance.

Virtual Backbone: A connected backbone formed by a subset of wireless nodes selected to perform communication tasks for the other nodes and the whole network.

Section IV
Security in Wireless PAN/LAN/MAN Networks

Chapter XLI
Evaluating Security Mechanisms in Different Protocol Layers for Bluetooth Connections

Georgios Kambourakis
University of the Aegean, Greece

Angelos Rouskas
University of the Aegean, Greece

Stefanos Gritzalis
University of the Aegean, Greece

ABSTRACT

Security is always an important factor in wireless connections. As with all other existing radio technologies, the Bluetooth standard is often cited to suffer from various vulnerabilities and security inefficiencies while attempting to optimize the trade-off between performance and complementary services including security. On the other hand, security protocols like IP secure (IPsec) and secure shell (SSH) provide strong, flexible, low cost, and easy to implement solutions for exchanging data over insecure communication links. However, the employment of such robust security mechanisms in wireless realms enjoins additional research efforts due to several limitations of the radio-based connections, for example, link bandwidth and unreliability. This chapter will evaluate several Bluetooth personal area network (PAN) parameters, including absolute transfer times, link capacity, throughput, and goodput. Experiments shall employ both Bluetooth native security mechanisms, as well as the two aforementioned protocols. Through a plethora of scenarios utilizing both laptops and palmtops, we offer a comprehensive in-depth comparative analysis of each of the aforementioned security mechanisms when deployed over Bluetooth communication links.

INTRODUCTION

Without doubt, the Bluetooth specification (IEEE 802.15) (Bluetooth SIG, 2003; IEEE, 2002) is gradually becoming the de-facto standard for replacing short range wired communications using radio technology. According to estimations, devices incorporating Bluetooth are predicted to quadruple in number between now and 2008, from under 100 million to about 440 million. Bluetooth enabled devices are used in several different environments and cover a wide range of applications. For instance, for mobile applications, the device periodically connects to the network to download music, to transfer files, or to synchronize with one's desktop on calendar and other files. Consequently, the safety and security of these applications, for instance, the security of the private information stored on the devices, becomes a major issue. By attacking actively or passively the communication link, aggressors could obtain personal and also important business data. However, security features (Gehrmann, Persson, & Smeets, 2004) must be carefully considered and analyzed in order to decide whether Bluetooth technology indeed provides the right answer for any particular task or application.

The Bluetooth standard has been long criticized for various vulnerabilities and security inefficiencies, as its designers are trying to balance between performance and complementary services including security. So far, both the Bluetooth Special Interest Group (SIG) (Bluetooth SIG, 2003) and several researchers have made significant contributions on Bluetooth security aspects, discovering numerous vulnerabilities and potential weaknesses and proposing solutions (Adam, 2003; Gehrmann, & Nyberg, 2002; Jacobson & Wetzel, 2001; Persson & Manivannan, 2003; Shaked & Wool, 2005). For example, the Bluetooth pairing procedure has been anticipated to be weak under certain circumstances. Moreover, other categories of threats, either active or passive, have also been investigated, including ad hoc security issues, malicious software like "Cabir," war-nibbling, and so forth.

An obvious choice for any Bluetooth application would be to use Bluetooth encryption provided at link layer. Virtually all Bluetooth devices support this feature, and it is, in most cases, considered to be adequately secure. However, this may not be applicable for all deployment scenarios. In order to establish a secure channel with another Bluetooth device, a preshared secret called PIN is required. A symmetric key is generated from this PIN. On customer devices this PIN typically consists of four or five digits. Supposing a whole piconet network would utilize this PIN to encrypt its communication, anyone acquiring this PIN could theoretically decrypt all communication. On top of that, in applications like VoIP that mandate IP connectivity to access points (APs), the encryption would end at the AP, which means that the AP, or any host that can manipulate the communication between the Mobile Device and the other end, can expose the data (see Figure 1). Thus, it is obvious that Bluetooth encryption is not well suited for all applications which may exploit Bluetooth connections.

Under these circumstances and for certain classes of security sensitive applications deployed in Bluetooth PAN networks, the investigation of complementary and advanced security protocols apart from Bluetooth's native security mechanisms, even if deployed as an interim countermeasure, is an interesting research issue. On the other hand, as Bluetooth wireless technology is targeting devices with particular needs and constraints (e.g., processing power and battery consumption) the trade-offs between security services and performance must be carefully considered. Furthermore, considering that radio links in general suffer from limited bandwidth and are unreliable by nature, performance issues must be thoroughly investigated to make a decision whether certain security protocols and their mechanisms are advantageous over Bluetooth connections, delivering robust and agile security services within tolerable service response times.

Figure 1. Sample scenario that mandates upper layer security

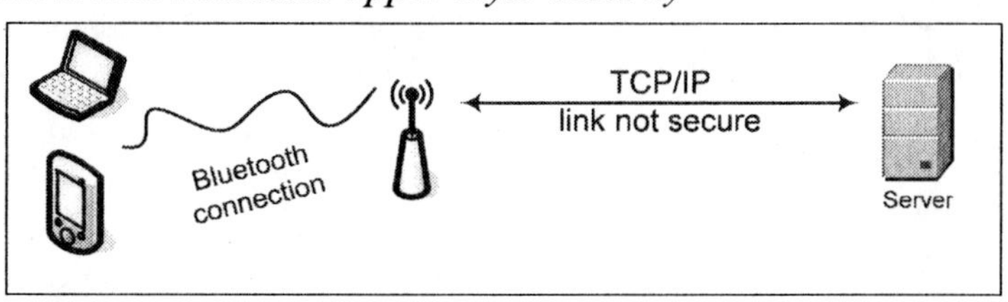

During the last few years, several researchers have examined various Bluetooth security parameters and some of them do explore performance parameters (e.g., Chakraborty, 2000; De Morais Cordeiro, Sadok, & Agrawal, 2001; Francia, Kilaru, Le Phuong, & Vashi, 2004; Golmie & Rebala, 2003; Howitt, 2002; Karnik & Kumar, 2000; Kitsos et al., 2003; Lim et al., 2001; Miorandi, Caimi, & Zanella, 2003; Wang, Arumugam, & Krishna, 2002). However, to the best of our knowledge, none of these works focus on performance evaluation comparing Bluetooth's native security mechanisms with well-respected, strong security protocols like IPsec and SSH.

The chapter will focus on the performance of existing protocols and mechanisms rather than on security itself, estimating the performance of both the built-in Bluetooth security mechanisms, namely security modes, and two other standard security protocols operating at different layers of the TCP/IP protocol suite, namely SSH and IPsec. Protocols like SSH and IPsec provide robust, flexible, costless, and easy to implement solutions for exchanging data over insecure communication links. However, although their deployment is a well established and accustomed practice in the wireline world, more research effort is needed for wireless links, due to the several aforementioned limitations. Depending on the scenario involved, the user may utilize SSH or IPsec security services, either individually or in combination with Bluetooth security modes, allowing applications to communicate securely, constructing a secure tunnel. Thus, in a sense, the whole procedure can also be seen as the deployment of small VPNs in Bluetooth PANs. Note however, that the efficiency of the SSH and IPSec depends mainly on the performance of the used end-system. On the contrary, Bluetooth security native modes utilize the hardware encryption of the Bluetooth chip, thus performance depends heavily on the chip per se. This situation will allow us to make several observations about different layer security mechanisms when deployed over dissimilar user devices.

Specifically, the chapter will evaluate several personal area network (PAN) parameters, including transfer times, link capacity, and throughput. Experiments shall employ both Bluetooth native security mechanisms as well as the two aforementioned protocols. Through a plethora of scenarios, utilizing both laptops and palmtops, we intend to offer a comprehensive in-depth comparative analysis of each of the aforementioned security mechanisms when deployed over Bluetooth communication links.

The rest of the chapter is structured as follows. The next section gives an overview of our experimental test-bed related parameters and procedures, while the third section presents the derived performance measurement results. The forth section offers an analytical discussion over the conducted results. The chapter finishes with some concluding thoughts and future directions of this work.

EXPERIMENTAL FRAMEWORK DESCRIPTION

The experimental topology consists of two pairs of machines. The first pair of Bluetooth devices employs a laptop and a palmtop machine, while the other consists of two similar laptop machines. The members of each pair are located at 10 meters apart and connected via Bluetooth adapters (or built in Bluetooth chip), thus forming a small two-member wireless PAN (WPAN) or piconet. The main components' characteristics, both software and hardware, are presented in Table 1. To estimate the performance of the Bluetooth network, the data were transmitted from one network node (server) to the other (client). Hence, in order to record the incoming and outcoming packets between the corresponding network entities and to calculate the network performance parameters we utilized on the server side the well known network analyzer "ethereal" (www.ethereal.com), version 0.10.12, which in turn uses the "tcpdump" tool. In addition, for the Linux environment, we employed the BlueZ official Linux Bluetooth protocol stack (www.bluez.org), which provides support for the core Bluetooth layers and protocols.

Bluetooth supports three different security modes called security modes I, II, and III, but in our tests we decided to use only security modes I and III. Security mode I offers no real security as authentication and confidentiality services are disabled. On the other hand, security mode II provides security services after the connection between the two devices has been established and only if a given application has requested them. Thus, the security services in mode II depend on the application running. The last security mode is the most powerful among the three modes because it mandates both authentication and confidentiality built-in mechanisms independently of the application running. These mechanisms are referred to as Bluetooth baseband security procedures, where the baseband layer deals with the SAFER+ algorithms (Massey, Khachatrian, & Kuregian, 1998). As implied, one of the terminals was acting as a client and the other one as the server. Therefore, the server should require security and the client should respond accordingly.

For IPsec, the engaged machines must have the same security policies in order to communicate securely. So, we configured Linux to use MD5 and SHA1 algorithms for data integrity and DES and 3DES algorithms for confidentiality in both machines by installing IPsec-tools (http://ipsec-tools.sourceforge.net/) and Openswan (www.openswan.org) as well. For SSH secured communication we used OpenSSH. In fact, many open-source projects exist. In addition to FreeSWAN and openswan which both enable IPsec in the Linux kernel, openvpn (http://openvpn.net/) can be used to create TLS-encrypted point-to-point connections. For SSH confidentiality services we chose four algorithms to test namely, 3DES, AES, Arcfour, and Blowfish. Finally, for both IPsec and SSH we employed only symmetric cryptography and manual keying procedures for the authentication of parties considering the fact that usually Bluetooth piconets are formed ad hoc and their users do not hold public key certificates.

PERFORMANCE MEASURES

As mentioned before, the experimental procedure consists of three main parts: evaluation of Bluetooth built-in security modes I (no security), and III

Table 1. Hardware and software characteristics of the engaged machines

First pair	Laptop Server	
	Processor	Intel Celeron M. – 1.4 GHz
	RAM	256 MB
	Operating System	SUSE Linux Ver. 10.0
	Bluetooth Adapter	Trust Bluetooth adapter Class 1
	Palmtop Client	
	Model	HP iPAQ h5400
	Processor	400 MHz Intel XScale PXA250
	RAM	64 MB
	Operating System	Familiar PDA OS 0.8.4
	Bluetooth Adapter	Bluetooth 1.1 compliant
Second pair	Laptop client and server	
	Processor	Intel Celeron M. – 1.4 GHz
	RAM	256 Mbytes
	Operating System	SUSE Linux Ver. 10.0
	Bluetooth Adapter	Trust Bluetooth adapter Class 1

(strong security), and estimation of the performance of IPsec and SSH mechanisms over Bluetooth links. In all scenarios we gathered measurements for the subsequent network performance parameters: absolute file transfer time (TT), achieved transfer rate (ATR), and throughput (THR). All measurements took place at the server node because of its processing power.

- The *Transfer_Time* represents the actual duration of transfers during a transaction.
- The *Achieved_Transfer_Rate* represents the actual transfer rate achieved during a transaction. In an ideal scenario, a constant data rate should be maintained between the two communication end-points. However, due to various reasons, mainly related to the wireless medium nature, this parameter is changing over time. We should underline the fact that bytes_sent and bytes_received could also contain retransmitted bytes.
  Achieved_Transfer_Rate(Kbps) = ((bytes_sent + bytes_received) * 8) / TT
- Throughput represents the percentage of Achieved_Transfer_Rate over the practical maximum_transfer_rate of the link, which in our case is 723 Kbps:
  Throughput(%) = achieved_transfer_rate / max_transfer_rate * 100
- Finally, *Achieved_Transfer_Rate_Improvement* is a comparison metric that indicates the improvement of the Achieved_Transfer_Rate with respect to the Bluetooth mode I achieved transfer rate Achieved_Transfer_Rate_B_I and is calculated as:
  Achieved_Transfer_Rate_Improvement(%) = (ATR - ATR_B_I) / ATR_B_I * 100

A positive value implies that the performance (or channel throughput) has increased compared to the Bluetooth mode I achieved transfer rate, while a negative one means that the performance has

Figure 2. Average metric values for network parameters measured/Bluetooth Modes I and III

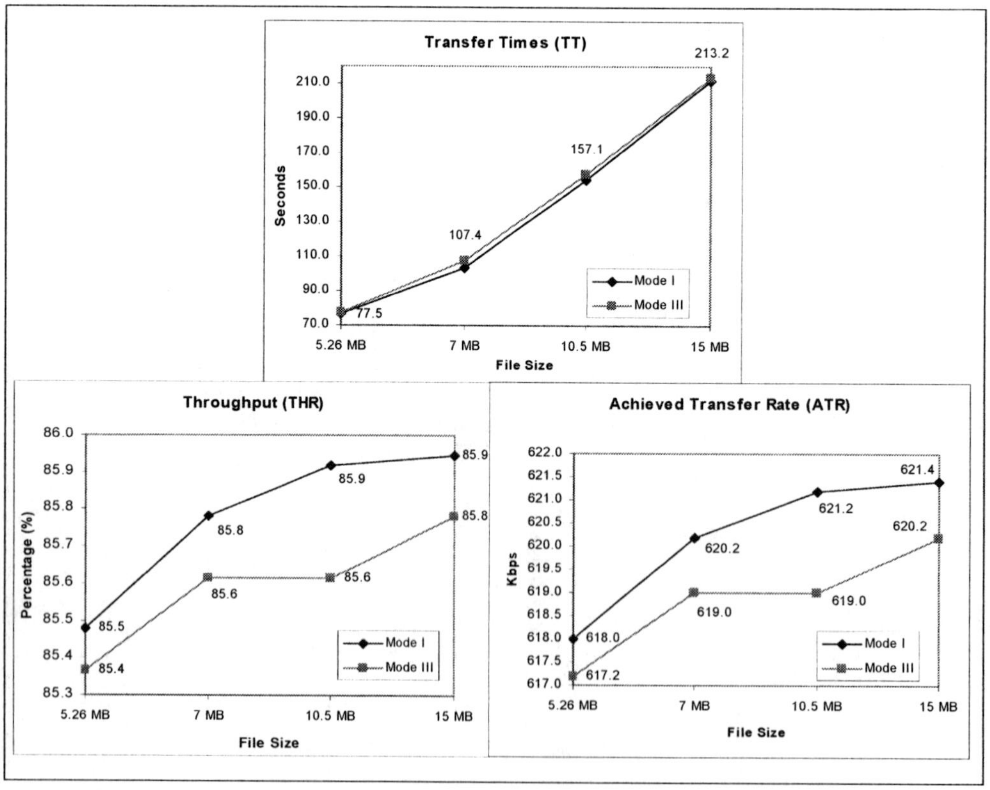

decreased. Measurements were gathered during repeated FTP file transfers, between the laptop server and the PDA client from the one hand and between the laptop client and server from the other. Each file was transferred twelve times and only average values were recorded. In all scenarios, the ping response times between client and server were varying among 19.7 and 21.8msecs. Due to space limitations, in the following first three subsections we present only the analytical results derived from the laptop server/PDA client, which is without doubt the most interesting one, while some indicative corresponding comparisons with the other laptop client–server pair is exhibited in the subsection titled "Comparison Between PDA and Laptop Clients."

Bluetooth Security Modes I and III Evaluation

Measurements for testing Bluetooth modes I and III were gathered by transferring four different files between each client–server pair. The files' sizes were 5.26, 7.0, 10.5, and 15 Mbytes, respectively. Figure 2 provides a graphical representation of these values comparing TT times achieved in the PDA client–laptop server piconet. As we can easily notice, the results are generally as expected, but there are some interesting points which need further analysis. At first, the TT metric is slightly higher for mode III, as well as the ATR is higher for mode I. This happens because mode III mandates authentication (handshake) at the beginning of each transaction. Keep in mind that the handshake time is included in TT too.

Moreover, encryption algorithms are applied during the transaction for mode III and as a result the overall transfer time is increased. We can also perceive that the larger the file size is, the longer the TT difference between mode I and mode III is expected to be. This situation is also depicted in the respective plot of Figure 2. In general, these measurements advocate that mode I utilizes the network better than mode III. Because of the volatile nature of the wireless link, we also report standard deviation (SD) for the measured values in Table 2.

Secure Shell (SSH) Evaluation

Experimental procedures for the SSH mechanism (IETF, 2006; OpenSSH, 2006) consider the transfer of the same four files, as before, between the client and the server. Table 3 displays the average times of all metrics used, while Table 4 presents the corresponding standard deviation values.

As we can notice, SSH gives highly increased transfer times when compared to Bluetooth security modes. For instance, we can spot a difference of +12.6 seconds to +13.4 seconds for the smallest file depending on the cipher used. Moreover, it is more than obvious that all the ciphers used are more or less of the same performance. This is easily proven if we examine for example the achieved transfer rates in each case, which shown very slight differences.

Another interesting assumption that we can make is that as the size of the file increases, the achieved transfer rate and the throughput become bigger. This happens because of the procedure of the authentication which takes place during the ini-

Table 2. Standard deviation for all Bluetooth scenarios

	MODE I			MODE III		
File Size (MB)	TT (sec)	ATR (Kbps)	THR (%)	TT (sec)	ATR (Kbps)	THR (%)
5.26	0.5	2.6	0.4	0.1	1.3	0.2
7	0.1	0.9	0.1	0.5	3.2	0.4
10.5	0.4	1.6	0.2	0.1	0.5	0.1
15	0.2	0.5	0.1	0.6	2.2	0.3

Table 3. Average values for network parameters measured (SSH)

	5.26 MB			7 MB		
	TT (sec)	ATR (Kbps)	THR (%)	TT (sec)	ATR (Kbps)	THR (%)
3DES	90.1	526.4	72.8	116.9	555.6	76.9
AES128	90.2	525.6	72.7	116.9	556.2	76.9
Arcfour	90.5	523.8	72.5	117.3	554.2	76.6
Blowfish	90.5	523.6	72.4	117.6	552.8	76.4
	10.5 MB			15 MB		
3DES	163.0	581.8	80.5	221.3	603.2	83.4
AES128	162.9	582.4	80.5	221.3	603.6	83.5
Arcfour	163.1	581.6	80.5	221.6	602.4	83.3
Blowfish	162.8	582.6	80.6	222.1	601.2	83.1

Table 4. Standard deviation for all SSH scenarios

	5.26 MB			7 MB		
	TT (sec)	ATR (Kbps)	THR (%)	TT (sec)	ATR (Kbps)	THR (%)
3DES	0.4	2.1	0.3	0.4	2.1	0.3
AES128	0.9	5.5	0.7	0.4	1.9	0.2
Arcfour	0.1	0.4	0.1	0.2	1.1	0.2
Blowfish	0.6	3.8	0.5	1.0	4.9	0.7
	10.5 MB			15 MB		
3DES	1.0	3.3	0.5	0.8	2.5	0.3
AES128	1.0	3.9	0.5	0.9	2.3	0.3
Arcfour	0.5	1.9	0.3	0.6	1.7	0.2
Blowfish	0.6	1.9	0.3	0.7	1.9	0.3

tial SSH handshake. In any case it should be noted that the improvement in the achieved transfer rates always compared to Bluetooth security mode I and induced by SSH, are negative for any scenario. This means that Bluetooth's native mechanisms offer better bandwidth and network utilization at almost all cases examined. This remark is confirmed by the values given in Table 5.

IPsec Evaluation

The procedure for the IPsec protocol (Kent & Atkinson, 1998a, 1998b) considers once again the transfer of the same four files between the client and the server. IPsec uses two mechanisms (protocols) that may be used independently or jointly to secure the outcoming traffic, namely authentication header (AH) offering data origin, connectionless

Table 5. %ATR deterioration for SSH

Size	Bluetooth Mode I	3DES	AES128	RC4	Blowfish
5.26	618.0	-14.8	-15.0	-15.2	-15.3
7	620.2	-10.4	-10.4	-10.6	-10.9
10.5	621.2	-6.3	-6.2	-6.4	-11.0
15	621.4	-2.9	-2.9	-3.3	-3.3

Table 6. Average values for network parameters measured (IPsec)

	5.26 MB			7 MB		
	TT (sec)	ATR (Kbps)	THR (%)	TT (sec)	ATR (Kbps)	THR (%)
AH_MD5	72.8	683.4	94.5	100.0	682.8	94.4
AH_SHA1	72.8	683.2	94.5	99.9	683.0	94.5
ESP_DES_MD5	74.4	681.0	95.0	102.0	686.6	95.0
ESP_3DES_MD5	73.8	681.0	95.7	102.2	685.2	94.8
ESP_DES_SHA1	74.2	680.0	95.2	102.0	686.6	95.0
ESP_3DES_SHA1	74.2	681.0	95.2	101.8	688.2	95.2
	10.5 MB			15 MB		
AH_MD5	145.9	682.6	94.4	205.2	683.4	94.5
AH_SHA1	145.7	683.4	94.5	205.1	683.8	94.6
ESP_DES_MD5	148.6	688.2	95.2	208.9	688.8	95.3
ESP_3DES_MD5	148.6	687.8	95.1	209.1	688.0	95.2
ESP_DES_SHA1	148.5	688.4	95.2	209.2	688.0	95.2
ESP_3DES_SHA1	148.6	688.0	95.2	210.5	683.6	94.6

Table 7. Standard deviation of measurements of all IPsec scenarios

	5.26 MB			7 MB		
	TT (sec)	ATR (Kbps)	THR (%)	TT (sec)	ATR (Kbps)	THR (%)
AH_MD5	0.0	0.5	0.05	0.1	0.8	0.12
AH_SHA1	0.1	0.4	0.1	0.1	0.0	0.05
ESP_DES_MD5	0.1	0.4	0.1	0.3	2.1	0.28
ESP_3DES_MD5	0.5	4.5	0.6	1.3	8.6	1.19
ESP_DES_SHA1	0.0	0.4	0.06	0.6	3.7	0.53
ESP_3DES_SHA1	0.0	0.4	0.06	0.1	0.4	0.1
	10.5 MB			15 MB		
AH_MD5	0.1	0.5	0.06	0.2	0.5	0.08
AH_SHA1	0.2	0.9	0.1	0.1	0.4	0.03
ESP_DES_MD5	0.1	0.8	0.09	0.1	0.4	0.04
ESP_3DES_MD5	0.1	0.8	0.08	0.1	0.0	0.03
ESP_DES_SHA1	0.0	0.5	0.02	0.3	1.0	0.13
ESP_3DES_SHA1	0.1	0.7	0.06	2.4	7.6	1.05

data integrity, and optionally replay protection, and encapsulating security payload (ESP) offering confidentiality and protection against traffic analysis. In our scenarios we utilized both mechanisms, using the MD5 and SHA1 algorithms for integrity and DES and 3DES to support confidentiality services. Note however that MD5 is not considered secure anymore and is reported here for the sake of completeness. In total, we deployed six scenarios as shown in Table 6.

First and foremost, all network metrics for IPsec are remarkably concentrated. Standard deviation values rendered in Table 7 confirm this remark. Surprisingly, IPsec gives better transfer times for all file sizes when compared to Bluetooth and SSH. This is also confirmed by %ATR improvement for IPsec shown in Table 8. In particular, all IPsec times are very close to those of Bluetooth's mode I, while at the same time are considerably better than SSH's. Note, that IPsec renders 210.5 seconds as the highest time duration for transferring the biggest file, while correspondingly SSH gives 222.1 seconds, mode III produces 213.2 seconds, and mode I 211.6 seconds. This is partially due to substantially increased (and highly stabilized) bandwidth that IPsec generates. The aforementioned observations are also confirmed by the fact that during IPsec measurements we had a very low rate of packet loss reported by the Ethereal utility. It is important to note that the throughput was better when using ESP. On the contrary, when using AH, the throughput for transferring the files was lower. This can be explained by the fact that authentication is applied in AH.

Comparison Between PDA and Laptop Clients

Considering the second experimental pair, which employs laptops for both the server and the client (see Table 1), TT times were better for all the corresponding scenarios, namely Bluetooth native security modes, SSH and IPsec. For instance, Bluetooth modes show a slight TT improvement ranging from 1 to 3 seconds depending on the file size. Specifically, TT for the 7 MB file was 102.8 and 106.5 for Bluetooth mode I and III, respectively. Approximately the same situation is reported for SSH and IPsec as depicted in Figure 3. This is expected as the laptop client incorporates a faster CPU and thus gains more in cryptographic operations that SSH and IPsec mandate. The same remark is applied for the other two network performance parameters, throughput and ATR. As in the PDA client case, IPsec continues to perform better under all circumstances for the laptop client due to its throughput optimization. However, IPsec

Table 8. % ATR improvement for IPsec

		AH_		ESP_DES_		ESP_3DES_	
File Size	Bluetooth Mode_I	MD5	SHA1	MD5	SHA1	MD5	SHA1
5.26	618.0	10.6	10.6	11.1	11.4	11.9	11.4
7	620.2	10.1	10.1	10.7	10.7	11.5	11.0
10.5	621.2	9.9	10.0	10.8	10.8	10.7	10.8
15	621.4	10.0	10.0	10.8	10.7	10.7	10.0

Figure 3. Comparison of network transfer times between Laptop and PDA clients

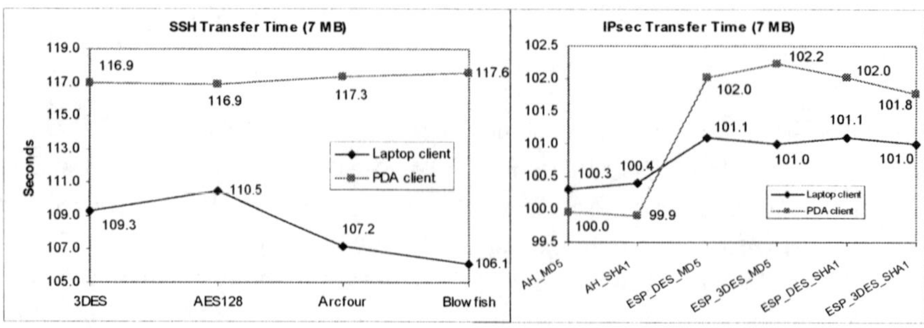

TT times remain very close to those of Bluetooth security modes. The same situation is confirmed by the minimum standard deviation values that characterize the IPsec case. Also in this case, SSH gives the worst performance compared with IPsec and Bluetooth native security modes.

COMMENTS ON THE RESULTS

This section provides a comparative view of the conducted results. Also, we attempt to provide a better explanation of the experiment outcomes. But before that we must shortly discuss important characteristics of Bluetooth connections that may affect the performance of the connection. Bluetooth employs frequency hopping spread spectrum (FHSS) to avoid interference. There are 79-23 in some countries-hopping frequencies, each having a bandwidth of 1MHz. Frequency hopping is assisted with fast automatic repeat request (ARQ), cyclic redundancy check (CRC), and forward error correction (FEC) to achieve high reliability on the wireless links. All the data/control packet transmissions are synchronized by the master. Slave units can only send in the slave-to-master slot after being addressed in the preceding master-to-slave slot, with each slot lasting 625 microseconds.

For real-time data such as video, synchronous connection oriented (SCO) links are used, while for data transmission, asynchronous connectionless link (ACL) links are employed. There are several ACL packet types, differing in packet length and whether they are FEC coded or not. The FEC coding scheme used in ACL DM mode is a shortened Hamming code, where each block of 10 information bits is encoded into a 15 bit codeword, and is capable of correcting single bit error in each block. Table 9 shows the different ACL packet types and their properties. The values in the table are theoretical without packet overhead. For example, over an ACL link using DH5, one can send about 300 to 320 kbit/s of UDP user data, while the theoretical limit is 433.9 kbit/s.

This means that in order to overcome the effect of low and varying link quality on throughput, the selection of the optimal link layer packet size, under estimated channel conditions, is crucial. Indeed some research work (Chen, Kapoor, Sanadidi, & Gerla, 2004) points this out by evaluating the "optimal" link layer packet size based on the current bit error rate of the channel. Moreover, in regions that Wi-Fi networks coexist with Bluetooth and because Wi-Fi and Bluetooth utilize spectrum in different ways, they can cause considerable interference between each other (depending on the relative location of the 802.11b and Bluetooth devices) (Yip & Kwok, 2004). By transmitting at the highest power level, Bluetooth class 1 devices would create more interference than Bluetooth's class 2 and class 3 devices, which transmit at lower power levels. Furthermore, because each Bluetooth PAN will occupy the entire ISM band, two or more coexisting Bluetooth PANs will occasionally collide, possibly causing loss of data packets. Of course, apart from implementation issues (e.g., protocol stacks), the aforementioned parameters are closely related and can affect real Bluetooth connections and the results gathered in this chapter. For instance, all experiments were conducted inside the coverage area of the University's hot-spot.

Table 9. Packet types for Bluetooth ACL Connections (theoretical values)

Mode	FEC	Packet (bytes)	Size (kbps)	Symmetric (kbps)	Asymmetric (kbps)
DM1	2/3	0-17	108.8	108.8	108.8
DM3	2/3	0-121	258.1	387.2	54.4
DM5	2/3	0-227	286.7	477.8	36.3
DH1	no	0-27	172.8	172.8	172.8
DH3	no	0-183	390.4	585.6	86.4
DH5	no	0-339	433.9	723.2	57.6

In the following, we present comparative graphs only for two of the three network parameters, transfer times, and throughput for the PDA client. As already noted in "Comparison between PDA and Laptop Clients," the laptop client scenario results are directly comparable with those of the PDA client and thus do not contribute further to this discussion. Consequently, Figure 4 illustrates a comparison of the transfer times for six selected scenarios in total. We easily spot that all times, especially for file sizes smaller than 10.5 MB, seem to be highly concentrated. This means that (excluding SSH ones) we have marginal differences between the performances' of the conducted scenarios. But, the bigger the size gets, the difference tends to slightly decrease. Apart from the fact that all tests have the Bluetooth link parameter in common, this can be explained by the fact that Bluetooth modes and IPsec utilize the network better.

On the downside, SSH does not always provide peak network performance because it traditionally has been more focused on providing security. In a nutshell, SSHv2 introduced an additional form of flow control that requires the receiver to ACK each packet before more packets can be sent. Most implementations seem to use packet sizes of 16K or occasionally 32K, with some going as low as 4K. This means that no matter how fast the link, every for example, at 16K the transmission stops for one round trip time awaiting the other side to send its ACK (referred to as a window adjust in SSHv2 terminology). In addition to the protocol-level handbrake, the SSH file transfer protocol (SFTP) that runs on top of SSH contains its own handbrake. This protocol recommends that reading and writing is limited to less than 32K of data, even though it is running over the reliable SSH transport which in turn runs over the reliable TCP/IP transport. One common implementation limits SFTP packets to 4K bytes, resulting in a mere 4% link utilization in the previously-presented scenario.

Figure 4. Comparison of network transfer times for six different scenarios (PDA client)

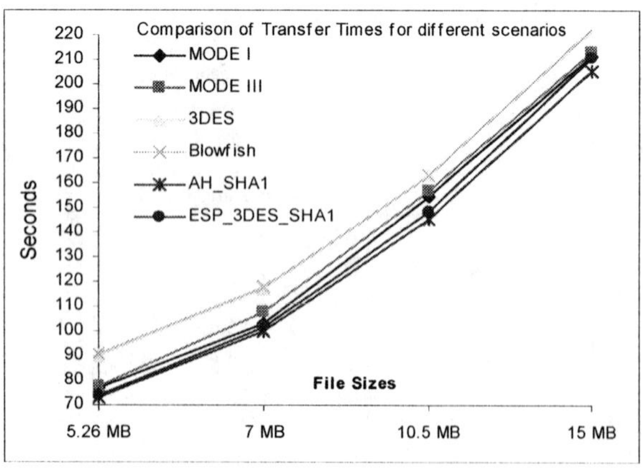

Finally, Figure 5 depicts a comparison of the achieved throughput for the specific six scenarios. This plot gives a clearer idea about the achieved network performance. In short, IPsec scenarios visibly have the best performance by far followed closely by the two Bluetooth's security modes. Moreover, we can make a very important observation about the SSH's performance. It is obvious that SSH's throughput increases as the file's size increases. This happens because of the handshaking phase which takes place during the initialization of each transaction. So, as the size of the transferred file increases, the impact of handshaking decreases and thus we notice an increase in the throughput. We should also report that the throughput of the other two scenarios remains more or less stable for all the file sizes we utilized. Another important issue is that during the experiments we observed a significant rate of packet loss for both Bluetooth security modes and SSH scenarios affecting their overall performance. Certainly, the main reason for this is the volatile nature of the wireless connection itself.

Additionally, it is well known that the addition of an IPSec header may cause IP fragmentation. However, the main concern in IPsec overhead is in the encryption, decryption, and authentication of the actual IPsec (ESP and/or AH) packets. Tunnel setup and rekeying occur much less frequently than packet processing and, except in highly unusual circumstances, their overheads are not worth worrying about. According to some other works (e.g., FreeSwan, 2002) utilizing low-end machines, a 60 MHz Pentium running a host-to-host tunnel to another machine shows an FTP throughput of slightly over 5 Mbit/s either way. Thereafter, we can conclude that in our case the IPsec mechanisms running on "relatively" low-end processors is not really a bottleneck. The overall performance is rather affected most by the quality of the Bluetooth link itself, meaning that due to better utilization of the link and possibly due to optimal ACL scheme and lower packet drop rate, IPsec performs slightly better than native Bluetooth modes do.

In Figure 6, we present some indicative ethereal screens that attest why in practice IPsec performs better from the other two in terms of the additional

Figure 5. Comparison of network throughput for six different scenarios (PDA client)

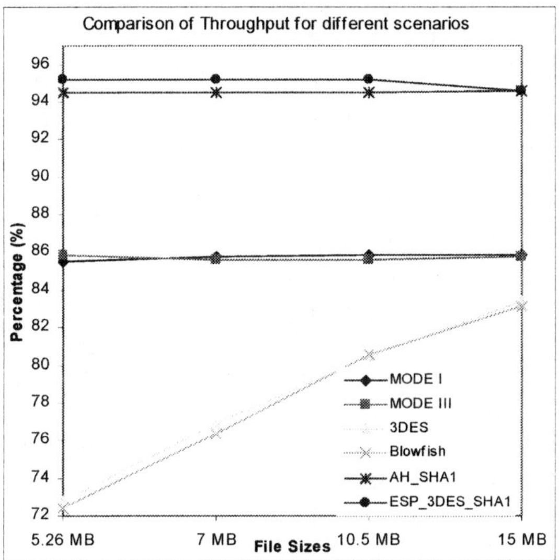

protocol overhead induced. These screens illustrate the overall network statistics for Bluetooth mode III and IPsec AH_MD5, respectively. The "Data" section corresponds to the overall percent of data that were sent from the server towards the PDA client for the 5.26 MB file. We observe that IPsec needs considerably lower percent of TCP data packets to complete the transaction (49.63%) than Bluetooth mode III which requires 66.24%. Note, that excluding ARP messages, the remaining percent corresponds to control information sent from the client to the server including ACKs, retransmissions, and so forth. Therefore, IPsec utilizes the link better, achieving higher performance.

Another important factor that may affect the conducted results is the operating system itself. For that we performed partial measurements using the Windows XP operating system in the laptop client, while keeping all the other test-bed parameters unchanged. Under this setting, we observed significantly lesser packet retransmissions and logged fairly better times. For example, for Bluetooth mode III and file size 10.5 MB we got an average transfer time of 150 seconds, namely 5 seconds better than Linux. One can presume that the Bluetooth stack is better implemented in Windows than in Linux or the Bluetooth adapters that we used perform better under Windows, perhaps due to their drivers' implementation. Nevertheless, a detailed analysis of this

Figure 6. Ethereal screens with protocol hierarchy statistics (PDA client)

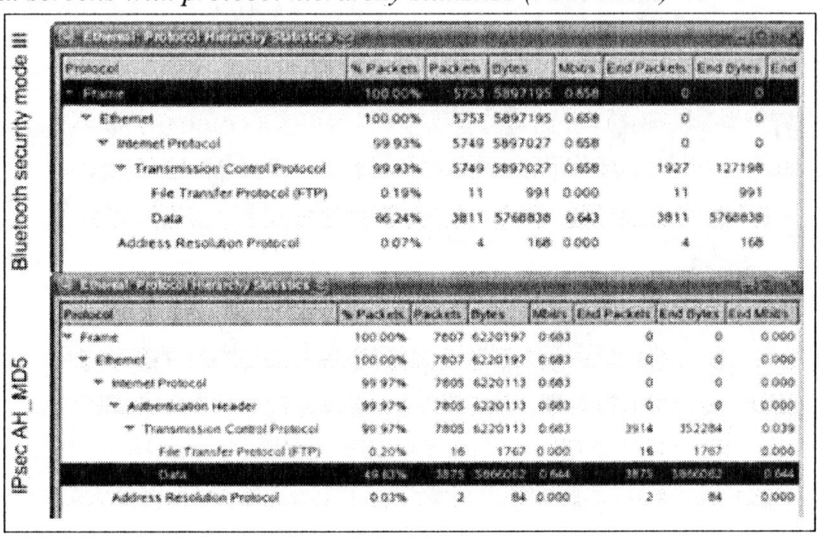

behavior between the two major operating systems is necessary but left for future work. An additional interesting research question is whether the recent updates of the Bluetooth specification to version 1.2 that have introduced significant changes in the Bluetooth protocol stack, including optional flow control, can affect the performance of the security mechanisms under investigation (Misic, Chan, & Misic, 2005). However, this is out of scope of the current chapter.

For the laptop client we also provide some indicative metrics concerning the physical memory consumption for the three categories of scenarios. More specifically, memory consumption for Bluetooth modes I and III was 552 KB, which is the "pand" daemon. For SSH we have an additional 1920 KB, thus in total 2472 KB ("sshd" and "pand" deamons), and finally for the IPsec case we have 4027 KB ("pluto" and "pand" daemons).

CONCLUSION AND FUTURE WORK

This chapter addresses performance issues for Bluetooth host-to-host connections. Three distinct categories of scenarios were used to test whether well respected security mechanisms of Internet and application layers of the TCP/IP suite are advantageous when deployed over Bluetooth PANs compared to Bluetooth native security modes. The results disclose that IPsec better utilizes the wireless link and thus provides radically improved transfer times when compared with SSH. Native Bluetooth modes service times are close to those of IPsec's thus significantly better from SSH ones. On the other hand, there is an important disadvantage which is the high amount of the memory resources IPsec consumes.

As future work we would like to expand this study, investigating the performance of asymmetric cryptography mechanisms, for example, public key certificates, and to support authentication services in the context of such protocols that promote automatic keying. Another direction is to detect how much energy is required for this sort of secure connections, as mobile devices can not afford batteries with unlimited capacity.

ACKNOWLEDGMENT

We would like to thank Mr. Alexis Andreadis and Mr. Paganos Charalampos for helping us with the network measurements.

REFERENCES

Adam, L. (2003). *Serious flaws in Bluetooth security lead to disclosure of personal data.* Retrieved October 14, 2007, from http://www.bluestumbler.org

Bluetooth SIG. (2003, November 1). *Specification of the Bluetooth system: Architecture & technology overview* (Version 1.2). Retrieved October 14, 2007, from http://www.bluetooth.com

Chen, L., Kapoor, R., Sanadidi, M. Y., & Gerla, M. (2004). Enhancing Bluetooth TCP throughput via link layer packet adaptation. In *Proceedings of the IEEE ICC '04* (Vol.7, pp. 4012-4016).

De Morais Cordeiro, C., Sadok, D., & Agrawal, D. P. (2001). Modeling and evaluation of Bluetooth MAC protocol. In *Proceedings of Tenth International Conference on Computer Communications and Networks* (pp. 518-522).

Francia, G., Kilaru, A., Le Phuong, & Vashi, M. (2004). An empirical study of Bluetooth performance. In *Proceedings of the 2nd Annual Conference on Mid-South College Computing, ACM International Conference Proceeding Series* (Vol. 61, pp. 81-93).

FreeSwan. (2002). *Performance of FreeSwan.* Retrieved October 14, 2007, from http://www.freeswan.org/freeswan_trees/ freeswan-1.95/doc/performance.html

Gehrmann, C., & Nyberg, K. (2002). *Enhancements to Bluetooth baseband security.* Ericsson Mobile Communcations AB, Ericsson Research.

Gehrmann, C., Persson, J., & Smeets, B. (2004). *Bluetooth security.* Artech House Publishers.

Golmie, N., & Rebala, O. (2003). Techniques to improve the performance of TCP in a mixed Bluetooth and WLAN environment. In *Proceedings of IEEE International Conference on Communications, ICC*, Anchorage, AK, (pp. 1181-1185).

Howitt, I. (2002). Bluetooth performance in the presence of 802.11b WLAN. *IEEE Transactions on Vehicular Technology, 51*(6), 1640-1651.

IEEE. (2002). Wireless PAN medium access control MAC and physical layer PHY specification. *IEEE standard 802.15.* New York: IEEE. Retrieved October 14, 2007, from http://www.ieee802.org/15/

IETF. (2006). *IETF secure shell (secsh) working group.* Retrieved October 14, 2007, from http://tools.ietf.org/wg/secsh/

Jacobson, M., & Wetzel, S. (2001). Security weaknesses in Bluetooth. In *Proceedings of the Conference on Topics in Cryptology: The Cryptographer's track at RSA* (LNCS 2020, pp. 176-191).

Karnik, A., & Kumar, A., (2000). Performance analysis of the Bluetooth physical layer. In *Proceedings of IEEE International Conference on Personal Wireless Communications* (pp. 70-74).

Kent, S., & Atkinson, R. (1998a). *IP authentication header* (AH) (IETF RFC 2402).

Kent, S., & Atkinson, R. (1998b). *IP encapsulating security payload* (ESP) (IETF RFC 2406).

Massey, J., Khachatrian, G., & Kuregian, M. (1998). Nomination of SAFER+ as candidate algorithm for the advanced encryption standard (AES). In *Proceedings of the 1st Advanced Encryption Standard Candidate Conference.* Retrieved October 14, 2007, from www.ee.princeton.edu/ ~rblee/safer+

Miorandi, D., Caimi, C., & Zanella, A. (2003). Performance characterization of a Bluetooth piconet with multi-slot packets. In *Proceedings of the WiOpt' 03.*

Misic, J., Chan, K. L., & Misic, V. B. (2005). TCP traffic in Bluetooth 1.2: Performance and dimensioning of flow control. In *Proceedings of WCNC '05* (pp. 1798-1804).

OpenSSH. (2006). *OpenSSH project home page.* Retrieved October 14, 2007, from http://www.openssh.org

Persson, K., & Manivannan, D. (2003). Secure connections in Bluetooth scatternets. In *Proceedings of the 36th Annual Hawaii International Conference on System Sciences (HICSS '03)* (p. 314b).

Shaked, Y., & Wool, A. (2005). Cracking the Bluetooth PIN. In *Proceedings of the 3rd ACM International Conference on Mobile Systems, Applications, and Services* (pp. 39-50). ACM Press.

Wang, F., Arumugam, N., & Krishna, G. H. (2002). Performance of a Bluetooth piconet in the presence of IEEE 802.11 WLANs. In *Proceedings of the 13th IEEE International Symposium on Personal, Indoor and Mobile Radio Communications* (Vol. 4, pp. 1742-1746).

Yip, H. K., & Kwok, Y-K. (2004). A performance study of packet scheduling algorithms for coordinating colocated Bluetooth and IEEE 802.11b in a Linux machine. In *Proceedings of the 7th International Symposium on Parallel Architectures, Algorithms and Networks* (ISPAN'04).

Yujin, L., Jesung, K., Sang, L. M., & Joong, S. M. (2001). Performance evaluation of the Bluetooth-based public Internet access point. In *Proceedings of the 15th International Conference on Information Networking* (pp. 643-648).

KEY TERMS

Bluetooth: An industrial specification for wireless personal area networks (PANs). Bluetooth provides a way to connect and exchange information between devices such as mobile phones, laptops, PCs, printers, digital cameras, and video game consoles via a secure, globally unlicensed short-range radio frequency.

Goodput: The application level throughput, that is, the number of useful bits per unit of time forwarded by the network from a certain source address to a certain destination, excluding protocol overhead retransmissions, and so forth.

IEEE 802.15: The IEEE 802.15 WPAN working group focuses on the development of consensus standards for personal area networks or short distance wireless networks. These WPANs address wireless networking of portable and mobile computing devices such as PCs, PDAs, peripherals, cell phones, pagers, and consumer electronics, allowing these devices to communicate and interoperate with one another. The IEEE Project 802.15.1 has derived a wireless personal area network standard based on the Bluetooth v1.1 Foundation Specifications.

IPsec: IPsec (IP security) is a suite of protocols for securing Internet protocol communications by encrypting and/or authenticating each IP packet in a data stream. IPsec also includes protocols for cryptographic key establishment. There are two modes of IPsec operation: transport mode and tunnel mode. IPsec is implemented by a set of cryptographic protocols for securing packet flows. Specifically, the authentication header (AH) protocol provides authentication, payload (message), and IP header integrity (with some cryptography algorithm also nonrepudiation). On the other hand, the encapsulating security payload (ESP) protocol provides data confidentiality, payload (message) integrity, and with some cryptography algorithm also authentication.

Network Performance: The level of quality of service of a telecommunications resource, protocol, or product.

Secure Shell or SSH: A set of standards and an associated network protocol that allows establishing a secure channel between a local and a remote computer. It uses public-key cryptography to authenticate the remote computer and to optionally allow the remote computer to authenticate the user. SSH provides confidentiality and integrity of data exchanged between the two computers using encryption and MACs.

Throughput: The amount of digital data per time unit that are delivered to a certain terminal in a network, from a network node, or from one node to another, for example, via a communication link.

Chapter XLII
Bluetooth Devices Effect on Radiated EMS of Vehicle Wiring

Miguel A. Ruiz
University of Alcala, Spain

Felipe Espinosa
University of Alcala, Spain

David Sanguino
University of Alcala, Spain

AbdelBaset M.H. Awawdeh
University of Alcala, Spain

ABSTRACT

The electromagnetic energy source used by wireless communication devices in a vehicle can cause electromagnetic compatibility problems with the electrical and electronic equipment on board. This work is focused on the radiated susceptibility (electromagnetic susceptibility [EMS]) issue and proposes a method for quantifying the electromagnetic influence of wireless radio frequency (RF) transmitters on board vehicles. The key to the analysis is the evaluation of the relation between the electrical field emitted by a typical Bluetooth device operating close to the automobile's electrical and electronic systems and the field level specified by the electromagnetic compatibility (EMC) directive 2004/104/EC for radiated susceptibility tests. The chapter includes the model of a closed circuit structure emulating an automobile electric wire system and the simulation of its behaviour under electromagnetic fields' action. According to this a physical structure is designed and implemented, which is used for laboratory tests. Finally, simulated and experimental results are compared and the conclusions obtained are discussed.

INTRODUCTION AND BACKGROUND

In the current vehicle coexist electronic and communications systems whose advantages are clear for the user but whose possible problems are not contrasted. The increasing use of radio frequency transmitters by automobile users makes it necessary to evaluate the risk caused by the coexistence of information and communication technologies in the reduced space inside the vehicle. In this context, the present work appears in order to bring up methods and results that contribute to establishing the possible risks limit of the use of wireless devices inside the automobile, and more precisely those based on Bluetooth technology.

To centre the problem, it is mentioned the tendencies in the automobile field that bet for the incorporation of new electrical and electronic systems (X-by-Wire technology) (Leen & Hefferman, 2002; Mazo, Espinosa, Awawdeh, & Gardel, 2005) front of the current mechanical systems, aspects of automotive electromagnetic compatibility (EMC) standard 2004/104/EC (2004) for evaluation of susceptibility/immunity in vehicles are detailed, it is justified the interest to focus the study on the extended Bluetooth wireless communication technology. However there are nonregulated questions by the 2004/104/EC concerning the use of Bluetooth devices what rise uncertainties around the risk derived from its use.

To get a better knowledge of this issue, we lay a few questions regarding the increase of the electronic equipment role in the automobile, the characteristics of commercial Bluetooth devices, some notes about the EMC European Directive involved in vehicles, and last but not least, some of the directive gaps concerning Bluetooth wireless devices in this context.

The Increase in Electrical and Electronic Components in Automobiles

It is clear that nowadays on board electronic components play an important role on vehicles (Bannatyne, 2000; Leen & Hefferman, 2002; Mazo et al., 2005), as much for the increase in the number of electronically controlled units (ECUs) as for the complexity of the communication system (field buses) implemented.

Continuous development in the industrial automobile sector means that dynamic systems that have traditionally been of a mechanical and hydraulic nature, such as the steering, braking, and acceleration are being replaced by electronic ones, which leads to the proposal of networks such as X-by-Wire with its own protocol (Mazo et al., 2005).

Taking advantage of the trend towards the use of DC voltage supplies of 36-42 volts instead of the 12-14 volts currently used, an increase in electronics is being adopted to control key elements of the automobile such as the steering, braking, and acceleration. For example, the car uses a range of electric actuators and also has an innovative driver interface. The driver has all the vehicle functionality in a special steering wheel, which is used for acceleration and braking as well as for steering and gear shifting. The vehicle uses a conventional engine for propulsion but electromechanical actuators for braking, clutching, and gear shifting (Larses, 2003).

With the progress of X-by-Wire technology, in-vehicle data traffic is always growing. Conventionally, individual wire harnesses were used for data transfers between control units and their associated sensors or display devices. As the number of control units and associated devices increase, the number of wire harnesses and interconnections required is swelling. The in-vehicle local-area network (controller area network [CAN], local interconnect network [LIN], and FlexRay) provides an answer to this problem: it minimises the use of individual wire harnesses for data exchanges and reduces both interconnections and vehicle weight, trying to improve consumption, power, security, and comfort.

However, associated with these electronic and communication innovations new sources of potential equipment failure appear, leading to the necessity to continue working on both diagnosis and prognosis in the automotive sector.

Bluetooth Devices and Applications in Automobiles

The presence of radio frequency transmitters in automobiles as a way for multiple wireless communication appliances continue to grow. Apart from the well known uses for the assistance and entertainment (GPS, laptops, PDAs, digital cameras, portable multimedia devices CD/DVD, etc.), others such as remote diagnosis, traffic control, accident assistance, and so forth are being promoted (Campos, Mills, & Graves, 2002; Mazo et al. 2005).

There are several wireless technologies (WiFi, DSRC, Zigbee, etc.) available to automobile manufacturers and users, but at present the most widely used is Bluetooth. Although the functionality and operativity of each technology is different, they have in common the incorporation of a transmitter or an electromagnetic energy source in the environment in which they operate. This extra energy can cause any kind of failure on equipment situated close to the transmitter, as is the case of ECUs on board a vehicle where the driver introduces several wireless devices. At the same time, the metal cage of the vehicle can act as a concentrating reflector, amplifying radio frequency (RF) density emitted by different radiation sources to higher and potentially more dangerous levels.

Bluetooth is an open technology that works with low power and is designed for short range (10 m-100 m), leading to being widely used in transport applications in general and in automobiles in particular. The operating frequency range is within the industrial, scientific, and medical (ISM) bandwidth used of 2.4 GHz to 2.4835 GHz. The frequency range is divided into 79 individual RF channels, each one separated by 1MHz. The output levels are divided into three classes (SIG, 2006): class I (100 mW, +20 dBm), class II (2.5 mW, +4 dBm) and class III (1 mW, 0 dBm).

The equation that determines the frequency for each one of the channels is as follows:

$$F(MHz) = 2402 + k \qquad \text{where, } k = 0.....78$$

In order to comply with out of band regulations in each country, a lower guard band of 2 MHz and an upper guard band of 3.5 MHz are used. The protocol uses a spread spectrum, or in other words, the transmission frequency changes randomly 1,600 times per second, reducing this way the possible interferences created by different transmitters working at the same time in the same frequency range.

Equipments transmit and receive using a time division multiplex (TDM). In addition, spread spectrum TDM provides a higher degree of security against eavesdropping and provides resilience to ambient noise. GFSK modulation is used in Bluetooth technology, where a logic 1 level is represented by a positive frequency shift and a logic 0 level is represented by a negative frequency shift. Keeping all this in mind, a Bluetooth transmitter, from an EMC viewpoint, can be considered as an interfering RF source in the 2.4 to 2.4835 Ghz frequency band.

Two levels of Bluetooth technology application can be considered inside an automobile: Bluetooth integrated into the vehicle at a system level and Bluetooth at a user device level. From a user device level point of view, Bluetooth technology allows connecting inside the vehicle electronic mobile devices such as PDAs, laptops, GPSs, handsfree sets, or cell phones, as seen in Figure 1.

The concept of 'Bluetooth integrated into the vehicle at a system level' is used when a Bluetooth network can provide a functionality and versatility similar to a vehicle control cabled network (e.g., CAN bus) which is nowadays the most widely extended solution (network and protocol) in vehicles.

Directive 2004/104/CE for the Assessment of EMC in Vehicles

In Europe, EMC activity in automobiles is regulated by the recent directive on electromagnetic

Figure 1. Typical applications of Bluetooth in vehicles

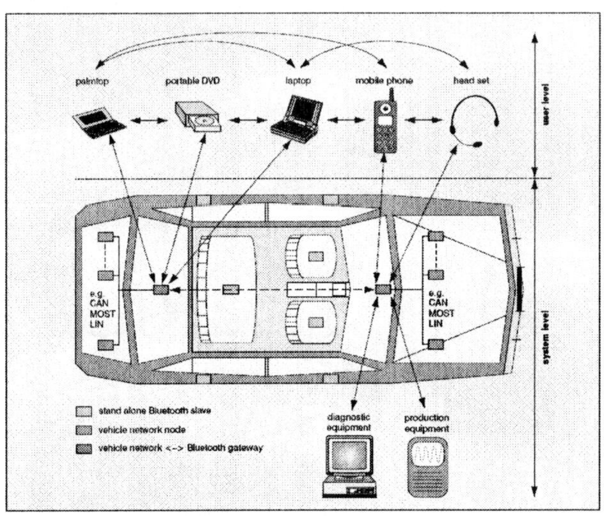

compatibility (2004/104/EC EMC, 2004), which since July 1, 2006, substitutes the earlier directive (95/54/EC EMC, 1995). The new directive requires tests to be carried out at both component level and on the vehicle as a whole. The range of required tests includes broadband and narrowband radiated emissions (CISPR 12, 2001; CISPR 25, 2002; Kerry, 2003), radiated susceptibility (ISO 11452-2, 2004), as well as conducted susceptibility and emissions along supply lines of electrical and electronic subcomponent (ISO 7637-2, 2004).

The present article focuses on the radiated susceptibility test in accordance with regulation ISO 11452-2 as this test allows determining the electromagnetic immunity of a device or electronic component on board the vehicle in proximity to RF transmitters.

Radiated Susceptibility Test According to ISO 11452-2

The before mentioned ISO 11452-2 describes a possible method for the radiated immunity test accepted by the EMC directive (2004/104/EC EMC, 2004) for the fulfilment of electromagnetic immunity requirements of electric and electronic components on a vehicle.

Following the standard, the electromagnetic susceptibility (EMS) test must be done in a semi-anechoic chamber. The electromagnetic field is generated by an antenna connected to a RF amplifier. To monitor the electric field intensity level (V/m) inside the semianechoic chamber, an isotropic probe must be used.

Figure 2 shows the setup as explained in the standard for the radiated test on a vehicle device

Figure 2. Setup of the radiated susceptibility test

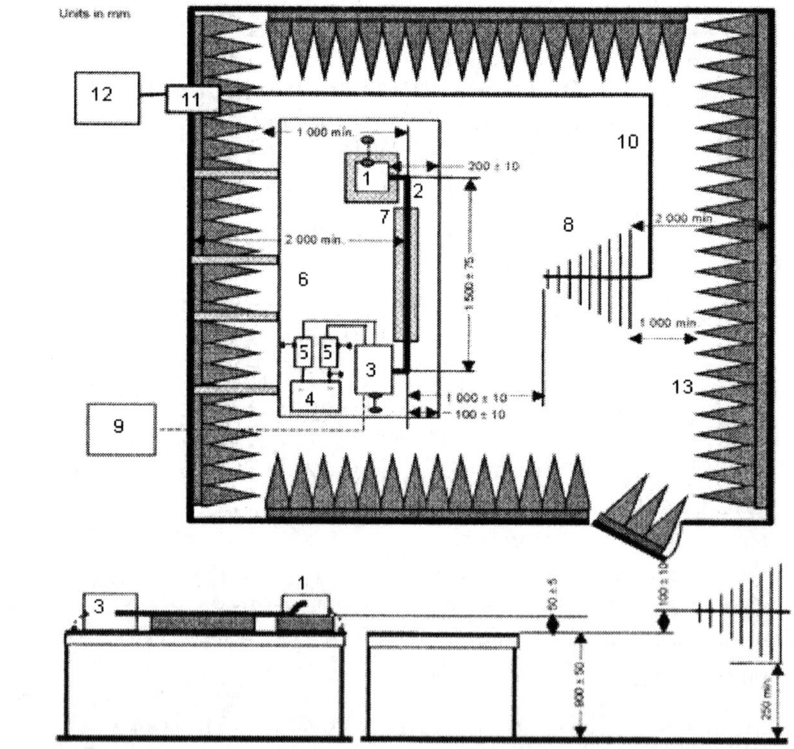

1. EUT (Equipment Under Test)
2. Test harness.
3. Load simulator.
4. Power supply.
5. Artificial network (AN de 50 Ω)
6. Ground plane (bonded to shielded enclosure).
7. Low relative permittivity support ($\varepsilon_r \leq 1.4$)
8. Antenna.
9. Stimulation and monitoring system.
10. High quality double-shielded coaxial cable (50 Ω)
11. Bulkhead connector.
12. RF signal generator and amplifier.
13. RF absorber material.

(equipment under test [EUT]). A metallic (copper or galvanised steel) ground plane of a minimum of 0.5 mm thickness and 1000 x 2000 mm (WxL) area has to be located 900±100 mm above the floor.

Each one of the power supply cables must be connected to the EUT through an artificial network (AN) [5] of 5 μH/50 Ω to get a reference impedance (usually 50 Ω). The ANs should be placed over the ground plane and connected to it.

The electric or electronic equipment under test has to be placed on a dielectric material [7] of low permeability ($\varepsilon_r \leq 1.4$) and 50±5 mm thickness. One of the EUT faces has to be placed 200±10 mm from the edge of the ground plane. The cables connected to the EUT are exposed along 1500±75 mm to the electromagnetic radiation generated by the antenna. They are placed on the same dielectric material as the EUT 100±10 mm away from the edge of the ground plane.

The antenna that generates the electric field has to be located at a 100±10 mm height above the ground plane, that is 1000 mm above the floor and also 1000 ± 10 mm away from the EUT cables.

The test procedure can be divided in two steps:

- A first one where the electric field level calibration is done (without EUT, cables nor ANs).
- A second in which the test is taken place based on the levels obtained in the preceding step.

In the calibration stage, an isotropic probe 150 ± 10 mm above the ground plane and 100 ± 10 mm away from the edge is used. The calibration is done for both horizontal and vertical electric field.

Aspects of Bluetooth Devices that are not Considered in Directive 2004/104/EC

Having mentioned some of the properties of Bluetooth, as well as the EMC regulation applicable to the automobile context, and focusing the study on the assessment of the susceptibility of the electrical and electronic components on board to radiations from different wireless devices, the following observations remain to be made:

- The specifications of the radiated susceptibility test, mentioned in the directive using the semianechoic chamber method to carry it out, determine that the range of frequencies to be tested is from 20 MHz to 2000 MHz. Therefore, the directive does not make it compulsory to test electrical and electronic automobile equipment at frequencies higher than 2 GHz. Bluetooth works at frequencies between 2.400 and 2.4835 GHz, and hence an electronic subsystem or component that complies with the directive does not guarantee electromagnetic compatibility in the presence of a Bluetooth device.
- The electrical field levels specified by the directive to be tested in the 20 to 2000 MHz range are of 30 V/m for 90% of the frequency band and 25 V/m for the whole frequency band. It is foreseeable that in the near future the directive will be modified to increase the range of frequencies to at least include the operating frequencies used by the wireless devices available on the market to automobile users.
- The test method specified in the directive corresponds to a situation in which the transmitter is not situated close to the equipment being studied. This leads to the use of a plane wave in the test setup, which requires one or more transmitter antenna working in far field. However, in this particular case it is easy to find Bluetooth transmitters within the automobile's own electrical and electronic system or another ones (introduced by users) operating a few centimetres away from the electronic systems and wires of the vehicle's electrical installation.

With this background, the present work is developed with the aim of determining whether a device that complies with the requirements of the EMC automobile directive presents any possible electromagnetic compatibility risks to Bluetooth transmitters located a short distance away. In addition a measure procedure is proposed for assessing the degree of interrelation between the electronics on board and the Bluetooth devices incorporated by vehicles' users.

Related Published Works

In the technical literature, negative examples of vehicle-communication system interaction can be found, as in the case of 'Project 54' (Kun, Lenharth, & Millar, 2004), in which the origin and possible solutions to random signal reception by appliances normally used by traffic police officers are analysed. There are other more complex cases, such as the one stated by Tatoian (2005), in which the possibility of equipping the police with electromagnetic systems in order to block cars in conflictive traffic conditions is assessed. The impact of the transient surrounding perturbations (especially due to electromagnetic interferences) on the dependability of systems distributed on TDMA-based networks in automotive domain is analysed in by Campos et al. (2002).

All of this justifies the interest of automobile manufacturers in regulating the incorporation of new information and communication technologies. In Australia for example, exists the FCAI (1997) initiative, in which the automobile industry and the nation's government are working together to establish the emission and susceptibility limits to which new vehicles must conform in order to guarantee the compatibility of the electronics on board the vehicle with the multimedia equipment for drivers available on the market. EMC centres work along the same lines in association with automobile manufacturers such as Audi or Renault (Renault, 2006).

On the other hand, there are several previous research works related to this subject. Stadtler Schoof, and Haseborg (2002) calculate that a 100 mW Bluetooth transmitter in far-field (1 m) generates a electric-field level of 2.45 V/m, that means a quite lower level to the one used in EMC test according to the 2004/104/EC standard. Nevertheless, simulations results presented by Schoof, Stadtler, and Haseborg (2003) inside a cockpit vehicle with a 100 mW Bluetooth transmitter achieved electrical-field levels of 25 V/m, which is close to the limit level indicated by EMC standard.

PROPOSED METHOD FOR ASSESSING THE POSSIBLE EFFECTS OF BLUETOOTH DEVICES USED INSIDE VEHICLES

Taking into account previous published studies (Schoof et al., 2003; Stadtler et al., 2002) and the EMC specifications in the automotive context, certain questions must be made in relation with the incorporation of Bluetooth transmitters in automobiles by either the manufacturer or the users of the vehicle. As mentioned earlier, the EMC directive (2004/104/EC EMC, 2004) does not require radiated susceptibility tests above 2 GHz and restricts the field level of the equipment under test to 25 or 30 V/m. Moreover, in present day traffic conditions, it is easy to find several Bluetooth transmitters inside the cabin of the vehicle and within a few centimetres of the vehicle's cables and electronic systems.

Fundament of the Proposed Measure

The setup for the radiated susceptibility test for an automobile component in accordance with ISO regulation 11452-2 (ISO 11452-2, 2004) was represented in the previously in the chapter. This setup contains similarities to the actual layout of the components inside a vehicle. For example, the equipment under test [1], wiring [2], simulators [3], and power supply [4], are placed on a ground plane that emulates the chassis of the vehicle. The length of wire exposed to the radiation is 1.5 m, being the usual length of cable on board a vehicle.

In this context, a study is made of the radio frequency current that is induced in the cable [2] when it is submitted to the action of a Bluetooth transmitter in near field, that is to say with a few centimetres between transmitter and cable.

Once the current induced in the EUT cable by the Bluetooth transmitter has been determined, the electrical field level that must be applied during the radiated susceptibility test in order to induce a current value identical to that induced by the Bluetooth transmitter a few centimetres away is analysed. If the electric field level required to induce the current value is under the 25 or 30V/m specified by 2004/104/EC EMC, it will confirm that all equipment that fulfil the EMC directive should not present compatibility problems. However, if the electric field level is similar to or higher than the one specified by EMC directive, there is no guarantee that the automotive component will not have electromagnetic compatibility problems in close proximity to a Bluetooth transmitter.

PRACTICAL IMPLEMENTATION AND RESULTS

Following the guidelines indicated by Stadtler et al. (2002), the setup shown in Figure 3 is used for the present research work. The impedance presented by the EUT [1] between the cable and the ground plane [6] is modelled as an ideal impedance of 50 Ω. At the other end of the cable an ideal impedance of 50 Ω represents the one corresponding to the artificial network [5] or to other auxiliary equipment.

In the first approach at validating the proposed thesis the electromagnetic simulation tool FEKO (2005) is used. In the laboratory experimental phase, a R&S ESIB 26 spectrum analyser syntonised to the transmission frequency of the device is used to measure the induced current. The resistance of 50 Ω that corresponds to the EUT is provided by the spectrum analyser input; as an impedance of 50 Ω at the other end of the cable, a load 50 Ω with an N connector is used. The analyser will register the voltage value at its input terminals and by direct relation the value of current induced in the cable is determined.

Design of the Interference Pattern

An electromagnetic radiation source in the 2.400 to 2.483 GHz range has been designed with adjustable power between 1 and 100 mW, emulating the behaviour of class I, II, and III Bluetooth transmitters. The radiation source consists of an antenna connected to a R&S SMR20 RF generator. The antenna design is based on a commercial radio frequency module (SparkFun, 2005), simulated using FEKO and implemented on a PCB.

Elements of the Setup

Figure 4 shows the setup used to measure the current induced in the cable when the Bluetooth transmitter is situated a short distance away. The right hand side of the cable is loaded with an impedance of 50 Ω, while the impedance of 50 Ω on the left hand side is provided by the spectrum analyser input (R&S ESIB 26), which is outside

Figure 3. Setup diagram of the test used to determine the current induced by a transmitter in near field

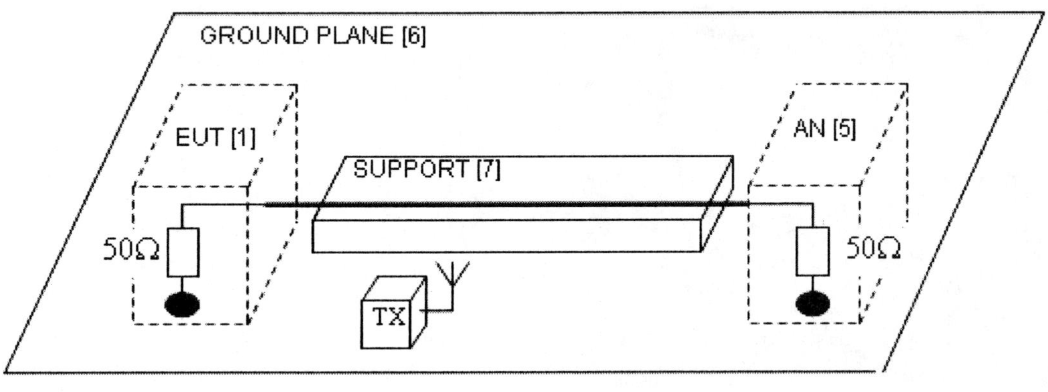

the semianechoic chamber (Space Saver of ETS) during the test and is connected by means of an RG214 cable. The attenuation caused by the RG214 cable is corrected by the spectrum analyser.

To measure the induced current, the radiation source is placed in different positions with respect to the 1.5 m long cable. The measurements are made with the transmitter facing the cable and in various positions along its length. The transmitter is placed at distances of 2 cm, 5 cm, and 8 cm from the cable and at heights with respect to the ground plane of 0.6 cm and 3.7 cm.

To determine the value of the electric field intensity (V/m), the setup represented in Figure 4 is used, corresponding to the radiated susceptibility test for automobile components (2004/104/EC EMC, 2004). The electric field level is registered by means of an isotropic electric-field probe

Figure 4. Setup of the test used to measure the current induced by a transmitter in near field (top). Setup used to determine the electric field level (down)

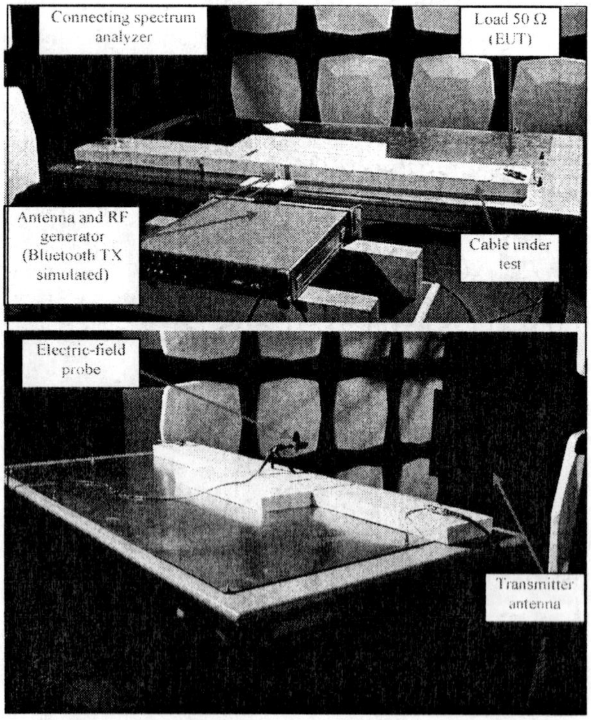

(FP6001 AR) placed at a height of 10 cm above the ground plane and 10 cm from the edge facing the antenna. The value of the current induced by the radiation of the AT4000 AR antenna situated at a distance of 1 m is constantly measured on the spectrum analyser. The power transferred to the antenna is varied until the induced current values are identical to those obtained when the Bluetooth transmitter was situated a few centimetres from the same cable. This is the way to determine the electric field level that induces the same current as a Bluetooth transmitter in the conditions previously described.

Results

In the following section, some of the results about the setups proposed in previous sections obtained by both simulations and practical measurements made in the laboratory are given, with the principal aim of determining the electromagnetic compatibility risks caused by commercial Bluetooth transmitters in automobiles.

The FEKO tool is used to simulate a ground plane with a 150 cm cable above it at a height of 5 cm, with both ends loaded with a resistance of 50 Ω. A monopole antenna connected to a generator was used as a transmitter in the simulation. The simulations are made with the antenna transmitter situated in the centre of the 1.5 m cable structure and at distances of 2 cm, 5 cm, and 8 cm and at heights above the ground plane of 0.6 cm and 3.7 cm. In addition, the simulations are carried out taking into account the different power types (I, II, and III) specified by the Bluetooth technology.

Tables 1 and 2 represent a comparison between the results obtained with the FEKO simulation tool and those obtained in laboratory tests. First of all, the results belong to a transmitter working at 2.425 GHz and at a height above the ground plane of 0.6 cm are presented. The table shows the variation in the induced current as a function of the distance that separates the transmitter from the cable, and for three different power transmission (+20, +4, and 0 dBm). For example, in case the class I transmitter is separated a distance of 2 cm from the cable, the simulated current value is 1990 μA in contrast with the value of 1870 μA experimentally obtained.

Besides, one can see in Table 2 the comparison between simulated and experimental induced current when the emission frequency is changed for three Bluetooth devices (class I, II, and III) at a distance of 5 cm and a height of 0.6 cm.

On the other hand, Table 3 shows some of the measurements obtained in the laboratory corresponding to the current induced by the transmitter located at a distance of 2 cm and 5 cm from the cable, and at a height above the ground plane of 0.6 cm. The same table shows the increase in the induced current due to the effect of different power class transmitter (class I, II, and III).

To conclude, Figure 5 shows the electric field levels that the structure being tested is submitted to in order to induce the same RF currents as those produced if a Bluetooth transmitter is situated in near field. The setup used for the test is the one

Table 1. Values obtained by simulation and experimentally of the induced current as a function of the transmitter distance. (frequency 2425 MHz and height 0.6 cm)

Wire Induced Current			
Power transmission Bluetooth devices	Distance (cm)	Simulation (μA)	Measurement (μA)
+20 dBm (Class I)	2	1990	1870.0
	5	879	715.3
	8	337	378.0
+ 4 dBm (Class II)	2	315	319.5
	5	139	123.0
	8	53.3	62.2
0 dBm (Class III)	2	200	203.4
	5	87.7	78.8
	8	33.5	43.3

Table 2. Values obtained by simulation and experimentally of the induced current as a function of the transmitter frequency (distance 5 cm and height 0.6 cm)

Induced Current			
Power transmission Bluetooth devices	Frequency (MHz)	Simulation (μA)	Measurement (μA)
+20 dBm (Class I)	2400	921	827.0
	2425	879	715.3
	2450	941	604.0
	2475	1060	645.0
+ 4 dBm (Class II)	2400	145	142.8
	2425	139	123.0
	2450	148	104.8
	2475	167	111.7
0 dBm (Class III)	2400	91.1	90.8
	2425	87.7	78.8
	2450	93.8	67.5
	2475	105	71.28

Table 3. Measurement of the induced current as a function of the frequency and of the Bluetooth transmitter location (height 0.6 cm)

Measurement of induced current			
Power transmission Bluetooth devices	Frequency (MHz)	Distance 2 cm (μA)	Distance 5 cm (μA)
+20 dBm (Class I)	2400	1974	827.0
	2425	1870	715.3
	2450	1772	604.0
	2475	1862	645.0
+ 4 dBm (Class II)	2400	335.7	142.8
	2425	319.5	123.0
	2450	301.6	104.8
	2475	331.5	111.7
0 dBm (Class III)	2400	213.3	90.8
	2425	203.5	78.8
	2450	193.0	67.5
	2475	203.0	71.28

Figure 5. Identical induced current on the cable under test by the intensity of electric field (V/m) according to the described test as well as Bluetooth transmitters working with variable distance and power (dBm)

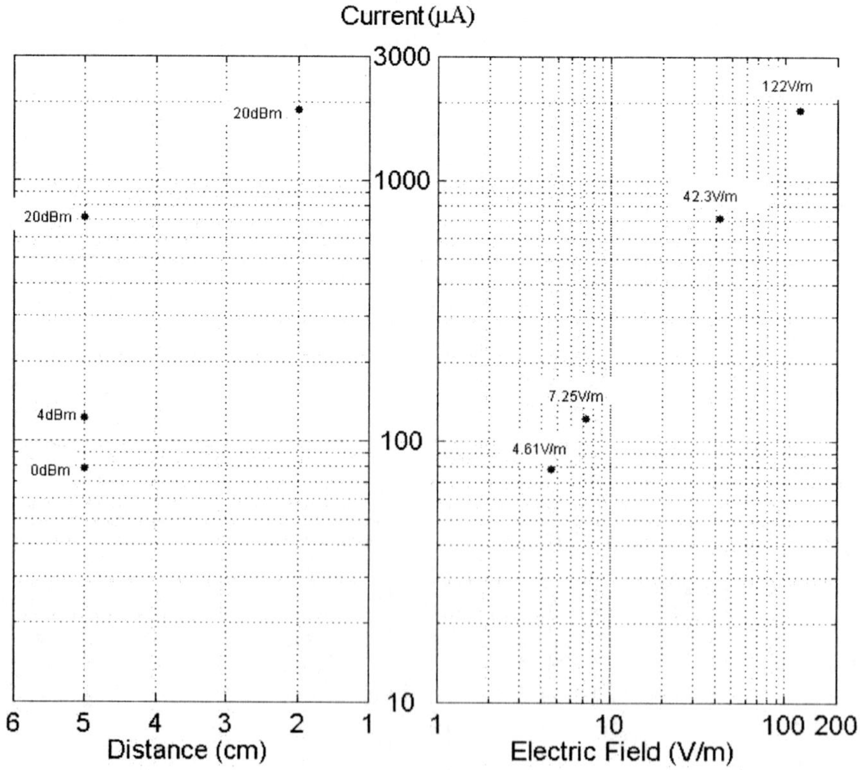

shown in Figure 4 (down). For example, a class I transmitter (+20 dBm) located at a distance of 5 cm and at a height of 0.6 cm induces a current of 715 μA. In the same way, this transmitter situated at a distance of 2 cm induces a current of 1870 μA. Identical current values are induced when the wire loaded by resistances of 50 Ω is exposed to a uniform plane wave with an electric field level of 42.3 V/m and 122 V/m, respectively.

FUTURE WORKS

Once the above shown results are analyzed, the authors suggest to keep on evaluating the electromagnetic field generated from these kinds of wireless communication devices and others alike, varying the setup conditions (relative cable and antenna location, cables, different antennas transmitting simultaneously, etc.). All this is done comparing the results obtained from the simulation tools as well as from the experimental tests in the EMC laboratory.

It would also be interesting to study and evaluate the amplifying effect due to the metallic structure of the cabin, measuring inside and outside the vehicle.

CONCLUSION

From the simulated and experimental results obtained by this work, it can be deduced that the electromagnetic interference supported by the cable structure under study, when situated a few centimetres from a commercial Bluetooth transmitter, is similar to the action of a plane wave with electric field levels superior to those specified by directive 2004/104/EC (25 or 30 V/m).

Comparing the magnitude of the electric fields obtained in the present analysis with the real values at which on board electronic components are tested in accordance with the EMC directive, it can be deduced that Bluetooth transmitters of 20 dBm can cause electromagnetic susceptibility problems in the vehicle's electronic and electrical systems, which would not be detected during the radiated susceptibility test according to a valid EMC directive for automobiles.

In short, more consideration should be given to the electromagnetic interference generated by Bluetooth devices as they get closer to electrical and electronic circuits whose performance they can affect, and even more so in confined spaces where multiple sources of interference coexist, as is the situation with automobiles. The effect of increasing the power of the transmitter or reducing the distance between it and the wired elements of the automobile is equivalent to submitting them to increasing electric far-field levels in radiated susceptibility tests in accordance with the 2004/104/EC EMC directive, which increases the risk of a failure in the system.

This work leads to support the need for the prevailing EMC directive to be modified in order to assess and ensure the electromagnetic compatibility of automobiles' on board systems in the presence of wireless devices with a frequency range above 2.0 GHz.

ACKNOWLEDGMENT

This work has been possible thanks to the support of the Centre of High Technology and Homologation (CATECHOM) at the University of Alcala (UAH), as well as the COVE Project funded by the Spanish Science and Education Ministry TRA2005-05409/AUT and TRA2006-12105/TAIR.

REFERENCES

2004/104/EC EMC. (2004). *Directive relating to the radio interference of vehicles*. Commission of the European Communities.

95/54/EC EMC. (1995). *Directive relating to the radio interference of vehicles*. Commission of the European Communities.

Bannatyne, R. (2000, May). The sensor explosion and automotive control systems. *Sensors Magazine, 17*(5).

Campos, F.T., Mills, N.W., & Graves, M.L. (2002). A reference architecture for remote diagnostics and prognostics applications. In *Proceedings of the IEEE Autotescon* (pp. 842-853).

CISPR 12. (2001). *Vehicles, boats and internal combustion engine driven devices. Radio disturbance characteristics. Limits and methods of measurement for the protection of receivers except those installed in the vehicle/boat/device itself or in adjacent vehicles/boats/devices.* The International Special Committee on Radio Interference (CISPR).

CISPR 25. (2002). *Radio disturbance characteristics for the protection of receivers used on board vehicles, boats, and on devices. Limits and methods of measurement.* The International Special Committee on Radio Interference (CISPR).

FCAI. (1997). *Federal Chamber of Automotive Industries (FCAI).* Retrieved October 15, 2007, from http://www.dcita.gov.au/Article/0,,0_4-2_4008-4_10465,00.html

FEKO. (2005). EM software & systems. *FEKO.* Retrieved October 15, 2007, from http://www.feko.info/

ISO 11452-2. (2004). *Road vehicles: Component test methods for electrical disturbances from narrowband radiated electromagnetic energy. Part 2: Absorber-lined shielded enclosure.* The International Organization for Standardization (ISO).

ISO 7637-2. (2004). *Road vehicles: Electrical disturbance from conduction and coupling. Part 2: Electrical transient conduction along supply lines only on vehicles with nominal 12V or 24V supply voltage.* The International Organization for Standardization (ISO).

Kerry, P.J. (2003). *EMC in the European Union.* IEEE. 0-7803-7779-6/03.

Kun, A., Lenharth, W., & Millar, W.T. (2004). *Project 54.* Dirham: University ofNew Hampshire. Retrieved October 15, 2007, from http://www.project54.unh.edu/

Larses, O. (2003). *Modern automotive electronics from an OEM perspective* (Tech. Rep. KTH S-100 44). Royal Institute of Technology, Mechatronics Lab, Department of Machine Design, Stockholm.

Leen, G., & Hefferman, D. (2002, January). Expanding automotive electronic systems. *IEE Computer, 35*(1), 88-93.

Mazo, M., Espinosa, F., Awawdeh, A.M.H., & Gardel, A. (2005). Automotive electronics diagnosis: State of the art and next tendencies. *FITSA.* Madrid. Retrieved October 15, 2007, from http://www.fundacionfitsa.org/fitsa/pub/Libro%20diagnosis%20electronica.pdf

Renault. (2006). *Renault EMC unit.* Aubevoye, France. Retrieved October 15, 2007, from http://www.worldcarfans.com/news.cfm/newsid/2060406.004/country/ecf/Renault-inaugurates-emc-unit

Schoof, A., Stadtler, T., & Haseborg, J.L. (2003, May 11-16). *Simulation and measurement of the propagation of Bluetooth signals in automobiles.* Paper presented at the 2003 IEEE International Symposium, EMC'03 (pp.1297-1300).

SIG. (2006). *Specification of the Bluetooth system.* Retrieved October 15, 2007, from http://www.bluetooth.com

SparkFun. (2005).*Transceiver MiRF - Miniature RF 2.4GHz.* Retrieved October 15, 2007, from http://www.sparkfun.com/commerce/product_info.php?products_id=153

Stadtler, T., Schoof, A., & Haseborg, J.L. (2002, September 9-13). *Electromagnetic compatibility of a system under the influence of a Bluetooth transmitter.* Paper presented at the Symposium EMC Europe 2002, Sorrento.

Tatoian, J. (2005). *Car chases zapped.* Pasadera, California: Eureka Aerospace. Retrieved October 15, 2007, from http://www.defensetech.org/archives/001369.html

KEY TERMS

Anechoic (Semianechoic) Chamber: An anechoic chamber is a room in which there are no echoes. This description was originally used in the context of acoustic (sound) echoes caused by reflections from the internal surfaces of the room but more recently the same description has been adopted for the radio frequency (RF) anechoic chamber. A RF anechoic chamber is designed to suppress the electromagnetic wave analogy of echoes: reflected electromagnetic waves, again from the internal surfaces. Both types of chamber are usually built, not only with echo suppression features, but also with effective isolation from the acoustic or RF noise present in the external environment. In a well designed acoustic or RF anechoic chamber the equipment under test will only receive signals (whether acoustic or RF) which are emitted directly from the signal source, and not reflected from another part of the chamber. The semianechoic chamber is a shielded room with radio frequency absorbing material on the walls and ceiling (not on the ground). This semi-anechoic chamber simulates an open field test site, and eliminates any ambient signals that may be present in an open field environment.

Bluetooth (Class I, II, and III): Bluetooth is the name of a wireless technology standard for connecting devices, set to replace cables. It uses radio frequencies in the 2.45 GHz range to transmit information over short distances of generally 33 feet (10 meters) or less. By embedding a Bluetooth chip and receiver into products, cables that would normally carry the signal can be eliminated.

There are currently three flavours or classifications of Bluetooth devices, relative to transmitting range. As the range is increased the signal used in the respective classification is also stronger. Note that class III devices are comparatively rare.

Class	Signal Strength	Range
Class I	100 mW (+20dBm)	Up to 328 feet (100 meters)
Class II	2.5 mW (+4 dBm)	Up to 33 feet (10 meters)
Class III	1 mw (0 dBm)	Up to 33 feet (10 meters)

CISPR: The Special International Committee on Radio Interference(abbreviated CISPR from the French name of the organization, Comitéinternational spécial des perturbations radioélectriques) is concerned withdeveloping norms for detecting, measuring and comparing electromagnetic interference in electric devices. CISPR's principal task is at the higherend of the frequency range, from 9 kHz upwards, repairing standards thatoffer protection of radio reception from interference sources such aselectrical appliances of all types, the electricity supply system,industrial, scientific and electro-medical RF, broadcasting receivers (soundand TV) and, increasingly, information technology equipment (ITE).

EMC-EMI-EMS: EMC is an abbreviation for electromagnetic compatibility. This means interoperability, or an electronic device's ability to operate in an electric environment without interfering other electronic devices (emission), and without being interfered by other devices in its vicinity (immunity).EMC is divided into two main areas: electromagnetic interference (EMI) and electromagnetic susceptibility (EMS). These two areas are again divided into two categories of phenomena: conducted phenomena and radiated phenomena. EMC testing comprises measurements of the emission generated on in- and outgoing cables, the emission generated as electric fields surrounding the device, immunity against several disturbance phenomena on in- and outgoing cables, immunity against electric fields generated by other electronic devices and radio transmitters, and immunity against electrostatic discharges generated by human intervention.

Near Field Communication (NFC): A short-range wireless connectivity standard (Ecma-340, ISO/IEC 18092) that uses magnetic field induction to enable communication between devices when they are touched together, or brought within a few centimetres of each other. Jointly developed by Philips and Sony, the standard specifies a way for the devices to establish a peer-to-peer (P2P) network to exchange data. After the P2P network has been configured, another wireless communication technology, such as Bluetooth or Wi-Fi, can be used for longer range communication or for transferring larger amounts of data.

RF: Short for radio frequency, any frequency within the electromagnetic spectrum associated with radio wave propagation. When a RF current is supplied to an antenna, an electromagnetic field is created that then is able to propagate through space. Many wireless technologies are based on RF field propagation, including cordless phones, radar, ham radio, GPS, and radio and television broadcasts. RF waves propagate at the speed of light, or 186,000 miles per second (300,000 km/s). Their frequencies however are slower than those of visible light, making RF waves invisible to the human eye.

WLAN: The acronym for wireless local-area network. Also referred to as LAWN. A type of local-area network that uses high-frequency radio waves rather than wires to communicate between nodes. LAN is a computer network that spans a relatively small area. Most LANs are confined to a single building or group of buildings. However, one LAN can be connected to other LANs over any distance via telephone lines and radio waves. A system of LANs connected in this way is called a wide-area network (WAN).

Chapter XLIII Security in WLAN

Mohamad Badra
Bât ISIMA, France

Artur Hecker
INFRES-ENST, France

ABSTRACT

The great promise of wireless LAN will never be realized unless there is an appropriate security level. From this point of view, various security protocols have been proposed to handle wireless local-area network (WLAN) security problems that are mostly due to the lack of physical protection in WLAN or because of the transmission on the radio link. The purpose of this chapter is (1) to provide the reader with a sample background in WLAN technologies and standards, (2) to give the reader a solid grounding in common security concepts and technologies, and (3) to identify the threats and vulnerabilities of WLAN communications.

WLAN STANDARDS AND TECHNOLOGIES, BENEFITS AND USE CASES

IEEE 802.11/wireless local-area network (WLAN) technologies (WLAN, 2003) have evolved phenomenally over the last few years. They have been widely deployed in a variety of network environments and they properly converge with actual Internet and 3G infrastructures.

IEEE 802.11 refers to a set of specifications for WLAN developed by IEEE. It specifies an over-the-air interface between a mobile station (STA) and a base station as well as between two mobile stations. Basically, WLAN networks can be seen as extensions of wired Ethernet networks. WLAN leverages on a set of newest digital communications technologies to make it possible to establish a local area network for computer communications without the use of cables.

IEEE approved the first 802.11 standard in 1997. This version is limited to only 1 and 2 Mbps data rates. Subsequently in 1999, 802.11a and 802.11b were approved, expanding to new radio bands (changing the usage of the 2.4 GHz ISM band and adding usage of the 5 GHz UNII band) and increasing the available data rates to 54 Mbps and 11 Mbps, respectively. Consequently, large deployments of

802.11 WLAN started being rolled out, especially in enterprises to replace or extend the wired local-area network (LAN) with an implementation of WLAN, and in airports and various business venues where they installed several WLAN access points offering a public Internet access (so-called hotspots), which can range from a small covered zone to many square miles of overlapping hotspots in metropolitan areas.

While the most obvious advantage of the WLAN is mobility, there are also other benefits:

- **Installing and maintaining flexibility:** Installation of a WLAN system is fast and easy and eliminates the terminal cabling costs. It extends to area where wires cannot be installed.
- **Apparent ease of use:** WLAN is easy for novice and expert users alike, eliminating the need of a large knowledge to take advantage of WLAN.
- **Transparency:** WLAN is transparent to a user network, allowing applications to work in the same way as they do in wired LANs.
- **Scalability:** WLANs are designed to be simple or complex; they range from networks suitable for a small number of nodes to full infrastructure networks of thousands of nodes and large physical area by adding access points to extend coverage and to provide users with roaming between different areas.

WLAN was developed to extend wired LAN wirelessly and therefore to minimize Ethernet cabling. It was designed to provide "data obscurity" equivalent to that provided by wired Ethernet with easier installation. However, there is some difference between WLAN and wired LAN due to constraints introduced by the first, especially the shared medium, interference, the collisions that cannot be detected reliably, the physical boundary that is difficult to control, and to the signal. These differences make the WLAN security harder to maintain in comparison to wired LAN. In WLAN, it is possible for an attacker to snoop on confidentiality communications or modify them to gain access to the network much more easily than the wired LAN. The open access to the networks permits malicious action at a distance and simplify passive interception. The temptation for unauthorized access and eavesdropping is also a reality (Khan & Khwaja, 2003) because an attacker could easily access the transport medium. This is not easy in wired LAN due to the physical access to the media. WLANs have introduced a new security threat, sometime referred to as parking lot attack (Arbaugh, 2003) (i.e., a person with a wireless computer and a makeshift antenna can gain access to your the WLAN from hundreds of feet away). Other security issues are mostly because of the lack of physical protection of the wireless network access or of the transmission on the radio that cannot be confined to the walls of an organization.

The original 802.11 standard defines authentication and encryption mechanisms based on the use of the wired equivalent privacy (WEP) protocol. Unfortunately, this protocol suffers from serious design flaws (Miller & Hamilton, 2002). Furthermore, it does not define a key management mechanism; it presumes that the secret key is conveyed between WLAN entities through a secure channel independent of 802.11 WLAN. As a result of different flaws discovered in WEP, the security of WLAN has been widely studied, and a set of standards have been developed by IEEE and IETF, especially 802.1X (802.1X, 2004), 802.11i (802.11i, 2004) and extensible authentication protocol (EAP) (Aboba, Blunk, Vollbrecht, Carlson, & Levkowetz, 2004). The 802.1X standard has been standardized by 802.1 working group. 802.1X was initially conceived to securely manage the access to different IEEE 802.1 networks. It is a framework for authenticating and controlling user traffic at the network level, as well as dynamically varying and exchanging encryption keys between a mobile station and an authentication server. By pushing the authentication method to the virtual layer, the 802.1X defines an open security architecture, which principally allows user authentication and, optionally, session key generation and derivation on a per-user and per-session basis. Because of this possibility for dynamic provisioning, 802.1X is used as the common base in the current WLAN security suites such as Wi-Fi protected access (WPA) (WPA, 2003) and IEEE 802.11i (802.11i, 2004).

The rest of the chapter presents a more detailed description of the various WLAN standards from the security perspective: challenges and possible attacks in WLAN security; WLAN infrastructure security; authentication, authorization, and access control; confidentiality and privacy; and key management and establishment.

WLAN MANAGEMENT FRAMES

A WLAN network is formed by entities called stations (STA). A WLAN can operate in two modes: infrastructure and ad hoc. In the ad hoc mode, each STA communicates directly with other stations. In the infrastructure mode, stations communicate with each other via a special STA, called access point (AP). Each AP additionally has a connection to the distributing system (DS), which can take different forms (wireless, wired, OSI layer, etc.). In this chapter, we focus on the infrastructure mode.

The infrastructure mode extends the range of the wired LAN. It introduces a notion of basic service set (BSS). Each BSS is formed by an AP and associated stations, and can be roughly understood as a WLAN equivalent of a cell (a base station and mobile nodes). It is uniquely identified by the medium access control (MAC) address of the STA of its AP, called BSSID. By using their DS connection, several APs can allow a station to move from one BSS to another. Several BSSs may be collected, constructing an extended service set (ESS). The identifier of the ESS is a case sensitive string of 32 bytes (ESSID), and can be roughly understood as a "network name." In the infrastructure mode, it is usually called SSID for convenience.

One of the primary services of WLAN management frames is to provide access control reliability. This is done originally based on a predetermined set of MAC address and improved later with 802.1X. The access control usually implements a way to provide authentication or authorization to a terminal attached to the network. WLAN uses a concept called port-based access control that is based on the notion of a port. The port-based access control blocks all traffic on a (logical) port until some condition is true. The condition for the port opening is a successful user association and authentication.

An association precedes each communication between the STA and the AP. The association is formed between a STA and an AP by exchanging messages, by the means of so-called management frames, allowing both STA and AP to create and to maintain the association states. WLAN defines three states: unauthenticated and unassociated, authenticated and unassociated, and authenticated and associated.

The management frames can be started by the STA sending a probe request management frame to find an AP affiliated with a selected ESSID, or scanning the beacon management frame broadcast by the APs at a fixed interval. As part of the association processes, the STA and the access point perform an authentication. IEEE 802.11 originally defines two authentication modes, the open system authentication (OSA), practically equivalent to no authentication, and shared key authentication (SKA), a simple challenge handshake protocol based on a preshared key between the STA and the AP and the specified WEP protocol. Furthermore, other methods can be used to restrict the access to an AP, such as classical MAC address filtering (whitelisting or blacklisting STA MAC addresses) and the suppression of service advertisement, usually called SSID hiding.

It must be noted that neither of these methods can be considered sufficiently secure given the current usage of the 802.11 technology. Since MAC addresses need to be transported in clear and can be easily changed, the MAC address filtering is not enough of a barrier. SSID hiding only can work as long as nobody uses the service, since the associating STA will try to solicit an AP under a given SSID, thus effectively disclosing this "secret." The included SKA scheme lacks mutuality and is way too static (no session key derivation, no key management) to be applicable in an operational industrial environment. Accidentally SKA was found to be misconceived, effectively rendering it useless. The details of these findings will be discussed in the next sections.

WLAN SECURITY ESSENTIALS

The 802.11 standard defines an optional encryption scheme to protected data streams exchanged over-the-air between the STA and the AP. This scheme, called WEP, was designed to prevent unauthorized access to the wireless LAN traffic. It uses the stream cipher RC4 for confidentiality equivalent to a traditional wired network, and a CRC-32 checksum for integrity protection. Shared key authentication type requires WEP support.

The 802.11 standard does not define how to distribute shared keys to the equipment in the network. In other words, WEP key management and distribution is outside the scope of 802.11.

WEP-Based Authentication

The first feature of WEP is to prevent unauthenticated users from gaining access to the WLAN network. STA attempting to gain access to the network must send an authentication frame containing, among others, its asserted identity to the AP, which replies with another authentication frame transporting a challenge text. The device encrypts the challenge text using WEP with the shared key and its own initialization vector (IV), concatenates the encrypted output to the IV, and sends the result to the AP:

STA→AP: Authentication Request (STA asserted identity)
AP→STA: Challenge
STA→AP: WEP(Challenge, IV, Key) = Challenge XOR RC4(IV | Key)
AP→STA: Success <or> Reject.

Using the same key, the AP decrypts the response and verifies that the decrypted text matches the challenge text it sent to the device, before accepting or denying device access to the network.

WEP Confidentiality and Data Integrity

WEP uses a 40-bit key that is concatenated to a 24-bit IV to form the traffic key. The resulting key is used as an input to the RC4 pseudo-random number generator (PRNG) to generate a pseudo-random key sequence.

When a device encrypts data using WEP, it calculates the integrity check value (ICV) over the data to be sent (ICV is implemented as a CRC-32-bits). The device concatenates the data and the ICV before the result is XORed with the key sequence (in a typical stream cipher manner).

Upon reception, the AP retrieves the IV from the arrived packet to generate the same pseudo-random key sequence. Then, the AP XORs the key sequence to the received frames and computes the ICV of the decrypted text, comparing it to the ICV of the received packet. If the two ICV do not match, the AP sends an error indication to the sender.

Note that due to interference, out-of-order or lost-packet rates are plausible within 802.11 communication channels. Therefore and in order to ensure the so-called self-synchronisation property, WEP uses a per-packet RC4 key and generates a separate keystream per packet. For that, WEP concatenates a per-packet IV to the WEP key.

ISSUES IN WLAN SECURITY

Due to the shared nature of the wireless medium, it is easy to create associations with unprotected wireless networks. Consequently, unauthorized STAs are able to launch attacks on a wireless network, for example, to affect the WLAN performance or to get an Internet access, as well as on another STA to eavesdrop on its established association. Attacks are classified as active and passive.

A passive attack is an attack where an unauthorized attacker monitors or listens on the communication between two or more parties. Active attacks have the possibility of inflicting undetected corruption on the data in transit by manipulating the cipher text in special ways that do not change its built-in cyclic redundancy checks.

WLAN devices broadcast their MAC addresses over-the-air and it is therefore easy to observe the MAC address for an associated mobile station and spoof it to masquerade as a legitimate device.

Due to the nature of WLAN, intruders can flood the open medium access and are able to execute denial-of-service attacks (DoS) to bring down WLAN access or services. An attacker may launch denial-of-service attacks by spoofing, replaying, or generating management frame packets.

Another problem related to the open medium is jamming WLAN frequencies. Jamming against WLAN is almost impossible to prevent and can be executed easily as noise or interference on channels that deliver WLAN services. For example, in a military environment, jammers are often located in helicopters as the line-of-sight propagation gives them an advantage over communication transmitters located on the ground (Stahlberg, 2000).

WLANs are also vulnerable to session hijacking attacks due to the lack of authentication of the management frames as well as to the WLAN state machines. Session hijacking is a combination of DoS and identity spoofing attacks and it can be launched by 1) eavesdropping on the medium to discover the MAC address of a legitimate station and/or of the AP, 2) deauthenticating the legitimate station to terminate its connection to the AP (spoofing STA or spoofing AP addresses), and 3) using the eavesdropped MAC to reauthenticate to a different or to the same AP on the same WLAN.

WEP Weaknesses

Shared key authentication was designed to help in reducing attacker activities against WLAN. Unfortunately, WEP has turned out to be much less secure than intended. Fluhrer, Mantin, and Shamir's (2001) paper entitled "Weaknesses in the Key Scheduling Algorithm of RC4" describes how an attacker can intercept transmissions and gain unauthorized access to wireless networks. Other problems are related to the insufficient IV length (thus permitting to decrypt frames without key knowledge), absent key management (on the one hand resulting in manual settings and typically weaker alphanumeric keys, and on the other hand directly exposing the long term secret), and to the absent message integrity checking (the available CRC32 integrity does not depend upon the keys and mainly targets transmission problems; it is therefore possible to alter a packet whose content was known even if it had not been decrypted). More information on WEP attacks may be found by Borisov, Goldberg, and Wagner (2001).

In a WLAN context, a passive attack takes advantage of several weaknesses in the key-scheduling algorithm of RC4. It could be done also by a comparison of the encrypted version of a known message (e.g., TCP fields) to repetitive IV-based encryption combinations of the known text and to reveal the secret key (Morrison, 2002). In fact, the 24-bit IV implies that 2^{24} packets can be protected with the same key, before changing the key. Because the IV is relatively short, and is transmitted in the clear text, it will be repeated with sufficient frequency that the rest of cipher can be relatively easily cracked. On the other hand, WEP by its design cannot efficiently reduce overhead of denial-of-service attacks. In particular, it does not protect beacon packets, or the part of the packet header, which includes the MAC address unencrypted. Consequently, it is not hard to infiltrate the WLAN using WEP.

Consequently, a dedicated task group called 802.11i has been set up by IEEE to create a replacement security solution. The released IEEE 802.11i amendment introduces an improved security mechanism called Wi-Fi protected access (WPA) to solve WEP-related authentication and confidentiality problems and to introduce an efficient frame integrity scheme. 802.11i security solution (called robust secure network or WPA2) uses a new counter-mode/CBC-MAC protocol (CCMP) cipher based on the advanced encryption standard (AES) instead of RC4.

802.1X, WPA, AND IEEE 802.11I (WPA2)

IEEE 802.11i is a dedicated task group to specify and to create a replacement security solution. It provides enhanced security services and mechanisms for the IEEE 802.11 medium access control beyond the features and capabilities provided by WEP. These security services are established by defining temporal key integrity protocol (TKIP) and counter-mode/CBC-MAC protocol (CCMP) that provide more robust data protection mechanisms than what WEP affords. 802.11i also introduces the concept of a security association and defines security association management protocols called the 4-way handshake and the group key handshake. Also, it specifies how IEEE 802.1X may be utilized by IEEE 802.11 LANs to effect authentication.

The IEEE 802.11i architecture usually contains or implements the following components:

- 802.1X for authentication, entailing the use of IETF's EAP and an authentication server.
- Robust security network (RSN) for keeping track of associations.
- AES-based CCMP to provide confidentiality, integrity, and origin authentication. Another important element of the authentication process is the four-way handshake, explained below.

WPA

Because WEP has been shown to be totally insecure and in order to strengthen the weak keys used by WEP, 802.11 Working Group has proposed a new WPA protocol called TKIP. This protocol is designed to strengthen the security of 802.1X networks and to leverage the existing WEP-enabled WLAN network interface card (NIC), while remaining backward compatible with existing hardware (no change in the hardware engine). This is done by distributing firmware/software upgrades including new algorithms to be added to WEP, such as message integrity code (MIC) and per-packet key mixing function.

TKIP uses a key scheme based on RC4, but unlike WEP that uses the master key for authentication and per-packet encryption, TKIP extends this key hierarchy to reduce the exposure of the master secret and to provide per-packet key mixing, a message integrity check as long as a rekeying mechanism. Consequently, TKIP ensures that every data packet is sent with its own unique encryption key. Moreover, it includes a key hash function to improve resistance against Fluhrer attacks (Fluhrer et al., 2001) and MIC and it uses 802.1X for key management and establishment. The MIC prevents forged packets from being accepted. Thanks to per-packet key mixing, it is very hard for an eavesdropper to correlate the IV and the per-packet key used to encrypt the packet (Chandra, 2005). More precisely, TKIP hashes the combination of the IV value, the data encryption key (derived from the master secret), and the MAC address. This mechanism addresses the WEP problem when concatenating the key with the IV to form the traffic key, and then reducing the ability of the related key attack.

Key Hierarchy

The master secret used in key hierarchy can be a preinstalled key or a per-session key. In fact, TKIP can be used with an IEEE 802.1X authentication server, which shares a master key with each user as a consequence of a successful authentication process as well as in a preshared key (PSK) mode where all authorized users share a PSK. These two modes target two distinct environments respectively, enterprise and home networking.

As we cited before, TKIP extends the WEP key hierarchy to reduce the exposure of the (long term) master secret and to provide per-packet key mixing, a message integrity check as long as a rekeying mechanism. This extension is shown in the following figure. At a given layer, the different keys are generated by applying the pseudo random function (PRF) on, among others parameters, the key of the upper layer and the MAC addresses of the two endpoints.

Preshared Key

As we cited before, 802.11i security solution uses 802.1X (see next section) that requires a logical authentication server entity. However, 802.11i defines the preshared key solution as an alternative to 802.1X-based master key establishment. This solution can be used for home or small networks

Figure 1. WPA Key hierarchy

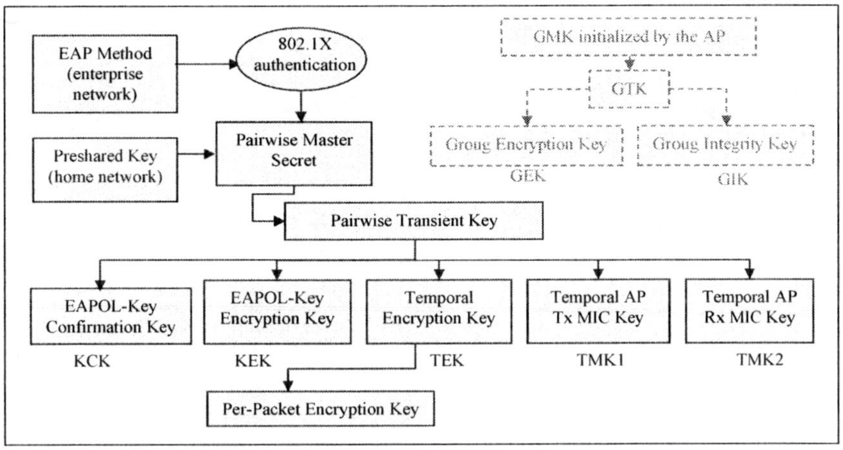

and does not require installation of an authentication server.

The PSK is 64 hexadecimal digits or a pass phrase 8 to 63 bytes long, in which each STA has its own PSK tied to its MAC address and uses it to get access to the network. The key hierarchy is showed in Figure 1. The PSK is however used directly to compute the pair-wise transient key (PTK). The rest of the key computation process remains unchangeable.

The PSK is a 256-bit random value or a pass phrase 8 to 63 bytes long, in which each STA has a PSK tied to its MAC address and uses it to get access to the network. The key hierarchy is showed in Figure 1. The PSK is however used directly to compute the PTK. The rest of the key computation process remains unchanged.

IEEE 802.1X

IEEE 802.1X is introduced for port-based network access control. It provides authentication to stations attached to a LAN port, establishing a point-to-point connection in case of success or preventing access from that port if authentication fails.

802.1X uses three terms:

- **The Supplicant:** A station that requests access to the network offered by the authenticator. It dialogue with the authentication server through the authenticator.
- **The Authenticator:** Typically a wireless access point that controls the state of each port (open/close) and mediates an authentication session between the supplicant and the authentication server.
- **The Authentication Server:** Typically a (remote authentication dial in user service) RADIUS server that performs the authentication process on behalf of the authenticator. The resulting decision consists of whether the supplicant is authorized to access the authenticator's network. Note that 802.1X does not require use of a central authentication server, and thus can be deployed with stand-alone bridges or access points, as well as in centrally managed scenario (802.1, 2004).

The most important component in 802.11i architecture is the IEEE 802.1X port access entity (PAE), which controls the forwarding of data to and from the MAC. A STA always implements a Supplicant PAE and implements EAP peer role, and an AP, acting as an Authenticator, always implements an Authenticator PAE and implements the EAP Authenticator role.

802.1X is based on EAP, which is a powerful umbrella that shelters multiple authentication

Figure 2. 802.1X messages exchange between a supplicant, an authenticator, and the authentication server

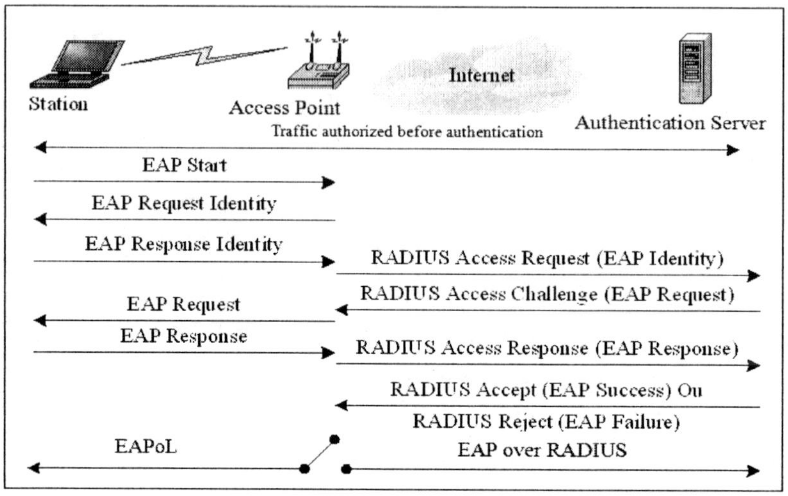

methods. When IEEE 802.1X authentication is used within 802.11 networks, EAP is used transparently between the station and the (usually remote) authentication server and relayed through the AP. 802.1X requires the cooperation between the authentication server and an EAP method. In the case of a wireless LAN, the EAP method is required to perform mutual authentication and key management and distribution\REF-RFC-REQ-EAP-WLAN. Using the flexibility proposed by the IEEE 802.1X architecture, multiple EAP-based security protocols and mechanisms such as EAP-SIM (Haverinen & Salowey, 2006), EAP-TLS (transport layer security) (Aboba & Simon, 1999), and protected-EAP (Palekar, Simon, Zorn, Salowey, Zhou, & Josefsson, 2004) are proposed. These EAP methods are used with the 802.11i (or WAP2) and WPA standards in order to establish authenticated access and key calculation and distribution.

IEEE 802.11i (WPA2)

The procedures defined in 802.11i adopt the key-hierarchy defined by WPA and provide fresh keys by means of protocols called the 4-way handshake and group key handshake. 4-way handshake is a pair-wise key management protocol used to confirm the mutual possession of a pair-wise master key (PMK) by two parties and to distribute a group temporal key (GTK). Several keys are established as a result of a successful authentication. The keys are derived from the PMK (in particular, the pair-wise transient key).

802.11i defines two key hierarchies: (a) pair-wise key hierarchy to protect unicast traffic and (b) GTK, a hierarchy consisting of a single key to protect multicast and broadcast traffic. Furthermore, it defines TKIP (uses existing hardware) and CCMP (needs additional hardware) to repair the problems caused by WAP. TKIP provides stronger security through a keyed cryptographic message integrity code (MIC), an extended IV space, and a key mixing function. And the CCMP is used to provide data confidentiality, integrity, and replay protection.

4-Way Handshake

Once the authenticator and the mobile station have agreed upon a shared PMK, they can begin a 4-way handshake: STA represents the station; STAA and AA, SNonce and ANonce, represent the MAC address and the nonce of the station and authenticator, respectively; SN is the sequence number; msg1, msg2, msg3, and msg4 are indicators of different message types; and $MIC_{EAPOL\text{-}KCK()}$ represents the

Figure 3. 802.11i 4-way handshake

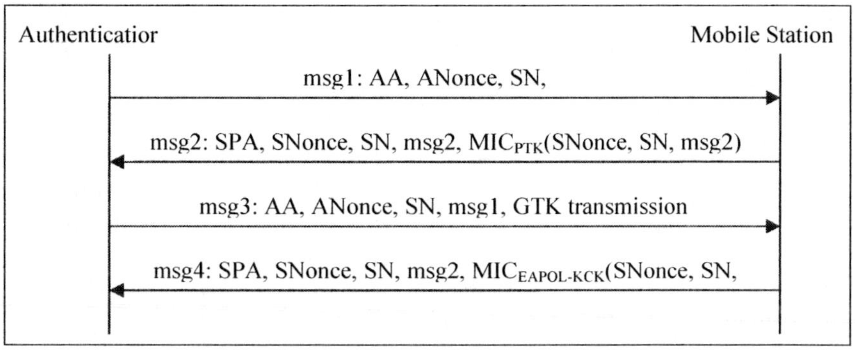

message integrity code calculated for the contents inside the bracket with the fresh PTK.

EAP

EAP is the IETF standard for extensible authentication for network access. It was designed to enable an extensible OSI layer 2 authentication before the IP configuration could be acquired. Originally developed for use with point-to-point protocol (PPP) (Simpson, 1994), it has subsequently also been applied to IEEE 802 wired networks (802.1, 2004) and wireless networks such as 802.11 (WPA, 2003).

EAP can be viewed as a transport framework and supports multiple authentications mechanisms such as EAP-TLS, EAP-SIM and EAP-AKA, without having to prenegotiate a particular one. Authenticators do not need to understand each request type and may be able to simply act as a pass-through agent for a "back-end" server on a host. The authenticator starts the EAP exchanges with a port closed state; it needs however to look for the success/failure sent by the authentication server to open/close the port.

EAP packets include all relevant information about the required authentication scheme, for example, authentication method and packet code (request, response, success, or failure), and allow for method negotiation (a special NAK type). The exact content of these packets is up to the chosen EAP authentication mechanism. The progression of an authentication procedure also depends on the chosen authentication mechanism.

Authentication Server

Typically, 802.1X performs authentication and key management through a server, such as AAA RADIUS (authentication, authorization and accounting, remote authentication dial in user service) or DIAMETER. RADIUS (Aboba, 2006; Rigney, Willens, Rubens, & Simpson, 2000) is a widely deployed AAA protocol. As we cited, the EAP packets are carried by EAPOL directly over the wireless interface between the STA and the access point, and by EAP over RADIUS between the access point and the authentication server. This effectively creates an EAP conversation channel between the station and the authentication server, which allows the supplicant to authenticate. The authentication is realized by the chosen authentication mechanism. This phase will also generate a secret key that will be used in the key hierarchy. The generated key of EAP between the STA and the AAA server is therefore conveyed to the AP using the AAA protocol.

EAP-TLS

EAP-TLS defines the transport of transport layer security (Dierks & Allen, 1999) in EAP. TLS is one of the most deployed security protocols, which is mainly due to its integration in navigators. TLS is an IETF-standardized authentication method derived from the secure sockets layer (SSL) protocol.

TLS authentication within EAP is quite straightforward. The TLS handshake packets are encapsulated in an appropriate EAP form and transported between the station and the authentication server. Because of the size of the certificates exceeding typical link MTUs, EAP-TLS additionally defines fragmentation. When the TLS authentication dialog succeeds, the authenticator (host requiring the authentication on behalf of a supplicant) gets the authorization information delivered within the RADIUS access-accept message and access to the network is granted.

EAP-TLS defines the full Handshake phase that involves the exchange of X.509 certificates and the cryptographic information to allow peers to be authenticated. This step requires several operations. First, peers must verify the integrity of certificates and should generally support certificate revocation messages (the peer may not have Internet connexion and therefore it can use online certificate status protocol (OCSP) to obtain the revocation status of a certificate [Blake-Wilson et al., 2003]). Also, the certificate must be verified to ensure it is signed by a trusted certificate authority (CA). Finally, the client (i.e., station) should be able to view the information about the certificate and the CA root.

EAP-TLS indicates that a secure connection may be terminated and resumed later. This (abbreviated handshake) phase may be established if the client and the server agree. During this phase, the client and the server will use the master key, which is calculated during the last full handshake phase, to mutually authenticate and to calculate and generate new keys for the secure channel. After that, they verify the integrity of their exchanged handshake messages and then begin to exchange data over the secure connection.

The abbreviated handshake allows the client and the server to avoid several expensive cryptographic operations such as private key computations, client/server certificate decoding and verification, online consultation of certificate revocation list (CRL), and generation and encryption/decryption of the premaster secret key, which is used for generating the master secret key. Moreover, the abbreviated TLS handshake shortens the authentication delay and preserves the precious radio bandwidth.

Security Problems

802.1X is intended to provide strong authentication, as well as key management and distribution. However, 802.1X suffers from some security problems related to its use in conjunction with WLAN 802.11. This conjunction suffers from the absence of an ensured synchronization of the various state machines, causing potential attacks such as man in the middle, session hijacking, and DoS. Furthermore, others attacks are possible because EAP does not include any integrity information to its transported packets. For example, at the end of the authentication process, the authenticator will send an EAP notification message to indicate the success or the failure of that process. Since this notification does not include any integrity protection data, an attacker can easily replace an EAP failure with EAP success and deny the access to the WLAN network.

On the other hand, He and Mitchell (2004) demonstrate a DoS attack against a 4-way handshake of 802.11i. The attack can be realixed by impersonating the authenticator, composing a Message 1, and sending to the STA. The attacker sends a forged Message 1 to the STA after Message 2 of the 4-way handshake. The STA will calculate a new PTK corresponding to the nonces for the newly received Message 1, causing the subsequent handshakes to be blocked because this PTK is different from the one in the authenticator. The attacker can determine the appropriate time to send out Message 1 by monitoring the network traffic or just flooding Message 1 with some modest frequency.

Some available tools can be used to crack WAP in PSK mode, especially coWPAtty, which is a brute-force cracking tool, which means that it systematically attempts to crack the WPA-PSK by testing numerous passwords, in order, one at a time (Fogie, 2005).

Additional Security Needs: Privacy and Identity Protection

During the authentication and security association phases, almost all security protocols, including 802.1X and EAP methods, exchange identity related data in clear text and without any encryption. Therefore, security parameters flowing in the network could potentially be logged, archived, and searched.

Basically, certificates are issued by a trusted third party linking the identity of the certificate owner to the public key, whereas the shared secret is managed through its identifier. Certificate or shared key identifiers are usually sent in clear text and consequently, entities cannot protect their identities from eavesdropping. Thus, an intruder can learn who is reaching the network, when, and from where, and hence, track users by correlating client identity to connection location. Especially in WLAN, where the access medium is open to eavesdroppers, and the mobility is a reasonable service, the location tracking can be a serious security issue. The PEAP and EAP-TTLS authentication methods can be used to protect user identity. Both are two-phase protocols with the first phase used to establish a TLS with only server authentication and the second phase used to deliver, among others, the user identity.

Privacy and identity protection are increasingly required for 802.1X/EAP and consequently, research is being carried out to add credentials and identity protection to EAP methods, especially to EAP-TLS. In this latter method, the client certificate is sent in clear text and therefore, an attacker can easily sniff packets conveying the client credentials. To avoid sending identity information in clear text during the TLS session, Hajjeh and Badra (in press) extend TLS with an enhanced, completely backwards compatible mechanism. The client identity protection is provided by symmetrically encrypting the client certificate with a key derived from the TLS master secret,

Hardware Security in WLAN

Many agencies (GAO, 2001) require the use of smart cards to overcome the vulnerabilities of the storage of private and shared keys. In fact, without smart cards, unauthorized access can be easily established to an authorized device (e.g., station) to retrieve confidential and personal data stored on it.

A smart card is a portable and tamper-resistant computer. It provides data security, data integrity, and personal privacy and supports mobility. Furthermore, major application areas including mobile communication use smart card to convey user subscription and identification information as well as to provide user identity and to build computer and network access.

In the 802.1X/EAP context, (Urien & Pujolle, 2005) describes the interface of the EAP protocol in smart cards, which can store multiple identities associated to EAP methods and appropriate credentials. It presents implementations of the EAP-TLS smart cards, which securely stores TLS security parameters, such as client X509 certificate, client private RSA key, and CA public key. For more information regarding the EAP smart card configuration and test steps, please refer to the OpenEAPSmartCard (2006), which is an open Java card platform for authentication in Wi-Fi and WLAN networks.

THE UNIVERSAL ACCESS METHOD

A different approach to authentication and authorization for WLAN is that based on Web-based unlicensed mobile access (UAM), the most prevalent form of access to WLAN. This approach defines a sign-on usage model using the user navigator or Web browser and it is adopted by a number of WLAN hotspots providers. The Web-based UAM approach is very simple. When the user attends to get Internet access through a given hotspot, this latter will redirect the user's browser to a local Web server. After redirection, the user will be invited to be authenticated by entering its credentials (e.g., username, password). These credentials are tunnelled through a secure session, typically established using TLS.

UNLICENSED MOBILE ACCESS

UMA stands for Unlicensed Mobile Access; a technology provides access to GSM and GPRS mobile services over unlicensed spectrum technologies, including Bluetooth and WLAN. By deploying UMA technology, service providers can enable subscribers to roam and handover between cellular networks and public and private unlicensed wireless networks using dual-mode mobile handsets. With UMA, subscribers receive a consistent user experience for their mobile voice and data services as they transition between networks (UMA, 2005).

The UMA architecture uses the following standard IP-based protocols without any modifications. It uses IP/TCP to provide a tunnel for GSM/GPRS signalling and SMS, IPSec ESP to provide a secure tunnel for user and control plane traffic, IKEv2 (Kaufman, 2005), and EAP-SIM (Haverinen & Salowey, 2006), and EAP-AKA (Arkko & Haverinen, 2006) for authentication and establishing and maintaining a security association between MS and UNC.

EAP-SIM and EAP-AKA introduce the smart card use (e.g., SIM cards). Due to the smart cards advantages cited before, a potential attacker will not be able to access the smart card memory to spoof or retrieve the private and personal data. Moreover, the attacker will not be able to have these data in clear text outside the smart card since UMA is operating over IPSec ESP, which provides a strong authenticated and encrypted session. However, cares must be taken when implementing UMA technology on an open terminal. To have a focus study on the impact of open terminal platforms when UMA technology is implemented with GSM, please refer to Grech and Eronen's (2005) work.

SECURITY PERFORMANCES IN WLAN

Security mechanisms usually involve using of certificates, public-key infrastructures, symmetric encryption/decryption, digest computation, and so forth. Therefore, 802.1X/EAP will add a performance impact, varying upon the deployed security protocol. Several studies have evaluated the security performance of WLAN and the performance impacts of WEP, WPA, EAP-TLS, and other authentication protocols.

Baghaei (2003) provides a study comparison of the following eight security solutions used by 802.11:

1. No security: no security mechanism activated with default configuration.
2. MAC address authentication carried out at the AP.
3. WEP authentication.
4. WEP authentication with 40-bit WEP encryption.
5. WEP authentication with 128-bit WEP encryption.
6. EAP-TLS authentication.
7. EAP-TLS with 40-bit WEP encryption.
8. EAP-TLS with 128-bit WEP encryption.

This study comparison includes an analysis of the effect of different TCP and UDP packet sizes on performance of secure networks. It shows that WEP encryption significantly degrades the

Figure 4. Throughput of TCP, UDP traffic in a congested network

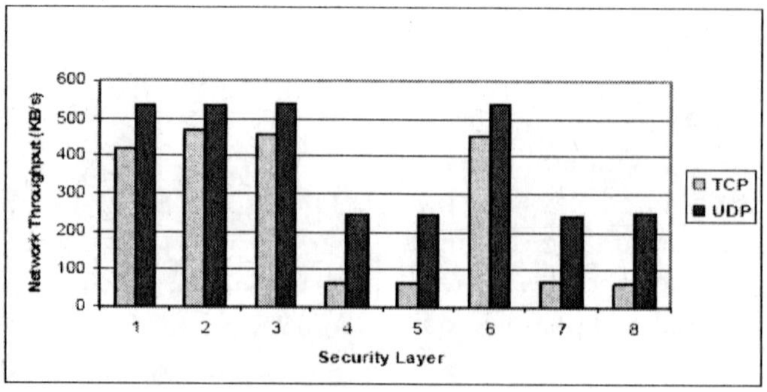

Figure 5. Computing times distribution for a smart card

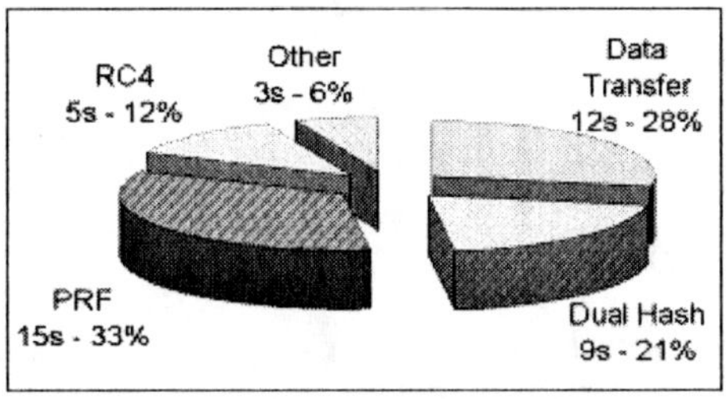

performance of congested wireless networks. Network performance degradation increased as the number of clients was increased under all security mechanisms.

On the other hand and in order to show the impact of smart cards use within 802.1X/EAP, we implemented EAP-TLS on smart cards, in which performance, benefits, and drawbacks are discussed and analysed by Urien, Badra, and Dandjinou (2004).

Figure 5 shows the repartition of computing times during the authentication phase. The smart card (10 MHz, 8 bits CPU, 2304 bytes RAM bytes, 96 Kbytes 32 Kbytes ROM, 32 Kbytes E2PROM) processes the EAP-TLS protocol in about 5 seconds (Urien & Badra, 2006). Note that benchmarks are performed on a 1 GHz Intel processor PC and only about 50 ms are required to execute an EAP-TLS session. This demonstrates the cost and performance influence of using smart cards, which are required for credentials and private data storing.

CONCLUSION

Wireless technologies have evolved phenomenally over the last few years. Wireless transmission has a big impact on new services and applications because it is the method for data communication for, among others, cellular phones, text pagers, and Wireless LAN 802.11. In this chapter, we focused on WLAN security threats, which extend on several levels, from the identity spoofing to the traffic analysis.

WLAN security risks have increased exponentially as wireless services have become more popular. The risks represent any malicious and undesirable event on the various applications, which possibly suffer from faults facilitating treat concretization. Risks can result in sniffing and hijacking of sensitive and personal data over the link for unprotected Internet access. The consequences are therefore variants (Hurley, 2002). It can eat up bandwidth, but it could pose a darker issue as virus writers can use the access to anonymously send viruses out.

In answer, WLAN defined, among other, the 802.1X standard, providing a framework for authenticating and controlling user traffic to a protected network, as well as dynamically varying and exchanging encryption keys between the wireless entity and the authenticator server. This is done using EAP methods, which are also deployed jointly with the 802.11i and WPA standards. Implementing WLAN technologies in a secure network requires on one hand a combination of these security measures. On the other hand, organizations need to adopt security measures and practices that help bring down their risks to a manageable level. In early 2006, therefore, ISO members voted the IEEE's 802.11i standard for adoption.

REFERENCES

802.1X. (2004). *IEEE Standards for local and metropolitan area networks: Port based network access control* (IEEE Std 802.1X-2004).

802.11i. (2004). *Institute of electrical and electronics engineers, supplement to standard for telecommunications and information exchange between systems: LAN/MAN specific requirements. Part 11: Wireless LAN medium access control (MAC) and physical layer (PHY) specifications: Specification for enhanced security* (IEEE 802.11i).

Aboba, B., Blunk, L., Vollbrecht, J., Carlson, J., & Levkowetz, H. (2004). *Extensible authentication protocol (EAP)* (RFC 3748).

Aboba, B., &. Calhoun, P. (2003). *RADIUS (remote authentication dial in user service) support for extensible authentication protocol (EAP)* (RFC 3579).

Aboba, B., & Simon, D. (1999). *PPP EAP TLS authentication protocol* (RFC2716).

Arbaugh, W., Shankar, N., & Wan, Y. (2001). *Your 802.11 wireless network has no clothe*. Retrieved October 16, 2007, from http://www.cs.umd.edu/~waa/wireless.pdf

Arkko, J., & Haverinen, H. (2006). *EAP-AKA authentication* (RFC 4187).

Baghaei, N. (2003). *IEEE 802.11 wireless LAN security performance using multiple clients* (Honours Project Report).

Blake-Wilson, S., et. al. (2003). *Transport layer security (TLS) extensions* (RFC 3546).

Borisov, N., Goldberg I., & Wagner D. (2001). *Intercepting mobile communications: The insecurity of 802.11 by.*

Chandra, P. (2005). *Bulletproof wireless security.* Elsevier.

Dierks T., & Allen, C. (1999). *The TLS protocol version 1.0* (RFC 2246).

Fluhrer, S., Mantin, I., & Shamir, A. (2001). *Weaknesses in the key scheduling algorithm of RC4* (LNCS 2259).

Fogie, S. (2005). *Cracking Wi-Fi protected access (WPA): Part 2.*

GAO (United States General Accounting Office). (2001). *Information security: Advances and remaining challenges to adoption of public key infrastructure technology* (Report to the Chairman, Subcommittee on Government Efficiency, Financial Management and Intergovernmental Relations, Committee on Government Reform, House of Representatives).

Grech, S., & Eronen, P. (2005). Implications of unlicensed mobile access (UMA) for GSM security. In *Proceedings of IEEE/Create-Net SecureComm 2005, The First International Conference on Security and Privacy for Emerging Areas in Communication Networks,* Athens, Greece, (pp. 3-12).

Hajjeh, I., & Badra, M. (in press). *Identity protection ciphersuites for transport layer security* (IETF Draft).

Haverinen, H., & Salowey, J. (2006). *EAP-SIM authentication* (RFC 4186).

He, C., & Mitchell, J. C. (2004). Analysis of the 802.11i 4-way handshake. In *Proceedings of the 2004 ACM Workshop on Wireless Security* (pp. 34-50). New York: ACM Press.

Hurley, E. (2002). *Company tackles wireless network security risks.* News Writer.

Kaufman, C. (Ed.). (2005). *Internet key exchange (IKEv2) protocol* (RFC 4306).

Khan, J., & Khwaja, A. (2003). *Building secure wireless networks with 802.11.* Indiana: Wiley.

Miller, B. R., & Hamilton, B. A. (2002). Issues in wireless security WEP, WPA and 802.11i. In *Proceedings of the 18th Annual Computer Security Applications Conference.*

Morrison, J. D. (2002). *IEEE 802.11 WLAN security through location authentication.* Naval Postgraduate School.

OpenEAPSmartCard. (2006). Retrieved October 16, 2007, from http://www.infres.enst.fr/~urien/openeapsmartcard/

Palekar, A., Simon, D., Zorn, G., Salowey, J., Zhou, H., & Josefsson, S. (2004). *Protected EAP protocol (PEAP) version 2* (IETF Draft).

Rigney, C., Willens, S., Rubens, A., & Simpson, W. (2000). *Remote authentication dial in user service (RADIUS)* (RFC 2865).

Simpson, W. (1994). *The point-to-point protocol (PPP)* (RFC1661).

Stahlberg, M. (2000). Radio jamming attacks against two popular mobile networks. In H. Lipmaa & H. Pehu-Lehtonen (Eds.), *Proceedings of the Helsinki University of Technology. Seminar on Network Security. Mobile Security.* Helsinki University of Technology.

UMA. (2005). Retrieved October 16, 2007, from http://www.umatechnology.org/

Urien, P., & Badra, M. (2006). Secure access modules for identity protection over the EAP-TLS - Smartcard benefits for user anonymity in wireless infrastructures. In M. Malek, E. Fernández-Medina, & J. Hernando (Eds.), *SECRYPT 2006, Proceedings of the International Conference on Security and Cryptography,* Setúbal, Portugal, (pp 157-163).

Urien, P., Badra, M., & Dandjinou, M. (2004). *EAP-TLS smartcards, from dream to reality.* Paper presented at the Fourth IEEE Workshop on Applications and Services in Wireless Networks.

Urien, P., & Pujolle, G. (2005). *EAP-support in smartcard* (IETF Internet Draft).

WLAN. (2003). *Information technology - telecommunications and information exchange between systems—local and metropolitan area networks—specific requirements Part 11: Wireless LAN medium access control (MAC) and physical layer (PHY) specifications* (IEEE Std. 802.11-2003).

WPA. (2003). *Wi-Fi protected access, version 2.0.*

Chapter XLIV
Access Control in Wireless Local Area Networks: Fast Authentication Schemes

Jahan Hassan
The University of Sydney, Australia

Björn Landfeldt
The University of Sydney, Australia

Albert Y. Zomaya
The University of Sydney, Australia

ABSTRACT

Wireless local area networks (WLAN) are rapidly becoming a core part of network access. Supporting user mobility, more specifically session continuation in changing network access points, is becoming an integral part of wireless network services. This is because of the popularity of emerging real-time streaming applications that can be commonly used when the user is mobile, such as voice-over-IP and Internet radio. However, mobility introduces a new set of problems in wireless environments because of handoffs between network access points (APs). The IEEE 802.11i security standard imposes an authentication delay long enough to hamper real-time applications. This chapter will provide a comprehensive study on fast authentication solutions found in the literature as well as the industry that address this problem. These proposals focus on solving the mentioned problem for intradomain handoff scenarios where the access points belong to the same administrative domain or provider. Interdomain roaming is also becoming common-place for wireless access. We need fast authentication solutions for these environments that are managed by independent administrative authorities. We detail such a solution that explores the use of local trust relationships to foster fast authentication.

INTRODUCTION

Wireless local area networks (WLAN) are rapidly becoming a core part of enterprise network access. The IEEE 802.11 standardization has lead to vendor interoperability and rapidly plummeting prices, making wireless access an economically tantalizing alternative to wired access. Currently, enterprise deployment incorporates support for mobility between access points (AP) as well as security and monitoring solutions. Mobility introduces a new set of problems, not present in a wired infrastructure, due to handoffs between network access points. The implications of frequent handoffs to different APs is that for communication security, the IEEE 802.11 standard requires that the mobile node (MN) has to undergo a full authentication process each time it wants to connect to a new AP. The recent security ratifications from the IEEE task group i (TGi) (IEEE 802.11i, 2004) defined several security remedies for WLANs in the standard IEEE802.11i. According to this standard, the complete (full) authentication process involves the use of 802.1X port-based access control architecture, and provides mechanisms for key management (IEEE 802.1X, 2001). An AAA server such as RADIUS (Rigney, Willats, Rubens, & Simpson, 2000; Rigney, Willats, & Calhoun, 2000) is to be used for authentication and key derivation. Following a successful authentication, the MN and the AP are to undertake a four-way handshake protocol for deriving various encryption keying material. Keying material derived in this way then is used in the encrypted (secure) communication sessions between an AP and the MN. Thus the four-way handshake, which does not involve the AAA server, is a *must* in each secure association of an MN to the AP and cannot be avoided.

However, the authentication process, suggested in the 802.11i ratifications using extensible authentication protocol (EAP) over transport layer security (TLS) can introduce significant handoff delays because it involves the exchange of a round of messages between the MN and the AAA server via the AP. It has been shown that a full EAP-TLS authentication (i.e., the full authentication) can take as long as *1.1* seconds (Mishra, Shin, & Arbaugh, 2004). The delay can only increase when the AAA server is located at the ISP's site, topologically far from the AP site. The longer the delay in handoffs, the longer the outage time experienced by applications. While this kind of delay is acceptable for applications with flexible response time requirements, emerging real-time applications, such as wireless voice-over-IP, have stringent delay requirements (Cisco IP phone). Thus, this kind of network delay and outages are detrimental for real-time applications, especially in frequent handoff scenarios, which hinders the success of wireless local networks to support such popular applications.

The aim of this chapter is, therefore, to provide readers with state-of-the-art knowledge on this significant issue, and solutions as found in the industry and literature. The mentioned issue arises from two directions: (1) intradomain handoffs and (2) interdomain handoffs. Thus solutions are needed for both. While various solutions have been mostly proposed for the first direction, we will show that interdomain, or interprovider handoffs are becoming a common place and need specific solutions that are different from the intradomain solutions because of the involvement of more than one administrative authority in the latter cases.

To reduce the handoff delays due to the exchanges of authentication messages when a MN hands off to a new AP (nAP), there have been several proposals from the industry and the research community. These solutions are targeted for providing fast access when changing APs belong to the same administrative network domain.

However, handoffs within a single domain might not always be the case. There are possible scenarios where different service providers need to collaborate to provide continuous connectivity to roaming users for supporting seamless services. In addition, IEEE 802.11 has lead to price levels suitable for the mass consumer market and small operators. This has caused an explosive trend in the deployment of residential gateways (RG) for home networking and wireless hotspots at city areas by various business owners and hotspot providers.

The capacity offered by these APs and residential gateways (RGs) at various sites may not be fully utilized since the traffic patterns typically vary considerably over the course of a day. Thus, there will be unutilized capacity that could be offered to active users despite not being their serving provider or AP.

In this model, each WLAN site that is connected to the Internet via an individual RG or AP can be considered as an individual domain as RGs are owned by individual residential consumers and wireless routers (or APs) at hotspots belong to individual providers or businesses. An example of a current operational commercial system building on this principle is the FON (FON Web site) community where individual subscribers share excess capacity with the global FON community and FON itself provides billing support so that used capacity is billed to a user's own account.

If the current deployment trend continues, in dense residential areas it will be common to find a substantially large number of RGs within range. This also provides the opportunity for load-sharing or load-balancing among the RGs by handing off some visiting connections to other RGs within range when the original RG's link utilization or load increases and affects its home traffic. Therefore, we see stationary handoff scenarios emerging in the multiowner RG access network architecture, which can also apply to the commercial city area hotspots. Depending on the load variation, there may be situations when during an active session a visiting mobile node will have to undergo frequent handoffs to many new RGs. The same applies to the city area hotspot architecture.

The previous proposals and implementations mentioned above aim at supporting a single domain where centralized control is an advantage. In the collaborating scenarios between service providers this is not the case since each domain has its own authentication mechanisms that are closed to other parties. In the RG access network or the city area hotspot architecture, such centralized control is not possible since each RG or AP is under its own authority and administration. Hence, a distributed approach is required for multiowner and multiprovider handoff scenarios.

The rest of the chapter is organized as follows. In the next section we provide background information on the port-based authentication-driven access control as suggested in IEEE802.11i, the security ratifications from the IEEE task group *i*. This section equips the reader with the fundamentals of the authentication-based access control process using 802.1X architecture. In the third section we elaborate fundamental proposals found in the literature that provide fast-authentication schemes applicable in the intradomain handoffs. This section also contains a subsection on comparative analysis of the presented proposals. In the fourth section we provide a possible direction that we have proposed for solving the interdomain handoff latency problem because of authentication delays. We sketch some open issues, and finally, we conclude this chapter.

IEEE 802.11I AUTHENTICATION PROCESS

In this section, we provide fundamental information on the authentication-based access control mechanism for IEEE802.11i, the security ratifications from IEEE802.11. IEEE 802.11i includes the use of the architectural framework of IEEE802.1X, the port-based network access control standard for different link layer technologies such as IEEE 802.3, FDDI, IEEE802.11, and so forth. In the standard, there are three entities involved in the authentication process: the supplicant or the user wireless device, the authenticator or the network port (wireless access point), and the authentication server such as the RADIUS server. Figure 1 shows this setup. The 802.1X standard uses *extensible authentication protocol* (Blunk, Vollbrecht, Aboba, Carlson, & Levkowetz, 2003) to support a variety of authentication mechanisms, of which the *transport layer security* (TLS) providing strong encryption and authentication at the transport layer is the most commonly used mechanism (EAP-TLS) (Aboba & Simon, 1999) within 802.11 networks. Next, we look at the functionalities of the entities of the 802.1X framework in light of the 802.11 network setting:

Figure 1. 802.1X setup

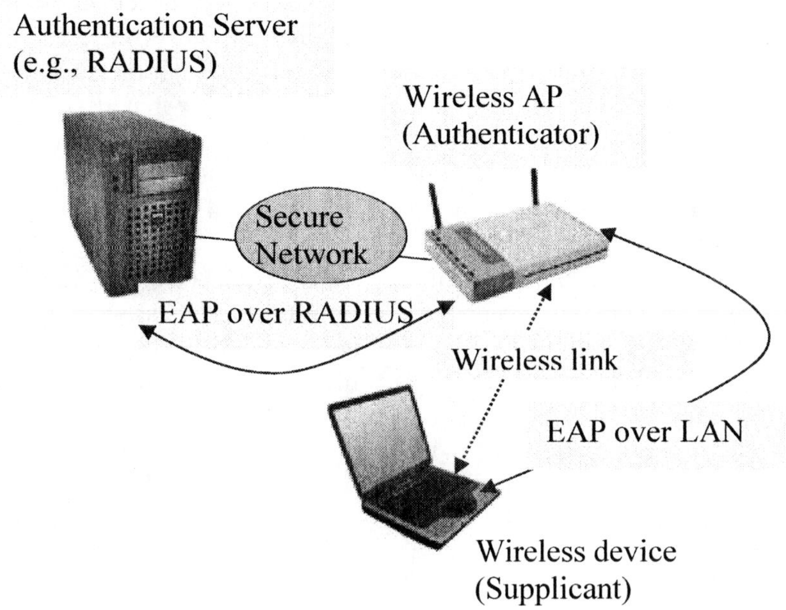

Figure 2. EAP-TLS authentication messages

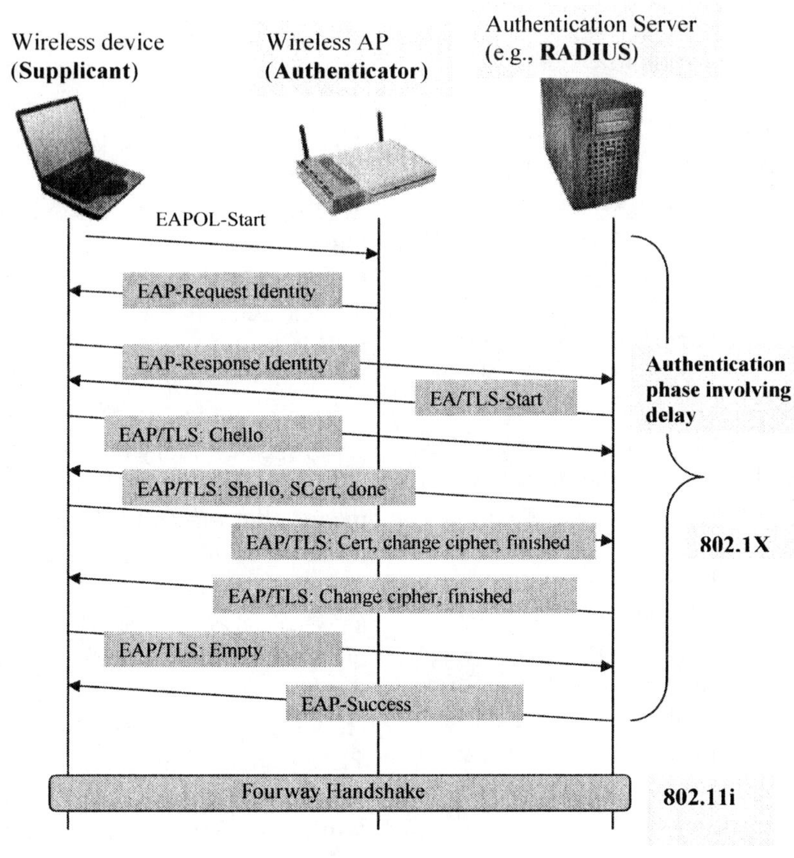

- **Supplicant:** This is a user device seeking link layer connectivity with a network so that it can use the services offered by the network.
- **Authenticator:** This is the wireless AP providing link layer connections to the user devices. In any network, typically there will be many APs. The authenticator liaises with the authentication server by relaying information to and from the supplicant. When the authenticator receives a success message from the authentication server, it allows the supplicant to establish a link layer connection.
- **Authentication server:** This is a central server which helps the authenticator with the authentication decision based on what it knows about the supplicant and the information supplied by the supplicant.

As EAP, and in particular EAP-TLS, serves as the authentication building block for strong authentication security, we discuss briefly these technologies.

A. EAP: EAP communications are in the form of challenge-response method. EAP uses four basic message types: EAP request, EAP response, EAP success, and EAP failure. After a few rounds of request and response exchange, the supplicant is notified of the outcome using EAP success or EAP failure. Regarding the transport of the authentication protocols used, as EAP does not have any addressing mechanism, EAP messages are encapsulated in EAP over local area network (LAN) (EAPOL) protocol between the supplicant and the authenticator, and as a RADIUS message (EAP over RADIUS) between the authenticator and the authentication server.

B. TLS: Specified in RFC2246, when used in 802.1X, the supplicant and the authentication server will undergo a mutual authentication process. At the root of this process are certificates at both of these entities from a common certificate authority (CA). The results of the mutual authentication using these certificates are a strong secret master key (MK), and an initial set of pseudo-random function which will be used to generate additional keying material. Using this function and the MK, a pair-wise maser key (PMK) is generated. The PMK further produces four pair-wise transient keys (PTKs) when used with particular cipher methods, and are used for origin authenticity and confidentiality of the four-way handshake procedure, as well as for data encryption.

Figure 2 shows the full EAP-TLS authentication steps and messages exchanged. At the end of a successful EAP-TLS authentication (EAP success message), there is a four-way handshake process which ensures that the AP and the MN are active, guarantees the freshness and synchronization of the shared encryption key, as well as binds the PMK to the medium access control (MAC) address of the MN.

INTRADOMAIN FAST AUTHENTICATION SOLUTIONS

One of the most significant features of wireless networks is that it does not restrict users to a fixed connection point (e.g., at a desk) while using the network. As long as the user is within the coverage area of a wireless access point or base station, the network connection keeps alive. The predominant mobility-friendly applications are voice and multimedia that can be supported when the user is continuously mobile. However, primarily due to this kind of mobility, the connections need to be switched from one access point to the next, a process called handoff, when the user crosses the coverage boundary of one AP and moves into the next one. Using 802.11i in wireless LANs, this means that the full EAP-TLS authentication has to be performed each time such handoff occurs. It is well-known that real-time applications, such as wireless voice over IP, have stringent delay requirements and can only tolerate moderate packet loss due to network outage occurring at handoffs. However, it has been reported that a full EAP-TLS authentication process can take as long as 1.1 second (Mishra et al., 2004), a number far too large to support the smooth operation of voice and multimedia applications in continuous mobility scenarios[1]. This number can only magnify when the RADIUS server is located topologically far from the AP. As the APs in wireless LANs have very small coverage[2], many APs are required to be installed to cover a certain geographical area of a network. Thus, continuous mobility implies that there will be many handoffs during an active real-time application session, even when the user is within the same network (domain). There needs to be mechanisms to cut down the authentication delay of 802.11i for this kind of intradomain handoffs. Below we discuss the IEEE 802.11i proposed solution, and those found in the literature to tackle this issue.

Preauthentication

This is the solution specified within the IEEE802.11i to support fast authentication at handoffs between APs in the same network domain or extended service set (ESS). In this solution, when an MN is connected with an old AP (oAP), it can initiate EAP-TLS authentication with a new AP (nAP) within the same ESS by sending an IEEE 802.X EAPOL-Start message via the oAP to the nAP. The nAP then may initiate the EAP-TLS authentication with the MN. The distributed system of the ESS has to be configured to forward the authentication messages to the oAP for the MN. While still connected with the oAP, preauthentication for the MN is performed by exchanging all the EAP-TLS authentication messages between the MN and the nAP. The process ends when after deriving the new PMK, the nAP sends the first message of the four-way handshake to the MN. The MN and the nAP must cache the new PMK to be used when the MN finally moves to the nAP. Preauthentication can be performed in advance to a group of APs that the MN may select from, for handing off in the future. At time of handoff, there will not be any more EAP-TLS exchanges, and the four-way handshake can be used straight away to resume the connection process.

While the preauthentication mechanism provides a great way to cut down the authentication delay necessary for supporting real-time applications in wireless LANs, in the current form, no mechanism has been used to select the most likely handoff candidate APs. Thus, there will probably be many instances of preauthentications that will not be utilized at all. This is a waste of resources. Also, when there is a large number of candidate APs, this mechanism does not scale and, in addition, puts extra loads on the AAA server. It is to be noted that the scope of the preauthentication is, however, limited to a single network domain or ESS, making it inapplicable in interdomain roaming scenarios.

Proactive Key Distribution

Proactive key distribution has been proposed as a mechanism to provide fast authentication at handoffs within the *same administrative domain*, by *predistributing* the keys to candidate APs in a *neighbor graph* (Mishra et al., 2004). Thus, this scheme avoids the involvement of the AAA or the RADIUS server for distributing the keys to the nAPs *during* handoffs. When the MN will finally move to the nAP, the key will be already there and the local handshake protocol (four-way handshake) can be used to establish the radio link between the MN and the nAP.

The most important concept of this proposal is the use of the neighbor graph. The neighbor graph is the dynamic identification of the mobility topology of the network: a set of APs that the mobile user device potentially could reassociate to. The authors suggest that this set is typically a small subset of all the APs in the wireless network. By selecting the possible candidate APs for handoffs by a particular MN, the cost of proactively distributing the key to these APs are justified and minimized. The scheme utilizes the concept of a reassociation relationship by which the authors mean that two APs have this relationship if it is physically possible for a given MN to handoff from one to the next. Thus, this relationship depends on factors such as physical distance between two APs and placement of the APs. The authors suggest that the neighbor graphs can be autonomously learned and maintained by the wireless network, and can be maintained either in a centralized or distributed manner. In their implementation, the authors have stored this information in the centralized manner, in the RADIUS server.

The authors propose that instead of distributing the original PMK to all the neighbor graph APs, the PMK is used to derive PMKs depending on the instance of reassociation (e.g., n^{th} reassociation) using a proposed equation. Special RADIUS messages have been also introduced to aid the key distribution process: NOTIFY-REQUEST, NOTIFY-ACCEPT, and ACCESS-ACCEPT. Once the MN completes a full EAP-TLS authentication, the AAA server sends a NOTIFY-REQUEST message to all the APs in the neighbor graph. This message informs the APs that a given MN may roam to their coverage. It is up to the APs to decide whether they want the security information (the PMK) for the MN. If the AP decides to get the security information at this stage, it sends a NOTIFY-ACCEPT message to the AAA server, and the AAA server sends an ACCESS-ACCEPT message in return to the AP containing the appropriate PMK and an authorization for the MN to remain connected to the network. From the experimental results, it has been shown that the average latency of the full authentication reduces to around 50ms from that of 1.1 second.

The scheme provides a practical and feasible way for maintaining the quality of real-time applications while the MN moves about in the same network. However, this imposes extra functionality and loads on the AAA server, because it has to send requests to candidate APs asking if they want the security key for the MN before it hands off to the APs. This centralized approach where a single AAA server controls and manages the key distribution will suit well the scenarios where the WLAN sites are all under the tight control of one central AAA server such that the server can derive and decide on the candidate APs for the MN's next move. This proposal will not be directly applicable to interdomain roaming scenarios.

The proposal from Mishra et al. (2004) has similarity with preauthentication proposal from IEEE 802.11i in the sense that (some) steps of the authentication process is initiated even before the MN moves to the nAP, that is, the proactive nature of the schemes. The two proposals differ in the sense that in preauthentication, it is up to the MN to choose (using no particular guideline) APs in the network to complete authentication before it performs the next handoff, but in the case of proactive key distribution scheme, *only* the APs the neighbor graph can get the PMK (some APs in the neighbor graph may decide not to ask for the key at this stage). Also, the predistribution of the PMK scheme does not involve the MN in the process of distributing the PMKs to the neighbor graph APs, whereas the preauthentication scheme involves the MN to complete the preauthentication process with the nAPs.

Proactive Key Caching

An industry solution, namely proactive key caching (PKC), is an extension of Airespace Inc.'s[3] wireless enterprise platform, developed along with Funk Software[4] and Atheros Communications (Atheros Communications). In PKC, the MN can use the same master key to roam across an *Airespace network*, visiting one AP to the next. This eliminates the need for RADIUS authentication at each handoff; only the four-way handshake will be required. Airespace has a centralized policy engine for creating and maintaining security parameters across the *entire enterprise*. The use of the central policy engine in the network also leads this solution to be centralized and suitable only for a single administrative domain.

Predictive Authentication

This proposal from Pack and Choi (2002) is a predictive-authentication scheme based on the selection of a frequent handoff region (FHR) which works in a centralized manner. The main idea is to formulate a FHR consisting of a number of APs in a public access LAN by using a FHR selection algorithm, and taking into account the user mobility and traffic pattern. The FHR APs are the ones that the MN is likely to associate with in the near future. The MN is preauthenticated to all the APs within the FHR so that when the MN handoffs from one AP to the next within that FHR, there is no time wasted in communicating with the RADIUS server.

The authors use the notion of movement ratio between APs which determines the handoff probability of a particular AP in the network. The movement ratio is affected by the user mobility and AP location. Movement ratio between APs can be measured by using an event logging system which logs the handoff information including login time and handoff times to different APs. Using a given equation and the log information, the handoff ratio is then calculated. To determine the FHR APs, the users' service level is considered as well. If the user can tolerate service disruptions during handoffs, less APs can be included in the FHR, and vice versa. Obviously, if there are more APs in the FHR, there will be more resources used for preauthenticating to them than if there were less APs in the FHR. Thus this differentiation based on service level serves an important purpose.

The key distribution in IEEE 802.1X has been modified to suit this scheme. Although the MN sends an authentication request to the AAA server via its current AP, the server sends the authentication response message (EAP success message) to APs in the FHR. The FHR APs keep the authentication information for the MN in soft state for a certain period, let us call it preauthentication validity period. If the MN does not handoff within that period, the information is no longer useful and the MN will then have to perform a full authentication if it handoffs after that period to that AP. If the MN moves to an AP in the FHR within the preauthentication validity period, the reassociation with the new AP is fast as it only uses the exchange of a couple of messages locally between the AP and the MN.

This scheme has much similarity with the preauthentication scheme, with the difference being that the predictive authentication proactively authenticates to a *group of APs* that the MN is more likely to handoff to, rather than just any new or next *APs* in the network selected by the MN. Predictive authentication is also a centralized solution in that the MN does not have to decide which AP(s) to preauthenticate to; the network (the AAA server) takes charge of that decision.

INTERDOMAIN FAST AUTHENTICATION

Up until now, we have discussed proposals that are designed for reducing the authentication delay of the radio link layer establishment when a mobile device moves from one AP to the next within the same network administrative domain. While these are the most common handoffs, roaming such as in wireless hotspot areas served by multiple providers is becoming more common these days, especially in the CBD areas of big cities. For example, the CBD areas may be covered by small business owners such as Starbucks, big providers such as T-mobile, and nonprofit providers such as the city council. The wireless networks (hotspots) from these entities belong to different administrative domains. Consider the following scenario. There is an overlap between a public network in a coffee shop and a council-operated open wireless mesh network. The council is providing a free service to the public and the coffee shop provides access to paying customers. Despite the business model, for the success of these networks, they must consider supporting real-time session continuation, thus we would need solutions to make the authentication speedy in these multidomain handoff scenarios. Other network setups that would require this kind of solutions can be residential neighborhood wireless networks, also known as community wireless networks as mentioned in the introduction section, where the neighbors want to share their broadband capacity over the wireless access networks they have at their individual premises. Such sharing has been envisioned by various researchers (e.g., Landfeldt, 2006; Thompson, 2006; Raniwala, 2005).

The solutions discussed so far cannot be directly applied to the interdomain handoff scenarios. The main reason is the tight administrative control that the individual domains operate in. The previous solutions also use central decision engines or centralized servers for key distribution, and so forth. Individual domains have their own mechanisms to manage handoffs, and would rather keep it to themselves. One network does not necessarily trust the other when it comes to access control. To ad-

Table 1. A comparison of the fast authentication schemes

Scheme Name	Initiated by	# of nAPs considered	nAP involvement	Applicability
Preauthentication	MN	Variable, selected by the MN	Decided by the MN	Intradomain handoffs
Proactive Key distribution	RADIUS server	A subset of APs in the network; determined by the neighbor graph	Decided by the nAP concerned	Intradomain handoffs
Predictive Authentication	RADIUS server	A subset of APs in the network; determined by the FHR	Decided by the RADIUS server	Intradomain handoffs
Proactive Key Caching	Centralized policy engine	All APs in the network	Centralized policy engine	Intradomain handoffs

dress this gap, we have proposed a "trust-cloud" key sharing model (Hassan & Landfeldt, 2006).

Trust-Cloud Key Sharing

According to our interdomain fast-authentication scheme based on a concept of "trust clouds," a trust cloud is formed among neighboring access points based on a relationship among the owners of the access points. The scheme enables fast and simple authentication for mobile devices that move between access points belonging to different administrative domains such as different ISPs. Used together with an appropriate routing scheme, the scheme enables continuous service of delay sensitive flows even while roaming between different access providers. We define the following terms:

Trust Link: A trust link defines the trust relationship between any two given RG[5]. RG_i and RG_j have a trust relationship between them if they agree to take part in key sharing for visiting mobile nodes between them.

Trust Cloud: A trust cloud is a collection of trust links for a given RG. Every RG has a different trust cloud. One RG can appear in many trust clouds, depending on its relationship with other RGs.

The model is a security key-sharing scheme which works on the basis of AP-to-AP (or RG-to-RG, network-to-network/hotspot-to-hotspot) trust. Unlike the implicit trust among the APs within a single administrative domain or an ESS, this trust is not implicit and is a translation from the trust among the AP-owners through a relationship with a third party such as an ISP or indeed through personal relationships if the community does not operate with a subscription-based model. For example, in community networks, the network operation is dictated by personal preferences, thus even if two AP-owners (or WLAN owners) share the same ISP, there is no guarantee that they would trust each other. This is the difference from neighborhood networks with federated networks such as FON.

In our model, the serving AP[6] of a visiting mobile node (VN) will share the key of the MN that is currently attached with it, within its trust cloud. So, depending on the number of APs in the serving AP's trust cloud, some of the APs in the hotspot area will have the key of the VN ready to be utilized for fast authentication when the VN hands off to one of these APs, and that AP will share the key further among its trust cloud APs. In our model of interdomain access points, provider-provider (or AP-AP, or RG-RG) trust is not necessarily transitive: if RG X trusts RG Y and RG Y trusts RG Z, it does not necessarily mean that RG X trusts RG Z. Moreover, as this trust may have to do with personal preferences, it is not necessary to be symmetric: RG X trusts RG Y does not necessarily mean that RG Y trusts RG X. Initially, we have simulated symmetry in the trust relationships between a given RG pair, and also that trust is not transitive as it depends on the relationship or understanding between any given pair of RG (or RG-owners). This means that if RG X trusts RG Y, RG Y also trusts RG X. However, we have also simulated with the symmetry being relaxed thus two RGs may have uni- or bi-directional trust relations, thus we deviate from a nondirected trust graph to a directed one. By using the concept of trust clouds in the area, we will see pockets of fast authentication enabled coverage area, and *not* an entire coverage area of federated fast authentication areas. Therefore, we would still require strong authentication mechanism provided by the EAP-TLS in this setup as not all the handoffs will be able to utilize fast authentication.

The fast-authentication for interdomain sites is achieved through cooperation among the trust cloud members. The approach is distributed without a central authentication server being involved in distributing the security master key to the access points belonging to the trust cloud. We have proposed two algorithms for mobile visiting nodes to select RGs to perform authentication at handoffs: trust-aware and trust-unaware. In the trust-aware handoff algorithm, the MN needing to handoff to a new RG actively seeks to handoff to an RG that is trusted by its prior-move RG, thus it has to keep track of which RGs are trusted by its prior-move RG. In the trust-unaware handoff though, the MN just seeks to handoff to a suitable RG (e.g., an RG that has low load and can accept more connections) but does not care about the fast authentication possibility as the RG it hands off to may or may not be trusted by its prior-move RG.

Performance Evaluation

We have carried out simulation-based performance evaluation. The scenarios we model are a VN trying to complete a series of communication sessions by utilizing the unused capacity of nearby RGs within its wireless communication range (RG hotspot). There are a total of N RGs in the hotspot area. The VN can sense the current load of each RG from their beacons, and can only associate with an RG that is lightly loaded. An RG is modeled as a two-state Markov chain where the states of an RG alternate between heavily loaded and lightly loaded. The time spent in each state is exponentially distributed with means (L) selected to obtain a given fraction of time an RG spends in the heavily loaded state7.

If an RG switches its state from lightly loaded to heavily-loaded while a VN session is in progress through that RG, the VN session will have to handoff to another lightly-loaded RG using one of the two trust cloud handoff algorithms, or the trustless one described in the previous section. If no lightly loaded RGs are available, the session is prematurely terminated.

The activity of the VN is modeled using the well known on-off process. When the VN completes a session, or a session is prematurely terminated, the VN enters a silence mode before initiating another session. The session and silence mode durations are exponentially distributed. Mean session duration is denoted by S. Once the VN enters the silence mode, its security association with a given RG becomes invalid (an inactivity timer is implemented within each RG, upon expiration of which the security associations of the VN become invalid). Consequently, the VN must go through the full security association process (full authentication involving the AAA server) at the start of each *new* session, even if it continues with the current RG.

The primary performance variable that we measure is the number of times a full authentication is needed for a session on average, since the goal is to reduce this variable. This number is basically one (for the initial association) plus the number of handoffs that require full authentication.

Figures 3 and 4 are two representative graphs from our simulation studies. First of all, we see that our trust-based handoff schemes, be it aware or unaware, achieves much lower per session full authentication than the usual no-trust or trustless

Figure 3. Full authentication vs. mean session time (S)

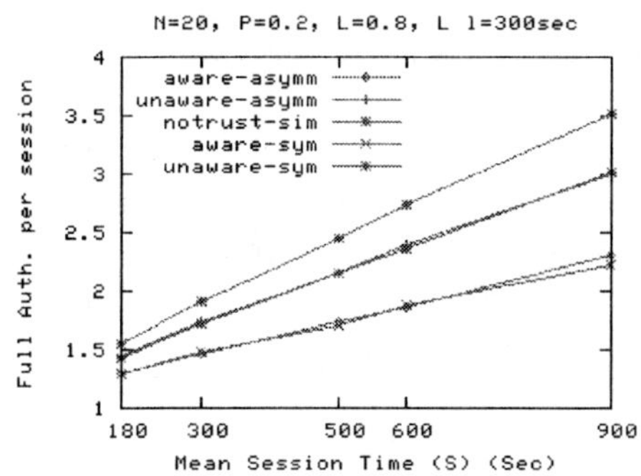

Figure 4. Full authentication vs. mean session time (S): Trust symmetry relaxed

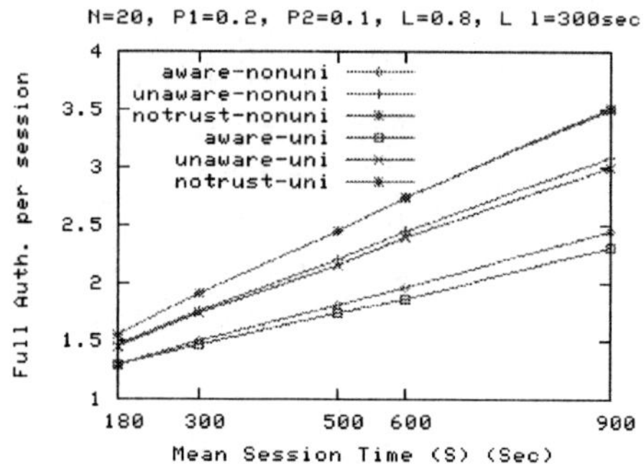

cases. Further more, Figure 3 shows a comparison of symmetric and symmetry-relaxed trust relations in a hotspot area covered by 20 RGs where links may or may not be symmetric. In this simulation, we have used the same trust probability value (P) of 0.2 for both symmetric and asymmetric scenarios (uniform trust probability). As this value is the same for all these cases, we see no difference in the performance. The impact of asymmetric trust relations comes in play when we consider different probability values in deciding trust links in the forward and backward directions for a given pair of RGs (nonuniform). The results then deviate from the case of symmetric trust model. We can observe this in Figure 4 where we have selected two different trust probabilities for the forward direction (P1=0.2) and the backward direction (P2=0.1). In this figure, as the two trust probability values are very close to each other, although we see a difference between the asymmetric uniform and nonuniform cases, the differences are not large. We further simulated an asymmetric nonuniform trust model which we found achieved much lower full authentication per session. In the uniform cases, we used P = 0.2, while in the nonuniform cases, we used P1=0.2 and P2=0.8. Thus we see that the handoffs benefit from the nonuniform asymmetric trust model as the probability values are different and the trust clouds improve because of the higher probability value P2.

Open Issues

The trust-cloud model provides a conceptual way forward in solving the interdomain lengthy authentication issues. However, to make this concept a reality, much more work is required. For example, work is needed to solve issues and answer questions such as what governs trust between providers or APs when we are talking about fast connection administering from neighboring providers or APs, how to formulate the trust groups automatically and in a scalable manner, what is the feasibility of implementation of such solutions, and so on. We are currently focusing on these.

CONCLUSION

Current access control mechanisms in the standard for IEEE802.11 wireless local area networks cannot support continuity of real-time streaming applications in mobile environments where the session has to handoff from one AP to the next. Handoffs involve delay in a few steps, one being the authentication process. In this chapter, we have provided a comprehensive guide on leading proposals for reducing the authentication delay. We have covered both inter and intradomain fast-authentication solutions. Fast-authentication solutions are an integral part of making the AP-switching fast enough to support delay-constrained popular applications.

REFERENCES

Aboba, B., & Simon, D. (1999, October). *PPP EAP TLS authentication protocol* (RFC 2716).

Atheros Communications Inc. Retrieved February 9, 2007, from http://www.atheros.com/

Blunk, L., Vollbrecht, J., Aboba, B., Carlson, J., & Levkowetz, H. (2003, September). *Extensible authentication protocol (EAP)* (Internet Draft draft-ietf-eap-rfc2284bis-06.txt).

CISCO IP Phone. *Cisco unified wireless IP phone 7920*. Retrieved February 1, 2007, from http://www.cisco.com/en/US/products/hw/phones/ps379/ps5056/index.html

Hassan, J., & Landfeldt, B. (2006, June). *Fast authentication in a collaborative wireless access network*. Paper presented at the IEEE International Conference on Communications (ICC).

IEEE 802.1X. (2001, June). *IEEE standard for local and metropolitan area networks: Port based network access control* (IEEE Std. 802.1X-2001).

IEEE 802.11i. (2004, June). *IEEE 802.11i: Wireless LAN medium access control (MAC) and physical layer (PHY) specifications: Medium access control (MAC) security enhancement.*

Juniper Networks. Retrieved February 9, 2007, from http://www.juniper.net/

Mishra, A., Shin, M. H., & Arbaugh, W. A. (2004, February). Pro-active key distribution using neighbor graphs. *IEEE Wireless Communications, 11*(1), 26-36.

Pack, S., & Choi, Y. (2002, October). *Pre-authenticated fast handoff in a public wireless LAN based on IEEE 802.1X model*. Paper presented at the IFIP TC6/WG6.8 Working Conference on Personal Wireless Communications.

Rigney, C., Willats, W., & Calhoun, P. (2000a, June). *Radius extensions* (IETF Internet RFC 2869).

Rigney, C., Willats, S., Rubens, A., & Simpson, W. (2000b, June). *Remote authentications dial in user service (RADIUS)* (IETF Internet RFC 2865).

KEY TERMS

FHR: A group of wireless access points in a public access LAN to whom the predictive authentication will be performed (Pack, 2002). FHR is selected by using a FHR selection algorithm, and taking into account the user mobility and traffic pattern.

Handoffs: Changing network link-layer connection from one network access point or network port to another one.

IEEE 802.11: Also known as Wi-Fi, this is a set of standards for WLANs from the IEEE 802 working group 11.

IEEE802.11i: An amendment to standard 802.11 to specify security mechanisms for Wi-Fi networks.

Neighbor Graph: A collection of APs that the mobile device is likely to handoff to in its next moves (Mishra, 2004).

Network Access Control: Used for security purposes. Network access control determines who (or which device) to give access to the network.

Trust Cloud: A trust cloud is a collection of trust links for a given access point or residential gateway (RG) (Hassan, 2006).

Trust Link: A trust link defines the trust relationship between any two given RG (Hassan, 2006).

Wireless Networks: Networks (of computers) that allow network nodes (e.g., user devices) to connect to the network infrastructure without any wire, typically using short range radio.

WLANs: Wireless local area networks. Local area networks that allow every computer to use a wireless LAN card with which it can communicate with other systems.

ENDNOTES

1. Typically, the overall latency of handoffs should not exceed 50ms.
2. For IEEE 802.11b, the coverage range is no more than 100-200 feet, as compared to the cellular coverage area in cities which is around 2640 feet, and more in the rural areas.
3. Airespace later was acquired by Cisco Systems (Cisco Systems Web site)
4. Funk Software has now been acquired by Juniper Networks (Juniper Networks)
5. In this section, RG, AP and wireless routers can be treated equally to mean wireless AP-type devices not belonging in the same domain, but to different domains.
6. In the interdomain handoff model, especially the trust-cloud model, the APs (or residential gateways-RGs, in the case of community networks) belong to different owners, and domains. APs and RGs are also used interchangeably here.
7. $L = \frac{L_h}{L_h + L_l}$, where L_h and L_l are the mean values for the sojourn times in the heavily and lightly loaded states, respectively.

Chapter XLV
Security and Privacy in RFID Based Wireless Networks

Denis Trček
University of Ljubljana, Slovenia

ABSTRACT

Mass deployment of radio-frequency identification (RFID) technology is now becoming feasible for a wide variety of applications ranging from medical to supply chain and retail environments. Its main draw-back until recently was high production costs, which are now becoming lower and acceptable. But due to inherent constraints of RFID technology (in terms of limited power and computational resources) these devices are the subject of intensive research on how to support and improve increasing demands for security and privacy. This chapter therefore focuses on security and privacy issues by giving a general overview of the field, the principles, the current state of the art, and future trends. An improvement in the field of security and privacy solutions for this kind of wireless communications is described as well.

INTRODUCTION

Radio-frequency identification (RFID) has its roots in WWII when it was used for the first time to distinguish British from German aircrafts. An aircraft was challenged to communicate a certain piece of information and on this basis a decision was made on whether to attack it or not.

This principle is the core of contemporary RFID technology, although, of course, the implementation technology is significantly different. It is now based on low-cost integrated circuits (ICs) called *tags*. Due to the ability to currently store up to two kilobytes of data on these tags, they constitute a very attractive technology in many areas. These include manufacturing, supply chain management, inventory management, healthcare applications, air-transportation, and so forth. All items (in containers) can be scanned together, while each item can be uniquely identified and traced. These properties give RFID technology significant advantages over existing bar-code systems that currently serve for low level, operational acquisition of data in the above mentioned business environments.

These appealing properties also have drawbacks, many of them in the area of security and privacy. But as RFID is already finding its place in contemporary information systems (ISs), these issues need to be addressed seriously, which is the goal of this chapter. In the second section, the background of RFID technology is given. In the third section, threats are described and countermeasures are given. In the fourth section anticipated future trends are discussed. There is a conclusion in the fifth section, while the chapter ends with references and key definitions.

BACKGROUND OVERVIEW

Some definitions have to be given first. One basic definition in the area of computer (communications) security states that security means minimization of vulnerabilities of assets and resources (ISO, 1989). *Wireless security* thus means minimization of vulnerabilities of assets and resources when communicating information in electro-magnetic media through a free-space environment. Finally, *RFID technology* will be defined as wireless identification technology which operates on radio frequencies and deploys low-cost ICs.

A model of RFID environment is described in Figure 1. It consists of *tags* (also called responders) and *readers* (also called transceivers). This is the front-end of RFID applications, which have their back-end in database management systems, where they are integrated with the rest of the IS (see Figure 1). It is generally assumed that RFID security and privacy is concerned with the front-end part (the left-hand side of the dashed vertical line in Figure 1). This is actually the part that is covered by the reader's signal; the tag's signal usually falls within its range.

Tags consist of a microchip and an antenna, both encapsulated in polymer material. The microchip has encoded data, called *identification* (ID), which typically include the manufacturer, brand, model, and serial number. Communication takes place on radio-frequencies, for example, from 125 kHz to 134 kHz for security cards and from 800 MHz to 900 MHz for retail applications (Roussos, 2006). However, increasing the frequency means increased accumulation of signal in bodies containing large quantities of water or in metal.

Communication is achieved by electromagnetic coupling between readers and tags. A reader transmits a signal, which induces a voltage in the tag's antenna. This coupling provides sufficient power

Figure 1. A model of the RFID security and privacy environment

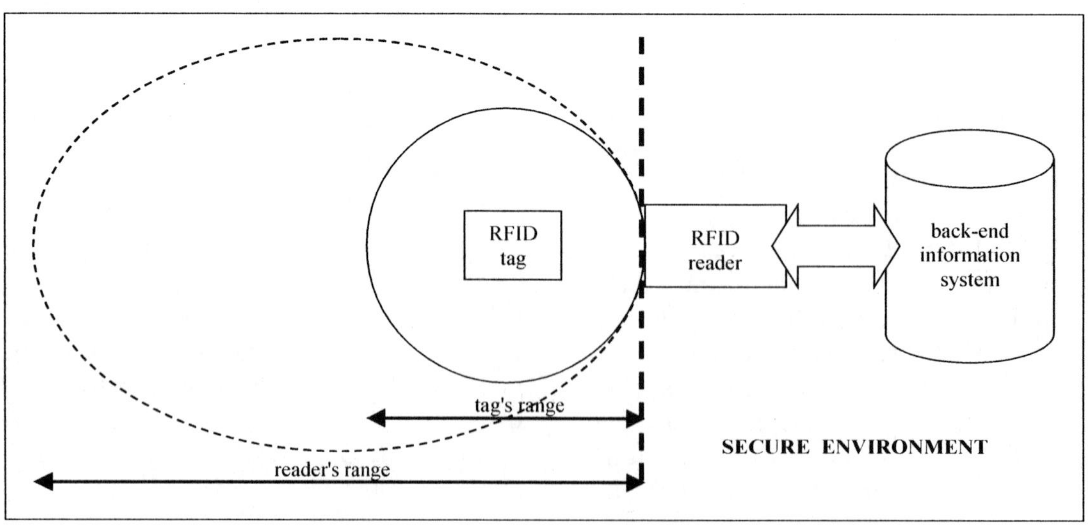

for a tag to respond (after performing some calculations if required). If a tag is powered through this coupling, it is called a *passive tag*. However, if a tag has some source of energy, for example, a battery, it is called an *active tag*. Each type has certain advantages and disadvantages. Passive tags are cheap, but remain active until being explicitly destroyed. They have a low operating perimeter (typically 3 meters) with a relatively high error rate. In contrast, active tags have a greater operating perimeter (up to a few hundred meters), lower error rate, and cease functioning when the source of power is exhausted. However, they are significantly more expensive. Both kinds of tags can be read only, write once-read many, or rewritable.

The main barrier to mass-deployment of RFID tags is their price. A wish-price is limited by five cents, but depending on quantities and using current technologies, many application niches can already be covered. The total cost consists mainly of cost of an antenna, which can be from €/US\$ 0.01 to €/US\$ 0.02, cost of silicon, and IC production; silicon typically costs €/US\$ 0.04/mm^2 (Weis, 2003), while IC production depends on the number of logical gates, that is, technology. But roughly, the cost ranges from €/US\$ 0.025/mm^2 with 1500 gates/mm^2 to €/US\$ 0.08/mm^2 with 60.000 gates (Weis, 2003).

A typical communication channel with a passive RFID is asymmetric. This means that forward communication, that is, communication from a reader to a tag, has one order of magnitude larger in range than backward communication, that is, from the tag to the reader. In the former case this is typically up to 100 meters, while in the latter case this is typically up to 3 meters. The reason, of course, is the power consumption constraint, which means that practical applications are limited to a range of up to 3 meters.

Thus, the cost factor dictates that a typical RFID, or a reference RFID implementation, is currently expected to have the following characteristics. It is passively powered and has 96 bits of read-only memory. These standardized bits serve to carry the *tag's identity*, which is unique for each tag (these IDs are stored in silicon by an imprinting process). A chip operates at 20,000 clock cycles, providing 200 read operations per second. An algorithm to respond to read primitives from a reader may be probabilistic (e.g., Aloha (Prasad & Rugierre, 2003) or deterministic (e.g., a binary walking tree) (Juels, Rivest, & Szydlo, 2003). With such algorithms, a single tag can be identified and isolated. The related process is called *singulation*. Finally, the number of available gates that can be devoted to security operations is in the range of 400 to 4,000.

The above estimates are based on figures from Weis (2003) by applying Moore's law, which states that for the same price the available processing power doubles every year and a half. It is therefore clear that processing resources to support security in RFID environments are very limited and lightweight cryptographic solutions thus provide an answer to this problem.

Moore's law also implies that there is always a point where "ordinary" cryptographic algorithms become feasible for computationally weak devices. An example of a thick RFID implementation, which is based on AES to provide authentication, can be found in the work of Feldhofer, Dominikus, and Wolkerstorfer (2004). Despite this, a permanent need exists for lightweight cryptographic protocols and also algorithms. One main reason is the gap between ordinary devices where space and power consumption are not a serious concern (e.g., tag readers, desktop systems), and weak devices with limited space and power consumption (e.g., RFID tags, smart-cards). This gap means that increased processing power affects both kinds of devices equally; in the case of a cryptographic algorithm, the key-length of this algorithm is extended.

As a consequence, weak devices are again less protected because they cannot deploy such intensive computations with enlarged keys. Further, if the above use of a cryptographic algorithm can be seen as a kind of variable cost (the longer the key, the higher the processing overhead), cryptographic protocols can be seen as a fixed cost. Note that cryptographic protocols are ordinary communication protocols that deploy cryptographic algorithms, and cryptographic protocols are often referred to as *security services*, while cryptography algorithms are referred to as *security mechanisms*. Both kinds of costs contribute to the total processing power requirements, and have to be kept low while at the same time enabling a comparable level of security to weak devices. This leads to a whole new research area (Juels, 2004).

RFID Threats and Countermeasures

The very basic threat to each and every tag is that it remains active when it is no longer supposed to be active. To counter this problem, RFID logic may implement *kill operation*, which means that upon receipt of a certain communication primitive, the tag becomes permanently inoperative by, for example, blowing a fuse in its circuitry. A more bullet-proof solution is exposure of RFID to microwave radiation that melts its metalized layer.

Risk management drives each and every provision of security and privacy in ISs. A typical process is depicted in Figure 2. It starts with the identification of assets A ($A = \{a_1, a_2, \ldots, a_n\}$) and threats T ($T = \{t_1, t_2, \ldots, t_m\}$) to those assets. For each asset and threat, that is, Cartesian product $A \times T = \{(a_1, t_1), (a_1, t_2), \ldots, (a_n, t_m)\}$, related vulnerabilities are identified together with the likelihood of a threat to get into interaction with the asset during a certain period of time. On this basis, the estimated damage $D(a_i, t_j)$ caused by interaction between asset a_i and threat t_j during this period is calculated. The result presents the upper bound for investment in safeguards. A certain degree of risk, called residual risk, is usually accepted and taken into account. This often makes sense economically. But in the majority of cases, a threat cannot be completely neutralized (Trček, 2006).

The challenging parts of this process are identification of threats and their probability. For identification of threats in RFID environments a comprehensive taxonomy from Garfinkel, Juels, and Pappu (2005) can be used. The first four threats are related to corporate security, and the rest to personal privacy:

- **Corporate espionage threat:** Tagged products may enable remote acquisition of supply chain details like logistics details, volumes, and so forth.
- **Competitive marketing threat:** Tags may enable access to customers' preferences and use the data gathered for competition.
- **Infrastructure attacks threat:** Where RFID is central to a competitor's advantage; disruption of RFID operations becomes an important point for attack.

Figure 2. Risk management process

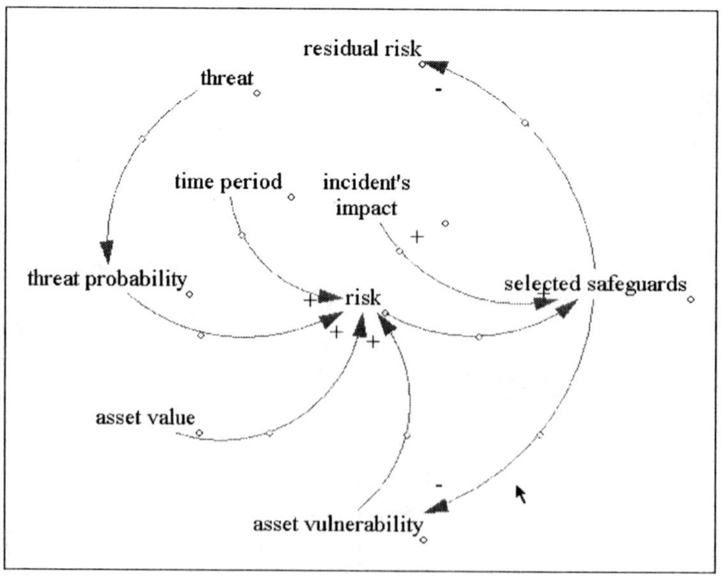

- **Trust perimeter threat:** Gathering additional volumes of data through RFID introduces new challenges related to sharing information in a trustworthy way.
- **Action threat:** Individuals actions may be monitored.
- **Association threat:** When tagged products are associated with an individual's ID (e.g., loyalty programs), these persons can be associated not only with the type of product, but with the exact product, due to its unique ID.
- **Location threat:** Tags can be triggered by covert readers at various locations to reveal a person's location.
- **Preference threat:** Tags disclose preferences of customers and help to identify, for example, more wealthy ones.
- **Constellation and transaction threats:** Constellation threat is similar to location threat, but in this case the identity of a customer is not known. Despite this, a particular person can be spotted and traced. Further, chaining one constellation threat with another, a whole chain of actions, or transactions, becomes traceable.
- **Breadcrumb threat:** When products are disposed with their original tags, an attacker may use them and is tracked with falsified identity. This is actually just another kind of identity theft.

On top of all this, a fundamental threat exists, called tag cloning, and such cloning has been successfully demonstrated (Bono, Green, Stubblefield, Juels, Rubin, & Szydlo, 2005). What countermeasures are at our disposal?

The basic option was mentioned at the beginning with the physical destruction of a tag (e.g., by exposure to microwaves or implementation of a logical kill command that makes chip inoperable). But the fact is that the latter approach often has flaws in implementations: logically killed tags may remain active or be reactivated (Roussos, 2006). In many situations, it might be even beneficial to keep these tags active; for example, tagged items may be used for smart-home applications or to help disabled people.

The most common approach to security and privacy is by deploying cryptography. Using cryptographic mechanisms (e.g., symmetric and asymmetric cryptographic algorithms, strong one way hash functions), the following cryptographic services can be implemented (ISO, 1995):

- **Authentication:** This ensures that the peer communicating entity is the one claimed.
- **Confidentiality:** This prevents unauthorized disclosure of data.
- **Integrity:** This ensures that any modification, insertion, or deletion of data is detected.
- **Access control:** This enables authorized use of resources.
- **Nonrepudiation:** This provides proof of origin and proof of delivery, such that false denying of the message content is prevented.
- **Auditing:** This enables detection of suspicious activities and analysis of successful breaches. It provides evidence when resolving legal disputes.

In case of RFID tags, authentication, confidentiality, and access control can be applied to counter threats described at the beginning of this section. But to make these security services operational, key management (i.e., handling of cryptographic algorithms' keys) has to be resolved (Trček, 2005). This is a complex issue in open environments and has been known as such for almost two decades. Suffice it to say that only very simple key management schemes are acceptable for RFID environments.

With regard to security and privacy, it is required that authentication, and consequently access control, is provided only to legitimate readers. Further, rogue readers should not be disclosed a tag's ID, but should also be prevented from tracing a tag, regardless of the inaccessibility of its ID. Put another way, when rogue readers interact with a tag, it should be practically impossible (i.e., computationally difficult) to link the multiple manifestations of a tag to this very tag.

An example of recent solutions that meet these requirements is the YA-TRAP protocol (Tsudik, 2006). However, to demonstrate a typical simple authentication RFID protocol, an example by Weis (2003) is given. In this case, RFID contains a derivative of some key, from which it is computationally hard to obtain the real key. This derivative, called metaID, can be a cryptographically strong hash of the key. The tag authenticates the reader as follows:

1. The reader queries a tag to send its metaID.
2. After obtaining metaID, the reader looks up the internal table to find the corresponding key, which is unique for each tag. It sends this key to the tag.
3. After obtaining the key, the tag hashes this key and compares it with its metaID. If values match, the tag provides its full functionality.

It can readily be observed that confidentiality is not used in the above scenario. This means that the communicated data are exchanged in plain and can be read by an adversary. In the above scenario, it suffices to intercept the key, and afterwards to falsely authenticate to tag.

This is not the only threat to confidentiality. Confidential data are stored on RFID, the most sensitive piece being its ID. Due to processing requirements and key-management problems, the tendency is to store data in plain. In this case, many other kinds of attacks can be applied that exploit a tag's tamper resistance, and may consequently lead to the tag's cloning (Anderson & Kuhn, 1996). Such attacks can be prevented by careful circuitry design principles that are common in smart-card design (e.g., scrambling of memory addresses, proper positioning of the memory layer within the integrated circuit, and inclusion of dummy components).

But, as is always the case with security and privacy in ISs, one should not rely solely on cryptography. One alternative architectural approach is the use of tag pseudonyms (Garfinkel et al., 2005). In this approach, each tag is given a cycling sequence of pseudonyms that are chosen to reply to reader requests. But by repeatedly scanning the same tag the whole cycling sequence would be discovered. A solution is to throttle the queries, that is, to use each pseudonym for some extended period of time.

Another option is blocker tags that are based on the already mentioned tree-walking algorithm. The singulation process deploying this algorithm uses a k-bit identifier represented as a binary tree. Each leaf in this tree is the tag's ID. The algorithm goes as follows:

1. The reader starts at the root (depth d=0). To decide whether to proceed along the subtree with 0 or the subtree with zero, it first emits 0.
2. If all tags broadcast 0, the reader proceeds by stepping one level down to the subtree beginning with 0 (depth d=1). If all tags broadcast 1, it steps one level down to the subtree beginning with 1 (depth d=1). If the reader receives both zeros and ones, it proceeds down both trees.
3. The procedure is repeated at each level (the total depth d of the tree equals the number of bits that are used for identifiers, that is, d=k), until only one singular tag is responding.

To spoof the reader, a blocker tag simply emits both bits 0 and 1, which means traversing the complete tree. This reply actually forces the reader to do an exhaustive search on the whole set of 2^{96} combinations, which is impossible. Such a tag would, of course, disrupt all operations. Thus only a subtree of the whole ID space can be allocated to privacy protecting bits (e.g., only that subtree that starts with bit 1). When a reader would enter such a subtree, it would cease the process of further singulation.

Another option is to regulate the range of emitted signal from a tag (Fishkin & Roy, 2003). Based on the strength of a signal, a tag can calculate an estimated distance to the reader and adapt its emitting power to the level, at which signal-to-noise ratio is such that a reader is able to read the reply. Other devices that are out of this range are automatically disabled from tracing the response.

Finally, we propose a new hybrid technique called one-time pseudonyms. This technique deploys the idea of tag pseudonyms and one-time password principle, which was proposed for the first time by L. Lamport (1981). The principle goes as follows. Take a seed value S_0 and apply a strong one-way hash function to obtain the value S_1 = hash(S_0). Next, calculate S_2 = hash(S_1), S_3 = hash(S_2), and so on. Now, when a tag is to be identified by a reader, the reader sends an integer that denotes the index of the iteration to be done on the seed. Of course, in the case of RFIDs, the unique ID is assumed to be used as a seed.

From the point of background (secure environment) operations, these one-time pseudonyms introduce a workload comparable to that for ordinary pseudonyms. The advantage is that there is no need for a logic circuit to throttle the queries (i.e., to change pseudonyms only every few minutes). Further, the cycle of one-time pseudonyms is not deterministic. Thus it is much harder for an attacker to follow the tag to collect the complete sequence, which is the case with the basic tags pseudonyms technique. There is a slight drawback with this technique if a large number of iterations are required. A straightforward possibility would be to limit the highest index (number of iterations), while another approach would be to store in an RFID, for example, each 20th iteration of its hashed ID. This would also significantly expand the range of applicable iterations, because the upper limit for this technique presents a computation load that is related to the response of a tag and to power consumption.

FUTURE TRENDS

With regard to cryptography, it is anticipated that encryption of RFIDs content will become the norm in providing confidentiality. Basically, in many scenarios it is not necessary that a tag be capable of performing strong encryption of its data. Once it has successfully authenticated a reader, changes to relevant data can be encrypted by the reader and written (back) to the tag. Permanent sensitive data can be also decrypted at the reader's side. But this requires efficient lightweight authentication protocols, and this area is likely to get further attention from researchers.

However, cryptography is no panacea. One basic reason, described in the second section, is the gap between ordinary computational devices (devices without stringent limitations on semiconductor area and power consumption) and weak processing devices. Thus noncryptographic solutions, as described in the previous section, may become more important.

It should be mentioned that no IS security and privacy can be fully exercised if there is no legal coverage. Although security and privacy related regulation is becoming broad and diverse, the RFID area has many specific issues that require tailored legislation. But this topic is outside the scope of this chapter.

CONCLUSION

A ubiquitous and pervasive computing paradigm is almost inherently tied to wireless communications. In addition, this paradigm rests on massive deployment of computing devices, which therefore have to be cheap. RFID technology is the special kind of such paradigm. Due to its specific properties it places further constraints on wireless security and privacy. These specific properties and available existing solutions have been discussed in detail in this chapter.

The majority of solutions that support security and privacy are built on cryptography, while some promising attempts are emerging that are not tied to cryptography (e.g., blocker tags, antenna-energy analysis, tag pseudonyms). Further, a new hybrid technique for tags authentication has been proposed that deploys the principles of tags pseudonyms and one-time passwords. It compensates the basic drawback of the pure tag pseudonyms approach, while the price to be paid (in terms of additional calculations and slightly prolonged response times) may be acceptable for many environments.

There is no doubt that RFID devices are about to become one kind of mainstream wireless technology. Currently being implemented mostly in supply chains and retail, their deployment possibilities are numerous, for example:

- Health care information systems can be improved for better handling of patients' data.
- Elderly and disabled people can benefit from new applications that are based on RFID.
- New applications for intelligent, smart-homes can be built on top of RFID.
- Scientific research of certain species can be improved, and so forth.

As for every technology, RFID technology has its advantages and disadvantages. To properly ensure security and privacy related measures, a proper risk management strategy has to be taken. Only the big picture of security and privacy assures to benefit from the use of this kind of wireless technology.

REFERENCES

Anderson, R., & Kuhn, M. (1996). Tamper resistance: A Cautionary Note. In *Proceedings of* the *USENIX Workshop on Electronic Commerce* (pp. 1-11). Berkeley: USENIX.

Bono, S., Green, M., Stubblefield, A., Juels, A., Rubin, A., & Szydlo, M. (2005). Security analysis of a cryptographically-enabled RFID device. In *Proceedings of the 14th USENIX Security Symposium* (pp. ?). Berkeley: USENIX.

Feldhofer, M., Dominikus, S., & Wolkerstorfer, J. (2004). Strong authentication for RFID systems using the AES algorithm. In *Proceedings of the 6th International Workshop Cryptographic Hardware and Embedded Systems* (LNCS 3156, pp. 357-370). Heidelberg: Springer.

Fishkin, K.P., & Roy, S. (2003). *Enhancing RFID privacy via antenna energy analysis* (Memo IRS-TR-03-012, Intel Research). Santa Clara: Intel.

Garfinkel, S.L, Juels, A., & Pappu, R. (2005). RFID privacy: An overview of problems and proposed solutions. *IEEE Security and Privacy, 3*(3), 34-43. Los Alamitos: IEEE.

International Standards Organization (ISO). (1989). *Information processing systems: Open systems interconnection - basic reference model, security architecture, part 2* (ISO 7498-2). Geneva: ISO.

International Standards Organization (ISO). (1995). *IT, open systems interconnection: Security frameworks in open systems* (IS 10181/1 thru 7). Geneva: ISO.

Juels, A. (2004). Minimalist cryptography for RFID tags. In C. Blundo & S. Cimato (Eds.), *4th Conference on Security in Communication Networks* (pp. 149-164). Heidelberg: Springer Verlag.

Juels, A., Rivest, R., & Szydlo, M. (2003). *The blocker tag: Selective blocking of RFID tags for consumer privacy.* Paper presented at the 8th ACM Conference on Computer and Communications Security (pp. 103-111). New York: ACM.

Lamport, L. (1981). Password authentication with insecure communication. *Communications of the ACM, 24*(11), pp. 770-772. New York: ACM.

Prasad, R., & Ruggiere, M. (2003). *Technology trends in wireless communications.* London: Artech House.

Roussos, G. (2006). Enabling RFID in retail. *Computer, 39*(3), 25-30. Los Alamitos: IEEE.

Trček, D. (2005). E-business systems security for intelligent enterprises. In M. Khosrow-Pour (Ed.), *Encyclopedia of information science and technology* (Vol. 2, pp. 930-934). Hershey, PA: IGI Global, Inc.

Trček, D. (2006). *Managing information systems security and privacy.* Heidelberg/New York: Springer.

Tsudik, G. (2006). *A-TRAP: Yet another trivial RFID authentication protocol.* Paper presented at the International Conference on Pervasive Computing and Communications PerCom 2006 (pp ?-?). Los Alamitos: IEEE.

Weis, S.A. (2003). *Security and privacy in radio-frequency identification devices.* Unpublished master's thesis, Cambridge MIT.

KEY TERMS

Active Tag: A tag that has some source of energy, for example, a battery.

Kill Operation: Upon receipt of a certain communication primitive a tag becomes permanently blocked, for example, by blowing a fuse in its circuitry.

Passive Tag: A tag that is powered through electromagnetic coupling and obtains power from the reader.

Reader: A device that queries tags to obtain their IDs and that is connected to the back-end part of information systems.

RFID Technology: Wireless identification technology that operates in radio frequencies and deploys low-cost ICs.

Security Mechanism: A basis for a security service, where using a particular security mechanism (e.g., cryptographic algorithm) enables the implementation of security service.

Security Service: A service provided by an entity to ensure adequate security of data or systems in terms of authentication, confidentiality, integrity, and nonrepudiation.

Singulation: A process (an algorithm) that enables isolation and identification of a single tag.

Tag: A small, low cost IC with unique ID and computational capabilities to support identification processes.

Tag's Identity (ID): This is a unique number for each tag that is stored in silicon with imprinting process.

Wireless Security: Minimization of vulnerabilities of assets and resources when communicating information in electromagnetic media through a free-space environment.

Chapter XLVI
Security and Privacy Approaches for Wireless Local and Metropolitan Area Networks (LANs & MANs)

Giorgos Kostopoulos
University of Patras, Greece

Nicolas Sklavos
Technological Educational Institute of Mesolonghi, Greece

Odysseas Koufopavlou
University of Patras, Greece

ABSTRACT

Wireless communications are becoming ubiquitous in homes, offices, and enterprises with the popular IEEE 802.11 wireless local area network (LAN) technology and the up-and-coming IEEE 802.16 wireless metropolitan area networks (MAN) technology. The wireless nature of communications defined in these standards makes it possible for an attacker to snoop on confidential communications or modify them to gain access to home or enterprise networks much more easily than with wired networks. Wireless devices generally try to reduce computation overhead to conserve power and communication overhead to conserve spectrum and battery power. Due to these considerations, the original security designs in wireless LANs and MANs used smaller keys, weak message integrity protocols, weak or one-way authentication protocols, and so forth. As wireless networks became popular, the security threats were also highlighted to caution users. A security protocol redesign followed first in wireless LANs and then in wireless MANs. This chapter discusses the security threats and requirements in wireless LANs and wireless MANs, with a discussion on what the original designs missed and how they were corrected in the new protocols. It highlights the features of the current wireless LAN and MAN security protocols and explains the caveats and discusses open issues. Our aim is to provide the reader with a single source of information on security threats and requirements, authentication technologies, security encapsulation, and key management protocols relevant to wireless LANs and MANs.

INTRODUCTION

The topic of this chapter is the security of 802.11 wireless local area networks (WLANs) and of 802.16 wireless metropolitan area networks (WMANs). These networks are based on the IEEE standards belonging to the 802 family, which include the much-beloved Ethernet (802.3) that is common today in homes and offices. Although the development of the 802.11 technology and standards have been ongoing sine the late 1990s, grassroots adoption of "wireless Ethernet" only began in the 2000-2001 timeframe when access point (AP) devices became cheap enough for the home user to obtain.

The convenience of having wireless access to the IP Internet is self-evident. The value proposition in terms of employee productivity has been so compelling that many enterprises began also to introduce the technology into their corporate networks. This enterprise adoption, however, was prematurely halted when security flaws in the wired equivalency privacy (WEP) algorithm were discovered and published. Various temporary patches were then suggested in order to support existing enterprise investments in WLAN equipment, with the IPsec-VPN (e.g., over the wireless segment) as the most common approach. The IEEE standards community completed the revision of the security-related components of 802.11 in 2004, with conforming products scheduled to be shipped in 2005.

This chapter is not a user guide to specific WLAN or WMAN products, and intentionally avoids specific references to such products. It is also not a thesis on the various engineering solutions that could have been applied to solve the Wi-Fi security problem. Instead, the chapter attempts to explain what current approaches and solutions have been adopted, and why these were chosen.

The contents of the chapter are arranged in four parts, where each part groups together topics and issues that are closely related. These parts roughly cover the topics of WLAN authentication and authorization, WLAN security algorithms and protocols, security in WLAN roaming, and security in WMANs. These are described in more detail next.

BACKGROUND

Traditionally, the term *authentication* in the context of computer and network security concerns the ability of a verifier (or prover) entity to ascertain the correct *identity* of another entity claiming to be that identity. Thus, the aim of authentication is for one entity to prove its identity to another based on some *credentials* possessed by that first entity. Examples of credentials include passwords, digital certificates, or even physical keys. The outcome of an authentication process is typically binary, namely success or fail. The process is typically defined and implemented as one or more *protocols*.

The term *authorization* pertains to the rights, privileges, or permissions given to an authenticated entity in relation to some set of resources. In practice, authorization for an entity to take actions (e.g., access network, read files) is preconditioned on a successful authentication. The functions of authentication and authorization are often accompanied by *accounting* (or auditing), with the three loosely referred to as *AAA*.

The level of authorization assigned to an entity when it seeks access to resources is often tied to the type and strength of the authentication protocol used and the type of credential possessed by the authenticated entity. Hence, differing levels of assurance or certainty regarding the outcome of an authentication process can be gained by using different credentials and authentication protocols.

For example, when a password (as a credential) is used with a weak protocol (e.g., plaintext challenge-response), then a low or weak level assurance is obtained as both the credential and the authentication protocol are weak. In contrast, a strong credential such as a digital certificate when combined with a strong authentication protocol, such as SSL or transport layer security (TLS), achieves a higher level of assurance regarding the identity of the authenticated entity.

In today's complex computer and network systems, multiple credentials might be needed for an entity to access multiple resources, each access instance of which may be governed by separate sets of privileges. Thus, often the term *layer* (of authentication and authorization processes) is used to describe complex situations.

One of the earliest questions with regards to authenticating dial-up users was how to run an authentication protocol with a (dial-up) client when it did not yet have an assigned IP address. This issue was of particular concern since the assignment of an IP address was subject to a successful authentication. However, most of the existing authentication protocols, such as IKE or SSL, were designed to be run over the IP layer with the end points possessing known source/destination IP addresses. This apparent chicken-and-egg problem was solved with the introduction of the extensible authentication protocol (EAP), first published as an Internet standard in 1998 in RFC2284 and more recently revised in RFC3748 (2004).

In the last couple of years EAP has come to the forefront of discussions, this time on 802.11 WLAN security and 802.1X. This is due to the similarity of the chicken-and-egg problem found in WLANs, namely the question of how a server on an IP network can authenticate an 802.1X supplicant when that supplicant does not yet have an IP address, and whose IP address assigned is in fact subject to a successful authentication by the server.

For the purposes of WLAN security EAP itself can be viewed as a framework within which a security protocol must be instantiated "inside" (on top of) EAP. Thus, with the emergence of EAP came the definition of a number of *EAP methods* which load EAP with the appropriate security protocol. One important EAP method is EAP-TLS, which is an instantiation of the TLS (or SSL) protocol inside EAP. More stringent that plain TLS in terms of certificate usage, EAP-TLS mandates the use of digital certificates at both the client and server side.

SECURITY ISSUES IN WIRELESS LANS AND MANS

Wireless networks make it easy for an attacker to read, modify, or drop packets, as the attacker often needs only to be in the vicinity of the communicating entities. In contrast, an adversary may need to gain entry to a physical access controlled building, wiring closet, or a network device to attack a wired network. The IEEE 802.11 standard attempts to emulate the physical attributes of wired medium in designing the WEP suite of security protocols and data transforms.

WEP consists mainly of an entity authentication protocol and a data security transform. The authentication protocol has two modes: open-system and shared-key authentication subtypes. The open-system authentication protocol is a 2-way handshake initiated by the entity requesting service. It consists of the requester asserting its identity and the responder returning with "successful" or "unsuccessful" only on the basis of whether open-system authentication is supported or not. In other words, open-system authentication protocol does not provide any security whatsoever. The shared-key protocol is a 4-way exchange, also initiated by the entity requesting service. It consists of request, challenge, challenge-response, and result messages in that order. The responder challenges the requester to provide proof of possession of the shared secret. The requester encapsulates the challenge-response message using the WEP transform to prove that it knows the WEP secret key. The responder decapsulates the message using its local copy of the shared secret key and compares the decrypted text with the original challenge text. If there is a match, the requester is granted a connection.

The data security transform, comprising the WEP encapsulation and decapsulation processes, is the other component of an 802.11 security solution. The standard recommends that the authentication protocol be used in conjunction with the data security transform.

WEP uses RC4 (Ron's code: a proprietary stream cipher designed by Ron Rivest of RSA Labs and later released into public domain) as the cipher to protect 802.11 frames. The WEP encapsulation process is designed to be "reasonably strong," self-synchronizing (considering that medium access control [MAC] protocol data units [MPDUs] may be dropped or arrive out of order, the receiving entity must be able to decapsulate an MPDU independent of prior or future MPDUs), efficient, and easy to implement in hardware or software. Thus, an initialization vector (IV) for the RC4 cipher accompanies each MPDU, along with a key ID (shared-key ID used to encapsulate the current MPDU), and an integrity checksum to protect against MPDU modification en route. Briefly, the key and the IV are input to the RC4 encryption algorithm to generate a pseudorandom key stream for MPDU encapsulation. A checksum is computed over the MPDU using cyclic redundancy check (CRC-32) for integrity protection. The checksum is then appended to the MPDU and is XORed with the keystream to generate the ciphertext. WEP decapsulation follows the reverse process, where, upon decryption of an MPDU, the received checksum is compared with the locally computed checksum to verify the MPDU's integrity.

WEP design contains several well-publicized flaws. The nature of the flaws will be clear as we delve into the design details. The design goals themselves are suspect: the stated goal is wired equivalency privacy, which in itself is hard to capture. The choice of cipher, and especially the use of a stream cipher for encapsulating packets, also makes it easy for an attacker to cryptanalyze WEP encapsulated MPDUs. The choice of integrity algorithm, CRC-32, which is linear, makes it easy for an attacker to modify encrypted MPDUs without the receiver being able to verify whether the packets are legitimate or not.

Wireless LAN devices or wireless stations (STA) are considered as logically external entities to an enterprise network. Radio frequency (RF) waves in most deployment scenarios do not have physical boundaries and thus STAs should be allowed to access the corporate network only after going through a similar authentication procedure as in the case of remote access. For remote access, enterprises typically use IPsec for general purpose access to their intranets, and SSL for access to e-mail and other similar applications. In both cases, remote access servers and clients mutually authenticate each other, and arrive at a common security association (SA). Use of keys within that SA is proof of authentication for accessing the enterprise intranet via the remote access server.

STAs establish a robust security network association (RSNA) with an AP using IEEE 802.1X and EAP for authentication and key distribution. From the resulting master key, the AP and the STA engage in a 4-way exchange to derive session keys. STAs can only communicate to other entities in the wired or wireless network via an authenticated secure channel (using CCMP or TKIP as the security protocols). Thus, an AP enforces access control to the wired network and provides a means for authenticated and confidential communication between the STA and other entities in the network.

The primary reason WLANs were developed was to allow untethered connections between a client and an 802.11 access point (AP), as a basis for further access to resources and services on the Internet. The next step in this process is wireless roaming, in which a client can move across multiple APs in one administrative domain and across multiple APs across differing administrative domains. Currently, the most prevalent model for wired roaming consists of a dial-up connection from a client (e.g., a laptop) through an Internet service provider (ISP), to a home domain (e.g., corporate network). This model presumes the prior existence of a business relationship between the client (or its corporation) and one or more ISPs.

The term *Wi-Fi roaming* can be loosely defined as the set of services supporting the deployment and management of 802.11 WLAN access at public venues or public *hotspots,* where the customer of one service provider can obtain services (e.g., IP connectivity) from a different (visited) service provider. The term *service provider* (SP) here is intentionally left abstract since in today's Internet a number of entities can take the role of providing one or more services relating to Wi-Fi roaming. It is important to note that Wi-Fi roaming involves the crossing of both network-administrative boundaries and security-administrative boundaries. Therefore, on-campus WLAN access at different remote locations (e.g., offices, buildings) under the same administrative jurisdiction is not considered here as Wi-Fi roaming.

The business case for Wi-Fi roaming is self-evident: consumers with laptops or handheld devices are willing to pay for IP connectivity through hotspots located throughout the world, provided that Wi-Fi access is easy to use and secure.

This desire is already true today, as seen in the case of dial up IP services. Many traditional ISPs see Wi-Fi roaming as providing a new business opportunity, by extending their edge services to a new kind of access point, namely, the public hotspot, while retaining as much as possible their investment in their existing backend authentication, authorization, and accounting (AAA) infrastructure.

For some *mobile network operators* (MNO) and carriers, the case for Wi-Fi roaming can even be considered imperative, as they are seeking to augment and extend existing mobile-related services to their customers at affordable prices. Mobile handsets that can make use of Wii hotspots—with speeds of 11 to 50 Mbps—could generate new business opportunities by providing users with higher-quality content and a higher level of interactivity. The case for Wi-Fi roaming is of particular interest to MNOs that have invested heavily in the recent acquisition of three-generation (3G) licenses.

Given the increasing mobility of the workforce, providing *secure* Wi-Fi roaming is an important challenge today. Corporations see remote access as a given fact of life and expect services from their ISPs supporting remote access. This is true in dial-up today, and it is something expected of Wi-Fi roaming in the near future.

Wi-Fi roaming has recently taken an interesting direction in North America due to the entrance of a number of MNO into this space. These MNOs want to enhance their 2G and 2.5G (and later their 3G) offerings with Wi-Fi related services. Many MNOs already perceive that in practice a universal mobile telecommunications system (UMTS) may not reach its theoretical data rates of 2 Mbps. Thus, Wi-Fi at hotspots—with speeds of up to 11 Mbps in 802.1lb and up to 54 Mbps in 802.1la—may provide a solution for the need for higher data rates complementing their 2.5G and 3G offerings. From a content perspective, the marriage of GSM/UMTS and Wi-Fi roaming makes very good sense. The ability of 802.11 Wi-Fi hotspots to provide high-speed connectivity to the Internet makes it attractive for downloading richer content for mobile devices (e.g., PDAs and GSM phones) beyond the ring tones of today. Such content may include MP3 music files, interactive online games, and MPEG4 video clips, depending on the capabilities of the device.

The authentication protocol used in GSM has been the *subscriber identity module* (SIM), while for UMTS it will be *UMTS subscriber identity module* (USIM). Both take the physical form of a *universal integrated circuit card* (UICC), with the next generation based on tamperproof smartcard technology. In the context of 3G-WLAN roaming, the proposed protocols (EAP methods) for authentication are EAP-SIM for SIM-based users and EAP-authentication and key agreement (AKA) for USIM-based users.

The choice of the SIM and AKA methods for authentication has been dictated by the need of the MNOs to keep as much as possible their back-end AAA infrastructure unmodified for Wi-Fi usage. However, since the SIM and AKA protocol were designed for GSM/UMTS networks, they are not transferable to the IP world without introducing some vulnerabilities. Thus, the "naked" SIM or AKA exchange needs protection while in the IP segment of the end-to-end handshake between the SIM/USIM card (in the UE) and the HLR/HSS at the home network. One possible solution around this problem is to wrap the SIM/AKA exchange within a TLS layer, which can be done by layering the SIM or AKA handshake above (wrapped within) protected extensible authentication protocol (PEAP) or tunneled TLS (TTLS).

However, in addition to these issues that are specific to EAP-SIM and EAP-AKA, the 3GPP-WLAN interworking security specifications have also outlined a number of other issues and requirements. Some of these are as follows:

- **Mutual authentication:** In addition to the user authenticating itself to the home network, the network must in turn authenticate itself to the user. The EAP-AKA protocol provides this feature.
- **Signaling and user data protection:** Subscribers should have at least the same security level for WLAN access as for their current cellular access subscription. This require-

ment translates to the need to protect their interfaces or connections between the Wi-Fi hotspot (namely, the WLAN access network) and the 3GPP network, between the 3GPP AAA proxy to the 3GPP AAA server, and between the 3GPP server and the HSS. For the connection between the Wi-Fi hotspot and the 3GPP network, most likely the protocol used will be RADIUS or Diameter.

- **Identity privacy:** The user's privacy while roaming from one Wi-Fi hotspot in another needs to be guarded. That is, when the user is assigned a temporary identifier (or pseudonym), it should be infeasible for an attacker to reverse this process and correlate the pseudonym with the actual user identifier. Naturally, it should also be infeasible for an attacker to generate a valid pseudonym note that temporary identifiers can be used within EAP-AKA.
- **Protection of the interface between UIC and WLAN access devices:** Here, the concern relates to the wish of operators to reuse existing UICC and GSM SIM cards in laptops and PDAs, which may have different physical security measures than mobile handsets. Thus, the UE is perceived to possibly have a functional split implemented over several physical devices/components, where one device holds the UICC/SIM card, while another device provides WLAN access.

The interface across this functional split needs to be protected. There is little point in providing a near tamper-free UICC or SIM card, when the WLAN-access device at the user (e.g., radio circuitry or WLAN software/hardware) can be manipulated by an attacker to obtain PS from the home network or visited network, as these services are core to the business of MNOs.

The aim is to provide protection to the level where attacking the PS domain (in UMTS network) by compromising the WLAN access device is at least as difficult as attacking the PS domain by compromising the card-holding device.

It is for this reason that there is currently interest in providing EAP functionality on board the UICC, thereby achieving true end-to-end EAP-AKA (or EAP-SIM) exchange between the network and the UICC (instead of the laptop hosting the UICC).

The area of broadband Internet has gained a lot of interest in recent years due the exciting business opportunities enabled by high-speed Internet connectivity to homes and businesses. Content owners (e.g., movie studios and record labels) and content providers/distributors (e.g., music and MPEG4 download services) see broadband Internet to the home as crucial to providing the next source of revenue, as it solves the difficult "last mile access" problem. Thus, if "content is king" as the saying goes, last mile access is the "queen" that enables content to flow to the consumer.

Today, only a fraction of U.S. households have broadband Internet in the form of cable modem services or DSL services. The mid-2004 subscriber numbers indicate that there are just over 18 million subscribers to broadband Internet. This is due, among others, to the difficulty in installing cables for those services in dense areas with old buildings and infrastructures, despite the fact that the two technologies have reached maturity. Furthermore, in many areas in the United States consumers who do have cable modem services available are unable to choose among service providers because a virtual monopoly has been established by one (or two) provider(s). The opportunity to remedy the situation is somewhat better in countries that are still developing their physical infrastructures today since they are able to build broadband Internet into their infrastructure designs.

It is with this background that *wireless metropolitan area networks* based on the 802.16 technology have recently gained a lot of interest among vendors and ISPs as the possible next development in wireless IP offering and a possible solution for the last mile access problem. With the theoretical speed of up to 75 Mbps and with a range of several miles, 802.16 broadband wireless offers an alternative to cable modem and DSL, possibly displacing these technologies in the future.

Wireless MAN security architecture has two main design goals: to provide controlled access to the provider's network, and to provide confidentiality, message integrity protection, and replay protection to the data being transmitted. WMAN communications can be one-to-one or one-to-many. In one-to-one communication, typically users are interested in protecting their data, and service providers in controlling access to their networks. In one-to-many communication, service or content providers encrypt data and provide keys to their subscribers; thus content access control is the only goal in this case.

For access control, one may use asymmetric (digital certificates) or symmetric (e.g., preshared keys, SIM cards) authentication methods; the revised 802.16 specification allows the use of either of these two classes of authentication methods. From a provider's perspective, an SS authenticating itself to a BS is sufficient for enforcing controlled access to the provider's network. However, for user data confidentiality, the one-way authentication is not sufficient. Consider, for example, that SS to BS authentication alone will not help detect an adversary claiming to be a BS and thereby launching a man-in-the-middle attack.

The revised IEEE 802.16 specifications update the cryptographic algorithms used for encryption and integrity protection, increase key lengths, and add replay protection. The revised key management protocol design consists of robust protection again replay attacks.

A few further additions to the 802.16 security architecture facilitate symmetric key-based authentication, and more importantly mobility. Specifically, a key hierarchy is defined for fast keying when a mobile SS (MS) associates with a new BS.

Controlled or metered access to the WMAN or any content disseminated via the WMAN is the foremost requirement of service providers. This basic requirement typically translates into many components.

First, any service provider's BS must be able to uniquely identify an MS that wants to get access to the network. The MS may identify itself to a BS using digital certificates or indirectly to a legacy authentication server (AS) (e.g., AAA server) in conjunction with a symmetric authentication method. In the latter case, the MS does not need to perform expensive computations as would be the case with digital certificates. Furthermore, in most cases, the BS only forwards authentication protocol messages to the backend AS for authentication. After verifying the MS's credentials, the AS informs the BS of the result—authentication success or failure—and securely transfers the master session key (MSK).

The second component of enforcing access control is key distribution. The BS must be able to uniquely and easily identify packets from authorized MSs so it can enforce authorized access to the WMAN. Thus, after successfully authenticating an MS, the BS establishes a secret key with the MS. The MS must include a proof of possession of the secret key with each packet. The most common way to achieve this is to compute a cryptographic integrity checksum with each packet, and include it with the packet. In WMANs, the BS and MS may derive the keying material as part of the authentication protocol, or the BS may supply the key(s) to the MS.

Third, the IEEE 802.16 specification defines a multicast and broadcast service (MBS). This allows WMAN service providers to distribute content efficiently via multicast to relevant subscribers. The provider enforces controlled access to the content by distributing a per-group secret key to the subscribers who paid for the additional services.

In addition to covering service providers' requirements, the security sublayer addresses WMAN users' requirements. User requirements are typically to protect the confidentiality and integrity of the data. In simpler terms, users want to ensure that a third party cannot read their communications, their data are not modified en route, and that no one injects or drops packets without being detected. It is quite difficult, if not impossible, to protect against an adversary dropping packets; the other requirements are fairly easy to achieve and the 802.16 standard specifies how to in the WMAN context. Specifically, in addition to encryption, WMAN secure encapsulation provides per-MPDU integrity protection as well as replay protection.

Within the extended security sublayer there are two versions of the PKM protocol: version 1 is quite similar to the basic security sublayer, except that it supports new ciphers including 3DES-ECB and AES-ECB for confidentiality of key material, and AES-CCM for MPDU confidentiality. HMAC-SHA-1 protects the integrity of the key management messages. PKMv2 comparatively has many more desirable properties, including mutual authentication using various combinations of RSA-based and EAP-based authentication protocols, additional message integrity algorithms, and key management protocols.

Before we delve into that discussion, a bit of context of the design motivation for PKMv2 is in order. PKMv2 is part of a specification to add mobility extensions to the base 802.16 standard. When MSs are mobile, it may be desirable that they preauthenticate with a BS they plan to associate with, to reduce any potential for interruption in service, be it access to the provider's network, or a-multicast/broadcast content delivery service. Thus preauthentication is one of the additional features in PKMv2. Similarly a key hierarchy is defined to allow an MS to authenticate itself to the backend AAA server once, irrespective of any number of BSs it may associate with. Along with these extensions for mobility, the new specification includes several enhancements to the WMAN security protocols. In the rest of this chapter, we discuss these additional features and their advantages and shortcomings.

EAP-based mutual authorization in PKMv2 alone can support mutual authentication (indirect mutual authentication via a proof of possession of a key, if a back end AS is involved). However, a combination of RSA authorization followed by an EAP authentication may also be used in WMAN access. In that case, the RSA authorization is considered to provide device mutual authentication, whereas the EAP authentication is user authentication (which is especially true if a SIM card is involved in authentication).

EAP authentication in PKMv2 is similar to that in the 802.1 X/EAP-based authentication of 802.11i STAs: the MS authenticates to an AS via an authenticator. The BS in 802.16 networks serves as the authenticator, although in some architectures the functionality of the authenticator and the BS might be separated (this model of separating the BS and the authenticator needs a further review before being considered secure). EAP authentication follows the steps below:

- The authenticator or the BS initiates the EAP authentication process. Note that in the public-key-based authentication protocol, the MS requests authentication. The BS sends an EAP request message to the MS. This is typically an EAP identity request encapsulated in MAC management protocol data unit (PDU) (i.e., the secondary management channel carries the EAP messages).
- The MS responds to the request with an EAP response message. The authenticator and the MS continue the EAP exchanges until the authentication server determines whether the exchange is a failure or a success. The exact number of the EAP messages depends on the method used for authentication.
- An EAP success or an EAP failure terminates the EAP authentication and authorization process. At the end of the protocol run, the BS and the MS have the primary master key (PMK).

If the EAP exchange follows and RSA authorization exchange, the EAP messages are protected using the EAP integrity key (EIK) derived as a result of the RSA authorization exchange. The EAP messages contain an AK sequence number (the AK and the EIK are derived from the RSA exchange) for replay protection and an OMAC digest, computed using the EIK, for integrity protection.

If a backend AS is involved in the EAP authentication process, the AS delivers the PMK to the authenticator or the BS after the EAP exchange is complete. The BS and the MS then engage in a 3-way exchange to prove to each other that they possess the PMK. The 3-way exchange can be run several times under the protection of the PMK to amortize the cost of the EAP authentication exchange.

IMPLEMENTATION ISSUES

IEEE 802 standards committee formed the 802.11 Wireless Local Area Networks (WLAN) Standards

Working Group in 1990. IEEE 802.11 standard does not provide technology or implementation, but introduces the specifications for the physical and the MAC layers. 802.11 is the wireless protocol for both ad hoc and client/server networks. The users' acceptance of this protocol is high. Although, the security of the transmission channel is a matter of special attention that always has to be considered.

The WEP scheme has been adopted by IEEE 802.11 standard to ensure security for the transmitted information. The basic two components of WEP are the pseudorandom number generator (PRNG) and the integrity algorithm. The PRNG is the most valuable component because it actually is the original encryption core. WEP adopts RC4 cipher as the PRNG unit and CRC-32 as the integrity algorithm. Although WEP is a good security scheme, the offered security in some cases can not satisfy the user demands. In order a higher security level to be ensured, 802.11i working group introduced, as protocol's security scheme, the advanced encryption standard (AES).

In this section the hardware implementation cost of both WEP and AES schemes is presented. In order to have a fair and detailed comparison between the two schemes, the same implementation platform has been used (i.e., the same field programmable gate array [FPGA] device).

For the AES scheme, a compact VLSI architecture is presented. The implementation of this architecture minimizes the allocated area resources. The area-optimized design does not sacrifice the system performance in a restricted way. The throughput of the design is much higher than the required by the IEEE 802.11 standard. Both WEP and AES schemes are compared in terms of implementation performance: allocated area resources, operating frequency, throughput, and power consumption. Aspects of the supported security of these encryption schemes are discussed and security level strength comparisons are given.

The proposed architecture for the implementation of the WEP scheme is illustrated in Figure 1.

In order to implement in hardware the cyclic redundancy check (CRC-32), a shift register of 32 flip-flops (F/Fs) and a number of XOR gates are used. So, a linear feedback shift register (LFSR) design is produced by using the F/Fs chain with

Figure 1. WEP scheme architecture

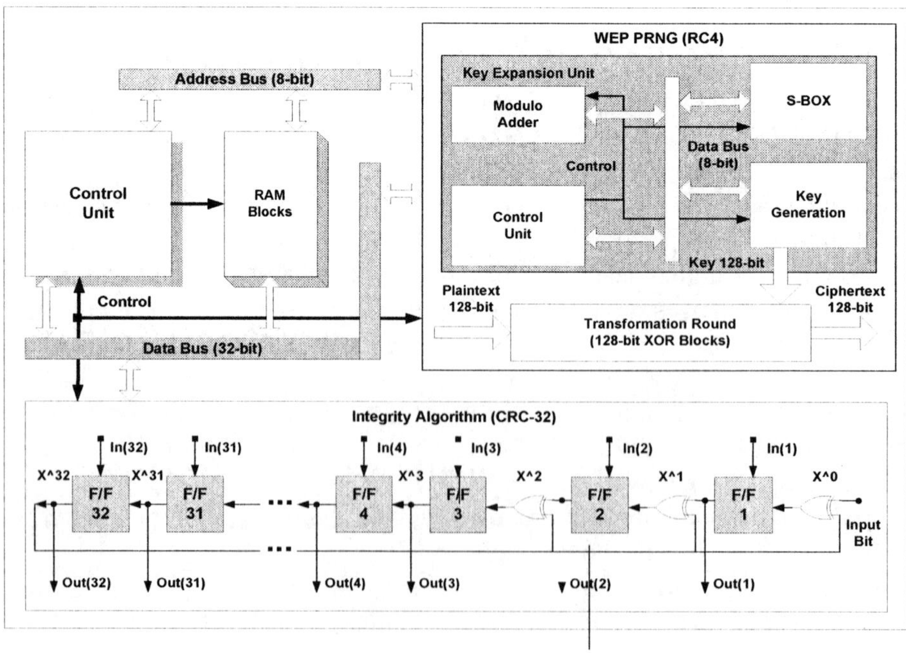

the XOR gates. The characteristic polynomial of this LFSR is:

$$(X) = X^{32}+X^{26}+X^{23}+X^{22}+X^{16}+X^{12}+X^{11}+X^{10}+X^{8}+X^{7}+X^{5}+X^{4}+X^{2}+X+1$$

The presence/absence of an XOR gate in CRC-32 architecture corresponds to the presence/absence of a term in G(X) polynomial. The required output message is the content of the LFSR after the input message last bit is sampled.

RC4 is a variable key-size stream cipher and operates on one plaintext block at a time. RC4 architecture consists of the key expansion unit and the transformation round. The key expansion unit is mainly a S-Box component. For the S-BOX implementation a 256-byte RAM memory block is used and another similar memory block is needed for the key array. The transformation round is a simple bit-by-bit XOR between the plaintext and the key.

IEEE 802.111 and Advanced Encryption Standard (AES)

AES proposed architecture operates in counter mode with cipher block chaining–message authentication code (CCMP). According to IEEE 802.11i working group, this operation mode is used to ensure, at the same time, integrity and privacy. The proposed AES architecture is shown in Figure 2.

The AES scheme architecture operates each time on a column of 32-bit data. It needs 41 clock cycles to complete the transformation of a 128-bit plaintext block. The column subunit is composed of four basic building blocks: S-Box, DataShift, MixColumn, and KeyAddition. The RAM-based design for the S-BOXes ([256x8]-bit) guarantees high performance. This "column"-based architecture minimizes the area resources compared with "state"-based architectures.

Figure 2. Advanced encryption standard (AES) scheme architecture

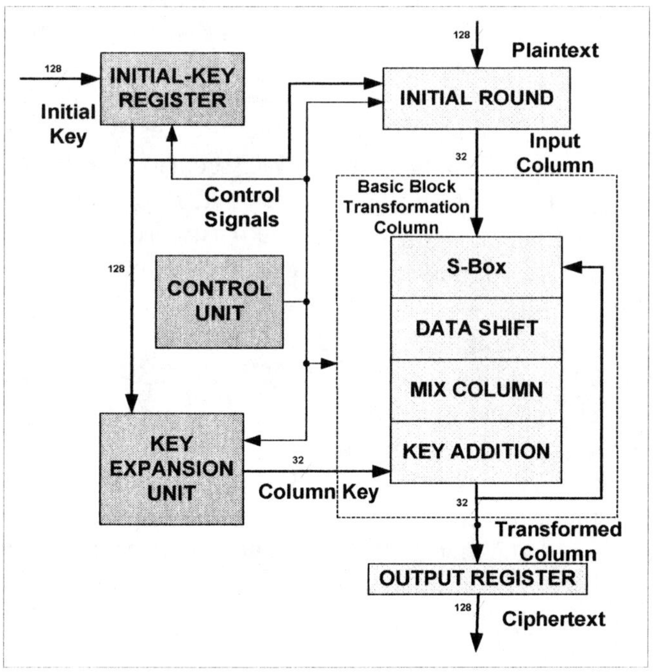

Figure 3. Implementations comparisons

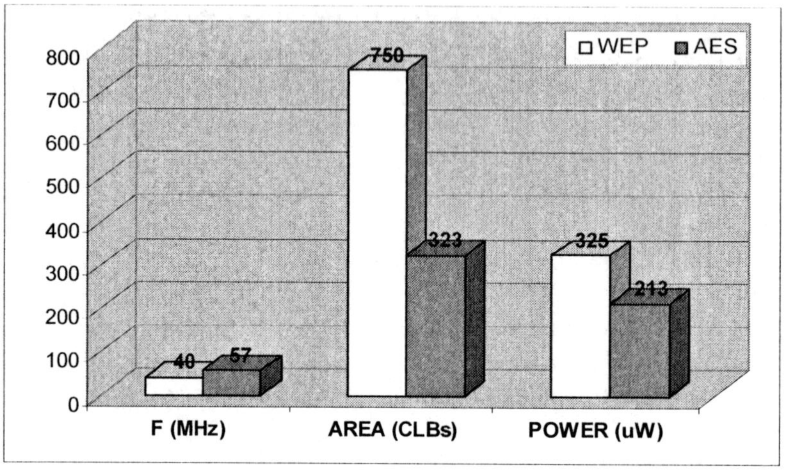

Implementation Cost and Performance Evaluation

In Figure 3, the synthesis results for both WEP and AES implementations are illustrated. For the hardware integration the FPGA device Xilinx Virtex (2V250fg256) (2003) has been used. For power consumption estimation, the Xilinx tool was used.

Based on the synthesis results regarding the area resources, the utilization of both implementations are: AES, 323 CLBs (allocated) + 1213 CLBs (unused) = 1526 CLBs (available in the FPGA device), and for WEP, 750 CLBs (allocated) + 776 CLBs (unused) = 1526 CLBs (available in the FPGA device). The AES implementation performs better compared with WEP implementation. The minimized area resources of AES do not sacrifice the system performance, which reaches throughput value 177 Mbps. On the other hand, RC4 is a more "heavy" design for mobile devices hardware implementation. This is due to the specified S-Boxes and the key expansion unit specifications. RC4 performance is the bottleneck for WEP throughput which reaches the value of 2.22 Mbps. The main RC4 implementation disadvantages, compared with AES, are: (1) more required silicon area resources, (2) higher power consumption, and (3) lower operating frequency. Concluding and based on the above comparison results, the proposed AES implementation is proposed for applications with special needs in both area resources and operation frequency.

Concerning security aspects, AES offers, at the same time, privacy and integrity. On the contrary, WEP scheme needs two different algorithms in order to support bulk encryption and data integrity. In some cases, where AES security is unbreakable, WEP security could be broken. These comparisons give AES advantages and make it an efficient and trustworthy solution for the next years' IEEE 802.11 networks.

CONCLUSION AND OUTLOOK

Generally speaking, the entire field of the "wireless Internet"—namely, wireless connectivity to the IP network—is still relatively new and may take a few more years to reach the level of ubiquity comparable to other access technologies such as dial-up over PSTN. What is evident, however, is that user mobility is a crucial aspect of the next generation Internet services where the ordinary user will expect connectivity to be something that is permanently available, much like electricity that is "always on."

There are a number of emerging technological trends today that may influence the future of WLANs and WMANs. We summarize these in the following and describe possible outcomes and developments in this exciting field.

- **The transparent "always on" Internet:** Increasingly the details of the operations of the IP Internet will become transparent or removed from the ordinary user. In the past, many early-home adopters of WLANs have had to familiarize themselves with important IP networking concepts (such as IP addresses, ports on switches/routers, and so forth) in order to set-up a home WLAN. Today, through improved quality and ease of use of WLAN products, most products are essentially plug-and-play. This human aspect is important because it contributes to the user's expectations of an always-on Internet, whether they access it from a home WLAN, a Wi-Fi hotspot, a wireless broadband (WMAN) provider, or from a 3G/GPRS provider. Increasingly, the lay user will not care how connectivity is provided, but will expect high-bandwidth connectivity to be ubiquitous. This expectation will in turn influence how service providers establish seamless services between the IP Internet and the 2G/3G mobile networks.
- **Increased adoption of 802.1X:** The 802.1X approach for WLAN authentication is increasingly being adopted by enterprises, due in part to its support within the Microsoft Windows family of products. Although various networking vendors have been touting proprietary security products for WLAN authentication, the completion of the revision to the IEEE 802.1X standard together with recent progress in the Internet Engineering Task Force (IETF) on EAP-related standards should lead to the strengthening of 802.1X in the market. In the context of Wi-Fi roaming, technically 802.1X provides better security than the UAM Web-based approach. Thus, one possible development is for MNOs to also begin adopting 802. IX for their Wi-Fi hotspots, possibly reusing their SIM-based authentication with 802.IX (e.g., using an EAP method such as EAP-SIM).
- **Enterprise adoption of "IPsec everywhere":** Many enterprises who were early adopters of intracampus WLANs solved the 802.11 WEP security problem by running IPsec connections internally within the enterprise network. Although the "IPsec everywhere" approach was initially promoted as a temporary patch over the insecure WLAN segment of the network, increasingly some enterprises have continued to use IPsec for other purposes (e.g., establish virtual LANs). There are a number of possible consequences—intended or unintended—of using IPsec in this manner.

One possible effect of the widespread use of IPsec within internal corporate networks—both LANs and WLANs—is to bring the connectivity layer one step higher, introducing a new *IPsec layer* in the stack. Thus, here IPsec could be seen as the network layer transport (instead of the plain IP at ISO/OSI layer 3), where the actual IP addresses of the endpoints become less important than the identity and IPsec credentials (e.g., digital certificates or shared secrets) of those endpoints. This deemphasizing of the IP layer and increased focus on the IPsec layer may force networking hardware vendors to introduce richer security functionality into their hardware. Examples would be routers that can route based not only on IP addresses, but also on other characteristics of the IPsec connection (e.g., IKE and IPsec security associations).

REFERENCES

Agis, E., Mitchel, H., Ovadia, S., Asisi, S., Bakshi, S., Iyer, P., et al. (2004, August). Global, interoperable broadband wireless networks: Extending WiMax technology to mobility. *International Technical Journal, 8*, 173-187.

Arkko, J., & Haverinen, H. (2004, December 21). *Extensible authentication protocol method for 3rd generation authentication and key agreement (EAP-AKA).* Retrieved October 22, 2007, from <draft-arkko-pppext-eap-aka-15.txt>

Cooklev, T. (2004). *IEEE wireless communication standards: A study of 802.11, 802.15, and 802.16.* New York: IEEE Press.

Eklund, C., Marks, R. B., Stanwood, K. L., & Wang, S. (2002, June). IEEE standard 802.16: A technical overview of the wirelessMAN air interface for broadband wireless access. *IEEE Communications Magazine*, 98-107.

Haverinen, H., & Saloway, J. (2004, December 21). *Extensible authentication protocol method for GSM subscriber identity modules (EAP-SIM).* Retrieved October 22, 2007, from <draft-haverinen-pppext-eap-sim-16.txt>

Hwang, G. J., Tseng, J. C. R., & Huang Y. S (2002). I-WAP: An intelligent WAP site management system. *IEEE Transactions on Mobile Computing, 1*(2).

IEEE. (2001). IEEE standard 802.16-2000 standard for local and metropolitan area networks. *Air interface for fixed broadband wireless access systems part 16.* New York: IEEE Press.

IEEE Standard for Local and metropolitan area networks-Port-Based Network Access Control. *ANSI/IEEE IEEE Std 802.1X-2001.*

Karri, R., & Mishra, P. (2003). Optimizing the energy consumed by secure wireless sessions-wireless transport layer security. *Journal of Mobile Networks and Applications, 8*, 177-185. Kluwer Academic Publishers.

RFC 2865. (2000, June). *Remote authentication dial in user service (RADIUS).*

RFC 2869. (2000, June). *RADIUS extensions.*

RFC 3580. (2003, September). IEEE 802.1X remote authentication dial in user service (RADIUS). *Usage guidelines.*

RFC 3748. (2004, June). *Extensible authentication protocol (EAP).*

Sklavos, N., & Koufopavlou, O. (2002). Architectures and VLSI implementations of the AES-proposal Rijndael. *IEEE Transactions on Computers, 51*(12), 1454-1459.

Sklavos, N., & Koufopavlou, O. (2003). Mobile communications world: Security implementations aspects-a state of the art CSJM journal. *Institute of Mathematics and Computer Science, 11*(2).

Wireless LAN Medium Access Control (MAC) and Physical Layer (PHY) specifications. (1999). *ANSI/IEEE Std 802.11: 1999 (E) Part 11, ISO/IEC 880211.*

Xilinx Inc. (2003). *Virtex, 2.5 V field programmable gate arrays, & a simple method of estimating power in XC40000X1/EX/E FPGAs application brief XBRF 014 v1.0.* San Jose, California. Retrieved October 22, 2007, from www.xilinx.com

KEY TERMS

Access Point (AP): The network access device for an 802.11 wireless network. It contains a radio receiver/transmitter. It may be an 802.1x authenticator.

Certification Authority (CA): An entity that issues digital certificates (especially X.509 certificates) and vouches for the binding between the data items in a certificate.

Extensible Authentication Protocol (EAP): A protocol used between a user station and an authenticator or authentication server. It acts as a transport for authentication methods or types. It in turn may be encapsulated in other protocols, such as 802.1x and RADIUS.

EAP-AKA: This document specifies an extensible authentication protocol (EAP) mechanism for authentication and session key distribution using the authentication and key agreement (AKA) mechanism used in the 3rd generation mobile networks universal mobile telecommunications system (UMTS) and CDMA2000. AKA is based on symmetric keys, and runs typically in a subscriber identity module (UMTS subscriber identity module [USIM], or removable user identity module [RUIM], a smart card like device).

EAP-LEAP: Lightweight extensible authentication protocol is a Cisco proprietary EAPType. It is designed to overcome some basic wireless authentication concerns through mutual authentication and the use of dynamic WEP keys.

EAP-PEAP: Protected extensible authentication protocol is a two-phase authentication like EAP-TLS. In the first phase the authentication server is authenticated to the supplicant using an X.509 certificate. Using TLS, a secure channel is established through which any other EAP-Type can be used to authenticate the supplicant to the authentication server during the second phase. A certificate is only required at the authentication server. EAP-PEAP also supports identity hiding where the authenticator is only aware of the anonymous username used to establish the TLS channel during the first phase but not the individual user authenticated during the second phase.

EAP-SIM: EAP-SIM is an authentication mechanism that makes use of the SIM card to perform authentication within the 802.1x framework for WLAN.

EAP-TLS: Transport layer security is an EAP-Type for authentication based upon X.509 certificates. Because it requires both the supplicant and the authentication server to have certificates, it provides explicit mutual authentication and is resilient to man-in-the-middle attacks. After successful authentication a secure TLS link is established to securely communicate a unique session key from the authentication server to the authenticator. Because X.509 certificates are required on the supplicant, EAP-TLS presents significant management complexities.

EAP-TTLS: Tunneled TLS is an EAP-type for authentication that employs a two-phase authentication process. In the first phase the authentication server is authenticated to the supplicant using an X.509 certificate. Using TLS, a secure channel is established through which the supplicant can be authenticated to the authentication server using legacy PPP authentication protocols such as PAP, CHAP, and MS-CHAP. EAP-TTLS has the advantage over EAP-TLS that it only requires a certificate at the authentication server. It also makes possible forwarding of Supplicant requests to a legacy RADIUS server. EAP-TTLS also supports identity hiding where the authenticator is only aware of the anonymous username used to establish the TLS channel during the first phase but not the individual user authenticated during the second phase.

European Telecommunications Standards Institute (ETSI): ETSI is a multinational standardization body with regulatory and standardization authority over much of Europe. GSM standardization took place under the auspices of ETSI. ETSI has taken the lead role in standardizing a wireless LAN technology competing with 802.11 called the high performance radio LAN (HIPERLAN).

Integrity Check Value (ICV): The checksum calculated over a frame before encryption by WEP. The ICV is designed to protect a frame against tampering by allowing a receiver to detect alterations to the frame. Unfortunately, WEP uses a flawed algorithm to generate the ICV, which robs WEP of a great deal of tamperresistance.

Institute of Electrical and Electronics Engineers (IEEE): A worldwide professional association for electrical and electronics engineers that sets standards for telecommunications and computing applications.

Initialization Vector (IV): Generally used as a term for exposed keying material in cryptographic headers; most often used with block ciphers. WEP exposes 24 bits of the secret key to the world in the frame header, even though WEP is based on a stream cipher.

Medium Access Control (MAC): The function in IEEE networks that arbitrates use of the network capacity and determines which stations are allowed to use the medium for transmission.

MPDU: MAC protocol data unit is a fancy name for frame. The MPDU does not, however, include PLCP headers.

MSDU: MAC service data unit is the data accepted by the MAC for delivery to another MAC on the network. MSDUs are composed of higher-level data only. For example, an 802.11 management frame does not contain an MSDU.

OFDM: Orthogonal frequency division multiplexing is a technique that splits a wide frequency band into a number of narrow frequency bands and inverse multiplexes data across the subchannels. Both 802.11a and the forthcoming 802.11g standards are based on OFDM.

Open Systems Interconnection (OSI): A baroque compendium of networking standards that was never implemented because IP networks actually existed.

Request for Comments (RFC): A series of numbered documents (RFC 822, RFC 1123, etc.), developed by the Internet Engineering Task Force (IETF) that set standards and are voluntarily followed by many makers of software in the Internet community.

Wireless Application Protocol (WAP): A standard for providing cellular telephones, pagers, and other handheld devices with secure access to e-mail and text-based Web pages. Introduced in 1997 by Phone.com, Ericsson, Motorola, and Nokia, WAP provides a complete environment for wireless applications that includes a wireless counterpart of TCP/IP and a framework for telephony integration, such as call control and telephone book access. WAP features the wireless markup language (WML), which was derived from Phone.com's HDML and is a streamlined version of HTML for small-screen displays. It also uses WMLScript, a compact JavaScript-like language that runs in limited memory. WAP also supports handheld input methods, such as a keypad and voice recognition. Independent of the air interface, WAP runs over all the major wireless networks in place now and in the future. It is also device-independent, requiring only a minimum functionality in the unit to permit use with a myriad of telephones and handheld devices.

Chapter XLVII
End-to-End (E2E) Security Approach in WiMAX:
A Security Technical Overview for Corporate Multimedia Applications

Sasan Adibi
University of Waterloo, Canada

Gordon B. Agnew
University of Waterloo, Canada

Tom Tofigh
WiMAX Forum, USA

ABSTRACT

An overview of the technical and business aspects is given for the corporate deployment of services over worldwide interoperability for microwave access (WiMAX). WiMAX is considered to be a strong candidate for the next generation of broadband wireless access; therefore its security is critical. This chapter provides an overview of the inherent and complementary benefits of broadband deployment over a long haul wireless pipe, such as WiMAX. In addition, we explore end-to-end (E2E) security structures necessary to launch secure business and consumer class services. The main focus of this chapter is to look for a best security practice to achieve E2E security in both vertical and horizontal markets. The E2E security practices will ensure complete coverage of the entire link from the client (user) to the server. This is also applicable to wireless virtual private network (VPN) applications where the tunneling mechanism between the client and the server ensures complete privacy and security for all users. The same idea for E2E security is applied to client-server-based multimedia applications, such as in Internet protocol (IP) multimedia subsystem (IMS) and voice over IP (VoIP) where secure client/server communication is required. In general, we believe that WiMAX provides the opportunity for a new class of high data rate symmetric services. Such services will require E2E security schemes to ensure risk-free high data-rate uploads and downloads of multimedia applications. WiMAX provides the capability for embedded security functions through the 802.16 security architecture standards. IEEE 802.16 is further subcategorized as

802.16d (fixed-WiMAX) and 802.16e (mobile-WiMAX). Due to the mobility and roaming capabilities in 802.16e and the fact that the medium of signal transmission is accessible to everyone, there are a few extra security considerations applied to 802.16e. These extra features include: privacy key management version 2 (PKMv2), PKM-extensible authentication protocol (EAP) authentication method, advanced encryption standard (AES) encryption wrapping, and so forth. The common security features of 802.16d and 802.16e are discussed in this chapter, as well as the highlights of the security comparisons between other broadband access, third-generation (3G) technologies, and WiMAX.

INTRODUCTION

The E2E security structure is transparent from the user's point of view and requires dedicated overhead and processing power. In the case of Wi-Fi, the overhead is a relatively large percentage of the total bandwidth, which makes Wi-Fi infeasible for most E2E security structures. However, in worldwide interoperability for microwave access (WiMAX), the security overhead is nominal and may not be an issue.

Today's enterprise customers are forced to use dedicated physical circuits such as leased lines to realize business class E2E security. With inherent WiMAX security features, a secured virtual private network (VPN) can easily be achieved over public networks. Instead of such dedicated leased line circuits, WiMAX users could enjoy VPN connectivity with up to 10 Mbps bandwidth to access the public backbones.

Personal broadband access technologies have undergone many challenges, one of which was digital subscriber line (DSL). DSL is a high-speed connection that utilizes the same wiring system as a regular telephone line uses. The advantages of DSL include: voice/data on the same line and higher data rates than regular modems. There are, however, a few downsides to DSL, including distance dependence (between users and the service provider) of data rate, unbalance rates for uploading and downloading of data, and having no complete physical area coverage.

All of the downsides of DSL technology appear in other personal broadband products. This is due to the physical limitations of wired technologies. WiMAX, on the other hand, is a wireless technology with very high bandwidth for voice/data applications, which does not appear to have any of the downsides of the wired technologies. WiMAX also has advantages over Wi-Fi technology in terms of longer range and larger bandwidth. This allows WiMAX to support a variety of broadband services.

Wi-Fi technology was not suited for personal broadband services due to a number of limitations, especially security. WiMAX, on the other hand, enjoys an all-IP open platform infrastructure with the benefit of its inherent security functions and features. This allows for faster and inexpensive provisioning of E2E secured services based on open standards. In addition WiMAX can be configured for self-installed services of multimedia VPN with enhanced end-to-end user control signalling.

The security aspect of WiMAX is an important issue: this includes state-of-the-art security mechanisms, such as very strong authentication with per station keys and higher-level security mechanisms. WiMAX's security strength is normally found in add-on products, such as in wired VPNs and virtual local area networks (VLANs), which are usually built into each of the WiMAX's base stations (BSs).

This chapter will present the characteristics of WiMAX security and how it fits into both consumer and business class structures. We believe that strong E2E security can be achieved with WiMAX without compromising performance.

Why Wireless Networks Could not Provide the Required Security

There were two main reasons why wireless was never considered as a secured high-performance backbone option for business and corporate applications. The first issue was the bandwidth limitations of wireless links and the second issue was the high security requirements of VPNs and IMS applications. The 802.11-based systems have an upper limit on bandwidth of 54 Mbps for 802.11g, however in real-world applications, this rate seldom tops more than 20-25 Mbps due to the overhead in the medium access control (MAC) layer. It is also very difficult to have a minimum guaranteed bandwidth for real-time applications such as VoIP and videoconferencing.

The current Wi-Fi security standard is presented in 802.11i, which contains many fixes for the security concerns in 802.11. However 802.11i has not been widely implemented and distributed among end-users and WiMAX is expected to dominate the market before 802.11i can affect the market. Therefore the main security comparisons are between Wi-Fi (802.11a/g) and 802.16. The main reasons for this weakness can be categorized as follow (Gast, 2004):

Problem #1: Easy Access
Since Wi-Fi networks generate beacon frames containing the network parameters all of the time, attackers with high gain antennas can find networks and launch attacks. With the inherent and add-on security features, WiMAX is expected to be resilient against such attacks.

Problem #2: "Rogue" Access Points
Anyone can have access to an inexpensive access point (AP) and get connected to a corporate network and bypass authorization. In WiMAX networks, an E2E security scheme can protect APs against such a scenario.

Problem #3: Unauthorized Use of Service
Nearly all APs have default configurations with wired equivalent privacy (WEP) or with a default key used in WEP by all the vendor's products. Without WEP, a network can be accessed by any anyone. Even with WEP enabled, a network is not considered to be secure nowadays.

Problem #4: Performance and Service Constraints
802.11b and 802.11g both have limited transmission capacities (11 and 54 Mbps) and due to MAC-layer overhead, the actual effective throughput is close to half of that rate. In addition, bandwidth is not guaranteed.

Problem #5: MAC Spoofing and Session Hijacking
802.11 networks do not authenticate frames and there is no protection against a forgery of the frame source address attack. Here, attackers can use spoofed frames to redirect traffic and corrupt address resolution protocol (ARP) tables. Station MAC addresses could easily be observed and engaged in malicious transmissions. Any user with a strong transmitter can be situated in the middle of a new session and potentially steal credentials and gain access through a man-in-the-middle (MITM) attack.

Problem #6: Traffic Analysis and Eavesdropping
802.11 is totally vulnerable to passive attacks. There is no security of the header information, thus, no protection against eavesdropping. Frame headers are always "in the clear" and sender-receiver pairs are vulnerable to traffic analysis.

Problem #7: Higher Level Attacks
Once an attacker gains access (either through session-hijacking, MITM, spoofing attacks, or through breaking the WEP secure key), it is possible to use that AP to launch attacks on other systems, which are within the trusted domain of the initially attacked AP.

The main reason for the failure of security in wireless networks is the fact that there are many weaknesses in the mechanisms and protocols used in the architecture.

WIMAX SECURITY LAYERS

Transmission control protocol (TCP)/IP protocol stacks have currently dominated the data traffic transmitted between transmitters and receivers attached to the backbone of the Internet. The same situation also applies to WiMAX infrastructures, therefore it is vital to study the performance and security aspects of WiMAX systems in a similar layered fashion. In principle, communication systems are based on the seven-layer OSI model. However, most systems communicating on the Internet's backbone obey the five-layer architecture and WiMAX security protocol foundation is based on the lower-layers (e.g., the MAC layer), which provides extra capabilities in constructing security functions.

WiMAX enjoys the inherent security features with an open system platform, the all-IP structure, with options to enhance security in different layers of the WiMAX open architecture. All of these features have contributed to the strength of WiMAX security, which potentially enables secured applications such as VoIP and content streaming. The E2E security scheme also plays a vital role in adding an extra security feature to enable secure connections for seamless roaming for wireless broadband technologies across any network supporting TCP/IP.

This layered approach to security is further discussed in sections 3.1, 3.2, and 3.3.

Physical Layer Security

IEEE 802.16 is a MAC-layer-based protocol and its security schemes are mainly situated in the security sublayer of the MAC layer, where most of the algorithms and security mechanisms initially work. Here, physical (PHY) and MAC layers are closely related to one another. The basic security functions at the physical layer are in the form of key-exchange, encryption, and decryption. These mechanisms are however controlled at the MAC layer. Therefore the main objective of this section is to understand the MAC layer security mechanisms.

Another aspect related to the PHY layer is the transmission power. Unlicensed WiMAX has the same inherent security capabilities as compared to the licensed WiMAX. They do differ, however, in the amount of transmission power (unlicensed WiMAX carriers having a lower maximum power) which limits the range and also the possibility for interference.

Some other security implementations at the physical layer exploit the fact that modulation is done at this layer. Some transmitters may use frequency inversion as a security deterrent (Chandra, 2002). For example, the transmitter may divide the spectrum into various frequencies and use the different frequencies in a predetermined fashion. Obviously, this requires that both the sender and the receiver share a frequency hopping pattern. This is a form of spread spectrum communication.

Spread-spectrum systems also have an inherent security mechanism since data meant for a particular receiver, cannot easily be intercepted by other receivers if they do not possess the frequency hoping order which is controlled by the key.

Physical layer security implementations do not provide robust protection against attacks as they are prone to attacks such as the disruption of service (denial-of-service [DoS]). Other passive and active attacks include cross-connects and adjacent channel interference.

Therefore parts of the encryption/decryption mechanisms (which are mainly controlled at the MAC layer) that deal with the physical act of hiding information from the intruders' eyes are part of the PHY layer security schemes.

MAC Layer Security

IEEE 802.16 specifications for security mainly fall within the MAC layer. Figure 1 shows the protocol layering of 802.16 and the MAC layer's security implementation. The separate security sublayer provides authentication, secure key exchange, and encryption.

Security within the MAC layer is called the *security sublayer*. Its goal is to provide access control and confidentiality of the data link.

When two parties establish a link, they are protected via a set of protocols that ensure confidentiality and unique access of the authorized parties. The unique handshaking between the two

Figure 1. IEEE 802.16 lower layers (Adapted from "Part 16," 2004)

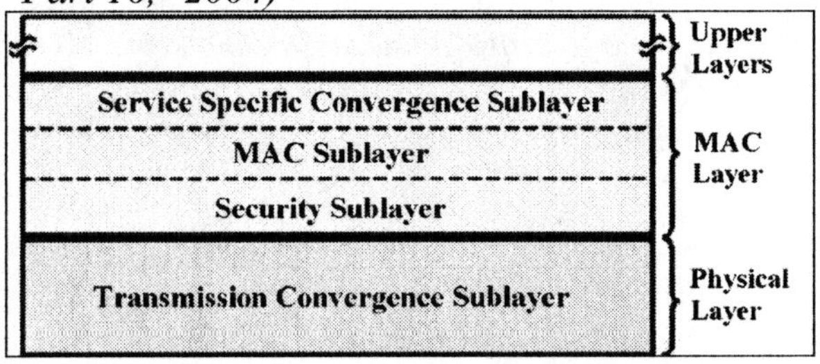

entities; namely BS and subscriber station (SS), is done at the MAC layer through security sublayer, which has five entities (Chandra, 2002):

- **Security associations:** A security association (SA) is a set of security information parameters that a BS and one or more of its client SSs share in order to support secure communications. Three types of SAs are defined as (Johnston & Walker, 2004) *primary, static,* and *dynamic* (Figure 2), which define the security keys and associations established between a SS and a BS during the authorization phase.
- **X.509 certificate profile:** This defines a digital certificate to verify the identity of subscribers and prevents impersonation (unauthorized SS or BS)
- **PKM authorization:** The privacy key management (PKM) protocol is responsible for privacy, key management, and authorizing an SS to the BS. The initial draft for WiMAX mandates the use of PKMv1 (Johnston & Walker, 2004), which is a one-way authentication method. PKMv1 requires only the SS to authenticate itself to the BS, which poses a risk for a MITM attack. To overcome this issue, PKMv2 was proposed (later adopted by 802.16e), which uses a mutual (two-way) authentication protocol. Here, both the SS and the BS are required to authorize and authenticate each other
- **Privacy and key management:** The privacy of the communications between the SS and the BS is achieved through the PKM protocol. Phifer, L 2. (2003, September). Applying RADIUS to Wireless LANs, using RADIUS For WLAN Authentication, Part I, from http://www.wi-fiplanet.com/tutorials/article.php/10724_3114511_1
- **Encryption:** The data communication between each SS and BS is encrypted using the advanced encryption standard (AES), with at least 128 bit keys. According to FIPS 140-2, AES-128 is computationally secure for data up to SECRET level for the next 10 years.

According to the initial drafts of WiMAX, the security sublayer provides enough security mechanisms to provide privacy, authentication, and encryption over the airlink. However, in order to achieve maximal security strength, true end-to-end security is required for a corporate wireless backbone, which enhances the security mechanisms specified by the initial drafts.

Security at Upper Layers

IEEE 802.16's main focus on the security issue is at the MAC layer, therefore WiMAX has the

Figure 2. Security model of the privacy sublayer (Adapted from Barbeau, 2005)

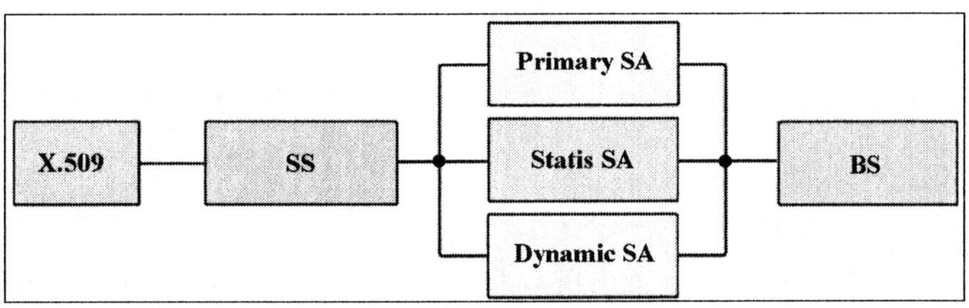

freedom to adopt the strong security measures for upper layers (network, transport, session, and application layers). Upper layer security options, such as Internet protocol (IP) security protocol (IPSec) and transport layer security (TLS), are examples of the current security schemes for upper layers. Through this freedom of choice, the security strength of WiMAX is comparable to the most secure networks in the market.

Lawful Interception (LI) or Lawful Legal Interception (LLI) (Baker, Foster, & Sharp, 2004; Brown, 2006; Mulholland, 2006)

Since WiMAX-enabled nodes will be connected as parts of the worldwide telecommunications networks and the telecommunications infrastructure of the world, the need for law enforcement access is required. Standards for access to the IP-based networks such as WiMAX have already been developed and are available from various standards and government bodies worldwide.

What follows is a discussion that focuses on two major standards bodies, the ETSI (European Telecommunications Standards Institute) and the IETF (Internet Engineering Task Force).

ETSI Approach to Lawful Interception

The Technical Committee on Lawful Interception (TCLI) is the leading body for lawful interception standardization within ETSI. Lawful interception standards have also been developed by ETSI technical bodies: AT, TISPAN (SPAN and TIPHON™), TETRA, and by 3GPP™. European governments might expect WiMAX vendors to provide this law enforcement access. Examples of ETSI standards include: "

- *ES 201 671: Lawful Interception (LI); Telecommunications Security; Handover Interface for the Lawful Interception of Telecommunications Traffic (revised version).*
- *ES 201 158: Lawful Interception (LI); Telecommunications Security; Requirements for Network Functions*
- *TS 102 234: Lawful Interception (LI); Telecommunications Security; Service-specific details for internet access services.*
- *TS 102 233: Lawful interception (LI); Telecommunications Security; Service-specific details for e-mail services.*
- *TS 102 232: Lawful Interception (LI); Telecommunications Security; Handover Specification for IP Delivery.*
- *TS 101 671: Lawful Interception (LI); Telecommunications Security; Handover interface for the lawful interception of telecommunications traffic.*
- *TS 101 331: Lawful Interception (LI); Telecommunications Security; Requirements of Law Enforcement Agencies.*
- *TR 102 053: Lawful Interception (LI); Telecommunications Security; Notes on ISDN lawful interception functionality.*
- *TR 101 944: Lawful Interception (LI); Telecommunications Security; Issues on IP Interception.*
- *TR 101 943: Lawful Interception (LI); Telecommunications Security; Concepts of Interception in a Generic Network Architecture."* (copied from Arend, 2007).

IETF Decision on the Lawful Interception

The IEFT has yet to consider wiretap requirements as part of their standards. The reasons for this decision are:

- Inappropriate in global standards – legal and privacy requirements are too varied
- Would increase protocol complexity and decrease security
- End-to-end security makes LI unworkable
- Other standards are already available

The IETF believes that designed mechanisms, which facilitate or enable wiretapping, or methods of using other facilities for such purposes, should be described openly, so as to ensure the maximum review of the mechanisms and to ensure that they adhere as closely as possible to their design constraints. This is considered by Cisco (Figure 3) for

LI in IP networks (RFC 3924) with the following requirements (Mulholland, 2006).

Carriers should be able to provide the following:

- **Content of the communication**
  - Audio content of the voice call
  - Packets to and from the subject
- **Communication-identifying information (CmII)**
  - Dialed digits in voice calls
  - Subject login information
  - Network addresses data

LI should not be detectable by the intercept subject and should include the followings:

- Knowledge of wire-tapping is limited to authorized personnel.
- Ability to correlate communication identifying information with the content of the communication.
- Confidentiality, authentication, and integrity of the CmII.
- Requirements vary between different agencies, regions, and countries.

Lawful access (LA) requirements are:

- Invisible to unauthorized personnel and other interceptors.
- Undetectable to the subject.
- Any available decryption keys should be provided to the authorities.
- Only authorized information should be provided.

SECURITY OF IMS AND WIMAX

The IP multimedia subsystem (IMS) uses a standardized next generation networking (NGN) architecture for wireline as well as wireless systems. This is particularly important for WiMAX backbones as they offer the required bandwidth for such multimedia traffic. More importantly is the fact that WiMAX comes in several flavors, some of which may coexist in a single network: fixed, portable, nomadic, and mobile. Therefore WiMAX covers wide areas of broadband access for personal and cellular communications, inline with the IMS coverage.

IMS provides new services as well as current and future Internet related services. This includes end-users ability to execute all related commands and functions even when they are far from their home networks, roaming through foreign networks. In order for IMS to achieve these goals, the architecture of IMS uses the open standard IP protocols, which is defined by the IETF and is enhanced by the 3rd Generation Partnership Project (3GPP). There are three variations of how

Figure 3. Lawful intercept architecture reference model (Adapted from Mulholland, 2006)

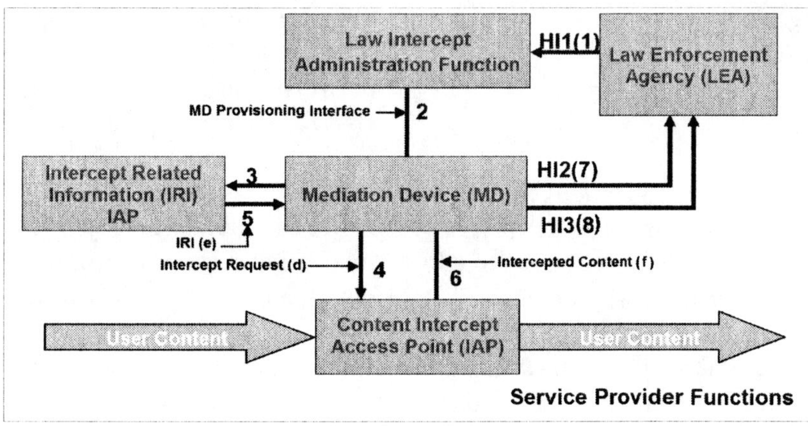

IMS works: session initiations between two IMS users, between an IMS user and a user on the Internet, or between two users on the Internet. IMS uses similar protocols for such initiations. Furthermore, service developers use IP protocol stack for the interfaces, which is why IMS can truly merge the Internet with the cellular world. This merge is done by using the cellular and mobile technologies, which provide ubiquitous access and Internet connections, which provides appealing services. Accordingly, WiMAX enjoys one of the most enhanced cellular technologies, which could work in the most efficient method delivering IMS data and applications.

In regards to the IMS security requirements, WiMAX security mechanisms are there to ensure all communicating parties, which gain access to the media, are legitimate and all parties wishing to gain access are thoroughly authenticated through the authentication and authorization protocols. This has to be done before any access is permitted. An ongoing mutual authentication mechanism ensures no illegitimate entity can highjack a session and abduct an already authenticated link and take over the communications at any points.

IMS is designed to work on either fixed or mobile systems. Since WiMAX offers most of the advantages of fixed networks, it is expected that IMS is going to be offered on a pure WiMAX backbone to address corporate and end-user requirements. The fact that WiMAX is based on an all-IP core structure makes it a perfect match for IMS, with its so many IP-based services in use. These services include voice over IP (VoIP), push to talk over cellular (POC), multiparty games, videoconferencing, messaging, community services, presence information, and content sharing.

Security of VoIP

One of the most important applications of IMS is the VoIP that runs over the standard IP. A VoIP system uses protocols, such as, H.323, MGCP, MEGACO, and/or session initiation protocol (SIP) for signaling, and real time protocol/real time control protocol (RTP/RTCP) for media transport and control. The threats for this type of scenario (client/server), as well as in IMS/WiMAX applications, including (Ramana Mylavarapu, 2005):

- Client impersonation (unauthorized client seeks access)
- Server impersonation (unauthorized server pretend to be authorized)
- Message tampering (additions, deletions, or delay of the message contents)
- Session tampering/hijacking (once the session between a legitimate client and server is established, an unauthorized entity takes the session)
- Signaling requests resulting in DoS attacks

To protect against any of the aforementioned vulnerabilities, an extensive two-way authentication method is used to ensure both the client's and the server's right of access and the establishment of IPSec security associate with the IMS terminal. This prevents the mentioned vulnerabilities as well as snooping attacks and replay attacks and to protect the privacy of every individual user.

Security issues in regards to SIP could also be summarized as follow (Access security for IP-based services, 2002):

- Protection mechanism of SIP signaling between the IMS server and the subscriber
- Subscriber's self authentication mechanism
- Subscriber's authentication mechanism to the IMS server

The reactive and proactive security measures are the encryption/decryption of SIP messages and deploying interconnection border control function (IBCF). IBCF is used as a gateway to external networks and provides network address translation (NAT) and firewall functions (Mylavarapu, 2005), two-way authentication-authorization schemes, and secure tunneling.

To enhance the deployment of IPSec, it is recommended to deploy IPv6 (Saito, 2003), which is the next generation Internet protocol. The important factor of IPv6 is its mandate for utilizing IPSec. Using a two-way IPSec connection (two one-way IPSec patterns) is required for an end-to-end security scheme (Saito, 2003).

END-TO-END APPROACH IN WIMAX

As mentioned earlier, the E2E scheme will ensure that the entire link from the user to the server is protected. E2E security is a major issue that could be addressed either in a peer-to-peer basis or in a multilayer manner. The E2E approach discussed here will not offer a comprehensive solution to the multilayer E2E; rather, it will present a peer-to-peer approach (i.e., BS to SS).

The heart of the airlink security scheme in WiMAX is the privacy key management version 2 (PKMv2), which offers a mutual authentication method to authenticate both the SS and the BS.

The following attacks could be mounted if the PKMv2 is not deployed:

- **BS and SS impersonations:** BS and SS should be able to authenticate the other party and find the unauthorized entity. The use of mutual authentication through PKMv2 (Figure 4) with suitable credentials. PKMv2 supports two authentication protocol schemes: Rivest-Shamir-Adleman (RSA) and extensible authentication protocol (EAP). EAP is mandatory for all devices.
- **Man-in-the-middle-attack:** This type of attack happens when one of the communication parties is not forced to authenticate itself. The same as the BS-SS impersonation. The use of PKMv2 will solve this problem.
- **Key exchange issue:** To encrypt and decrypt information between two parties, temporal keys and sessions keys are used. The key distribution in the initial draft uses triple data encryption standard (3DES) for exchanging keys. A 2-key 3DES based-key wrap is currently used for temporal encryption key (TEK) exchange. TEKs should not be used more than one time and there should be a mechanism to ensure that TEKs do not repeat. Otherwise this suffers from replay attacks as there are no dynamic components in the key exchange protocol and it also suffers from the man-in-the-middle attacks. To avoid this problem, various TEKs should be used and the EAP authentication framework and the AES-counter for cipher-block-chaining message authentication code (CCM) cipher suite should be used with PKMv2.

PKMv2 authentication/authorization method is shown in Figure 4.

WIMAX VS. 3G TECHNOLOGIES

In this section, 3G cellular technologies such as global system for mobile communications (GSM), universal mobile telecommunications system (UMTS), and coded division multiple access (CDMA) are compared with Mobile-WiMAX (802.16e), as they all fall into the cellular technology category. Mobile-WiMAX is a good alternative to the current 3G technologies.

Figure 4. The 2-way authentication and authorization of PKMv2 (Adapted from Adibi, Bin, Ho, Agnew, & Erfani, 2006)

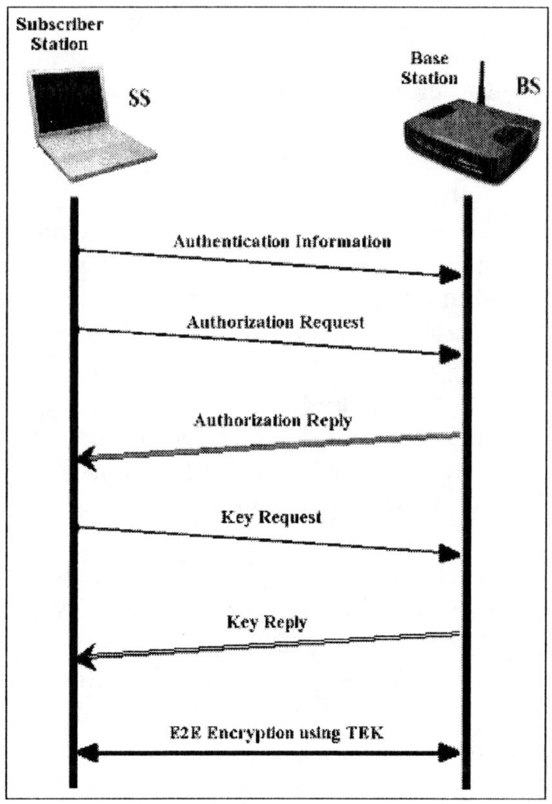

Security Breaches in 3G Technologies

GSM has been around for quite some time and the security mandates in GSM were designed according to the security requirements of when it was designed. Therefore GSM networks suffered several security issues (i.e., one-sided authentication mechanisms). As the technology evolved and matured, GSM/UMTS and CDMA provided the market with stronger security options.

Here are short descriptions of the security problems associated with 3G technologies:

- **Subscriber identity module (SIM) forgery:** SIM cards, mostly used in GSM and GPRS systems, are subject to security threat of forgery due to one-way authorization techniques.
- **Wireless application protocol (WAP) is insecure:** GSM uses WAP for data security, which is considered insecure.
- **Communication signaling in the clear:** Most GSM communication signaling is in the clear with no protection or encryption. This makes it prone to a variety of attacks.
- **Insecurity of base station:** GSM base stations are prone to man-in-the-middle attack scenarios, due to the one-way nature of the authentication scheme.
- **Encryption disability:** UMTS systems are susceptible to a downgrade attack, which eliminates the encryption. An attacker could disable the encryption and trap a legitimate user in a false base station scenario.
- **International mobile subscriber identity (IMSI) security issue:** For first-time registration of users, the IMSI is sent in clear text and an illegitimate entity could take over the session
- **Authentication key agreement (AKA) issue:** Both UMTS and CDMA use AKA. AKA is based on a challenge protocol, which is an unbalanced technique and AKA relies on the availability of a tamper-resistant smart-card in the device, which is also considered to be breakable
- **Cellular authentication and voice encryption (CAVE) issue:** CDMA uses CAVE, which is based on 64-bit authentication key (A-key) and an electronic serial number (ESN). CAVE and ESN are consider computationally weak when a brute force attack is launched against them

Mobile-WiMAX (802.16e)

The WiMAX specification mandates AES-CCM (Barbeau, 2005) encryption (equivalent to FIPS 140-2) between customer premises equipments (CPEs) and the base stations, protecting both the MAC and the PHY layers. The device key management is based on X.509 digital certificates public key, which uses RSA as the public encryption algorithm and other security measures (i.e. confidentiality) are based on AES.

A true E2E security scheme, which is very hard to achieve in 3G technologies, is also available in 802.16e through the use of PKMv2. Therefore Mobile-WiMAX outperforms the strongest members of the 3G family.

CONCLUSION

WiMAX has both a sophisticated set of security protocols in its security suite and advanced bandwidth allocation mechanisms, which makes it a suitable candidate for enterprise applications. The E2E security scheme is capable of providing maximum security for all data and control signals between SSs and BSs. This chapter was intended to take a closer look at the E2E security scheme for WiMAX and to address corporate security requirements. These requirements, which could be addressed by WiMAX, are as follows.

Multi-Level Security and Control ("Product Overview," 2006)

Corporate servers are usually located in highly secured data centers. All data frames should be protected with 128-bit AES encryption technique on an end-to-end basis. Multiple levels of password and authentication methods can be used. These requirements are supported by the security mandates of WiMAX, specified by the current standards.

End-user Remote-Access

End-users are able to connect remotely to a far-away server location using a secure wireless tunnel and access multimedia data and transmit private information. Remote-access is the basic requirement of a VPN and through the deployment of WiMAX, a secure and efficient VPN is achievable.

End-User Security Through Encryption (Data Security)

- AES with 128-bit keys protects the data stream automatically.
- Authentication using dual passwords and end-to-end user authentication.
- BS and CPE Security: BS, CPE, and other devices could be protected through ad-on security features.

Device Authentication

- Devices connected to the backbone of an enterprise will be authenticated, authorization, and protected using authentication, authorization, and certification techniques (i.e., PKMv2, X.509, etc.).
- There should be a complete logging of authorized and unauthorized devices. This will allow tracking of any security violations.

Secure IMs For Fixed And Mobile Applications

Using both fixed- and mobile-WiMAX on the backbone of an enterprise or a corporate server, users will have access to the variety of IMS applications and data, including secure VoIP applications and other VPN access techniques.

Security at the CPE

The security implementations at the customer premises equipments are required to be very high as they are the gateways between the subscriber stations and the services provider.

WIMAX vs. WI-FI

Fixed- and especially mobile-WiMAX outperform the security strength in the latest version of the 802.11 family (802.11i), though 802.16e and 802.11i have many features in common.

WIMAX and IMs Security

The security features in WiMAX are mostly applied at the MAC layer (layer II), where the security sublayer is located. However, WiMAX has the option to adopt very strong security features implemented at the higher layers (i.e., application layer) to meet minimum security requirements for IMS applications.

Security in Fixed- and Mobile- WIMAX

Even though security options built into the mobile-WiMAX are stronger, due to the physical variations and conditions, there has been enough security built into both fixed and mobile WiMAX to ensure complete security from the end-user and VPN applications.

REFERENCES

Adibi, S., Bin, L., Ho, P. H., Agnew, G. B., & Erfani, S. (2006, May). *Authentication authorization and accounting (AAA) schemes in WiMAX.* Paper presented at the Conference on Electro/Information Technology, EIT'06.

Arend, P. V. D. (2007, March 19). *Lawful interception.* Retrieved October 23, 2007, from http://portal.etsi.org/li/Summary.asp

Baker, F., Foster, B., & Sharp, C. (2004, October). *Cisco architecture for lawful intercept in IP networks* (RFC 3924). Retrieved October 23, 2007, from http://www.educause.edu/ir/library/powerpoint/NMD0613B.pps

Barbeau, M. (2005, October 13). *WiMax/802.16 threataAnalysis.* Paper presented at the Q2SWinet'05. School of Computer Science Carleton University.

Brown, I. (2006). *The Internet standards process.* Retrieved October 23, 2007, from http://www.cs.ucl.ac.uk/staff/I.Brown/infosoc-course/internetstandards.ppt

Chandra, P. (2002, July 30). Securing WLAN links: Part 3. *Telogy networks.* Retrieved October 23, 2007, from http://www.CommsDesign.com

Gast, M. (2004). *The top seven security problems of 802.11 wireless* (Airmagnet technical white paper).

Johnston, D., & Walker, J. (2004). Overview of IEEE 802.16 security. *International Journal.*

Mulholland, C. (2006, February 8). *Cisco systems lawful intercept capabilities.*

Mylavarapu, R. (2005, August 1). Security considerations for WiMAX-based converged network. *RFDESIGN.*

Part 16: *Air Interface for Fixed Broadband Wireless Access Systems, IEEE Std 802.16-2004* (http://standards.ieee.org/getieee802/download/802.16-2004.pdf)

Product Overview. (2006). *Citrix GoToMyPC corporate.* Retrieved October 23, 2007, from https://www.gotomypc.com/downloads/pdf/m/GoToMyPC_Corporate_Product_Overview.pdf

Saito, Y. (2003, December). IPv6 and new security paradigm. *NTT communications.*

Technical Specification Group Services and System Aspects; *3G Security; Access security for IP-based services* (Release 5). ARIB STD-T63-33.203, 2002-06

Chapter XLVIII
Evaluation of Security Architectures for Mobile Broadband Access

Symeon Chatzinotas
University of Surrey, UK

Jonny Karlsson
Arcada University of Applied Sciences, Finland

Göran Pulkkis
Arcada University of Applied Sciences, Finland

Kaj Grahn
Arcada University of Applied Sciences, Finland

ABSTRACT

During the last few years, mobile broadband access has been a popular concept in the context of fourth generation (4G) cellular systems. After the wide acceptance and deployment of the wired broadband connections, such as DSL, the research community in conjunction with the industry have tried to develop and deploy viable mobile architectures for broadband connectivity. The dominant architectures which have already been proposed are Wi-Fi, universal mobile telecommunications system (UMTS), WiMax, and flash-orthogonal frequency division modulation (OFDM). In this chapter, we analyze these protocols with respect to their security mechanisms. First, a detailed description of the authentication, confidentiality, and integrity mechanisms is provided in order to highlight the major security gaps and threats. Subsequently, each threat is evaluated based on three factors: likelihood, impact, and risk. The technologies are then compared taking their security evaluation into account. Flash-OFDM is not included in this comparison since its security specifications have not been released in public. Finally, future trends of mobile broadband access, such as the evolution of WiMax, mobile broadband wireless access (MBWA), and 4G are discussed.

INTRODUCTION

During the last decade, wireless network technologies have greatly evolved and have been able to provide cost-efficient solutions for voice and data services. Their main advantages over wired networks are that they avoid expensive cabling infrastructure and they support user mobility and effective broadcasting. As a result, mobile wireless networks have managed to take over a large percentage of the "voice" market, since the global system for mobile communications (GSM) cellular technology has promoted the worldwide expansion of mobile telephony. Furthermore, nowadays broadband Internet has become a necessity for many home and business users. Moreover, in the context of all-IP network convergence, an increasing share of telephony subscribers is migrating towards VoIP solutions mainly due to the decreased cost compared to fixed telephony. Therefore, the main challenge is to find spectrum- and cost-efficient solutions for the provision of mobile broadband services. In this direction, a large research community of academic and industrial origin has dedicated considerable effort on designing, implementing, and deploying systems for mobile broadband access, such as Wi-Fi, universal mobile telecommunications system (UMTS), WiMax, and flash-orthogonal frequency division modulation (OFDM). According to the predictions, in the years to come, more and more of our voice samples and data packets will be carried over wireless broadband links through the Internet. Therefore it becomes imperative that these messages are secured from malicious eavesdroppers and attackers. Especially in applications such as e-banking, e-commerce, and e-government the revelation of sensitive data to unauthorized persons, unauthorized data submission, and/or the interruption of system availability can cause financial damage, user preferences' surveillance, industry espionage, and/or administrative overhead.

The purpose of this chapter is to analyze and compare the security architectures of the dominant mobile broadband technologies. More specifically, the objectives are to:

- Describe and analyze the security architectures of mobile broadband technologies.
- Identify the strong and weak points of each technology in terms of access control based on authentication, confidentiality, integrity, and physical layer resilience.
- Compare the investigated security architectures based on a risk evaluation of the identified security vulnerabilities.

MOBILE BROADBAND TECHNOLOGIES

This section discusses the mobile technologies Wi-Fi, UMTS, WiMax, and flash-OFDM. Authentication performance, confidentiality, and integrity mechanisms for each technology are analyzed.

Wi-Fi

Wi-Fi was the first widely-deployed technology for wireless computer networks. It was originally designed to provide portability support in local area networks (LANs). However, Wi-Fi has also been utilized in other scenarios, such as wireless metropolitan area networks (WMANs), since it was the first wireless technology with support for mobile communication and for a wide range of portable and mobile devices.

The Wi-Fi radio interface is based on the IEEE 802.11 standard and is available in three versions:

- **802.11a**
    - **Frequency:** 5.5 GHz,
    - **Modulation:** OFDM
    - **Bandwidth:** 54 Mbps
- **802.11b**
    - **Frequency:** 2.4 GHz
    - **Modulation:** Direct sequence spread spectrum (DSSS)
    - **Bandwidth:** 11 Mbps
- **802.11g**
    - **Frequency:** 2.4 GHz
    - **Modulation:** OFDM
    - **Bandwidth:** 54 Mbps

In this context, Wi-Fi alliance is an organization testing products in order to evaluate that they correctly implement the set of standards defined in the IEEE 802.11 specification. After the products have successfully passed these tests, they are allowed to use the Wi-Fi logo.

Security Architecture

Wi-Fi security standards include wired equivalent privacy (WEP), Wi-Fi protected access (WPA), and WPA2. WEP was the first introduced security standard. WPA was designed to be a security protocol that corrects the security deficiencies of WEP and to be backward compatible with existing hardware. The last development in Wi-Fi security is the WPA2 standard which was published in June 2004 by the IEEE 802.11i group. WPA2 was designed to offer a further improved security scheme (Edney & Arbaugh, 2003). The aforementioned security specifications are analyzed and compared in the following paragraphs.

Authentication

Authentication services are utilized to allow a client to communicate with the serving access point. After successful authentication, a session is initiated and it can be terminated by either the client or the access point. Wi-Fi provides the following link-layer authentication schemes:

- Closed system authentication
- Media access control (MAC) filtering
- WEP suthentication—Shared RC4 key
- WPA and WPA2 authentication—802.1X/extensible authentication protocol (EAP)

Closed system authentication, MAC filtering, and WEP authentication are not recommended due to their well-known serious security flaws (Borisov, Goldberg, & Wagner, 2001; Lynn & Baird, 2002; Welch & Lathrop, 2003).

WPA and WPA2 security schemes have some major design differences from WEP, since the authentication and the confidentiality processes operate totally independently from each other (Baek, Smith, & Kotz, 2004). The authentication process of WPA and WPA2 adopts the three-entity model of IEEE 802.1x which was originally designed for the point-to-point protocol (IEEE, 2001). The three entities involved in this protocol are the client, the access point (AP), and the authentication server (AS). First, the client request to obtain access to the network. The AP acts as a network guard, allowing access only to the clients that the AS has authenticated. Finally, the AS is responsible for deciding whether the client is allowed to access the network. These three entities utilize EAP to exchange communication messages in order to coordinate the authentication process (Stanley, Walker, & Aboba, 2005).

In addition, there is a lighter version of WPA, called WPA-preshared key (WPA-PSK). This version is based on a shared secret key or passphrase in order to authenticate the wireless clients. As a result, an attacker can use a wireless sniffer to capture the 4-way WPA handshake, log the packets, and then try a brute force attack using a dictionary file (Van de Wiele, 2005). Thus, if WPA-PSK is deployed, the robustness of the network security totally depends on the length and the complexity of the secret key.

Encryption

Encryption services are utilized to provide confidentiality over wireless communication links. In Wi-Fi networks the following encryption schemes are available:

- WEP based on the RC4 (Ron's Code 4) stream cipher
- WPA encryption based on the temporal key integrity protocol (TKIP)
- WPA2 encryption based on the advanced encryption standard (AES)

WEP is a weak implementation of the RC4 stream cipher and WEP encryption is thus not recommended (Borisov et al., 2001; Stubblefield, Ioannidis, & Rubin, 2002; Welch & Lathrop, 2003).

WPA encryption is based on TKIP. It incorporates the basic functionalities of WEP, but improvements have been made to address the security flaws. The length of the initialization vector (IV) has been increased from 24 bits to 48 bits and therefore the possibility of reused keys has been significantly decreased. Furthermore, WPA does not directly utilize the master keys. Instead it constructs a hierarchy of derived keys to be utilized in the encryption process. Finally, WPA dynamically cycles keys while transferring data. Since keys are regularly changed, a malicious user has a very short time window to attempt an attack.

WPA2 was designed from scratch taking the vulnerabilities of the previous security architectures into account. WPA2 allows various network implementations, but the default configuration utilizes the advanced encryption standard (AES) and the counter mode CBC MAC protocol (CCMP). AES is a block cipher, operating on blocks of 128 bit data, and is a replacement of the RC4 algorithm used by WPA. AES is much more robust since it has already been tested in various security architectures without revealing serious vulnerabilities. CCMP comprises of two main parts. The first is the counter mode (CM) which is responsible for the privacy of the data in combination with AES. The second is the cipher block chaining message authentication code (CBC-MAC) providing data integrity checking and authentication.

Integrity

Integrity services are responsible for making sure that transmitted information is not replayed or modified during transmission. The following techniques are applicable in Wi-Fi networks:

- WEP cyclic redundancy heck 4 (CRC-32) Checksum
- WPA Integrity
- WPA2 Integrity

WEP checksum is a noncryptographic linear function of the plaintext. This means that multiple messages may correspond to a single 32-bit number. Hence, an experienced intruder could modify the plaintext in such a way that the checksum remains unchanged. Furthermore, due to the linearity of both the RC4 stream cipher and the CRC-32 checksum, the attacker is able to change the message even when he does not know the plaintext (Welch & Lathrop, 2003).

WPA has incorporated mechanisms for the prevention of replay attacks. More specifically, the TKIP sequence counter (TSC) based on the IVs is utilized, so that the receiver can identify and reject "replayed" messages. Furthermore, WPA uses an improved integrity mechanism in order to generate the message integrity check (MIC). This mechanism, called Michael, is able to detect possible attacks and deploy countermeasures to prevent new attacks.

WPA2 utilizes CCMP for providing integrity services. CCMP generates a MIC using the CBC-MAC method. In this method, even the slightest change in the plaintext will produce a totally different checksum.

Security Vulnerabilities

Although the Wi-Fi security architecture has been greatly improved since WEP, there are still vulnerabilities which cannot be addressed by WPA2. These vulnerabilities can lead to a number of link layer denial-of-service (DoS) attacks (Van de Wiele, 2005). All the DoS techniques described here are fairly easy to use with freely available tools found on the Internet. In most of the cases, the attacker will use different forged MAC addresses to mount DoS attacks. These attacks can be detected by specialized hardware (e.g., air monitor, security aware access point) which can detect the misuse of the infrastructure. Furthermore, this specialized hardware can notify the people responsible for the follow-up of a DoS incident and give an estimate on where the attacker is located by considering the signal and noise levels.

Disassociation Storm

Before any wireless communication can occur, a client has to send an association frame to the access point asking to join the network. Similarly, after the end of the wireless session, the access point or client has to send a disassociation frame to terminate the connection. The frames of these messages are broadcasted and can be sniffed by an attacker. The attacker can then flood the network with spoofed disassociation frames every time the client tries to join the network, thus disrupting the association process and the network access.

Authenticated / Deauthenticated Storm

The aforementioned principle can be exploited in order to disconnect a client and try to keep the client disconnected. This technique starts by sending a spoofed deauthentication frame followed by a disassociation frame in order to make sure that the client has disconnected from the legitimate access point. In a more advanced version of this attack, a fake probe request and some beacon frames are transmitted in order to force the client to connect to a rogue access point which ignores or monitors the client's traffic.

UMTS

Universal mobile telecommunications system (UMTS) is one of the third generation (3G) wireless cellular technologies for mobile communication. Mobile devices like smartphones, laptops, and handheld computers can be used. UMTS is standardized by the 3G partnership project (3GPP) and it is mainly deployed in Europe and Japan. Theoretically UMTS supports up to 1920 Kbps data transfer rates, but currently the real world performance can reach 384 Kbps. It uses the W-code division multiple access (CDMA) technology over two 5 MHz channels, one for uplink and one for downlink. The specific frequency bands originally defined by the UMTS standard are 1885-2025 MHz for uplink and 2110-2200 MHz for downlink.

In UMTS network topology, a mobile station is connected to a visited network by means of a radio link to a particular base station (Node B). Multiple base stations of the network are connected to a radio network controller (RNC) and multiple RNCs are controlled by a general packet radio service (GPRS) support node (GSN) in the packet-switched case. The visitor location register (VLR) and the serving GSN keep track of all mobile stations that are currently connected to the network. Every subscriber can be identified by its international mobile subscriber identity (IMSI). In order to protect against profiling attacks, this permanent identifier is sent over the air interface as infrequently as possible. What is more, locally valid temporary mobile subscriber identities (TMSI) are used to identify subscribers whenever possible. Every UMTS subscriber has a dedicated home network with which the subscriber shares a long term secret key Ki. The home location register (HLR) keeps track of the current location of all subscribers of the home network. Mutual authentication between a mobile station and a visited network is carried out with the support of the current serving GSN (SGSN) or the mobile switching center (MSC)/VLR respectively.

The new series of 3.5G mobile telephony technologies, known as high speed packet access (HSPA), will provide more bandwidth to the end-user, improved network capacity to the operator, and enhanced interactivity for data applications. HSPA refers to the improvements made in the UMTS downlink, known as high speed downlink packet access (HSDPA), and the UMTS uplink, usually referred to as high speed uplink packet access (HSUPA) but also referred to as enhanced dedicated channel (E-DCH).

HSDPA provides a bandwidth of 14.4 Mbps/user. For multiple-input-multiple-output (MIMO) systems up to 20 Mbps can be achieved. Both HSDPA and HSUPA can be implemented in the standard 5 MHz carrier of UMTS networks and can coexist with original UMTS networks. As HSPA specifications refer only to the access network, there is no change required in the core network (CN) except from the high data-rate links required to handle the increase in clients' traffic generated by HSPA.

Security Architecture

The 3G security architecture is based on GSM, but certain improvements are added in order to correct the described security vulnerabilities.

Authentication

Authentication and key agreement (AKA) is the main security protocol of UMTS in the 3GPP specification. According to AKA, a mobile device and a base station have to authenticate each other. Figure 1 provides an overview of the AKA process. The authentication vector includes the following components:

a. A random number (RAND)
b. An expected response (XRES)
c. A cipher key (CK)
d. An integrity key (IK)
e. An authentication token (AUTN)

RAND and XRES are utilized by the network to authenticate the mobile station (MS), whereas AUTN is utilized by the MS to authenticate the network. After the mutual authentication, the two communicating parties can agree on the CK and the IK which will be used throughout the rest of the session.

Confidentiality and Integrity

UMTS employs the UMTS encryption algorithm (UEA) in order to provide information confidentiality. The encryption process of UEA is based on the f8 algorithm. One of the main improvements of UMTS is that the link layer encrypted channel is established between the MS and the GSN instead of the BS, as in GSM. Furthermore, UEA is utilized to protect not only the data channels but also certain signalling channels.

Figure 1. 3GPP authentication and key agreement (AKA)

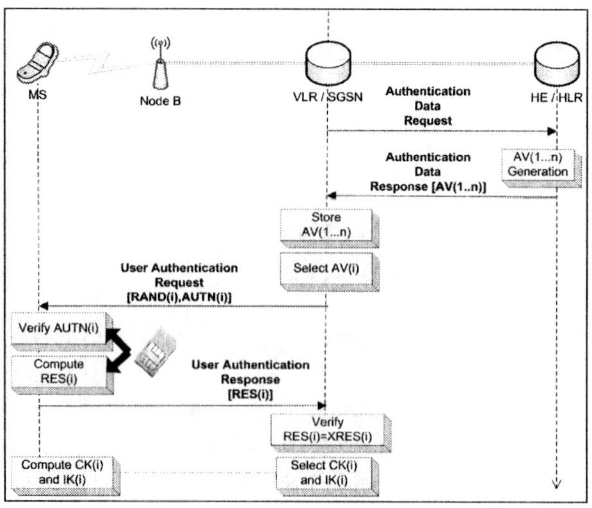

For user confidentiality UMTS utilizes the same mechanism as GSM. Instead of the IMSI, a temporary identity (TMSI) assigned by VLR is used to identify the subscriber in the communication messages exchanged with the BS. However, the IMSI is still transmitted in clear-text over the air while establishing the TMSI. This has been proved to be a starting point for security attacks against UMTS.

Data integrity in 3GPP is assured explicitly through the UMTS integrity algorithm (UIA). The UIA operation is based on the f9 algorithm. UIA is utilized to protect both communication and signalling. UEA and UIA are presented in Figure 2.

GSM Compatibility

UMTS has been designed to be backwards compatible with GSM. It includes standardized security features in order to ensure world-wide interoperability and roaming. More specifically, GSM user parameters are derived from UMTS parameters using a set of predefined conversion functions. However, GSM subscribers roaming in 3GPP networks are supported by the GSM security context, which is vulnerable to the aforementioned GSM vulnerabilities.

Security Vulnerabilities

3G security has been significantly improved compared to GSM. However, there are still vulnerabilities related to the backwards compatibility with GSM. Meyer and Wetzel (2004a, 2004b) present a man-in-the-middle attack which can be mounted even if the subscriber utilizes a 3G enabled device within a 3G base station coverage. The described attack goes far beyond the anticipations of the 3GPP group. UMTS subscribers are vulnerable

Figure 2. UMTS encryption and integrity algorithm

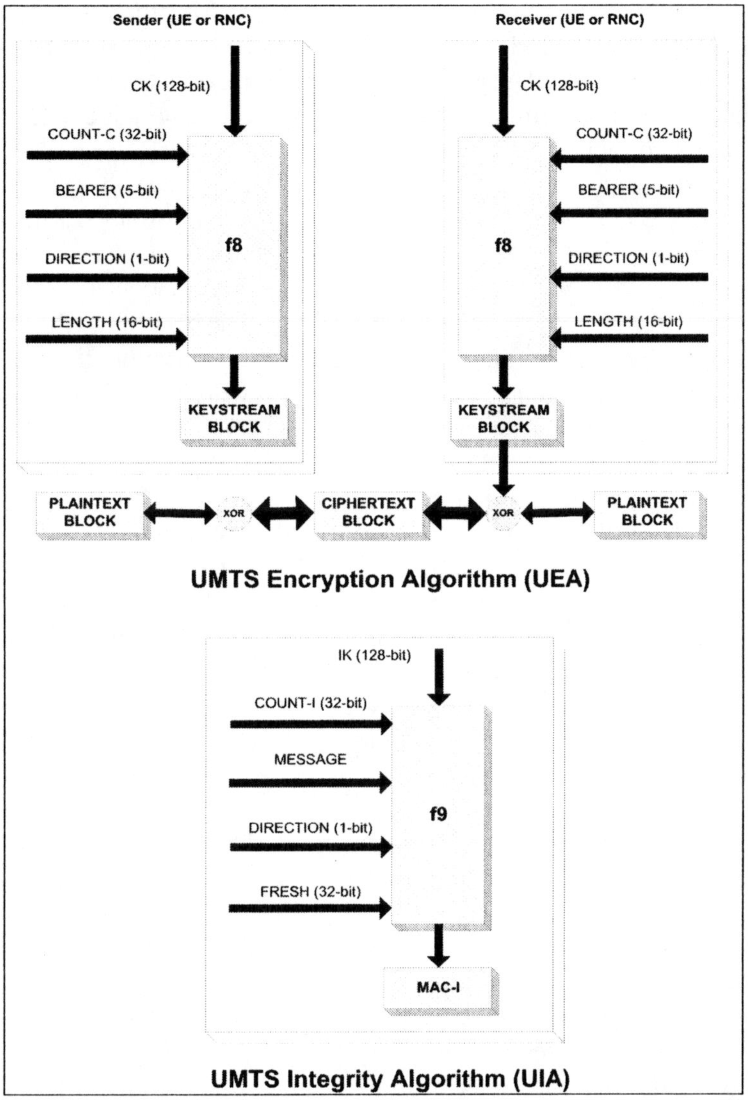

to what 3GPP calls a "false base station attack" even if subscribers are roaming in a pure UMTS network and even though UMTS authentication is applied.

This attack can be categorized as a "roll-back attack." This category of attacks exploits weaknesses of old versions of algorithms and protocols by means of the mechanisms defined to ensure backward compatibility of newer and stronger versions. According to this technique, the attacker acts on behalf of the victim's mobile station in order to obtain a valid authentication token AUTN from any real network. It is assumed that the attacker has already retrieved the IMSI of the targeted subscriber, since the latter is sent in clear-text when establishing a TMSI. The attacker can capture the AUTN by initiating the AKA procedure with any legitimate network. The next step is to impersonate a valid GSM base station to the victim mobile station. The mobile station connects and verifies the rogue BS, since it possesses a valid AUTN. Subsequently, the rogue BS is configured by the attacker to utilize "no encryption" or weak encryption. Finally, the attacker can send to the mobile station the GSM cipher mode command including the chosen encryption algorithm. The man-in-the-middle attack is mounted and the attacker can use passive or active eavesdropping without being detected.

WIMAX

The IEEE 802.16 or broadband wireless access (BWA) Working Group was established in 1999 to prepare specifications for broadband wireless metropolitan area networks. The first 802.16 standard was approved in December 2001 and was followed by three amendments: 802.16a, 802.16b and 802.16c. In 2004 the 802.16-2004 standard (IEEE-SA, 2006) was released and the earlier 802.16 documents including the a/b/c amendments were withdrawn. An amendment to the standard 802.16e (IEEE-SA, 2006) addressing mobility was introduced in 2005. The main additions of the 802.16e were low density parity check (LDPC) codes at the physical layer, enhanced MIMO setup functions, new states for MS operation, parameter-defined power saving classes of mobiles, and enhanced FFT sizes for scalable OFDMA.

WiMax aims at providing high data rate triple-play wireless services to fixed users, to nomadic users, and to users of mobile devices. It is based on a low latency quality of service (QoS) architecture in order to provide real-time multimedia services. It operates on the 2-6 GHz (IEEE802.16e) and 10-66 GHz (IEEE802.16-2004) frequency bands and it uses the OFDMA technology for modulation and medium access.

Security Architecture

WiMax has been designed with security in mind, especially after the serious vulnerabilities discovered in the original Wi-Fi security protocol. The IEEE 802.16 specifications include a security sublayer within the MAC layer. The IEEE 802.16 security architecture is based on the following issues:

- **Authentication:** The baseline authentication architecture, by default, employs a public key infrastructure (PKI) based on X.509 certificates. The base station (BS) validates the client's certificate before permitting access to the physical layer (see Figure 3). First, the subscriber station (SS) sends to the BS an authorization request containing the certificate, the available security capabilities, and the security association identifier (SAID). The BS verifies the certificate and generates a 128 bit authentication key (AK). Then, the BS sends to the SS an authorization reply, which contains the AK encrypted with SS's public key, the AK's lifetime, the selected security suite, and an AK sequence number. The SS uses its private key to recover the AK, which can now be utilized as an authentication token in further communication.
- **Key exchange:** The SS and the BS can agree on a transport encryption key (TEK), which will be utilized for data encryption (see Figure 3).
  TEK is randomly generated by the BS. The AK established during authentication is used to derive two additional keys:
  - Message authentication key (HMAC key), which is utilized to provide message integrity and AK confirmation during the key exchange process.
  - Key encryption key (KEK), which is utilized for encrypting the TEK before sending it back to the SS. The modes for encrypting TEK are:
    a. 3DES with a 112 bit KEK
    b. AES with a 128 bit KEK
    c. RSA using SS's public key
- **Data encryption and integrity:** The modes for implementing data privacy are:
  - Data encryption standard (DES) with a 56 bit key and cipher block chaining (CBC), which utilizes the Initialization Vectors obtained during Key Exchange,
  - AES with a 128 bit key and counter mode with cipher block chaining message authentication code protocol, which

Figure 3. WiMax authentication and key exchange process

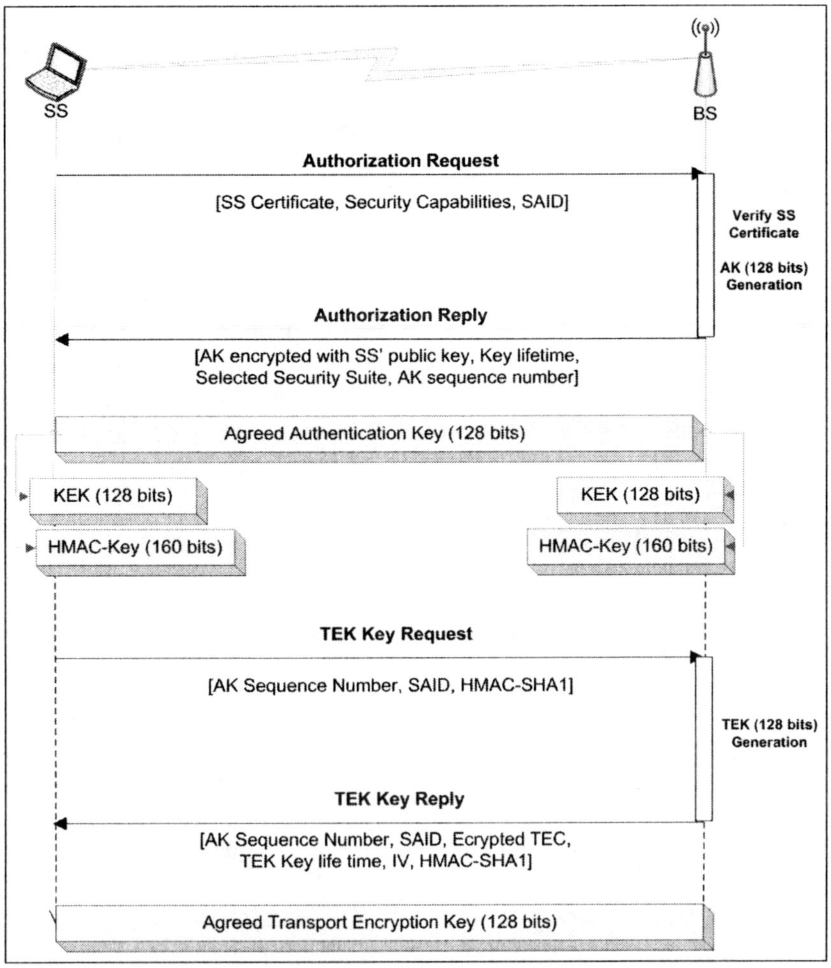

provides message integrity and replay protection.

Security Vulnerabilities

WiMax supports unilateral device level authentication (Barbeau, 2005), which can be implemented in a similar way as Wi-Fi MAC filtering based on the hardware device address. Therefore, address sniffing and spoofing make a MS masquerade attack possible. In addition, the lack of mutual authentication makes a man-in-the-middle attack from a rogue BS possible. However, a successful man-in-the-middle attack is difficult because of the time division multiple access (TDMA) model in WiMax. The attacker must transmit at the same time as the legitimate BS using a much higher power level in order to "hide" the legitimate signal. Furthermore, WiMax supports mutual authentication at user network level based on the generic extensible authentication protocol (EAP) (Aboba, Blunk, Vollbrecht, Carlson, & Levkowetz, 2004). EAP variants, EAP- transport layer security (TLS) (X.509 certificate based) (Aboba & Simon, 1999) and EAP-subscriber identity module (SIM) (Haverinen & Salowey, 2004), are supported.

In the data privacy domain, the main security threat is the transmission of unencrypted management messages over the wireless link. Eavesdropping of management messages is a critical threat for users and a major threat to a system. For example, an attacker could use this vulnerability to verify the presence of a victim at its location before perpetrating a crime. Additionally, it might be used by a competitor to map the network. Another major vulnerability is the encryption mode based on DES. The 56 bit DES key is easily broken by brute force with modern computers. Furthermore, the DES encryption mode includes no message integrity or replay protection functionality and is thus vulnerable to active or replay attacks. The secure AES encryption mode should be preferred over DES.

Finally, there is a potential for DoS attacks because authentication operations trigger the execution of long procedures. For example, a DoS attack could flood a MS with a high number of messages to authenticate. Due to low computational resources, the MS will not be able to handle a large amount of invalid messages, rendering the DoS attack successful.

FLASH-OFDM

Fast low-latency access with seamless handoff orthogonal frequency division multiplexing (flash-OFDM) is an OFDM-based proprietary system which specifies the physical layer, as well as higher protocol stack layers. It is an all IP technology and it aims to compete with GSM/3G networks. Already implemented flash-OFDM technology operating in the 450 MHz frequency band can offer a maximum download speed of 5.3 Mbps and an upload speed of 1.8 Mbps.

Design objectives have included design of a high capacity physical layer, a packet-switched air interface, a contention-free and QoS-aware MAC layer, and efficient operations using existing Internet protocols. The air interface is designed and optimized across all protocol stack layers. Fast hopping across all tones in a pseudorandom predetermined pattern is employed. Channel coding and modulation are carried out on a per-segment basis and can be individually optimized for each channel. The ability to send segments of arbitrary size enables the MAC layer to perform efficient packet switching over the air interface. Given segments can be dedicated for use with predefined functionality. Thus there is no need to send overheads, such as message headers. Therefore, network layer traffic experiences small delays and no significant delay jitter.

Security Architecture

The security relies on "defence in depth," that is, virtual private network (VPN) tunnelling and end-to-end encryption are used. Security specifications for flash-OFDM have not been presented in public (Lehtonen, Ahonen, Savola, Uusitalo, Karjalainen, Kuusela et al., 2006).

Security Analysis

A security analysis of the mobile broadband technologies Wi-Fi, UMTS, and WiMax is presented. Inclusion of flash-OFDM in this comparison is not possible because of the unavailability of public security specifications. Threats are analyzed with respect to the likelihood of occurrence, the impact on the network operation, and the global risk they represent. In the following paragraphs, we first describe in detail the evaluation and comparison methodology, and then a group of tables is presented in which the security threats of the investigated technologies are evaluated. Security threats are classified based on four main axes: authentication, confidentiality, integrity, and physical layer resilience. Finally, the security evaluations of the studied technologies are compared and presented in a concise overview table.

Methodology

The evaluation and comparison methodology was based on the method described by Barbeau, (2005) and ETSI (2003). More specifically, three main criteria are considered: likelihood, impact, and risk. "Likelihood" refers to the probability that an attack associated with a specific threat is successfully launched. In this context, two variables are considered:

a. The technical difficulties of mounting the attack in terms of the required software, hardware, and estimated time duration.
b. The attacker's motivation in terms of the level of network access or the severity of the system malfunction that the attack achieves.

Three levels of likelihood are available as described in Table 1. "Impact" refers to the consequences of an attack in terms of user and network security. The two variables of impact are:

a. User impact in terms of the severity of network access degradation.
b. System impact in terms of the severity of network degradation or outage.

Three levels of impact are available as described in Table 1. According to the level of likelihood and impact, numerical values from a predefined range are assigned to each criterion (see Table 1). For a specific threat, the "risk" refers to an overall threat level which is determined by the product of the likelihood value and impact value.

Security threats which result in a high evaluated risk value are critical and additional measures should be taken to protect the network perimeter, whereas threats which have a low risk can be tolerated without employing countermeasures.

In this point, it is worth noting that this quantitative ranking is subjective. However, this is a useful evaluation and comparison methodology which can stimulate a structured discussion based on the evaluation criteria, that is, likelihood, impact, and risk. The comparison axes are authentication, confidentiality, integrity, and physical layer resilience.

Table 1. Evaluation and comparison methodology

		Variables		
Criteria	**Cases**	**Difficulty**	**Motivation**	**Rank**
Likelihood	Unlikely	Strong	Low	1
	Possible	Solvable	Reasonable	2
	Likely	None	High	3
Impact		**User**	**System**	
	Low	Annoyance	Very limited outages	1
	Medium	Loss of service	Limited outages	2
	High	Long time loss of service	Long time outages	3

	Risk = Likelihood x Impact		
Risk	Minor	No need for countermeasures	1-3
	Major	Threat need to be handled	3-6
	Critical	High priority	6-9

Objective-Based Comparison

This section applies the aforementioned methodology on four main objectives of wireless security architectures: authentication, confidentiality, integrity, and physical layer resilience. For each objective, a thorough discussion describes the rationale behind the ranking of the security threats.

Authentication Evaluation

Wi-Fi includes four security threats which are all ranked to have a high impact on the system, since the attacker can exploit them to override the authentication checks or launch a combination of attacks which will grant him full network access. However, the likelihood ranking greatly varies. Closed system authentication and MAC filtering are very likely to be attacked by sniffing software which is readily available on the Internet. WEP attacks are more complicated, because a combination of software is required to induce and capture network traffic and then exploit the weak IVs in order to crack the key. WPA-PSK is even more difficult to break since it requires a brute force attack. The resilience of WPA-PSK is greatly dependent on the length and the complexity of the preshared key.

UMTS is far more resilient to authentication attacks, since most of the security gaps have been identified during the deployment of GSM and tackled in the specification design of UMTS. However, UMTS includes two main authentication vulnerabilities which can be exploited to launch a man-in-the-middle attack (high impact). The IMSI hijack threat refers to the deployment of a rogue BS in order to initiate an authentication procedure and steal the IMSI of a mobile user. The motivation for this attack is high, but the equipment is expensive and complicated to configure. AUTN capture is the second step of the attack and it refers to capturing an authentication token by masquerading a MS.

It assumes that the IMSI Hijack attack has been already successfully launched. However, this attack does not require the deployment of a rogue BS and therefore it is more possible to happen.

In the WiMax architecture, the main security threat is the device-level authentication mode. When this mode is utilized without certificate support, it is as vulnerable as MAC filtering and it can be exploited to launch MS or BS masquerading attacks. A less critical vulnerability is the DoS attack which can be launched by flooding authentication requests. This attack mostly affects the MS due to its limited processing resources, but it is not a major threat since it has a medium impact and a low motivation.

Confidentiality Evaluation

Wi-Fi includes some major vulnerabilities. It supports a null mode encryption which is configured as default in the majority of the commercial access points. WEP encryption can provide an elementary level of protection, but it is still too weak to keep the intruders out. WPA-PSK offers a satisfactory level of confidentiality, if long and complex keys are utilized. The ranking of the Wi-Fi confidentiality vulnerabilities is similar to authentication ranking, since both objectives are based on the same mechanisms.

UMTS incorporates strong encryption algorithms which have eliminated the deficiencies of its predecessor GSM. Nevertheless, the backwards compatibility with GSM can be exploited to compromise dual-band mobile devices by launching a man-in-the-middle attack. In this attack, the rogue BS can mandate the MS to use null mode encryption or one of the GSM encryption modes which can be easily broken (Biham & Dunkelman, 2000; Biryukov, Shamir, & Wagner, 2000). However, this is an unlikely attack since it requires the deployment of a BS and a prior successful launch of the IMSI hijack and AUTN capture attacks.

WiMax security architecture includes two main shortcomings. First of all, the DES encryption mode provides an inadequate level of confidentiality, since it can be easily broken. In addition, the eavesdropping of unencrypted management frames can be easily established, but it cannot greatly affect the system if robust authentication and integrity mechanisms have been deployed.

Integrity Evaluation

Wi-Fi supports null mode which leaves the messages totally unprotected against modification and replay attacks. WEP CRC-32 integrity mechanism provides a moderate level of protection, but there is no replay protection and the integrity protection can be overridden by an experienced attacker.

The UMTS architecture includes a major shortcoming, namely the inadequate replay protection of authentication tokens. This vulnerability can have a high impact since it allows the reuse of the token retrieved by an AUTH capture attack and the completion of the UMTS man-in-the-middle attack. However, it requires a prior successful launch of IMSI hijack and AUTN capture. Therefore it results in a high technical difficulty.

WiMax supports two modes that can greatly compromise information integrity. The first is the DES mode which does not support integrity and replay protection of data frames. The second is the null MAC mode for management frames, which can allow the intruder to inject modified management frames and affect the network operation.

Physical Layer Resilience Evaluation

The resilience of the physical layer of each technology is evaluated with respect to jamming and scrambling. Jamming is achieved by introducing a source of noise strong enough to significantly reduce the capacity of the channel. Scrambling is similar to jamming, but it takes place for short intervals of time and it is targeted to specific frames or parts of frames.

Wi-Fi comprises of the three different specifications IEEE 802.11a/b/g which all utilize random medium access techniques but operate on different physical channels. IEEE 802.11a/g operate on a 5 MHz OFDM channel, whereas IEEE 802.11b operates on a 5 MHz DSSS channel. The DSSS is more resilient to narrowband jamming than OFDM and therefore jamming has a higher impact on IEEE802.11a/g. However, if the attacker wants to jam all the channels, the attacker has to jam a bandwidth of 40 MHz, which is quite difficult. Scrambling is easier to launch because of the random medium access layer.

UMTS operates on two 5 MHz DSSS channels, one for the uplink and one for the downlink. It is resilient to narrowband jamming because of the DSSS modulation, but it is still vulnerable to scrambling because of the random access.

WiMax operates on a 1.25-20 MHz OFDM channel and it employs TDMA techniques. Thus, it can be vulnerable to jamming especially if it operates on a narrow channel, but it is resilient to scrambling due to the TDMA.

OVERALL COMPARISON

The results from authentication, confidentiality, integrity, and physical layer resilience evaluation are presented in Table 2.

As follows, the overall comparison results:

- **Wi-Fi:**
  - **Authentication:** 6.75
  - **Confidentiality:** 6
  - **Integrity:** 6
  - **PHY Resilience:** 5
  - **AVERAGE RISK:** 5.94
- **UMTS**
  - **Authentication:** 4.5
  - **Confidentiality:** 3
  - **Integrity:** 3
  - **PHY Resilience:** 3.5
  - **AVERAGE RISK:** 5.94
- **WiMax**
  - **Authentication:** 6.5
  - **Confidentiality:** 6
  - **Integrity:** 7.5
  - **PHY Resilience:** 3
  - **AVERAGE RISK:** 5.75

Wi-Fi has the highest average risk, which is quite reasonable because of the initial lack of security mechanisms in the Wi-Fi specification and the subsequent failure of WEP. WPA and WPA2 modes are much more secure, but the poor usability and the limited security awareness have constrained their wide deployment. UMTS proved to be quite robust by eliminating the security inefficiencies of its predecessor GSM. However, an attacker can still exploit some backward-compatibility issues to launch a man-in-the-middle attack. WiMax's performance was not satisfactory enough mainly due to the provision of weak security modes. Nevertheless, the practical performance is greatly dependent on the actual security decisions of the network operators. These decisions vary according to the provided service requirements.

Table 2. Security evaluation

AUTHENTICATION EVALUATION				
Technology	**Threat**	**Likelihood**	**Impact**	**Risk**
Wi-Fi	Closed System	3	3	9
	MAC Filtering	3	3	9
	WEP	2	3	6
	WPA-PSK	1	3	3
Average Risk				**6,75**
UMTS	IMSI Hijack	2	3	6
	AUTN Capture	1	3	3
Average Risk				**4,5**
WiMAX	Device-level Authentication	3	3	9
	DoS on MS	2	2	4
Average Risk				**6,5**
CONFIDENTIALITY EVALUATION				
Technology	**Threat**	**Likelihood**	**Impact**	**Risk**
Wi-Fi	Null	3	3	9
	WEP	2	3	6
	WPA-PSK	1	3	3
Average Risk				**6**
UMTS	Rogue BS – Null / Weak	1	3	3
Average Risk				**3**
WiMAX	DES mode	3	3	9
	Management Frames	3	1	3
Average Risk				**6**
INTEGRITY EVALUATION				
Technology	**Threat**	**Likelihood**	**Impact**	**Risk**
Wi-Fi	Null	3	3	9
	WEP	1	3	3
Average Risk				**6**
UMTS	AUTN Replay	1	3	3
Average Risk				**3**
WiMAX	DES mode – Null integrity	3	2	6
	Management Frame-Null MAC	3	3	9
Average Risk				**7,5**
PHYSICAL LAYER RESILIENCE EVALUATION				
Technology	**Threat**	**Likelihood**	**Impact**	**Risk**
Wi-Fi	Jamming (IEEE 802.11a/g)	2	3	6
	Scrambling (IEEE 802.11a/g)	3	3	9
	Jamming (IEEE 802.11b)	2	2	4
	Scrambling (IEEE 802.11b)	3	2	6
Average Risk				**5**
UMTS	Jamming	1	2	2
	Scrambling	2	2	4
Average Risk				**2,5**
WiMAX	Jamming	1	3	3
	Scrambling	1	3	3
Average Risk				**3**

FUTURE TRENDS

Broadband wireless access networking is presently a rapidly evolving ICT area. Three important development trends can be identified:

- WiMax evolution for long range broadband wireless access.
- Development of a broadband wireless access technology supporting high speed mobility.
- Emerging 4G wireless cellular technology.

WiMax Evolution

The WiMax standard was finalized in June 2004. WiMax has the potential to change telecommunications as it is known today. "It eradicates the resource scarcity that has sustained incumbent service providers for the last century. As this technology enables a lower barrier to entry, it will allow true market-based competition in major telecommunications services like voice, video and data" (Ohrtman, 2005).

WiMax can offer a point-to-point range of 50 km with a throughput of 72 Mbps. The WiMax technology will make personal broadband services profitable to service providers and will be available to business and consumer subscribers at affordable prices. The first mobile WiMax products are expected to be introduced into the market in the first quarter of 2007. New technologies such as MIMO and beam forming for higher throughput and capacity will be introduced in 2007 (WiMax Forum, 2006).

Mobile Broadband Wireless Access (MBWA)

The IEEE 802.20 (or MBWA) Working Group was established in December 11, 2002, with the aim to develop a specification for an efficient packet-based air interface that is optimized for the transport of IP based services. The goal is to enable worldwide deployment of affordable, always-on, and interoperable BWA networks. The group will specify the lower layers of the air interface, operating in licensed bands below 3.5 GHz and enabling peak user data rates exceeding 1 Mbps at speeds of up to 250 km/h. A draft version of the specification was approved in January 18, 2006.

4G – Future Wireless Cellular Technology

Frameworks for future 4G networks, which seamlessly integrate heterogeneous mobile technologies in order to provide enhanced service integration, QoS, flexibility, scalability, mobility, and security, are currently being developed. However, these frameworks raise security vulnerabilities. An international consortium presents requirements and recommendations for the evolving 4G mobile networking technology (Akhavan, Vivek Badrinath, & Geitner, 2006). The 4G technology, which is at its infancy, is supposed to allow data transfer up to 100 Mbps outdoor and 1 Gbps indoor. The International Telecommunications Union (ITU) defines 4G as downlink throughput of 100 Mbps or more, and corresponding uplink speeds of at least 50 Mbps.

The 4G technology will support roaming for interactive services such as video conferencing. The cost of the data transfer will be comparatively low and global mobility will be possible. The networks will be all IPv6 networks. WLAN, 2.5G, 3G, and other networks such as SATCOM, WiMAX, and Bluetooth will be integrated in 4G networks. The antennas will be much smarter and improved access technologies like OFDM and MC-CDMA will be used. More efficient algorithms at the physical layer will reduce the inter-channel interference and cochannel interference.

Security Issues

Seamless convergence of heterogeneous wireless networks provides new security challenges for the research community. Global authentication architectures are needed which can operate independently of the wireless physical protocol. In addition, specifications are needed for maintaining the confidentiality and the integrity of the communication data while the user terminal is in a hand-off state. In this direction, a forum of mobile operators called fixed mobile convergence alliance (FMCA) is working on defining specifications for the convergence of heterogeneous networks in the context of all IP 4G wireless systems.

Security policy issues are:

- The use of lightweight and flexible authentication, authorization, account, and audit (AAAA) schemes,
- The use of Trusted Computing (Reid, Nieto, & Dawson, 2003), and
- Different security polices for different services are recommended for 4G systems (Zheng, He, Xu, & Tang, 2005a).

Several security architecture proposals for 4G wireless systems have been made:

- Zheng, He, Yu, and Tang (2005b) propose a security architecture with:
  - Network access security features.
  - Network area security features for secure data exchange between network nodes.
  - User area security features for secure access to ME/USIM.
  - Application security for secure end-to-end data exchange.
- Integration of the SSL security protocol and a public key infrastructure is outlined and evaluated by Kambourakis, Rouskas, and Gritzalis (2004).
- A hierarchical trust model for 4G wireless networks is proposed by Zheng et al. (2005a).

CONCLUSION

In this chapter, the dominant mobile broadband technologies have been evaluated and compared based on their security performance. Three technologies were taken into consideration: Wi-Fi, UTMS, and WiMax. Their security architectures have been presented and analyzed in order to highlight the main security deficiencies. The evaluation and comparison methodology was based on assigning qualitative rankings to security threats with respect to the following criteria: likelihood, impact, and risk. The methodology was applied on four evaluation axes: authentication, confidentiality, integrity, and physical layer resilience. According to the comparison results, Wi-Fi is more liable to security attacks, followed by WiMax and UMTS. However, WiMax has not been widely tested under real-world systems due to its recent release. More security vulnerabilities may therefore be discovered in the future. Finally, the security architecture of UMTS is quite robust because of the lessons learned from GSM, but it is still not invincible against an experienced attacker with the right equipment.

REFERENCES

Aboba, B., Blunk, L., Vollbrecht, J., Carlson, J., & Levkowetz, H. (2004). *Extensible authentication protocol (EAP)* (IETF RFC 3748).

Aboba, B., & Simon, D. (1999). *PPP EAP TLS authentication protocol* (IETF RFC 2716).

Akhavan, H., Vivek Badrinath, V., & Geitner, T. (2006). *Next generation mobile networks beyond HSPA & EVDO* (White Paper.NGMN—Next generation mobile networks Ltd.) Retrieved April 24, 2007, from http://www.ngmn.org/

Baek, K., Smith, W., & Kotz, D. (2004). *A survey of WPA and 802.11i RSN authentication protocols* (Tech. Rep. TR2004-524). Dartmouth College, Computer Science.

Barbeau, M. (2005). WiMax/802.16 threat analysis. In *Proceedings of the 1st ACM Workshop on QoS and Security for Wireless and Mobile Networks (Q2SWinet),* Montreal, (pp. 8-15).

Biham, E., & Dunkelman, O. (2000). Cryptanalysis of the A5/1 GSM stream cipher. In *Proceedings of the First International Conference on Progress in Cryptology* (pp. 43-51).

Biryukov, A., Shamir, A., & Wagner, D. (2000). *Real time cryptanalysis of A5/1 on a PC.* Paper presented at the Fast Software Encryption Workshop 2000, New York.

Borisov, N., Goldberg, I., & Wagner, D. (2001). Intercepting mobile communications: The insecurity of 802.11. In *Proceedings of the 7th Annual International Conference on Mobile Computing and Networking*, Rome, (pp. 180-189).

Edney, J., & Arbaugh, W. A. (2003). *Real 802.11 security: Wi-Fi protected access and 802.11i* (1st ed.). Addison-Wesley Professional.

ETSI. (2003). *Technical specification ETSI TS 102 165-1 V4.1.1.*

Haverinen, H., & Salowey, J. (2004). *Extensible authentication protocol method for GSM subscriber identity modules (EAP-SIM)* (Internet draft [work in progress]). Internet Engineering Task Force.

IEEE. (2001). IEEE standards for local and metropolitan area networks: Standard for port based network access control. *IEEE Std 802.1x-2001.* Retrieved April 24, 2007, from http://standards.ieee.org/getieee802/download/802.1X-2001.pdf

IEEE-SA. (2006). IEEE 802.16 LAN/MAN broadband wireless LANS. *IEEE 802.16 standards.* Retrieved April 24, 2007, from http://standards.ieee.org/getieee802/802.16.html

Kambourakis, G., Rouskas, A., & Gritzalis, S. (2004). Performance evaluation of public key-based authentication in future mobile communication systems. *EURASIP Journal on Wireless Communications and Networking, 1*, 184-197

Lehtonen, S., Ahonen, P., Savola, R., Uusitalo, I., Karjalainen, K., Kuusela, E., et al. (2006, September). *Information security in wireless networks.* Ministry of Transport and Communication. Finland: LUOTI Publications. ISBN 952-201-783-3. Retrieved April 24, 2007, from http://www.luoti.fi/material/InfoSec_in_WNetworks_final.pdf

Lynn, M., & Baird, R. (2002). *Advanced 802.11 attack.* Paper presented at the Black Hat 2002 Conference, Las Vegas. Retrieved April 24, 2007, from http://www.blackhat.com/presentations/bh-usa-02/baird-lynn/bh-us-02-lynn-802.11attack.ppt

Meyer, U., & Wetzel, S. (2004a). On the impact of GSM encryption and man-in-the-middle attacks on the security of interoperating GSM/UMTS networks. In *Proceedings of IEEE International Symposium on Personal, Indoor and Mobile Radio Communications (PIMRC2004).*

Meyer, U., & Wetzel, S. (2004b). A man-in-the-middle attack on UMTS. In *Proceedings of ACM Workshop on Wireless Security (WiSe 2004).*

Ohrtman, F. (2005). *WiMax handbook. Building 802.16 wireless networks.* McGraw-Hill Communications.

Reid, J., Nieto, J., & Dawson, E. (2003). Privacy and trusted computing. In *Proceedings of the 14th International Workshop on Database and Expert Systems Applications* (pp. 383-388).

Stanley, D., Walker, J., & Aboba, B. (2005). *Extensible authentication protocol (EAP) method requirements for wireless LANs* (IETF RFC 4017).

Stubblefield, A., Ioannidis, J., & Rubin, A. (2002). *Using the Fluhrer, Mantin, and Shamir attack to break WEP.* Paper presented at the NDSS.

Van de Wiele, T. (2005). *Wireless security: Risks and countermeasures* (UNISKILL Whitepaper).

Welch, D. J., & Lathrop, S. D. (2003). *A survey of 802.11a wireless security threats and security mechanisms* (Tech. Rep. ITOC-TR-2003-101). United States Military Academy.

WiMax Forum. (2006). *Mobile WiMax—Part I: A technical overview and performance evalation.* Retrieved April 24, 2007, from http://www.wimaxforum.org/home/

Zheng, Y., He, D., Xu, L., & Tang, X. (2005a). Security scheme for 4G wireless systems. In *Proceedings of 2005 International Conference on Communications, Circuits and Systems* (Vol. 1, pp. 397-401).

Zheng, Y., He, D., Yu, W., & Tang, X. (2005b). *Trusted computing-based security architecture for 4G mobile networks.* Paper presented at the Sixth International Conference on Parallel and Distributed Computing, Applications and Technologies PDCAT 2005 (pp. 251-255).

KEY TERMS

Authentication: Verification of the identity of a user or network node who claims to be legitimate.

Broadband: A network connection with a bandwidth of about 2 Mbps or higher.

Confidentiality: A cryptographic security service which allows only authorized users or network nodes to access information content.

EAP: Extensible authentication protocol (EAP) is an authentication protocol used with 802.1X to pass authentication information messages between a suppliant and an authentication server.

Integrity: A security service which verifies that stored or transferred information has remained unchanged.

UMTS: Universal mobile telecommunication system (UMTS) is a global third generation wireless cellular network for mobile telephony and data communication with a bandwidth up to 2 Mbps which can be upgraded up to 20 Mbps with high speed packet access (HSPA).

Wi-Fi: Wireless local area networking based on IEEE 802.11 standards.

WiMax: Wireless metropolitan area networking based on IEEE 802.16 standards.

WPA, WPA2: Wi-Fi protected access (WPA) is a protocol to secure wireless networks created to patch the previous security protocol WEP. WPA implements part of and WPA2 implements the entire IEEE 802.11i standard. In addition to authentication and encryption, WPA also provides improved payload integrity.

Chapter XLIX
Extensible Authentication (EAP) Protocol Integrations in the Next Generation Cellular Networks

Sasan Adibi
University of Waterloo, Canada

Gordon B. Agnew
University of Waterloo, Canada

ABSTRACT

Authentication is an important part of the authentication authorization and accounting (AAA) schemes and the extensible authentication protocol (EAP) is a universally accepted framework for authentication commonly used in wireless networks and point-to-point protocol (PPP) connections. The main focus of this chapter is the technical details to examine how EAP is integrated into the architecture of next generation networks (NGN), such as in worldwide interoperability for microwave access (WiMAX), which is defined in the IEEE 802.16d and IEEE 802.16e standards and in current wireless protocols, such as IEEE 802.11i. This focus includes an overview of the integration of EAP with IEEE 802.1x, remote authentication dial in user service (RADIUS), DIAMETER, and pair-wise master key version (2PKv2).

INTRODUCTION

Extensible authentication protocol (EAP) is a universally accepted authentication mechanism, frequently used in different wireless technologies. Although the applications of EAP protocol are not limited to wireless local area networks (LANs), they could be used for authentication in wired-based LAN applications. However EAP is most often used in wireless LANs. The integrations of EAP and other security protocols and mechanisms often result in strong security frameworks.

These integrations are often established with other security protocols and mechanisms, such as transport layer security (EAP-TLS), message digest 5 (EAP-MD5), privacy key management (PKM-EAP), and so forth.

The organization of the sections of this chapter is as follows: Section II will discuss details about the EAP-IEEE 802.1x interactions. Section III is dedicated to remote authentication dial in user service (RADIUS) and DIAMETER in the authentication/authorization schemes. Section IV talks about the IEEE 802.1x-EAP functions implemented in Wi-Fi (IEEE 802.11i) and introductions to EAP-MD5, lightweight extensible authentication protocol (LEAP), EAP-TLS (TTLS) and protected extensible authentication protocol (PEAP). Section V presents the PKMv2-EAP scheme in worldwide interoperability for microwave access (WiMAX) (IEEE 802.16) followed by section VI, which is a configured testbed for a WiMAX system. Sections VII and VIII contains conclusions and references respectively.

EAP AND IEEE 802.1X

Based on RFC 3748 (Aboba, Blunk, Vollbrecht, Carlson, & Levkowetz, 2004), EAP runs on top of IEEE 802.1x (Figure 1), therefore 802.1x is the key issue to understanding the EAP. IEEE 802.1x offers a strong framework for authenticating and controlling user traffic for protecting networks. IEEE 802.1x also offers dynamically varying encryption keys. IEEE 802.1x uses EAP in both wired and wireless LANs and supports multiple authentication methods, such as Kerberos, one-time passwords, and public key certificates. Our main focus is on wireless technologies.

IEEE 802.1x initially starts the communications by an attempt to connect with an authenticator (i.e., an 802.16 or 802.11 access point [AP]) to authenticate an unauthenticated supplicant. The AP responds back by enabling a port for passing only EAP packets between the clients to the authentication server, which is usually located on the wired side of the AP. The AP blocks all other traffic (i.e., HTTP and dynamic host configuration protocol [DHCP] packets), until the AP (authenticator) is able to verify the client's identity using an authentication server (e.g., DIAMETER or RADIUS). Once authenticated, the AP opens the client's port for the rest of traffic types.

To better understand how 802.1x operates, the interactions mentioned in Table 1a usually happen between various 802.1x elements.

As showed in Figure 1, EAP is an important component of an 802.1x-based infrastructure. EAP improves the authentication scheme provided by the point-to-point protocol (PPP) (RFC 1661). EAP provides PPP with a generalized framework for

Figure 1. 802.1x authentication components (Adapted from Kwan, 2003)

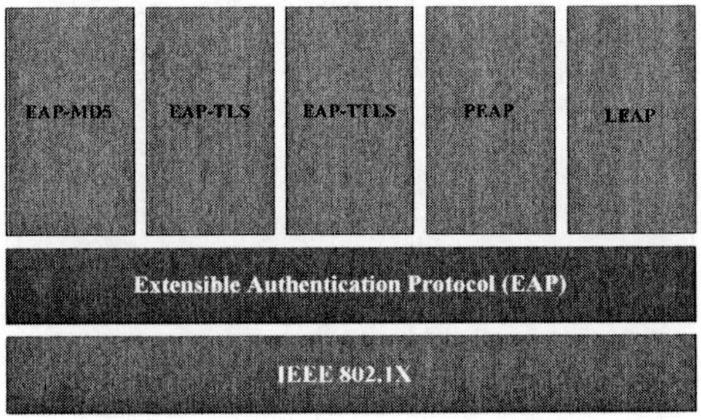

Figure 2. Different layers of 802.1x (Adapted from Leira, 2005)

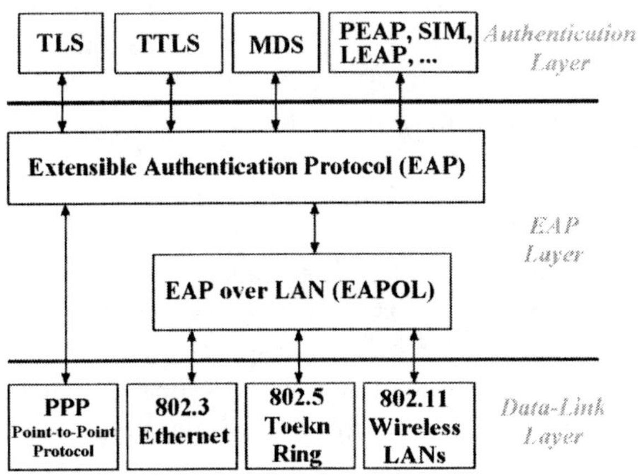

Table 1a. A summary of EAP messages in a supplicant-authenticator-authentication server scenario

	SUPPLICANT		AP (ACCESS POINT) AUTHENTICATOR		AUTHENTICATION SERVER
1.	Sends EAP-start message	→	Receives the message		
2.		←	Replies with an EAP-request ID		
3.	EAP-response with ID	→		→	Verifies the client's identity using Digital Certificates, etc
4.			Receives Accept/Reject message	←	Sends Accept/Reject to the AP
5.	Receives Accept/Reject note	←	Sends Accept/Reject notification		
6.	If accepted, port is open And messages are accepted	→		→	AP forwards the messages to the Authentication Server
7.	If rejected, the end				

various types of authentication schemes (Chen & Wang, 2005). The 802.1x standard includes a definition of EAP encapsulation for Ethernet packages used over LANs, which is called EAP over LAN (EAPOL). Figure 2 (Leira, 2005) shows various layers of selective authentication and network type 802.1x.

There are three main components found in 802.1 X-based systems:

- Supplicant, which is the client/user
- Authenticator, which is the mediator between the client and the Authenticator Server
- Authenticator Server, which determines if the client (supplicant) has the correct information for authentication. This could be a RADIUS or a DIAMETER server

In most cases, both supplicant and the authentication server have relatively more processing capabilities than the authenticator. The authenticator is mostly responsible for forwarding, therefore it requires less power as compared to the other two components. An AP can serve well as the role of an authenticator, which makes the system well suited for wireless networks.

Figure 3, which has more details compared to Table 1a, shows how the communication between the supplicant, authenticator, and the authentication server works. Initially the authenticator blocks all traffic except for the EAPOL-based traffic. The rest of the communication process is similar to that of Table 1a. As shown in Figure 3, the EAP scheme operates in the following fashion (Piscitello, 2005) (see Box 1).

In a true end-to-end secure wireless network, it is not only crucial that the authenticator and authentication server ensure user's legitimacy, but also the supplicant has to be confident that the authentication server and the authenticator are legitimate and not spoofing devices who try to

Figure 3. The supplicant-authenticator-authentication server relationship

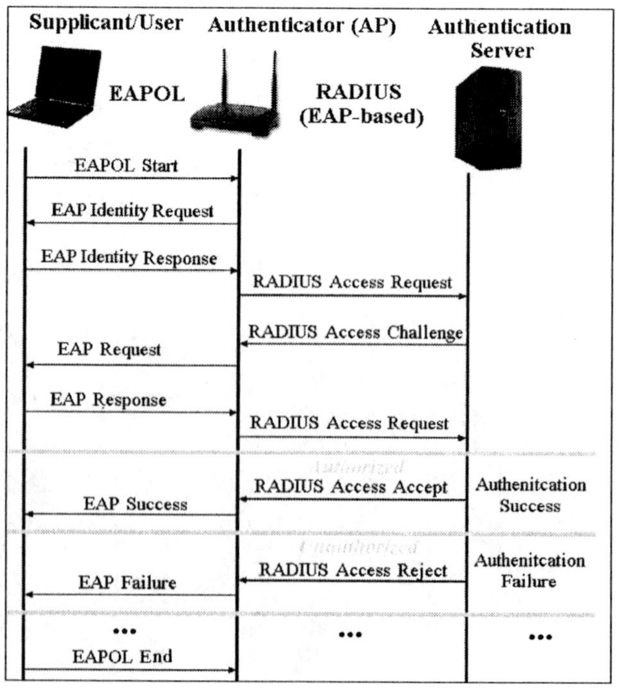

Box 1.

#	Process Taking Place	Message Transmitted/State
1.	Supplicant tries to connect to the authenticator (AP)	802.1x Associate Request
2.	Authenticator detects supplicant and enables client's port	Port set to Unauthorized
3.	Authenticator returns a response to supplicant and waits	802.1x Associate Response
4.	Supplicant transmits a message to authenticator	EAP-START
5.	Authenticator replies a message to supplicant, asks for identity	EAP-REQUEST IDENTITY
6.	Supplicant provides its identity to authenticator	EAP-RESPONSE
7.	Authenticator forwards EAP-RESPONSE to authentication server	FORWARD EAP-RESPONSE
8.	Authentication server authenticate clients	Authenticates via EAP-TLS, LEAP
9.	If accepted by authentication server, signals to authenticator	ACCEPT
10.	If rejected by authentication server, signals to authenticator	REJECT
11.	If authenticator receives acceptation, responds to supplicant Supplicant can use the wireless LAN	EAP SUCCESS Port set to AUTHORIZED
12.	If authenticator receives rejection, responds to supplicant Supplicant remain blocked from the wireless LAN	EAP FAILURE Port state no change
13.	If client succeeded, authenticator passes global key to client	Global Key Passed
14.	When client terminates session, it logs off	EAP LOGOFF

obtain the user name and password from the user. This scenario can be prevented by using a mutual authentication scheme where the authentication server and the authenticators also have to be authenticated by the supplicant. Examples of such mutual authentication schemes are used in TLS, tunneled TTLS (TTLS), LEAP, and PEAP.

IEEE 802.1x also provides a framework to reduce or eliminate the danger of session hijacking and man-in-the-middle (MITM) attacks, however it requires that the right type of authentication (mutual authentication) be used. Secure authentication does not yet imply secure communication. A strong encryption method is required to ensure data confidentiality. EAP enables the usage of different types of encryption with dynamic key distribution techniques.

RADIUS AND DIAMETER

Both RADIUS (Hill, 2001) and DIAMETER (Calhoun, Loughney, Guttman, Zorn, & Arkko, 2003) are authentication, authorization, and accounting (AAA) protocols for applications and mechanisms used in network access or Internet protocol (IP) mobility. They are intended to work in both local and roaming situations.

Many applications running through ISPs using modems, DSL, cable, or wireless connections require some sort of user name/password for access permission. This information is usually transmitted to a RADIUS server, over a network access server (NAS) device using the point-to-point protocol (PPP) and the RADIUS protocol. The RADIUS server verifies that the information is correct. This is done using authentication schemes, such as, password authentication protocol (PAP), challenge handshake authentication protocol (CHAP), or EAP. If authentication and authorization are accepted, then the server will authorize access to the ISP network and select an IP address and other access control parameters (L2TP parameters).

The RADIUS server is also notified of any session start-stop for related accounting, billing, and other statistical issues. RADIUS is an extensible protocol in which most RADIUS vendors have their own hardware and software implements.

The DIAMETER protocol is proposed to replace RADIUS and it is designed to be backward compatible in most cases. The main differences between DIAMETER and RADIUS protocols are, (see Box 2).

The message format and the authentication flows in DIAMETER EAP applications are given in Figures 4 and 5.

Applying RADIUS to Wireless LANs

In wireless-based networks that use 802.1x port access control, the wireless station is a remote user and the wireless AP behaves as the network access server (NAS) (Phifer, L 2., 2003). The IEEE 802.11-based protocols (a, b, or g) are used to associate the wireless stations to the wireless APs.

Once the client is associated, it transmits an EAP-Start message to the AP. The AP sends a request to the wireless station, asking for its identity and relays the message to an AAA server using a RADIUS-based access-request user name message.

As expected, through the AP, the wireless station and the AAA server establish the authentication process by exchanging RADIUS access-challenge and access-request messages. According to the specific EAP type, an encrypted TLS tunnel could be used to convey the messages inside of the tunnel.

If an access-accept message is sent by the AAA server, the wireless station and the AP establish a handshake. This generates session keys that are used by either temporal key integrity protocol (TKIP) or wired equivalence privacy (WEP) to encrypt data. At this point, the port is unblocked by the AP and the wireless station is able to send and receive data to and from the attached LAN.

If an access-reject message is sent by the AAA server, the client will be disassociated by the AP.

Box 2.

#	DIAMETER uses:	RADIUS uses:
1.	Reliable transport protocol (TCP or stream control transmission protocol [SCTP])	Uses an unreliable transport protocol (UDP)
2.	End-to-end transport level security protocols (IPSec or TLS)	End-users, such as, CHAP and PAP
3.	Transition support for RADIUS	No direct compatibility with DIAMETER
4.	Large address space for AVPs (attribute value pairs) – 32 bits	Smaller address space – 8 bits
5.	A peer-to-peer protocol scheme Server-initiated messages support	Client-server protocol scheme Request/response scheme only
6.	Both stateful and stateless models	Only a stateless model
7.	DNS (dynamic name system), SRV (generalized service location), and NAPTR (naming authority pointer), for dynamic discovery of peers	Static Discovery agents
8.	Capability Negotiation (version, applications, etc)	No such built-in capability
9.	Application layer acknowledgements and built-in Failover (device-watchdog request/ device-watchdog answer [DWR/DWA])	No such failover mechanism
10.	Error notification	No such notification
11.	Better roaming support	Average support for fixed and roaming users
12.	Better extended command and attributes	Average command and attributes
13.	Better Mobile-IP supports and stronger security	Average security options

Figure 4. DIAMETER message format (Adapted from Wu, Chen, Chen, & Fan, 2005)

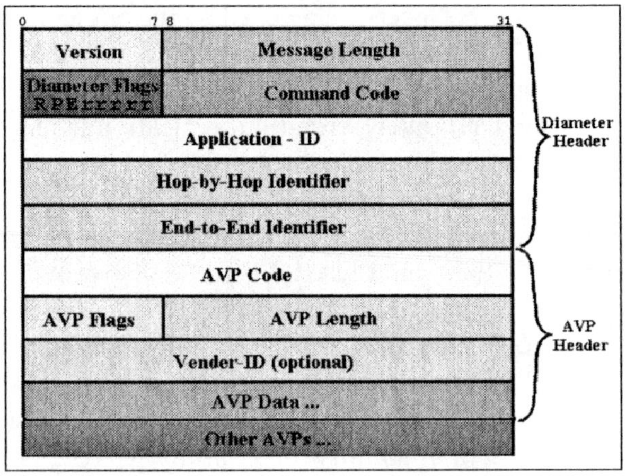

Figure 5. Authentication flows in diameter EAP applications (Adapted from Wu, Chen, Chen, & Fan, 2005)

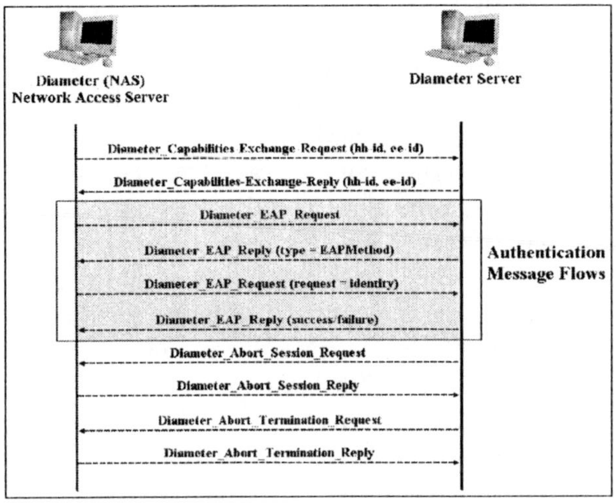

At this point, the failed supplicant can try the authentication process again, however it is prevented by the AP from data packet transmissions. It should be noted that the failed client is still able to listen to the transmitted data across the wireless channel. This raises the importance of encryption techniques for privacy over the air.

The AAA server uses the attribute-value pairs, which are included in the RADIUS messages. This is to deliver session parameters to the wireless station via the AP, such as, session-timeout or VLAN tag (Tunnel-Private-Group-ID=tag, Tunnel-Type=VLAN). The additional information, which can be delivered and used, depends on the AAA Server, AP, and the wireless station settings.

EAP and Different Authentication Methods

EAP by itself cannot protect the authentication message exchange between the client, authenticator, and authentication server. In order to secure the message exchange, an EAP authentication protocol is necessary. The commonly used EAP authentication protocols include (Kwan, 2003; Phifer, 2003; "What are Your EAP Authentication Options?," 2005):

EAP-MD5 (RFC 1994): The EAP-MD5 protocol lets a RADIUS server authenticate LAN stations through MD5 hash verification for each user/password. For a trusted Ethernet, this is a simple and reasonable choice where there is little risk of an outsider active attack or sniffing. EAP-MD5, on the other hand, is not suitable for wireless LANs or public Ethernets, since the station identities and password hashes are prone to easy outside sniffing. A man-in-the-middle attack or session hijacking could also be an issue. EAP-MD5 is able to protect the message exchange flow through creating a unique digital signature," which authenticates each packet using this to ensure authenticity for the EAP messages. EAP-MD5 has light computational weight and this increases it's timing performance, which makes it fairly easy to implement and configure. EAP-MD5 does not use public key infrastructure (PKI) certificates for validating clients nor does it provide strong encryption for protecting the authentication messages between the supplicant and the authentication server. EAP-MD5 is most suitable for the EAP message exchanges in wired networks where the EAP client is directly connected to the authenticator. In this case, the chances for message interception and eavesdropping are relatively very low. Therefore for wireless 802.1x authentication schemes, stronger and more robust EAP authentication protocols should be deployed.

EAP with transport layer security (EAP-TLS): EAP-TLS is discussed in RFC 2716, which is the only secured standard option (along with EAP-TTLS) designed for wireless LANs. EAP-TLS mandates a procedure in which the station and the RADIUS server are both required to prove their identities using public key cryptography (i.e., security tokens, smart-cards, or digital certificates). This procedure is secured by an encrypted TLS tunnel, which makes EAP-TLS very resilient to against dictionary, man-in-the-middle, and other types of attacks. However, the station's identity, which is the name attached to the certificate, can still be sniffed through eavesdropping. EAP-TLS is a very attractive candidate for large enterprises, which only use Windows (2000/2003/XP)-based applications with deployed certificates. EAP-TLS provides strong security schemes by requiring both client and authentication server (mutual authentication) to be authenticated and authorized by using PKI certificates. This works well within 802.1x authentication schemes as the TLS tunnel between the client and the authentication server protects the EAP messages from sniffing and eavesdropping. The only notable drawback of EAP-TLS is the requirement of PKI certificates on both sides (clients and authentication servers). This causes complications in roll-out and maintenance procedures and increases the amount of overhead to establish a secure link as certificates can be quite large. Figure 6 shows the EAP-TLS message flow.

EAP with tunnelled TLS (EAP-TTLS): EAP-TTLS is an extension of EAP-TLS, which provides the benefits of a strong encryption scheme without the complexity of mutual certificates on both sides (client and authentication server). Similar to the EAP-TLS scheme, EAP-TTLS scheme supports mutual authentication, however it only requires the authentication server to be validated to the client using a certificate exchange. EAP-TTLS allows the client to be authenticated by the authentication server through a user name/password process and only requires a certificate used by the authentication server. EAP-TTLS simplifies the roll out and maintenance procedures while retaining strong security and relatively strong authentication scheme. A TLS tunnel is used for protecting EAP messages and for reusing existing user credential services for 802.1x authentication, such as RADIUS, active directory, and LDAP. AP-TTLS also provides backward compatibility for other authentication protocols, such as, PAP, CHAP, MS-CHAP, and MS-CHAP-V2. If TLS tunnels are not used, EAP-TTLS is not considered secure and can be fooled into revealing identity credentials. EAP-TTLS is most suitable for infrastructures that require strong authentication without mandating the use of mutual certificates. Wireless 802.1x authentication schemes usually support EAP-TTLS.

Protected EAP (PEAP): PEAP is an Internet-draft (still not an RFC), which is similar to EAP-TTLS in terms of supporting mutual authentication. PEAP is currently being supported by Cisco Systems, RSA Data Security Inc., and Microsoft. PEAP is an authentication protocol alternative to EAP-TTLS, which overcomes EAP weaknesses through:

a. Protecting user credentials
b. Securing EAP negotiation flows
c. Standardizing key exchange flows
d. Supporting fragmentation and reassembly procedures
e. Supporting fast reconnects

PEAP allows the utilization of other EAP-based authentication protocols and securing the transmission through utilizing a TLS encrypted tunnel. PEAP relies on the TLS keying method for the key creation and exchange mechanisms. The PEAP client is authenticated directly with the back-end authentication server. The authenticator acts as a pass-through device, which does not require much processing power or manipulation and needs little understanding of the EAP authentication protocol mechanism. Unlike EAP-TTLS, PEAP does not support inherent username and password authentication against an existing user (unlike LDPA). To support this, every specific vendor has its own feature built on top of the protocol. PEAP is most suitable for infrastructures, which require strong authentication without the use of mutual certificates, similar to EAP-TTLS. Wireless 802.1x authentication schemes usually support PEAP.

Cisco's lightweight EAP (LEAP): LEAP goes beyond EAP-MD5 in addressing the security issues of wireless networks by delivering the keys used for WLAN encryption and requiring mutual authentication. Mutual authentication reduces the risk of an attacker posing as an AP (MITM attack). However, station identities and passwords remain vulnerable to dictionary sniffing attacks. LEAP is mostly used when Cisco-based APs and cards are involved. LEAP mandates mutual authentication between the client and the authenticator. The client first has to authenticate itself to the authenticator and then the authenticator should authenticate itself to the client. If the two authentication procedures are done successfully, a network connection is granted. Unlike EAP-TLS, LEAP is username/password-based and is not based on PKI certificates. This simplifies roll-out and maintenance procedures. Being the proprietary to Cisco is one of the drawbacks of LEAP, which is the reason it has not been widely adopted by other networking vendors. LEAP is most suitable for wireless scenarios that support Cisco AP's and LEAP compliant wireless NIC cards.

EAP-SIM: The EAP method for global system for mobile communications (GSM) subscriber identity module (SIM), or EAP-SIM, is an EAP-based mechanism used for authentication and session key distribution, which is used in the GSM-SIM. EAP-SIM is described in RFC 4186.

Tables 1b and 2 show summaries and comparisons between all mentioned EAP-based protocols.

Depending on the specific EAP authentication protocol used, IEEE 802.1x authentication protocol can help to solve the following security issues (Kwan, 2003):

- **Dictionary attack:** In this type of attack, the attacker obtains the challenge/response message exchange from a password authentication session and uses a brute force mechanism to find the password. IEEE 802.1x solves this type of attack by using TLS-based tunnels for protecting credential exchanges among authenticator and supplicant.
- **Session hijack:** In this attack, the attacker is able to sniff the packets passed between the client and the authenticator and to recover the client's identity information. This pushes the "legitimate" client out of the scope through a form of denial-of-service (DoS) attack and impersonates the client to continue the conversation with the authenticator (DoS and session hijacking). IEEE 802.1x can thwart the session hijacking through its ability to securely authenticate with dynamic session-based keys.
- **Man-in-the-middle:** The MITM attack happens in one-way authentication or unbalanced schemes, where the attacker obtains the necessary information from the client and/or

Table 1b. Comparison between different EAP methods in terms of client/server strength (Adapted from Phifer, 2003)

	EAP-MD5	EAP-TLS	EAP-TTLS	LEAP	PEAP
Supplicant Authentication	Password Hash	Public Key (Certificate or Smart Card)	CHAP, PAP, MS-CHAPv2, EAP	Password Hash	Any EAP, Public Key, or EAP-MS-CHAPv2
Server Authentication	None	Public Key (Certificate)	Public Key (Certificate)	Password Hash	Public Key (Certificate)
Dynamic Key Delivery	No	Yes	Yes	Yes	Yes
Security Risks	Identity exposure, Dictionary attack, Man-In-The-Middle (MitM) attack, Session hijacking	Identity exposure	Man-In-The-Middle (MitM) attack	Identity exposure, Dictionary attack	Man-In-The-Middle (MitM) attack

the authenticator and comes in the middle of the session and becomes the "middle man." Through IEEE 802.1x's authentication and dynamic session-based keys, the encryption of the data stream between the client and authenticator can prevent this type of attack.

IEEE 802.11I - WLAN SECURITY STANDARD IMPLEMENTATION

The IEEE 802.11i standard was designed to provide secure communications in wireless LANs, which is part of the IEEE 802.11 specifications. For many years, WEP was used as a WLAN security technology. However, WEP has been proven not to be secure with today's computational power due to the short period of the stream cipher used and how key stream reuse allows the data to be recovered (to name a few). IEEE 802.11i enhances the encryption, authentication, and key management schemes of WEP. IEEE 802.11i is based on a strong security scheme, the Wi-Fi protected access (WPA).

WPA in 802.11i

WPA is a subset of IEEE 802.11i, the standard for WLAN security, and consists of the followings:

- An authentication mechanism that uses IEEE 802.1x or pre-shared keys scheme.
- An encryption mechanism, which uses temporal key integrity protocol (TKIP), per IEEE 802.11i definition. TKIP could be software-based offered by products that support WEP.

Figure 6. Message flow of EAP-TLS

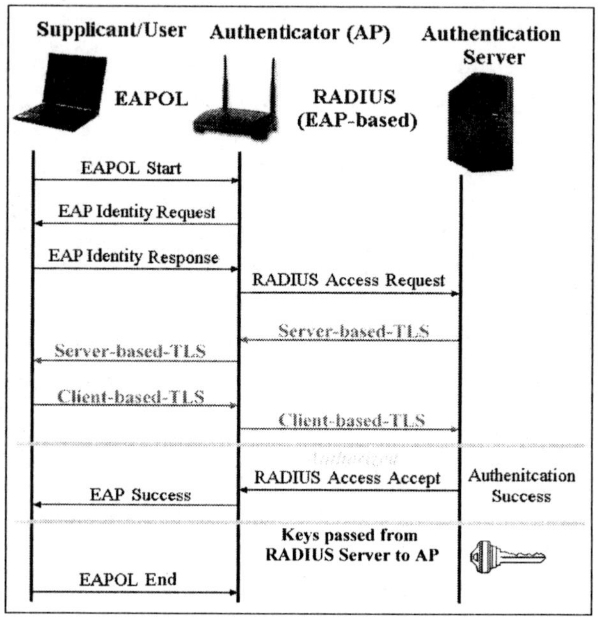

WPA2 and 802.11i

WPA2 is the second generation of WPA security introduced by the Wi-Fi Alliance (Lehembre, 2005). It is consisted of:

- An authentication mechanism that uses IEEE 802.1x or pre-shared keys scheme.
- An encryption mechanism, which uses advanced encryption standard (AES), per IEEE 802.11i definition.

WPA2 with AES is eligible for FIPS 140-2 (specified by the United States Government's National Institute of Standards and Technology [NIST] and uses four level of securities) compli-

Table 2. Comparison among various EAP methods in terms of wireless security strength (Adapted from "What are Your EAP Authentication Options?," 2005)

Protocol	The Implementation Procedure	Authentication Attributes	Deployment Difficulty	WEP Key Generated?	Wireless Security
EAP-MD5	Challenge-based password authentication	One-way	Easy	No	Poor
EAP-TLS	Certificate-based, two-way authentication (mutual)	Two-way (Mutual)	Difficult	Yes	Excellent
LEAP	Username/hashed password authentication	Two-way (Mutual)	Easy	Yes	Good. if strong passwords are used
PEAP or TTLS	Server authentication via certificates, client via other methods	Two-way (Mutual); Identity hiding (opt)	Moderate	Yes	Better than that of EAP-MD5

ance. WPA2 is a requirement for Wi-Fi compliance from 2006.

EAP Method Requirements for Wireless LANs

RFC 4017 (Stanley, Walker, & Aboba, 2005) specifies the requirements for EAP methods used in IEEE 802.11-based systems, which uses IEEE 802.11i for authentication and authorization. This in turn could be applied to IEEE 802.16 as well. 802.11i MAC security enhancements makes use of both IEEE 802.1x and EAP. Today's deployments of IEEE 802.11 wireless LANs are based on EAP, integrated with several EAP methods, namely: EAP-TLS, EAP-TTLS, PEAP, and EAP-SIM, which were discussed before. These methods support authentication credentials, including *digital certificates*, *secure tokens*, *usernames/passwords*, and *SIM secrets*.

IEEE 802.11i specifies the usage of EAP for both authentication and key exchange among the EAP peers and servers. RFC 3748 (RFC 3748 - EAP) outlines the EAP usage within IEEE 802.11i, which is subject to threats, given that WLAN provides ready access to any attacker within range.

The following four components are integral parts of IEEE 802.11i specifications (IEEE 802.11i: WLAN Security Standards," 2006):

- **Temporal key integrity protocol (TKIP):** TKIP is a protocol which uses an RC4 cipher for encryption of data and deals with confidentiality of data. TKIP improves the security weaknesses of WEP. It uses a message integrity code, called "TKIP-Michael algorithm," which authenticates end devices for legitimacy. TKIP utilizes a mixing function to overcome weak-key and brute-force attacks. TKIP is used in 802.11i during two phases:
    - **First phase:** In the first phase, TKIP is used together with an improved message integrity check (MIC). This is to stop data manipulation.
    - **Second phase:** In the second phase, TKIP and MIC are replaced with counter with cipher block chaining message authentication code (CCMP). CCMP uses the AES encryption scheme.

    TKIP offers three advantages over WEP:
    - Longer initialization vector (IV), which minimizes the chance session key reuse
    - Key hashing, which results in a different key used for each data packet
    - MIC, which ensures that the message is not altered during the communication between sender and receiver
- **Counter-mode/CBC-MAC protocol (CCMP):** CCMP is similar to TKIP, in which it deals with the confidentiality of data, as well as authentication and encryption. One of the differences between CCMP and TKIP is the fact that CCMP uses AES in counter mode for data confidentiality. The other difference is the usage of cipher block chaining message authentication code (CBC-MAC) for authentication and integrity. In the architecture of 802.11i, CCMP uses a 128-bit key scheme. CCMP provides protections for some fields, which are not encrypted through a mechanism, which is so-called additional authentication data (AAD). AAD protection includes a scheme which prevents attackers from replaying packets to various destinations.
- **IEEE 802.1x:** IEEE 802.11i is a wireless implementation of 802.1x, which offers an effective framework to authenticate and control user traffic and also offers dynamically varying encryption keys. Through this component (802.1x), 802.11i is able to get tied to EAP.
- **EAP encapsulation over LANs (EAPOL):** As discussed in Figure 2, EAP layer covers EAPOL, which is a key protocol in IEEE 802.1x for key exchange. Two main schemes covered in the EAPOL-key exchanges are defined in IEEE 802.11i, which are the 4-way handshake and the group key handshake.

PKMV2-EAP SCHEME IN WIMAX (IEEE 802.16)

WiMAX (IEEE 802.16) stands for worldwide interoperability for microwave access, which is maintained by the WiMAX Forum. WiMAX has similarities with Wi-Fi; however it claims to achieve higher bandwidth (up to 70 Mbps) over a 70 mile (+110 km) range, which outperforms Wi-Fi. There are also some similarities between the security schemes between WMAX's and IEEE 802.11i.

In this section, the security mechanisms for WiMAX are described. For an end-to-end authentication scheme, WiMAX uses extensible authentication protocol with privacy key management (EAP-PKM), which relies on the transport layer security (TLS) standard and public key cryptography ("WiMAX Technology," 2005). PKM is a protocol, which uses the Rivest, Shamir, and Adleman (RSA) public-key scheme, X.509 digital certificates, and a strong encryption scheme for the subscriber station (SS)-base station (BS) interactions. There are two PKM protocols supported in the IEEE 802.16 standard; PKM version 1 (PKMv1) and PKM version 2 (PKMv2). PKMv1, which is a one-way authentication method, is proven to be prone to variety of attacks and is not covered in this chapter. PKM supports two authentication protocol mechanisms:

1. RSA public key-based certificates, mandatory in all devices
2. EAP

Authorization via PKM RSA Authentication Protocol

Figure 7 shows the authorization and authentication processes of PKMv2 protocol using a request/grant access method. For a SS (PKM client) to have access to the BS network, the PKM server has to authorize the connection and the SS also needs to authenticate the BS; after that, the SS will have security features enabled. Once the SS associates with the BS, the SS shares a private encryption key with the BS and communication between

Figure 7. PKMv2 authentication and authorization process (Adapted from Adibi, Bin, Ho, Agnew, & Erfani, 2006)

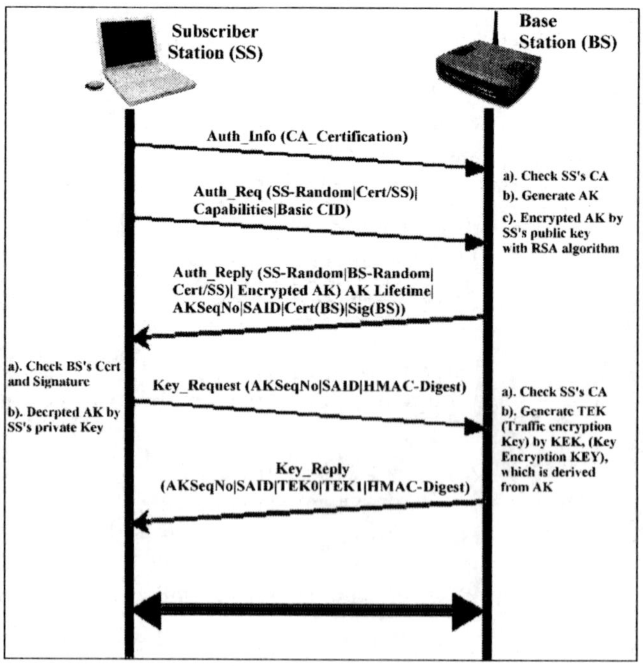

the BS and SS can be initiated using encrypted messages.

Authorization via PKM Extensible Authentication Protocol

After the SS is associated to the BS, the EAP authorization procedure starts. Figure 8 shows the EAP authorization and authentication flow steps:

Security Analysis of WiMAX Authentication

The EAP-PKM is intended to secure WiMAX clients and servers in a more robust way. The following list summarizes the strength of EAP-PKM:

1. PKMv2 supports mutual authentication, which can prevent man-in-the-middle attacks.
2. The X.509 digital certificate issued for each SS is unique and cannot be easily forged.
3. Each service has a unique security association identifier (SAID), therefore if one service is compromised, the other services are not affected.
4. The limited lifetime of authorization key (AK) provides key-refresh and periodic reauthorization, which prevents attackers from gathering enough data to launch cryptanalysis.
5. To correct replay attacks, it is recommended to add a random value transmitted from BS and SS for SA authorization.
6. WiMAX security supports two strong encryptions algorithms; triple data encryption standard (3DES) and AES, which are considered leading edge (AES in particular).
7. The ability of an SS to cache or transfer the master key to avoid a full reauthentication procedure.
8. EAP-PKM relies on the TLS standard that is based on public key cryptography, which is costly for some wireless vendors. Therefore, a high performance security processor is dedicated to BS in WiMAX, which enables the implementation of a complicated authentication system in WiMAX.

In this section, a WiMAX-based authentication using EAP-TLS and EAP-PKM were presented. This included the PKMv2 handshaking schemes. It is believed that WiMAX possesses more extensive security power compared to the ones in Wi-Fi, which in turn will favor WiMAX in the comparative market share.

Figure 8. 802.16e EAP authentication process (Adapted from Adibi et al., 2006)

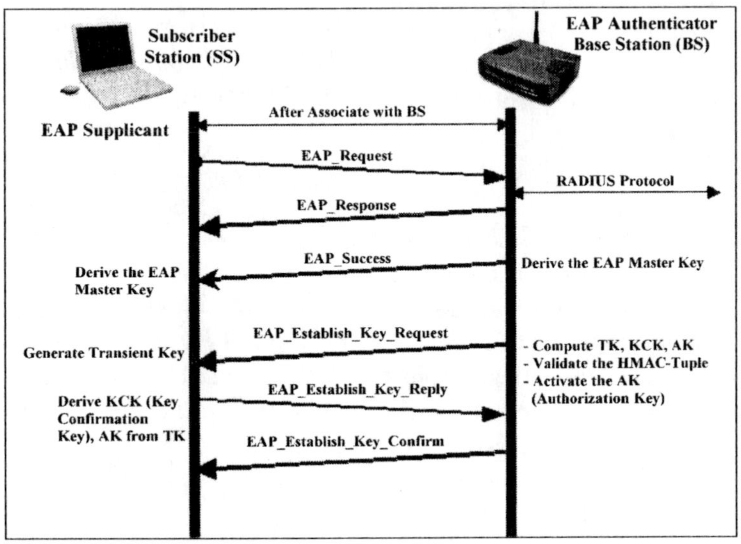

CONCLUSION

In this chapter, to help address the security issues of unauthorized access, The development of IEEE 802.1x was to provide a standard authentication mechanism in port-based scenarios. EAP, on the other hand, offers supports to a variety of standard authentication messaging protocols. EAP provides multivendor solutions to support network authentication framework. Additional EAP types, including EAP-SIM and EAP-SecurID (which supports hardware tokens), are also defined. EAP specifies the method in which supplicant/authenticator/authentication server interact and the type of standard messaging exchanged between them. EAP, however does not specify the actual authentication protocol. Therefore, EAP's advantages can be summarized as:

- EAP permits multiple authentication protocols without extra setup steps.
- EAP is flexible and supports multiple authentication protocols without the necessity of requiring to match an authenticator to a specific authentication mechanism. EAP permits the authentication server to selects the best suitable authentication protocols, which is supported on the client, as well as itself. This is usually done without the need for fully configuring the authenticator with the authentication protocol. In this scenario, the authenticator acts as a pass-through device (pass-through is optional).
- The authenticator has the ability to act as a pass-through device for non-local clients and at the same time, authenticate local clients using authentication protocols it may not support locally.
- The existence of a separate authenticator and authentication server operating in the pass-through mode, permits simplifications of the credentials and development of standard messaging protocols. The authenticator is responsible for determination of the outcome of the authentication from the access-accept or reject message provided by the authentication server. The outcome of the authentication is not affected by the EAP packet's contents.

This may pose as vulnerability manipulations and different attacks. Throughout this chapter, where appropriate, the application of EAP using different authentication/authorization methods for wireless applications, were discussed. Special attention was given to EAP-TLS and EAP-PKMv2 for 802.16e systems.

REFERENCES

Aboba, B., Blunk L., Vollbrecht, J., Carlson, J., & Levkowetz, H. (2004, June). *Extensible authentication protocol (EAP)* (RFC 3748).

Adibi, S., Bin, L., Ho, P. H., Agnew, G. B., & Erfani, S. (2006, May). *Authentication authorization and accounting (AAA) schemes in WiMAX.* Paper presented at the Conference on Electro/information Technology (EIT'06).

Calhoun, P., Loughney, J., Guttman, E., Zorn, G., & Arkko, J. (2003, September). *Diameter base protocol* (RFC 3588).

Chen, J. C., & Wang, Y. P. (2005, December). Extensible authentication protocol (EAP) and IEEE 802.1x: Tutorial and empirical experience. *43*(12), supl.26-supl.32.

Hill, J. (2001). *An analysis of the RADIUS authentication protocol.* InfoGard Laboratories.

IEEE 802.11i: WLAN Security Standards. (2006). *Javvin Technologies, Inc.* Retrieved October 25, 2007, from http://www.javvin.com/protocol80211i.html

Kwan, P. (2003, May). *802.1x authentication & extensible authentication protocol (EAP)* (White Paper).

Lehembre, G. (2005, December). *Wi-Fi security – WEP, WPA and WPA2.* Retrieved October 25, 2007, from http://www.hsc.fr/ressources/articles/hakin9_wifi/hakin9_wifi_EN.pdf

Leira, J. (2005, April 15). *WLAN - IEEE 802.1x.* The Norwegian Research Network (UNINETT).

Phifer, L. (2003, September). *Deploying 802.1X for WLANs: EAP types*. Retrieved October 25, 2007, from http://www.wi-fiplanet.com/tutorials/article.php/3075481

Phifer, L 2. (2003, September). *Applying RADIUS to Wireless LANs, using RADIUS For WLAN Authentication, Part I.* Retrieved from http://www.wi-fiplanet.com/tutorials/article.php/10724_3114511_1

Piscitello, D. M. (2005, April 16). *IEEE 802.1x and EAP primer*. Core Competence, Inc.

Stanley, D., Walker, J., & Aboba, B. (2005, March). *Extensible authentication protocol (EAP) method requirements for wireless LANs* (RFC 4017).

What are Your EAP Authentication Options? (2005, May). *InteropNet labs full spectrum security initiative*. Retrieved October 25, 2007, from http://www.opus1.com/nac/whitepapers-old/04-EAP_OPTIONS-LV05.PDF

WiMAX Technology. (2005). Retrieved October 25, 2007, from http://ww.hifn.com/docs/WiMAX_AB_1.4.pdf

Wu, W. T., Chen, J. C., Chen, K. H., & Fan, K. P. (2005). *Design and implementation of WIRE diameter*. Paper presented at the 3rd International Conference on Information Technology: Research and Education, ITRE 2005.

KEY TERMS

AP: Access point (or wireless access point) is a device that connects wireless devices (i.e., mobile users [MUs], laptops, etc.) together. APs are usually connected to another device called wireless controller (WC). A wireless network is usually comprised of a WC and a few APs, servicing MUs.

DIAMETER: DIAMETER is an authentication, authorization, and accounting (AAA) protocol, an updated version of RADIUS.

DHCP: Dynamic host configuration protocol is a protocol that automatically manages (temporarily assign and release) IP addresses to devices on the network (wireless and wired).

EAP: Extensible authentication protocol is a universally famous authentication protocol accepted framework for authentication. Its integration with other security schemes usually produces strong frameworks for various wireless and wired applications.

MD5: Message-digest algorithm 5 is a 128-bit hash function, which is a widely used cryptographic element. MD5 has shown some weaknesses; therefore it is not counted a robust scheme nowadays.

PEAP: Protected EAP is a security method which transmits authentication information, including passwords. PEAP can be used in variety of scenarios including wireless and wired topologies.

PKM: Privacy key management is a private key scheme used with EAP and TLS for providing end-to-end security schemes for wireless technologies.

RADIUS: Remote authentication dial in user service is an AAA protocol that works in a client/server application scenario. RADIUS oversees the authentication and authorization scheme of the session established between two entities. It is further updated by DIAMETER.

TLS: Transport layer security is used mostly in client/server applications, which require endpoint authentication and communications privacy, particularly over the Internet. This is mostly done using cryptographic measures.

WiMAX: WiMAX stands for worldwide interoperability for microwave access, which has been defined by the WiMAX Forum, formed in 2001. WiMAX is also known as IEEE 802.16 standard, officially titled WirelessMAN and is an alternative to DSL (802.16d) and cellular access (802.16e).

About the Contributors

Yan Zhang received the PhD degree in School of Electrical & Electronics Engineering, Nanyang Technological University, Singapore. From August 2004 to May 2006, he worked with the National Institute of Information and Communications Technology (NICT), Singapore. Since August 2006, he has worked with Simula Research Laboratory, Norway (http://www.simula.no/). He is on the editorial board of the *International Journal of Network Security*. He is currently serving as the Book Series Editor for the book series, *Wireless Networks and Mobile Communications* (Auerbach Publications, CRC Press, Taylor, and Francis Group). He is serving as co-editor for several books: *Resource, Mobility and Security Management in Wireless Networks and Mobile Communications; Wireless Mesh Networking: Architectures, Protocols and Standards; Millimeter-Wave Technology in Wireless PAN, LAN and MAN; Distributed Antenna Systems: Open Architecture for Future Wireless Communications; Security in Wireless Mesh Networks; Mobile WiMAX: Toward Broadband Wireless Metropolitan Area Networks; Wireless Quality-of-Service: Techniques, Standards and Applications; Broadband Mobile Multimedia: Techniques and Applications; Internet of Things: From RFID to the Next-Generation Pervasive Networked Systems; Unlicensed Mobile Access Technology: Protocols, Architectures, Security, Standards and Applications; Cooperative Wireless Communications; WiMAX Network Planning and Optimization; RFID Security: Techniques, Protocols and System-On-Chip Design; Autonomic Computing and Networking; Security in RFID and Sensor Networks;* and *Handbook of Research on Wireless Security.* He serves as industrial co-chair for MobiHoc 2008, program co-chair for UIC-08, general co-chair for CoNET 2007, general co-chair for WAMSNet 2007, workshop co-chair FGCN 2007, program vice co-chair for IEEE ISM 2007, publicity co-chair for UIC-07, publication chair for IEEE ISWCS 2007, program co-chair for IEEE PCAC'07, special track co-chair for "Mobility and Resource Management in Wireless/Mobile Networks" in ITNG 2007, special session co-organizer for "Wireless Mesh Networks" in PDCS 2006, and he is a member of Technical Program Committee for numerous international conference, including CCNC, AINA, GLOBECOM, ISWCS, ICC, and so forth. He received the Best Paper Award and Outstanding Service Award as Symposium Chair in the IEEE 21st International Conference on Advanced Information Networking and Applications (AINA-07). His research interests include resource, mobility, energy, and security management in wireless networks and mobile computing. He is a member of IEEE and IEEE ComSoc.

Jun Zheng received the BS and MS degrees in electrical engineering from Chongqing University, China, in 1993, 1996, respectively, the MSE degree in biomedical engineering from Wright State University, Dayton, Ohio, in 2001, and the PhD degree in computer engineering from University of Nevada, Las Vegas in 2005. Currently he is an assistant professor in the Department of Computer Science at Queens

College of The City University of New York. He is also a member of the faculty of the doctoral program in computer science at the Graduate School and University Center of The City University of New York. He is the co-editor for two books: *Security in Wireless Mesh Networks* and *Handbook of Research on Wireless Security*. He served as general co-chair for WAMSNet-07, track co-chair for ITNG 2007, and session co-organizer for PDCS 2006. He also served as TPC member for several international conferences. His research interests are mobility and resource management in wireless and mobile networks, media access control, performance evaluation, network security, computer architectures, fault-tolerant computing, and image processing. He is member of IEEE.

Miao Ma received the BEng. and MEng. degrees in electrical engineering from Harbin Institute of Technology, China, respectively, and the PhD degree in electrical and electronic engineering from Nanyang Technological University (NTU), Singapore. From August 2002 to December 2006, she worked at the Institute for Infocomm Research (I2R), Singapore. Since January 2007, she has been working at the Hong Kong University of Science and Technology (HKUST). She is a member of IEEE. Her research interests include media access control, cognitive radio, security, wireless communications, and networking.

* * * * *

Sasan Adibi received his BSc degree from Amirkabir University, Tehran, Iran, in 1996, first MSc degree from Brunel University, London in 1999, and second MSc degree from University of Windsor, Canada, in 2005. He is currently studying towards his PhD degree at the University of Waterloo and working as a contractor for Siemens Canada as a Verification System's Engineer. His areas of research include security (ad hoc networks, 3G, WiMAX, and Wi-Fi), quality of service (QoS) for MPLS- and wireless-based systems, and optimizations in ad hoc networks. His work experiences include extensive verification in routing (MPLS, OSPF, BGP), quality of service features (DiffServ, 802.11e, WMM, etc.), security and authentication (WPA v2, OKC, 802.1X, 802.11i, and RADIUS), Wi-Fi (802.11a/b/g), and VLAN (802.1D, 802.1Q, and 802.1p).

Gordon B. Agnew received his BASc and PhD in electrical engineering from the University of Waterloo in 1978 and 1982, respectively. He joined the Department of Electrical and Computer Engineering at the University of Waterloo in 1982. In 1984, he was a visiting professor at the Swiss Federal Institute of Technology in Zurich where he started his work on cryptography. Dr. Agnew's areas of expertise include cryptography, data security, protocols and protocol analysis, electronic commerce systems, high-speed networks, wireless systems, and computer architecture. He has taught many university courses and industry sponsored short courses in these areas as well as having authored many articles. In 1985, he joined the Data Encryption Group at the University of Waterloo. The work of this group led to significant advances in the area of public key cryptographic systems including the development of a practical implementation of elliptic curve-based cryptosystems. Dr. Agnew is a member of the Institute for Electrical and Electronics Engineers, a foundation fellow of the Institute for Combinatorics and its Applications, and a registered professional engineer in the Province of Ontario. Dr. Agnew has provided consulting services to the banking, communications, and government sectors. He is also a co-founder of CERTICOM Corp., a world leader in public key cryptosystem technologies.

Sheikh Iqbal Ahamed is an assistant professor in the Department of Mathematics, Statistics, and Computer Science at Marquette University and director of the Ubicomp Research Lab. His research focuses on pervasive security, trust, and privacy for pervasive computing. He received the PhD in computer science from Arizona State University and the BS from the Bangladesh University of Engineering and Technology.

Christer Andersson is a doctoral student at Karlstad University, Sweden, and his main research topic is designing and evaluating technologies for anonymous communication in mobile networks. He has proposed anonymity technologies for both infrastructured and infrastructureless mobile networks. He is furthermore interested in measuring the degree of anonymity and performance in mobile networks, as well as finding an appropriate trade-off between anonymity and performance. He holds a Licentiate in engineering degree from Karlstad University (2005), and a Master Degree in computer science (2002) from Linköping University, Sweden.

AbdelBaset M.H. Awawdeh received the BEng degree in industrial automation engineering from Palestine Polytechnic University in 1999 and the MSE and PhD degrees in electronics engineering from Alcala University in 2004. From 1999 to 2000 he joined the Department of Electrical and Computer Engineering at Palestine Polytechnic University, Palestine. Since 2004, he has held a researcher position in the Department of Electronics at University of Alcala, Spain. His technical interests include multiagent system interaction and design, vehicles on-board electronics, and vehicles fault diagnosis system.

Mohamad Badra is employed by the CNRS (National Center for Scientific Research, France) researching wireless networks security. Badra performs his research activities at the LIMOS Laboratory - UMR6158, University Blaise Pascal. He was a postdoctoral fellow at the Computer Science and Networks Department, ENST-Paris. His research interests include key exchange, wireless security, public-key infrastructures, smart cards, and security algorithms. Badra received a PhD in networks and computer sciences from ENST-Paris. He is a member of the IEEE and of ESRGroups.

Sungmin Baek received his BS degree in School of Information and Communication Engineering, Sung Kyun Kwan University in 2004 and an MS degree from Department of Computer Engineering and School of Computer Science and Engineering, Seoul National University in 2006. Currently, he is a research engineer in information and technology laboratory, LG Electronics Institute of Technology. His research interests include multimedia transmission over wireless network and wireless personal area networks.

Javier A. Barria received the BSc degree in electronic engineering from the University of Chile, Santiago, in 1980, and the PhD and MBA degrees from Imperial College London in 1992 and 1998, respectively. From 1981 to 1993, he was a system engineer and project manager (network operations) with the Chilean Telecommunications Company. Currently, he is a reader in the Intelligent Systems and Networks Group, Department of Electrical and Electronic Engineering, Imperial College London. His research interests include communication networks monitoring strategies using signal and image processing techniques, distributed resource allocation in dynamic topology networks, and fair and efficient resource allocation in IP environments. He has been a joint holder of several FP5 and FP6 European Union project contracts all concerned with aspects of communication systems design and management. Dr. Barria was a British Telecom Research Fellow (2001 – 2002) and a Tan Chin Tuan Research Fellow, NTU Singapore (2003 - 2004). He is a fellow member of IEE, member of IEEE, and a chartered engineer.

Paolo Bellavista graduated from University of Bologna, Italy, where he received PhD degree in computer science engineering in 2001. He is now an associate professor of computer engineering at the University of Bologna. His research activities span from mobile agent-based middleware solutions and pervasive wireless computing to location/context-aware services and adaptive multimedia. He is member of IEEE and Italian Association for Computing (AICA). He is an associate technical editor of the *IEEE Communication Magazine*.

Soong Boon-Hee received his BEng (honors I) degree in electrical and electronic engineering from University of Auckland, New Zealand and a PhD degree from the University of Newcastle, Australia in 1984 and 1990, respectively. He is currently an associate professor with the School of Electrical and Electronic Engineering, Nanyang Technological University. From October 1999 to April 2000, he was a visiting research fellow at the Department of Electrical and Electronic Engineering, Imperial College, UK under the Commonwealth Fellowship Award. He was awarded the Tan Chin Tuan Fellowship to visit the Centre for Advanced Computing and Communications, Duke University in June 2004. He also served as a consultant for mobile IP in a recent technical field trial of next-generation wireless LAN initiated by IDA (InfoComm Development Authority, Singapore). His area of research interests includes mobile ad hoc and sensor networks, mobility issues, mobile IP, optimization of wireless networks, routing algorithms, optimization and planning of mobile communication networks, queuing theory system theory, quality of service issues in high-speed networks, and signal processing. He has served as a reviewer for a number of IEEE top journals and international conferences. He has served on Technical Program Committee for IEEE Globecom 2004, 2005, 2006, and 2007, IEEE WCNC 2005, and IEEE ISWCS 2004, 2005, 2006, and 2007. He is currently organizing co-chair of IEEE Vehicular Technology Conference, Spring 2008 to be held in Singapore. He is currently on the technical committee ISO 204/WG16 that tracks developments in the intelligent transport sector. He is listed in Marquis Who's Who in Science and Engineering 2006-2007. He has published more than one hundred international journals and conferences. He is a senior member of IEEE and a member of ACM.

Noureddine Boudriga is a professor of telecommunication and director of the Communication Networks and Security (CN&S) Research Laboratory at the University of November 7th at Carthage, Tunisia. His research interests have covered several topics including communication networks, network engineering, optical networks, wireless networks, Internet work, and network security. He received his PhD in mathematics from the University of Paris XI and his PhD in computer science from the University of Tunis. Professor Boudriga is the recipient of the Presidential Award for Science and Research in Communication Technologies (Tunisia, 2004).

John Buford is a research scientist with Avaya Labs. Previously he was a lead scientist at the Panasonic Princeton Laboratory, VP of Software Development at Kada Systems, director of Internet Technologies at Verizon, and chief architect-OSS at GTE, Laboratories. Earlier he was tenured associate professor of computer science at the University of Massachusetts Lowell, where he also directed the Distributed Multimedia Systems Laboratory. He has authored or co-authored ninety refereed publications and the book *Multimedia Systems*. He is an IEEE senior member and is co-chair of the IRTF Scalable Adaptive Multicast Research Group. He holds the PhD from Graz University of Technology, Austria, and MS and BS degrees from MIT.

Mihaela Cardei is an assistant professor in the Department of Computer Science and Engineering at Florida Atlantic University, and director of the NSF-funded Wireless and Sensor Network Laboratory. Dr. Cardei received her PhD and MS in computer science from the University of Minnesota, Twin Cities, in 2003 and 1999, respectively. Her research interests include wireless networking, wireless sensor networks, network protocol and algorithm design, and resource management in computer networks. Dr. Cardei is a recipient of the 2007 Researcher of the Year Award at Florida Atlantic University. She is a member of IEEE and ACM.

Luca Caviglione (M.D. 2002, Ph.D. 2006) participated in several research projects funded by the EU, by ESA, and by Siemens COM AG. He is author and co-author of many academic publications about TCP/IP networking, P2P systems, QoS architectures, and wireless networks. He is a member of the Italian IPv6 Task Force and he participates in several TPCs and performance talks about IPv6 and P2P. He is with the Institute of Intelligent Systems for Automation (ISSIA) – Genoa Branch of the National Research Council of Italy.

Symeon Chatzinotas has a BSc in electrical and computer engineering from Aristotle University of Thessaloniki and a MSc in microwave engineering and wireless subsystem design from University of Surrey. Since 2006 he has been working on his PhD at the Centre for Communication Systems Research, University of Surrey. His current research interests include mobile networking, wireless security, and network information theory.

Hsiao-Hwa Chen is currently a full professor in Institute of Communications Engineering, National Sun Yat-Sen University, Taiwan. He received BSc and MSc degrees with the highest honor from Zhejiang University, China, and a PhD degree from the University of Oulu, Finland, in 1982, 1985 and 1990, respectively, all in electrical engineering. He worked with Academy of Finland as a Research Associate from 1991 to 1993 and the National University of Singapore as a lecturer and then a senior lecturer from 1992 to 1997. He joined Department of Electrical Engineering, National Chung Hsing University, Taiwan, as an associate professor in 1997 and was promoted to a full professor in 2000. In 2001 he joined National Sun Yat-Sen University, Tai-Wan, as a founding director of the Institute of Communications Engineering of the University. Under his strong leadership the institute was ranked second in the country in terms of SCI journal publications and National Science Council funding per faculty member in 2004. In particular, National Sun Yat-Sen University was ranked first in the world in terms of the number of SCI journal publications in wireless LANs research papers during 2004 to mid-2005, according to a research report (www.onr.navy.mil/scitech/special/354/technowatch/textmine.asp) released by The Office of Navel Research, USA. He was a visiting professor to the Department of Electrical Engineering, University of Kaiserslautern, Germany, in 1999, the Institute of Applied Physics, Tsukuba University, Japan, in 2000, Institute of Experimental Mathematics, University of Essen, Germany in 2002 (under DFG Fellowship), the Chinese University of Hong Kong in 2004, and the City University of Hong Kong in 2007.

Thomas Chen is an associate professor at Southern Methodist University. Prior to joining SMU, he worked on ATM research at GTE Laboratories (now Verizon). He has been the editor-in-chief of *IEEE Communications Magazine* since 2006. He also serves as senior technical editor for *IEEE Network*, and was the founding editor of *IEEE Communications Surveys*. He co-authored *ATM Switching Systems* (Artech House 1995). He received the IEEE Communications Society's Fred W. Ellersick Best Paper Award in 1996.

Yifan Chen received the BEng and PhD degrees in electrical and electronic engineering from Nanyang Technological University (NTU), Singapore, in 2002 and 2006, respectively. He is presently with the Biomedical Engineering Research Centre, NTU, as a research fellow. His current research interests involve ultra-wideband (UWB) radar system for biomedical applications including microwave imaging of human tissues and noncontact vital-signs monitoring, statistical modeling of mobile radio channels, UWB signal processing for wireless communications and geolocation systems, multiple-antenna system performance analysis, and wireless networks.

Zhijia Chen is currently a PhD student in Department of Computer Science and Technology, Tsinghua University. He is a visiting graduate student at School of Engineering of Stanford in Spring 2007. His research area is in P2P networking and media streaming. He has published four academic papers in area of P2P streaming, protocol modeling, and so forth. He is the International First Prize winner in American Mathematical Contest in Modeling (MCM 2004 Meritorious Winners). He is also the network session chair in 1st Beijing-Hong Kong Doctoral Forum on Network and Media 2006.

Yanghee Choi received BS in electronics engineering from Seoul National University, MS in electrical engineering from Korea Advanced Institute of Science, and Doctor of Engineering in computer science from Ecole Nationale Superieure des Telecommunications (ENST) in Paris, in 1975, 1977, and 1984, respectively. He was with the Electronics and Telecommunications Research Institute (ETRI) during 1977-1991. He is now leading the Multimedia and Mobile Communications Laboratory in Seoul National University. He is also director of Computer Network Research Center in Institute of Computer Technology (ICT). He is vice-president of Korea Information Science Society. His research interest lies in the field of multimedia systems and high-speed networking.

Mohammad M. R. Chowdhury is working toward the PhD degree in the University Graduate Center at Kjeller (UniK)/University of Oslo, Norway in the area of user mobility and service continuity. He received his MSc from Helsinki University of Technology in radio communications. His current areas of interest are identity and identity based service interactions, seamless user experience in heterogeneous wireless networks, and development of innovative service concepts for mobile operators.

Tomasz Ciszkowski received MSc degree in electronics and computer engineering from Faculty of Electronics and Information Technology of Warsaw University of Technology (WUT), Poland, in 2004. Currently, he is working toward a PhD degree in telecommunications at WUT on reputation service in anonymous ad hoc networks. Since 2004 he has been working for Polish Telecom in multimedia services division. His research activities are reflected in European research projects on next generation Internet (EuroNGI) and end-to-end QoS support over heterogeneous networks (EuQoS).

Amitabha Das obtained his BTech (honors) degree in electronics and electrical communication engineering from the Indian Institute of Technology, Kharagpur in 1985, and his PhD in computer engineering from the University of California, Santa Barbara, in 1991. Currently he is an associate professor in the School of Computer Engineering in Nanyang Technological University, Singapore. His research interests include wireless and mobile networks, network security, and intrusion detection. He is a senior member of IEEE.

Robert H. Deng received his Bachelor from National University of Defense Technology, China, and his MSc and PhD from the Illinois Institute of Technology. He has been with the Singapore Management University since 2004, and is currently a professor, associate dean for Faculty & Research, and director of SIS Research Center, School of Information Systems. Prior to this, he was principal scientist and manager of Infocomm Security Department, Institute for Infocomm Research, Singapore. He has 26 patents and more than 200 technical publications in international conferences and journals in the areas of computer networks, network security, and information security. He served as general chair, program committee chair, and member of numerous international conferences, including PC co-chair of the 2007 ACM Symposium on Information, Computer and Communications Security. He received the University Outstanding Researcher Award from the National University of Singapore in 1999 and the Lee Kuan Yew Fellow for Research Excellence from the Singapore Management University in 2006.

Mieso Denko is an associate professor of computing and information science at the University of Guelph, Ontario, Canada. He received his BSc degree in statistics and mathematics from Addis Ababa University. He received his MSc degree form the University of Wales, UK, and his PhD degree from the University of Natal, South Africa, both in computer science. His current research interests include wireless ad hoc networks, wireless mesh networks, wireless sensor networks, pervasive computing, and networking. He has published numerous research papers in international journals, conferences, workshops, and contributed to book chapters. Currently he is co-editing three books in the above areas. Dr. Denko has been actively involved in professional services as organizer or co-organizer of international conferences, symposiums, and workshops, as well as TPC member for a number of conferences and workshops. Among these, most recently he was the general co-chair of the IEEE PCAC-07, general vice-chair of ISPA-07 and program vice-chair of IEEE AINA-07. Currently he is a program vice-chair of the IEEE AINA-08, and co-organizer and program co-chair of the IEEE MHWMN-07 and IST-AWSN-07. Dr. Denko is a senior member of the ACM, a member of the IEEE, ACM SIGMOBILE, IEEE Communications Society, and IEEE Computer Society. Currently, he is an associate professor of computing and information science at the University of Guelph, Ontario, Canada.

Yacine Djemaiel holds a Master Degree in telecommunications and he is currently preparing his PhD thesis in telecommunications in the Engineering School of Communications (SUP'COM, Tunisia). He is conducting research activities in the area of intrusion detection and tolerance and digital investigation of security incidents. Since September 2006, Mr. Djemaiel has been a teacher assistant in telecommunications.

Felipe Espinosa got the BSc and MSc degrees in telecommunications from Polytechnics University of Madrid (Spain) in 1984 and 1991, respectively. He received the PhD degree in telecommunications from University of Alcala (Spain) in 1998. He was a lecturer from 1985 to 2000 and has been an associate professor since 2000, always in the Electronics Department at the University of Alcalá (Spain). His main research interests include electronic control and communication applied to cooperative guidance of robots and vehicles, as well as intelligent transportation systems.

Simone Fischer-Hübner has been a full professor at the Computer Science Department of Karlstad University since June 2000, where she is the head of the PriSec (Privacy & Security) research group. She received Doctoral (1992) and Habilitation (1999) degrees in computer science from Hamburg University. Her research interests include technical and social aspects of IT-security, privacy, and privacy-enhancing technologies. She was a research assistant/assistant professor at Hamburg University (1988-2000) and a guest professor at the Copenhagen Business School (1994-1995) and at Stockholm University/Royal Institute of Technologies (1998-1999).

J. Antonio Garcia-Macias holds a PhD from the Institut National Polytechnique de Grenoble (INPG), France. He is currently a researcher at CICESE Research Center, working in the Computer Science Department. His current research interests are wireless (ad hoc and sensors) networks, ubiquitous computing, next-generation Internet services and protocols, and distributed collaborative systems.

Kaj J. Grahn, Dr. Tech. from Helsinki University of Technology, is presently a senior lecturer in telecommunications at the Department of Business Administration, Media, and Technology at Arcada Polytechnic, Helsinki, Finland. His current research interests include mobile and wireless networking and network security.

Stefanos Gritzalis holds a BSc in physics, an MSc in electronic automation, and a PhD in informatics, all from the University of Athens, Greece. Currently he is an associate professor, the head of the Department of Information and Communication Systems Engineering, University of the Aegean, Greece, and the director of the Laboratory of Information and Communication Systems Security (Info-Sec-Lab). His published scientific work includes several books and more than 150 journal and international conference papers. The focus of these publications is on information and communication systems security. He was a member (secretary general, treasurer) of the Board of the Greek Computer Society.

Yong Guan is an assistant professor in the Department of Electrical and Computer Engineering at Iowa State University. He received his BS (1990) and MS (1996) in computer science from Peking University, China, and his PhD (2002) in computer science from Texas A&M University. His research interests are computer and network forensics, wireless and sensor network security, and privacy-enhancing technologies for the Internet. He received the Best Student Paper Award from the IEEE National Aerospace and Electronics Conference in 1998, won 2nd place in the graduate category of the International ACM Student Research Contest in 2002, and was named the Litton Assistant Professor by Iowa State University in 2007.

Mohamed Hamdi received his Engineering Diploma, Master Diploma, and PhD in telecommunications from the Engineering School of Communications (Sup'Com, Tunisia) in 2000, 2002, and 2005, respectively. From 2001 to 2005 he worked for the National Digital Certification Agency (NDCA, Tunisia) where he was head of the Risk Analysis Team. Dr. Hamdi was in charge in building the security strategy for the Tunisian Root Certification Authority and in continuously assessing the security of the

NDCAs networked infrastructure. He has also served on various national technical committees for securing e-government services. Currently, Dr. Hamdi is serving as a contract lecturer for the Engineering School of Communications at Tunis. He is also a member of the Communication Networks and Security Lab (Coordinator of the Formal Aspects of Network Security Research Team), where he is conducting research activities in the areas of risk management, algebraic modeling, relational specifications, intrusion detection, network forensics, and wireless sensor networks

Munirul M. Haque is currently a PhD student at Purdue University. He received the MS degree in computer science at Marquette University where he researched pervasive computing, security, and privacy in the Ubicomp Research Lab. He completed the BS in computer science and engineering from Bangladesh University of Engineering and Technology.

Jahan Hassan is a research fellow at the School of Information Technologies, University of Sydney. She received her PhD in 2004 from University of New South Wales, Sydney, and Bachelor degree in 1995 from Monash University, Melbourne, both in computer science. She is published widely in peer-reviewed conferences and journals. She was a member of the Technical Program Committee of IEEE LCN 2006, IEEE ICC 2007, IEEE ISWPC 2007, IADIS AC 2006, and IADIS WAC 2007. She served as a reviewer for many conferences and journals. Her research interests include mobile and wireless networking architectures and wireless network security. Her current project focuses on the fast authentication techniques for multiprovider access networks.

Artur Hecker received a diploma in computer science (Dipl.inform.) from the University of Karlsruhe (TH), Germany in 2001. In 2005, he received a PhD degree in computer science and networking from the ENST, France. After his thesis, he worked as CTO of Wavestorm SAS, which he co-founded in 2003. Since 2006, Dr. Hecker holds a position as associate professor at the INFRES department at the ENST. His present research interests are wireless access security, security assurance of complex systems, network and service management, and autonomous networking. Dr. Hecker is actively involved in several IST FP6 and EUREKA CELTIC research activities.

Silke Holtmanns received her PhD in mathematics from the University of Paderborn (Germany), Department of Computer Science and Mathematics. She has been a senior researcher at Nokia Research Center since 2004. Before that, she was working in Ericsson Research Lab Aachen (Germany) as a master research engineer and at the University of Paderborn as a scientific assistant. She has more then 30 publications and co-authored several books on mobile security. She is also rapporteur of six 3GPP security specifications and reports and involved in various standardization activities.

Ismail Khalil Ibrahim is a senior researcher and lecturer at the Institute of Telecooperation- Johannes Kepler University Linz, Austria, where he teaches, consults, and conducts research in mobile multimedia applications and services, agent technologies, and information integration. He received his MSc and PhD in computer engineering and information systems from Gadja Mada University, Indonesia. Dr. Ibrahim previously served as a research fellow at Intelligent Systems Group in the Netherlands and as project manager at the Software Competence Center Hagenberg, Austria. He is the editor-in-chief of *Advances in Next Generation Mobile Multimedia* book series and *Journal of Mobile and Multimedia Communications*, and co-editor in chief of the *International Journal of Web Information Systems* (*JWIS*) and the *Journal of Mobile Multimedia (JMM)*. His research interests also include business, social, and policy implications associated with the emerging Web technologies.

Biju Issac is a lecturer in the School of IT and Multimedia in Swinburne University of Technology (Sarawak Campus), Malaysia. He is also the head of Network Security Research Group in the Information Security Research Lab at Swinburne University Sarawak. He is an electronics and communication engineer with a post graduate degree in computer applications. Currently he is doing part-time PhD in networking and mobile communications in UNIMAS, Malaysia. His research interests are in wireless and network security, wireless mobility, and IPv6 networks.

Tao Jiang is a research scientist at the Department of Electronic and Computer Engineering, University of Michigan, Dearborn. He received BS and MS degrees in applied geophysics from China University of Geosciences, Wuhan in 1997 and 2000, respectively, and a PhD degree in information and communication engineering from Huazhong University of Science and Technology, Wuhan, P. R. China in April 2004. From August 2004 to August 2005, he worked at Brunel University, London, as an academic visiting scholar, and then moved to University of Puerto Rico in 2006. His current research interests include the areas of wireless communications and corresponding signal processing, especially for OFDM and MIMO systems, cooperative networks, cognitive radio, and ultra wideband communications.

John Felix Charles Joseph is currently pursuing PhD in computer science from Nanyang Technological University, Singapore. His research interests include security in wireless and ad hoc networks, computational intelligence, multicast routing security, and multimedia. He received his Bachelor in engineering, computer science from Madras University, India in 2002 and MS from Anna University, India in 2005. His current work involves design of an intrusion detection algorithm for mobile wireless ad hoc network environment.

Admela Jukan is a professor in electrical and computer engineering at the Technical University Carolo Wilhelmina in Braunschweig, Germany. Prior to coming to TU Braunschweig, she was with University of Illinois at Urbana Champaign (UIUC), Georgia Tech (GaTech), University of Quebec (EMT-INRS), and Vienna University of Technology (TU Wien). From 2002-2004, she served as program director in computer and networks system research at the National Science Foundation (NSF) in Arlington, VA. While at NSF, she was responsible for funding and coordinating US-wide university research and education activities in the area of network technologies and systems. She received the MSc degree in information technologies and computer science from the Polytechnic of Milan, Italy, and the PhD degree (cum laude) in electrical and computer engineering from the Vienna University of Technology (TU Wien), Austria. Dr. Jukan is the author of numerous papers in the field of networking, and she has authored and edited several books. She serves as a member of the Quality Assurance Committee for the EU Network of Excellence, ePhoton/One. Dr. Jukan has chaired and co-chaired several international conferences, including IFIP ONDM, IEEE ICC, and IEEE GLOBECOM. She serves on the editorial board of the IEEE Communications Surveys and Tutorials. She is a senior member of the IEEE.

György Kálmán is a graduate student at UniK, University Graduate Center in Kjeller, Norway. His research area covers personal and device authentication, security, and privacy in wireless systems. He got his MSc degree in the area of communication networks from the Budapest University of Technology and Economics. He was research fellow at Telenor R&I at the Media Platforms group.

Georgios Kambourakis received the diploma in applied informatics from the Athens University of Economics and Business and the PhD in information and communication systems engineering from the Department of Information and Communications Systems Engineering of the University of Aegean. He also holds a MEd from Hellenic Open University. Dr. Kambourakis is a lecturer in the Department of Information and Communication Systems Engineering of the University of the Aegean, Greece. His research interests are in the fields of mobile and ad hoc networks security, VoIP security, security protocols, and PKI, and he has more than 35 publications in the above areas.

Jonny Karlsson has a BSc in information technology from Arcada Polytechnic, Helsinki Finland. Since May 2002 he has been working at Arcada Polytechnic as a course assistant and course teacher in programming and network security related courses and as a research assistant. His current research interests include wireless and mobile network security.

Paris Kitsos received the BSc degree in physics in 1999 and a PhD in 2004 from the Department of Electrical and Computer Engineering, both at the University of Patras. Currently is research fellow with the Digital Systems & Media Computing Laboratory, School of Science & Technology, Hellenic Open University (HOU). His research interests include VLSI design, hardware implementations of cryptographic algorithms, and security protocols for wireless communication systems. Dr. Kitsos has published more than 60 scientific articles and technical reports, as well as is reviewing manuscripts for books, international journals, and conferences/workshops in the areas of his research. He has participated in international journals and conferences organization, as program/technical committee member and guest editor.

Giorgos Kostopoulos received his diploma in electrical and computer engineering from the Electrical & Computer Engineering Department, University of Patras, Greece in 2003. Since then he has been working as a researcher engineer in the Department of Electrical and Computer Engineering of the University of Patras. His research interests include security in wireless networks, new generation networks architectures, security management in new generation networks, and communication networks. Giorgos Kostopoulos has published more than 15 technical papers and book chapters in these areas. He has also participated as senior engineer in European Research Projects.

Zbigniew Kotulski received his MSc in applied mathematics from Warsaw University of Technology and PhD and DSc degrees from Institute of Fundamental Technological Research of the Polish Academy of Sciences. He is currently professor at IFTR PAS and professor and head of Security Research Group at Department of Electronics and Information Technology of Warsaw University of Technology, Poland. He is the author and co-author of three books and more than 150 research papers on applied mathematics, cryptology, and information security.

Odysseas Koufopavlou received the Diploma of Electrical Engineering in 1983 and the PhD degree in electrical engineering in 1990, both from University of Patras, Greece. From 1990 to 1994 he was at the IBM Thomas J. Watson Research Center, Yorktown Heights, NY. He is currently a professor with the Department of Electrical and Computer Engineering, University of Patras. His research interests include computer networks, high performance communication subsystems architecture and implementation, VLSI low power design, and VLSI crypto systems. Dr. Koufopavlou has published more than 150 technical papers and received patents and inventions in these areas. He has participated as coordinator or partner in many Greek and European R&D programs. He served as general chairman for the IEEE ICECS'1999.

Geng-Sheng (G.S.) Kuo worked with R&D laboratories of the communications industry in the United States, such as AT&T Bell Laboratories. In August 2000, he joined National Chengchi University, Taipei, Taiwan as a professor. Since 2001, he has been invited as chair professor of Beijing University of Posts and Telecommunications (BUPT) in Beijing, China. His current research interests include mobile communications, wireless communications, and IP-networks. From 2001 to 2002, he was editor-in-chief of *IEEE Communications Magazine*, whose impact factor in 2002 was 3.165. Currently, he is area editor for *Networks Architecture of IEEE Transactions on Communications*, editor and ComSoc representative to *IEEE Internet Computing*, editor of *European Transactions on Telecommunications*, and so forth.

Taekyoung Kwon is an assistant professor in Multimedia & Mobile Communications Lab., School of Computer Science and Engineering, Seoul National University He received his PhD, MS, and BS degrees in computer engineering from Seoul National University in 2000, 1995, and 1993, respectively. He was a visiting student at IBM T. J. Watson Research Center in 1998 and a visiting scholar at the University of North Texas in 1999. His recent research areas include radio resource management, wireless technology convergence, mobility management, and sensor network.

Pekka Laitinen received his MSc degree in information sciences in Helsinki University of Technology, Department of Engineering Physics and Mathematics. He is principal engineer in Nokia Research Center where he has been working since 1996. His research interests include identity management and applied security.

Björn Landfeldt received a BSc equivalent from the Royal Institute of Technology in Sweden. He received his PhD from The University of New South in 2000. Afterwards he joined Ericsson Research in Stockholm as a Senior Researcher. In 2001, Dr. Landfeldt took up a position as a CISCO senior lecturer in Internet Technologies at the University of Sydney. He has published more than 50 publications in international books, journals, and conferences. Dr. Landfeldt is serving on the editorial boards of international journals and as a program committee member of many international conferences. His research interests include wireless systems, systems modeling, mobility management, and QoS.

Peter Langendoerfer received his doctoral degree in 2001. Since 2000 he has been with the IHP in Frankfurt (Oder) where he is leading the mobile middleware group. He has published more than 55 refereed technical articles, filed seven patents in the security/privacy area, and worked as guest editor for the *Journal of Super Computing* (Kluwer), *Computer Communications* (Elsevier), *Wireless Communications and Mobile Computing* (Wiley), and *ACM Transactions on Internet Technology*. He is/was also a TPC member/chair of many conferences. His research interests include mobile communication (especially privacy and security issues), protocol engineering, and automated protocol implementation.

Shahram Latifi, an IEEE fellow, received the Master of Science degree in electrical engineering from Fanni, Teheran University, Iran in 1980. He received the Master of Science and the PhD degrees both in electrical and computer engineering from Louisiana State University, Baton Rouge in 1986 and

1989, respectively. He is currently a professor of electrical engineering at the University of Nevada, Las Vegas and director of the Center for Information and Communication Technologies (CICT). Dr. Latifi has designed and taught graduate courses on security, image processing, computer networks, fault tolerant computing, and data compression in the past 16 years. He has given seminars on the aforementioned topics all over the world. He has authored over 120 technical articles in the areas of image processing, document analysis, computer networks, fault tolerant computing, parallel processing, and data compression. His research has been funded by NSF, NASA, DOE, Boeing, Lockheed, and Cray Inc. Dr. Latifi is an associate editor of the IEEE Transactions on Computers and co-founder and general chair of the IEEE International Conference on Information Technology. He is also a registered professional engineer in the State of Nevada.

Bu-Sung Lee received his BSc (honors) and PhD from the Electrical and Electronics Department, Loughborough University of Technology, UK in 1982 and 1987, respectively. He is currently associate chair (research) with the School of Computer Engineering, Nanyang Technological University. He is also the founding president of Singapore Research and Education Networks (SingAREN). He has been an active member of several national standards organization such as the National Grid Pilot Project. His research interests are in network management, broadband, distributed, ad hoc and mobile networks, network optimization, as well as grid computing.

Supeng Leng is an associate professor in the School of Communication and Information Engineering, University of Electronic Science and Technology of China (UESTC). He received his BEng degree from UESTC in 1996, and PhD degree from Nanyang Technological University (NTU), Singapore in 2005. He has experience as a R&D engineer in the field of computer communications, and as a research fellow in the Network Technology Research Center, NTU. His research focuses on ad hoc/sensor networks, wireless mesh networks, and broadband wireless networks.

Mo Li received the BE from Beijing University of Posts and Telecommunications. Then, he worked for Lucent Technologies and Computer Associates (CA), where he has been involved in the design of system architectures for DWDM/SDH/IP Backbone O&M systems. He is currently working toward the PhD at the Faculty of Engineering, University of Technology, Sydney. His research interests include handover management and trust-assisted networking.

Xinghua Li obtained his ME and Ph D degrees in computer architecture and computer application from Xidian University (Xi'an) in 2004 and 2006, respectively. Currently, Xinghua Li is the lecturer of the School of Computer of Xidian University. His research interests include information and network security.

Zheng-Ping Li received the BE degree at Department of Electronics and Information, Lanzhou Railway University, Lanzhou, China, in 2000. He is currently working toward the PhD degree in the School of Telecommunication Engineering, Beijing University of Posts and Telecommunications, Beijing, China. His research interests include medium access control, routing, and intrusion detection in wireless mesh networks.

Shiguo Lian, member of IEEE, SPIE, and EURASIP, got his PhD degree in multimedia security from Nanjing University of Science and Technology in July 2005. He was a research assistant at City University of Hong Kong from March to June in 2004, studying on multimedia encryption. He has being with France Telecom R&D Beijing since July 2005, focusing on multimedia content protection, including digital rights management (DRM), image or video encryption, watermarking and authentication, and so forth.

Chuang Lin is a professor and the former head of the Department of Computer Science and Technology, Tsinghua University, Beijing, China. He received his PhD degree in computer science from Tsinghua University in 1994. Professor Lin is a senior member of the IEEE, the Chinese Delegate in TC6 of IFIP, and has served as associate editor for several journals. His current research interests include computer networks, performance evaluation, logic reasoning, and Petri net theory and its applications. He has co-authored more than 200 papers in research journals and IEEE conference proceedings in these areas and has published three books.

Bin Lu is an assistant professor in the Department of Computer Science at West Chester University of Pennsylvania. Dr. Lu received her BS (1996) and MS (1998) degrees in computer science from Harbin Institute of Technology, China, and her PhD (2005) in computer science from Texas A&M University. Her research interests include network security, quality of service, and wireless networks.

Jianfeng Ma received his BS degree in mathematics from Shaaxi Normal University (Xi'an) in 1985, and obtained his ME and PhD degrees in computer software and communications engineering from Xidian University (Xi'an) in 1988 and 1995, respectively. Professor Ma is a member of the executive council of the Chinese Cryptology Society. Currently, Professor Ma is the director of the Ministry of Education Key Laboratory of Computer Networks and Information Security, and he is the dean of the School of Computer of Xidian University. His research interests include information security, coding theory, and cryptography.

Ismat K. Maarouf obtained his BS degree in computer engineering in 2005 and an MS degree in computer networks in 2007 from King Fahd University of Petroleum and Minerals (KFUPM), Dhahran, Saudi Arabia. He is currently working as a research assistant in the Computer Engineering Department in KFUPM. His main research interests include mobile ad hoc and wireless sensor networks, computer networks security, reputation systems, and WLAN-Cellular networks integration.

Michael Maaser received his Master's degree in computer science from Brandenburg University of Technology Cottbus in 2004. After his thesis about negotiation of privacy he started as a research scientist at IHP. His research focuses on privacy preserving techniques mainly, but not limited to, the field of location based services. Throughout the recent 2 years he has seven publications thereof five in the area of privacy and two filed patents.

Ashraf S. Hasan received the BSc degree in electrical and computer engineering from Kuwait University in 1990, and the MEng in engineering physics (computer systems) from McMaster University, Hamilton, Canada in 1992. He received his PhD in systems and computer engineering from Carleton University, Ottawa, Canada in 1997. During 1997-2002, he was with Nrtel Networks Research and Development where he focused on development and evaluation of radio resource management algorithms for broadband and 3G networks. Since 2002, he has been with the Computer Engineering Department at King Fahd University of Petroleum and Minerals, Dhahran, KSA as an assistant professor. His research interests include radio resource management for 3rd and 4th G networks, wireless local area networks, and integration of heterogeneous networks.

Amel Meddeb Makhlouf received the engineering eegree (in 2001) and the Master degree in communications (in 2003) from the Engineering School of Communications (SUP'COM, Tunisia). She is member of the Communication Networks and Security (CN&S) Research Laboratory (University of November 7th, Carthage, Tunisia). Since September 2004, she has joined the Engineering School of Communications (SUP'COM, TUNISIA) as a teacher assistant in telecommunications.

Leonardo A. Martucci is a doctoral student at Karlstad University, Sweden, where he works with research on privacy enhancing technologies for wireless environments. He is involved in education, research, deployment, and industrial projects in the field of wireless network security and privacy since 2001. Mr. Martucci's research is focused especially in privacy problems in dynamic and distributed environments, such as mobile ad hoc networks. He holds a Licentiate in engineering from Karlstad University (2006), a Masters in electrical engineering (2002), and an electrical engineer degree (2000) from University of São Paulo, Brazil.

Geyong Min is a senior lecturer in the Department of Computing at University of Bradford, United Kingdom. He received the PhD degree in computing science from University of Glasgow, UK, in 2003. His research interests include performance modeling and simulation, network traffic engineering, mobile computing and wireless networks, multimedia systems, and information security. Dr. Min has published over 100 research papers in the well-established journals and conferences. Dr. Min serves on the editorial board of the *International Journal of Wireless and Mobile Computing* and *Journal of Simulation Modeling Practice and Theory*, and serves as the guest editor for 10 international journals.

Lawan A. Mohammed is currently an assistant professor in Computer Science and Engineering Technology Department at King Fahd University of Petroleum and Minerals (HBCC Campus), Saudi Arabia. His main research interests are in the design of authentication protocols for both wired and wireless networks, wireless mobility, group oriented cryptography, smartcard security, and mathematical programming.

Rebecca Montanari graduated from University of Bologna, Italy, where she received PhD degree in computer science engineering in 2001. She is now an associate professor of computer engineering at the University of Bologna. Her research primarily focuses on policy-based networking and systems/service management, mobile agent systems, security management mechanisms, and tools in both traditional and mobile systems. She is member of IEEE and AICA.

Luminita Moraru is currently a PhD candidate in the TCS-sensor lab of the Computer Science Department of the University of Geneva. She received a BS degree in electrical engineering and computer science from the Polytechnic University of Bucharest, in 2004, and a MS degree in computer science (embedded systems) from the University of Science and Technology of Lille, in 2005. Her research interests are in sensor networks, mobile ad hoc networks, security, and reputation based trust. Her current research focuses on security and QoS of routing protocols for sensor networks.

A. R. Naseer is an assistant professor in Department of Computer Engineering at King Fahd University of Petroleum and Minerals, Dhahran, KSA. He received the PhD degree in computer science and engineering from Indian Institute of Technology (IIT), Delhi, India, in 1996. He is a recipient of "Best Student Paper Award" at the IEEE/ACM 7th International Conference on VLSI Design, 1994 for his doctoral research paper in the area of FPGA based synthesis. His current research interests include wireless sensor networks security, reputation systems, computer networks, design automation of digital systems, FPGA based synthesis, computer architecture, parallel computing, and multicore processor architectures. He has published several refereed journal and conference papers on related topics.

Huansheng Ning received BS degree from Anhui University, China, in 1996, and a PhD degree from Beihang University, China, in 2001. From 2002 to 2003, he was the CTO of Aerospace Golden Card Company. Since 2004, he has been an associate professor in Beihang University. His current research interests include RFID, EM computing, ITS, and so forth.

Josef Noll holds a professor stipend from the University of Oslo in the area of mobile services. Working areas include mobile authentication, wireless broadband access, personalized services, and the evolution to 4G systems. He is also senior advisor in Movation, Norway's leading innovation company for mobile services. Previously he was senior advisor/group leader at Telenor R&I, project leader of "Operators' Vision on Systems Beyond 3G" and other international projects, use-case leader in the EU "Adaptive Services Grid (ASG)" project, and has initiated a.o. the EU's 6th FP ePerSpace and several ITEA and Eurescom projects.

Christoforos Ntantogian received his BSc degree in computer science and telecommunications from the Department of Informatics and Telecommunications, University of Athens, Greece. In 2006 he finished his postgraduate studies in computer systems technology in the same department and currently he is a PhD student. Since 2004 he has been working for the Communication Networks Laboratory of the University of Athens and he is a member of the Security Group.

Sangheon Pack received the BS (2000) and PhD (2005) degrees from Seoul National University, both in computer engineering. Since March 2007, he has been an assistant professor in the School of Electrical Engineering, Korea University, Korea. From July 2006 to February 2007, he was a postdoctoral fellow at Seoul National University. From 2005 to 2006, he was a postdoctoral fellow in the Broadband Communications Research (BBCR) Group at University of Waterloo, Canada. His research interests include mobility management, multimedia transmission, and QoS provision issues in next-generation wireless/mobile networks. He is a member of the ACM and the IEEE.

Luis E. Palafox received his BS in computer engineering from the University of Baja California in 1997. He also received his MS degree in digital systems from the National Polytechnic Institute of Mexico in 2002. In 2004, he enrolled in the PhD program in computer science program at the CICESE Research Center in Ensenada. He is a faculty member of the School of Chemical Science and Engineering at the University of Baja California since 1999. His areas of interest are computer networking, embedded systems, wireless sensor networks, and digital signal processing.

Cyrus Peikari, MD, is a practicing physician and author of several leading technical security books, including *Security Warrior* from O'Reilly and *Maximum Wireless Security* from SAMS. In his work with Airscanner Corporation he pioneered some of the first antivirus solutions for handheld computing devices. His main area of research is in reverse engineering of "airborne viruses." Dr. Peikari has been a popular speaker and keynote at several major security conferences.

Steffen Peter received his diploma in computer science from the Brandenburg University of Technology at Cottbus (BTU) in 2006. In 2006 he joined the IHP in Frankfurt (Oder), where he was also involved in developing a hardware TCP accelerator as a student. In his diploma thesis he was developing hardware cryptography accelerators. He is a member of the mobile middleware group, working on the research of solutions for security issues in wireless sensor networks. He has filed three patents and has authored two technical papers. His research interests include security and privacy in mobile environments focusing on efficient hardware implementation.

Krzysztof Piotrowski received his Master in computer science from the University of Zielona Gora (Poland) in 2004. Since 2004, he has been with the IHP in Frankfurt (Oder) where he is a member of the mobile middleware group. He published 15 refereed technical articles in the area of security and privacy. His research interests include mobile/wireless communication (focus on privacy and security issues), especially on resource-constrained devices (wireless sensor networks).

Olivier Powell is a senior researcher at the Computer Science Department of the University of Geneva in Switzerland. He was previously a Swiss National Research Foundation fellow at the Research and Academic Computer Technology Institute and the University of Patras in Greece. Previously, he was a post-doctoral research associate at the TCS-sensor lab of the University of Geneva. He received a PhD in computer science in the field of complexity theory from the University of Geneva and a MSc degree in mathematics from the same university. His current research interest is algorithmic aspects of wireless sensor networks.

Göran Pulkkis, Dr. Tech. from Helsinki University of Technology, is presently senior lecturer in computer science and engineering at the Department of Business Administration, Media, and Technology at Arcada Polytechnic, Helsinki, Finland. His current research interests are network security, applied cryptographic, and quantum informatics

Slim Rekhis holds a PhD and a Master degree in telecommunications from the Engineering School of Communications (SUP'COM, Tunisia). He is conducting research activities in the area of digital investigation of security incidents, formal modelling, intrusion detection and tolerance, and wireless security. Since September 2005, Dr. Rekhis has been an assistant professor in telecommunications.

Angelos Rouskas received the Diploma in Electrical Engineering from the National Technical University of Athens (NTUA), the MSc in communications and signal processing from Imperial College, London, and the PhD in electrical and computer engineering from NTUA. He is an assistant professor in the Department of Information and Communication Systems Engineering of the University of Aegean, Greece, and director of the Computer and Communication Systems Laboratory. Dr. Rouskas has been involved in several European and Greek funded research projects and has published extensively in the field of mobile and wireless communication networks.

Miguel A. Ruiz was born in Valdepeñas (Ciudad Real), Spain. He received the Technical Telecommunication Engineering and Telecommunication Engineering degrees from the Polytechnic School at the University of Alcala (Madrid), Spain, in 1999 and 2003, respectively. He is currently working toward the PhD degree in telecommunications at University Alcala. Since 2000, he has been working in the Electromagnetic Compatibility Laboratory as technical manager at the High Technology and Homologation Center (CATECHOM), research support center of the University Alcala. Furthermore, he is an assistant lecturer at the Electronic Department of the same university. His main research interest is EMC effect on electrical and electronic automotive systems.

Kumbesan Sandrasegaran holds a PhD in electrical engineering from McGill University (Canada) (1994), a Masters of Science degree in telecommunication engineering and information Systems from Essex University (UK) (1988), and a Bachelor of Science (honors) degree in electrical engineering (first class) (UZ) (1985). Dr Sandrasegaran is a professional engineer (Pr.Eng) (ECSA) and has more than 20 years experience working either as a practitioner, researcher, consultant, and educator in telecommunication networks. During this time, he has focused on the planning, optimization, forecasting, security, and network management of telecommunication networks. At present, he is program head of ICT Engineering at the Faculty of Engineering, University of Technology Sydney (UTS).

David Sanguino was born in Talavera de la Reina (Toledo), Spain. He received the technical telecommunication engineering degree from the Polytechnic School at the University of Alcala (Madrid), Spain, in 2004. He is currently working toward the telecommunication engineering degree at University Alcala (UAH). Since 2005, he has been working in the Electromagnetic Compatibility Laboratory as Technician at the High Technology and Homologation Center (CATECHOM), research support center of the University of Alcala.

Boot-Chong Seet received his PhD in 2005 from the School of Computer Engineering, Nanyang Technological University (NTU), where he is currently serving as an instructional faculty. Prior to joining NTU, he was with the Singapore-MIT Alliance (SMA), National University of Singapore, where he worked as a research fellow for a pilot project on adaptive location-aware computing. His current research interests include ad hoc, mesh, and sensor networks, mobile peer-to-peer computing, vehicular communications, and emerging broadband wireless technologies. He has over 20 refereed publications and one patent pending. He is a member of IEEE and ACM SIGMOBILE.

Jean-Marc Seigneur is a senior researcher and lecturer at the University of Geneva. He received his MSc and PhD in computer science from Trinity College Dublin. His more than 30 international scientific publications cover ubiquitous computing security, trust, reputation, and privacy. He is an international expert reviewer for French ANR security research projects and the European Commission. He worked in Hewlett-Packard in France and China. He leads the http://www.trustcomp.org online community on computational trust management with now more than 190 academic and industrial members. He has provided technical consulting and presentations to many companies, among them, Philips, Ericsson, SAP, and Amazon.

Moushumi Sharmin is currently a PhD student at University of Illinois. She received the MS degree in computer science at Marquette University where she researched pervasive computing, security, and privacy in the Ubicomp Research Lab. She completed the BS in computer science and engineering from Bangladesh University of Engineering and Technology.

Nicolas Sklavos received the PhD degree in electrical and computer engineering, and the diploma in electrical and computer engineering, in 2004 and 2000, respectively, both from the Electrical & Computer Engineering Department, University of Patras, Greece. His research interests include cryptography, wireless communications security, computer networks, and VLSI design. He holds an award for his PhD thesis on "VLSI Designs of Wireless Communications Security Systems" from IFIP VLSI SOC 2003. He was the general co-chair of MobiMedia'07. He has participated to international journals and conferences organization as program committee member and guest editor. Dr. N. Sklavos is a member of the ACM, IEEE, IEE, the Technical Chamber of Greece, and the Greek Electrical Engineering Society. He has authored or co-authored up to 90 scientific articles, books chapters, tutorials, and reports in the areas of his research.

Nilothpal Talukder is a graduate student in computer science at Marquette University where he researches pervasive computing, security, and privacy in the Ubicomp Research Lab. He completed the BS in computer science and engineering from Bangladesh University of Engineering and Technology.

Daniela Tibaldi graduated from University of Bologna, Italy, where she received her PhD degree in computer science engineering in 2006. Her research activity is focused on middleware solutions for supporting the secure service provisioning in mobile and heterogeneous environments. Since 2002 she works at the DSAW – Direction and Development of Web Activities of the University of Bologna with both technical and quality management responsibilities. One of the DSAW main tasks is to build the University Web sites, services, and the corresponding technological, informative, and organizational infrastructure to fully support University educational, academic, and administrative activities.

Tom Tofigh is a principal and technical member of the AT&T architecture team. He is responsible for architecture studies and vendors technology evaluation. Currently, he supports the AT&T labs advanced services and architecture group. Tom has worked in semiconductor companies as director of product management, director of software development, and has consulted and worked for a number of start-ups and had responsibility for architecture and developments of switches and access products. In addition Tom attended George Washington University and completed his doctoral course work in electrical engineering and computer science graduate school. Furthermore, Tom has a judicial doctoral degree from Northern Virginia Law School with emphasis in intellectual properties. Currently, Tom is the founder and chair of the WiMAX Forum's Application Architecture Working Group.

Alessandra Toninelli graduated from University of Bologna, Italy, where she is currently a PhD student in computer science engineering. Her research interests focus on semantic-based middleware supports for service provisioning, context-aware services, security solutions for pervasive environments, policy-based service management, and mobile agent systems. She is a member of IEEE and ACM.

Denis Trček is principal investigator at Jozef Stefan Institute and has been involved in the field of computer networks, security, and privacy for almost 20 years. He has taken part in various European projects, as well as domestic projects in government, banking, and insurance sectors. His bibliography includes over one hundred titles, including works published by renowned publishers like Springer and Wiley. D. Trcek has served (and still serves) as a member of various international boards, from editorial to professional ones. He is inventor of a patented family of light-weight cryptographic protocols. His interests include e-business, security, trust management, privacy, and human factor modelling.

Yu Wang received the PhD degree in computer science from Illinois Institute of Technology in 2004, and the BEng degree and the MEng degree in computer science from Tsinghua University, China, in 1998 and 2000. He has been an assistant professor of computer science at the University of North Carolina at Charlotte since 2004. His current research interests include wireless networks, ad hoc and sensor networks, mobile computing, and algorithm design. He has published more than 50 papers in peer-reviewed journals and conferences. Dr. Wang is a recipient of Ralph E. Powe Junior Faculty Enhancement Awards from Oak Ridge Associated Universities.

Yawen Wei is a PhD candidate in the Department of Electrical and Computer Engineering at Iowa State University. She obtained her BEng (2004) in electronic engineering from Tsinghua University, China. Since then she has been doing research on localization security issues and location-based services in wireless sensor networks.

Bing Wu is an assistant professor in the Department of Mathematics and Computer Science at Fayetteville State University. Dr. Wu received his PhD and MS in the Department of Computer Science and Engineering at Florida Atlantic University. His research interests include wireless ad hoc and sensor networks, mobile computing, and network security. He has worked as research assistant for four years at Motorola. He has published more than ten papers including refereed journal, book chapter, and conference proceedings. He is a member of IEEE.

Jie Wu is a distinguished research professor at the Department of Computer Science and Engineering, Florida Atlantic University and a program director at US National Science Foundation. He has published over 350 papers in various journals and conference proceedings. His research interests are in the areas of wireless networks and mobile computing, routing protocols, fault-tolerant computing, and interconnection networks. Dr. Wu was on the editorial board of *IEEE Transactions on Parallel and Distributed Systems* and was a co-guest-editor of *IEEE Computer and Journal of Parallel and Distributed Computing*. He served as the program co-chair for MASS 2004, program vice-chair for ICDCS 2001, and program vice-chair for ICPP 2000. He was also general co-chair for MASS 2006 and is general chair for IPDPS 2008. He is the author of the text *Distributed System Design* published by the CRC press. He was also the recipient of the 1996-97, 2001-02, and 2006-07 Researcher of the Year Award at Florida Atlantic University. Dr. Wu has served as an IEEE Computer Society Distinguished Visitor and is the chairman of IEEE Technical Committee on Distributed Processing (TCDP). He is a member of ACM and a senior member of IEEE.

Christos Xenakis received his BSc degree in computer science in 1993 and his MSc degree in telecommunication and computer networks in 1996, both from the Department of Informatics and Telecommunications, University of Athens, Greece. In 2004 he received his PhD from the University of Athens (Department of Informatics and Telecommunications). Since 1996 he has been a member of the Communication Networks Laboratory of the University of Athens and, currently, he is the head of the Security Group. In addition, he is a lecturer (faculty of the Department of Technology Education and Digital Systems) in the University of Piraeus, Greece.

Lu Yan is a research fellow at University College London and a Visiting Fellow at University of Cambridge. Previously, he was with Department of Information Technologies in Åbo Akademi University, Distributed Systems Design Laboratory in Turku Centre for Computer Science (TUCS), Institute of Microelectronics (IME) in Peking University. He holds visiting professor positions in both École Supérieure d'Ingénieurs généralistes (ESIGELEC) and École Supérieure de Commerce de Rouen (ESC).

Laurence T. Yang is a professor in computer science at St Francis Xavier University, Canada. His research includes high performance computing and networking, embedded systems, ubiquitous/pervasive computing, and intelligence. He has published around 250 papers in refereed journals, conference proceedings, and book chapters in these areas. He has been involved in more than 100 conferences and workshops as a program/general conference chair and more than 200 conference and workshops as a program committee member. He served as the vice-chair of IEEE Technical Committee of Supercomputing Applications (TCSA) until 2004. Currently he is on the executive committee of IEEE Technical Committee of Scalable Computing (TCSC), of IEEE Technical Committee of Self-Organization and Cybernetics for Informatics, of IFIP Working Group 10.2 on Embedded Systems, and of IEEE Technical Committee of Granular Computing. He is also the co-chair of IEEE Task force on Intelligent Ubiquitous Computing. In addition, he is the editors-in-chief of nine international journals and a few book series. He is serving as an editor for around 20 international journals. He has been acting as an author/co-author or an editor/co-editor of 30 books from Kluwer, Springer, Nova Science, American Scientific Publishers, and John Wiley & Sons. He has received three Best Paper Awards, as well as: the IEEE 20th International Conference on Advanced Information Networking and Applications (AINA-06); one IEEE Best Paper Award, 2007; one IEEE Outstanding Paper Award, 2007; Distinguished Achievement Award, 2005; Distinguished Contribution Award, 2004; Outstanding Achievement Award, 2002; Canada Foundation for Innovation Award, 2003; and University Research/Publication/Teaching Award 00-02/02-04/04-06.

Hao Yin, is currently an associate professor with the Department of Computer Science and Technology, Tsinghua University, Beijing, China He received Ph.D. degrees in electrical engineering from Huazhong University of Science and Technology, China in 2002. His research interests span broad aspects of network architecture, P2P technology, wireless network, video coding, multimedia communication over wireless network, and network security. He has published over 50 papers in refereed journal and conferences. He is on editorial boards of *Advances in Multimedia* and *AD HOC NETWORKS Journal*, and has been involved in organizing over 12 conferences.

Rong Yu was born in Guangdong, China, in 1979. He received his BE degree in communications engineering from Beijing University of Post and Telecommunications (BUPT), Beijing, China, in 2002. After that, he joined the Electronic Engineering Department of Tsinghua University, Beijing, China, where he received his PhD degree at July 2007. His research interests include protocol design and performance analysis of wireless sensor networks and board-band wireless multimedia networks.

Zhen Yu is a PhD candidate in the Department of Electrical and Computer Engineering at Iowa State University. He obtained his BEng (1995) and MEng (2001) in electrical engineering from Shanghai Jiao Tong University, China. He also received his MS in electrical engineering from Iowa State University in 2003. Since then he has been researching security issues in wireless networks and distributed systems.

Said Zaghloul is currently a PhD candidate at the Technical University Carolo-Wilhelmina in Braunschweig, Germany. Prior to his PhD studies, he was with Sprint-Nextel as a telecommunication design engineer mainly focusing on wireless IP infrastructures. During his employment at Sprint-Nextel, he submitted two patents in the area of telecommunication protocols and received excellence awards. In 2002, he received the first IEE award for his BSc graduation project in UMTS capacity planning. In 2003, he was granted a Fulbright Scholarship to pursue his MSc studies at the University of Kansas. In 2005, Mr. Zaghloul received his MSc degree with honors. His research interests include wireless protocols, IP technologies, and wireless communications.

Guo-Mei Zhu received the BE degree in communication engineering from ChongQing University of Posts and Telecommunications, Chongqing, China, in 2002. She is currently pursuing her PhD degree at the School of Telecommunication Engineering, Beijing University of Posts and Telecommunications, Beijing, China under the supervision of Professor Geng-Sheng Kuo. Her current research interests include distributed intrusion detection for wireless networks, cross-layer communication protocol design for wireless networks, next generation wireless networks, and wireless mesh networks.

Albert Y. Zomaya is currently the head of school and the CISCO systems chair professor of internetworking in the School of Information Technologies, The University of Sydney. He is the author/co-author of more than 300 publications and serves as an associate editor for several leading journals. Professor Zomaya is the recipient of the Meritorious Service Award (in 2000) and the Golden Core Recognition (in 2006), both from the IEEE Computer Society. He is a chartered engineer (CEng), a fellow of the American Association for the Advancement of Science, the IEEE, and the Institution of Electrical Engineers (U.K.), and a distinguished engineer of the ACM.

Aneta Zwierko holds MSc and PhD in telecommunications from Warsaw University of Technology, Poland. Her doctoral thesis "Cryptographic Protocols for Mobile Agent Systems with Applications" concerned application of cryptographic protocols in mobile environment for providing integrity, anonymity, and more complex services such as secure e-voting. Her current interest include zero-knowledge proofs and its application, identification, and authentication protocols, anonymity and privacy, security issues of the agent systems, E/M-voting protocols, electronic payments, and AI and its application in security.

Index

D

E

F

G

H

I

K

L

M

Q

R

S

T

U

V

W

X

Y

Z